Engineering Materials

Springer
Berlin
Heidelberg
New York
Barcelona
Hong Kong
London
Milan
Paris
Tokyo

M. Bengisu

Engineering Ceramics

With 243 Figures and 27 Tables

Springer

Murat Bengisu, Ass. Prof.
Eastern Mediterranean University, Department of Industrial Engineering
Famagusta TRNC, Via Mersin 10, Turkey
E-Mail: murat.bengisu@emu.edu.tr

ISBN 3-540-67687-2 Springer-Verlag Berlin Heidelberg New York

Library of Congress Cataloging-in-Publication Data

Bengisu, M. (Murat), 1963-
Engineering ceramics / M. Bengisu.
p. cm. - - (Engineering Materials)
Includes bibliographical references.
ISBN 3540676872 (alk. paper)
1. Ceramic materials. I. Title. II. Series.
TA455.C43 B46 2001
620.1'4- -dc21 2001020875

Springer-Verlag Berlin Heidelberg New York
a member of BertelsmannSpringer Science+Business Media GmbH

http://www.springer.de

Printed in Germany

Typesetting: Cameraready by author
Cover concept: eStudio Calamar Steinen
Cover production: de'blik, Berlin

SPIN: 10757845 57/3020/M - 5 4 3 2 1 0 – Printed on acid-free paper

To Pelin, Basak, and Beste and to my parents Nevin and Özdemir.

Preface

Today's rapidly advancing technology always demands materials with more stringent specifications for each new application. The industrial world asks for machines and electronic equipment with higher production rates, improved reliability, longer service life, higher precision, and resistance to more severe service conditions. Engineering ceramics are partly a result of this need and the developments in today's technology and industry. Scientists and manufacturers played a key role in the development of engineering ceramics in the past 50 years. Today, ceramics constitutes one of the most studied materials groups. Due to the very large number of publications in this domain, it takes a lot of skill to keep up with the development in ceramic materials, just as in any other field. Nevertheless, it is the responsibility of the student, technician, engineer, or scientist to be aware of major developments in their field. Books describing the state of art in the developing science and engineering fields are indispensable sources. Yet no book can be complete or final in that sense.

This book gives a brief introduction to the structure of ceramic materials and then follows a flow similar to that which a ceramic product experiences during its lifetime. It starts with the raw material, continues with the processing and consolidation of these materials, and ends with the basic properties, characterization, and applications. I hope that it will serve its purposes and be of some help to those who search for answers.

The project of writing the book called "Engineering Ceramics" started in 1989 when I was a Ph.D. student at the New Mexico Institute of Mining and Technology. I felt that there was a need for an up-to-date, comprehensive book that was easy to understand and at the same time detailed enough to answer the needs of a graduate student. I had written a few papers on different topics by then, a critique on a review paper, a few term papers, and a Master's Thesis. The feedback from my colleagues and professors was encouraging, so the process of collecting published material, reading, and writing began.

The books available were either on a specific subject or their contents were outdated. There was only one book that covered most of the subjects related to engineering ceramics, and it was quite up to date (*Richerson's Modern Ceramic Engineering*). *Engineering Ceramics* covers new and important subjects in this field, in addition to the essential subjects that are treated in *Modern Ceramic Engineering*. There was also Kingery, Bowen, and *Uhlmann's Introduction to Ceramics*, which was quite comprehensive but lacked new and important information. I realized the necessity to address various subjects that had not been treated before, and that some existing chapters already treated in other books could be improved by treating them from the perspective of recent developments.

Engineering Ceramics is aimed as an introductory course book for undergraduate and graduate students of materials science, chemistry, electrical and electronics engineering, and mechanical engineering. It is written so that the reader gets a clear idea of the topics covered, with application examples, tables and figures, where appropriate. The arrangement of the book will make it easy for the reader to understand: after a chapter on the structure of ceramics, the book continues by describing the raw materials for ceramic parts, their preconsolidation processing, consolidation, properties, evaluation, and applications. A final chapter on traditional ceramics is added so the the reader can understand how engineering ceramics are distinguished from conventional ceramics. The sequence of the book is not much different from an actual process of ceramic manufacturing. The ceramic part is shaped from raw materials and densified by certain means, after which certain properties are obtained, evaluated, and the part is used for a certain application. To focus on the subject, I did not include topics such as ceramic coatings, machining of ceramics, or ceramic joining. All of these subjects are covered in detail in separate books.

During my M.Sc., Ph.D., and postdoctoral research from 1984 to 1993, I worked on different aspects of ceramic materials, including sintering (plasma, microwave, rapid rate, conventional, hot pressing), toughening (particle, whisker, zirconia), hard coatings (plasma spraying, glow discharge deposition), self-propagating synthesis, dynamic shock wave treatment (effect on retained tetragonal zirconia, on consolidation behavior, direct explosive compaction), preconsolidation processes (spray drying, slip casting, dry pressing, ball milling, etc.), and erosion. I went through almost every step either by myself or as part of a team. This meant buying the necessary materials (ceramic powders, whiskers, laboratory equipment, etc.), designing the experiments, preparing a powder mixture, shaping the mixture, consolidation by different methods, sampling, evaluation of mechanical and physical properties, and microstructural analysis. I published various papers, including a review in the *Annual Review of Materials Science*. I attended various conferences and symposia. In all cases, I presented our research personally. With the book in mind, I collected information from various sources, including my colleagues' research areas and the literature. I browsed through all important journals in this field to see if there was anything I could use. After my postdoctoral period, I continued my literature search to refine and complete the book. Meanwhile, having taken part in the development and production of new FIAT cars, I had the opportunity to better understand some applications of ceramic materials in automotive parts and machining. My industrial experience allowed me to understand the requirements of certain ceramic parts, such as catalytic converters, their testing, and their performance. During my stay in Italy, for 1.5 years, I had the chance to browse through recent European literature on engineering ceramics, which also includes papers and books written in German, French, and Italian. I continued to follow the literature and developments in the field of ceramics and materials science.

I thank my wife, Pelin, for her full support of my project and for her patience during the time I spent with my second love, this book. I also thank the following, who have made important contributions in my life: my father Özdemir for inspiring me in this project with his well appreciated work both in engineering as well as in his other interests, my mother Nevin for always being there for me, and my M.S. and Ph.D. advisor, Prof. Osman Inal for his leadership in my academic and personal development. I thank all my friends who encouraged my efforts on this book.

Further thanks to Oguz Tosyali, Jon Donlon, Joseline Donlon, Malcolm Walsh, and Elvan Yilmaz for their feedback about the book and to all my colleagues who allowed me to use some of their illustrations and photographs. Thank you Basak Sila and Beste Gizem, for being such nice kids.

Famagusta, Northern Cyprus, *Murat Bengisu*
April 2001

Introduction

Ceramics make up a group of materials that has attracted a lot of interest in the last 30 years, especially in the last decade, although mankind has made use of it probably since its existence, shaping clay to make pots, dishes, brick, etc. Some of the raw materials necessary for ceramics such as silica, magnesia, and alumina are plentiful and cheap. The low toughness along with the high cost of finished ceramic parts, however, restricted their use as engineering materials until recently. One of the earliest industrial ceramics was silicon carbide, introduced by Carborundum Co. in the 1890s. During World War II, a major interest in research on ceramic materials was generated by scientists who realized their outstanding electrical properties and heat resistance. After the war, ceramics were studied to create advanced materials for use in computers, electronics, including capacitors, thermistors, varistors, piezoelectric devices, and other components. Ceramic cutting tools are used in industry for severe machining conditions. Ceramic matrix composites are relatively new materials with considerably higher fracture toughness and reliability compared to monolithic ceramics, making them potential alternatives to many structural engineering components.

One of this century's most important discoveries was also made possible by ceramic materials. The new high temperature superconductors are ceramic materials of complex structure and superior electronic properties. These developments and others that will be discussed throughout the book suggest that ceramic materials are among the most important materials groups that will have many applications in the future. Among the main barriers to their widespread use today, however, are their still relatively low toughness levels, limited reliability, problems encountered in their fabrication, and thus, their high price. Therefore, more research directed toward the solution of these problems and the development of an advanced understanding in related phenomena are necessary.

Chapter 1 of this book describes the structure of ceramic materials and the features that distinguish them from polymers and metals. It briefly describes the electronic structure of atoms, types of bonding, and crystalline and noncrystalline structures to serve as a basis for later discussions. A brief section on phase diagrams is added for completeness.

Chapter 2 gives some examples of conventional methods of high purity ceramic powder production and discusses modern powder production methods in detail. Whiskers and fibers are other raw materials used for ceramic parts. Growth methods of whiskers, so far known to us, are discussed with their pros and cons. Production methods for ceramic fibers and their relevant properties are discussed in detail. Single crystals other than whiskers can be regarded as final products on their own, but they are also included here to make the reader aware of their differences from whiskers and fibers. The sections were divided according to the fiber type in the case of fiber production methods since generally, specific methods have been devised for different types of fibers. On the other hand, in the case of powders, whiskers, and single crystals, one method is used to produce different materials. Therefore, sections on powders and whiskers were divided according to the method applied.

Chapter 3 discusses three related steps necessary for manufacturing a ceramic product for engineering applications: preconsolidation processing, green forming, and consolidation. Preconsolidation processing is the initial processing step in which the raw materials are conditioned for the green forming step. Detailed information is given on rheology and forces acting in a slurry to better understand preconsolidation processes using a slurry. Green forming is the stage where ceramic powders are given the desired near net shape before densification. Green forming methods discussed include current industrial practice, laboratory scale efforts, as well as new developments. Consolidation is the stage in which the final product is achieved, usually by firing at high temperature. Sintering mechanisms and factors affecting sintering are discussed in detail, since sintering is one of the most important consolidation processes for ceramics. Emphasis is given to alternative consolidation methods because detailed information is usually hard to find in other books.

Important mechanical, physical, electrical, and chemical properties are discussed in Chap. 4. The underlying theory for each property is discussed, and some specific examples are given in the related section in text, table, or schematic form. Methods for evaluating these properties and examples of properties quoted in the literature are given when appropriate. Mechanical and electrical properties are discussed in more detail since they are mostly the key properties sought in a ceramic application. Similarly, emphasis is given to corrosion with some specific sections for different ceramics. This emphasis is due to the lack of combined information on the corrosion behavior of ceramics in the literature. Some of these subsections, although not listed in the table of contents, are further divided when necessary. For example, the sections on ionic conduction and superconductors are further divided according to the material groups to explain the differences between them.

Chapter 5 describes various methods for sample preparation, optical and electron microscopic examination, nondestructive evaluation methods, and compositional analysis. The chapter presents important methods relevant to ceramic materials with examples specific to characterizing ceramic materials. Many new methods, which may provide alternative routes to conventional practice, are also discussed. The section on chemical and compositional analysis is quite voluminous since many analytical methods exist and new ones are being presented continuously.

Chapter 6 describes current applications of ceramic materials in structural, refractory, electronic, electrical, magnetic, and biotechnological components. Applications related to energy production include nuclear reactors, conventional energy production, and other approaches such as fuel cells and magnetohydrodynamic reactors. Potential applications are also discussed in each section. The advantages of employing ceramic materials instead of polymers or metals for these applications are explained.

The final chapter (Chap.7) very briefly describes conventional ceramics, how they are produced, and how they differ from engineering ceramics. It gives an idea of conventional ceramics so that the reader can distinguish between engineering and conventional ceramics. The prevalent structures in conventional ceramics and related properties are described.

Contents

1 Structure of Ceramic Materials

1.1 Electronic Structure of Atoms

This section will give a brief description of the electronic structure of atoms to serve as a basis for later discussions such as various types of bonds and electronic properties of ceramics.

Democritus (ca. 470-ca. 380 B.C.) imagined the atom as a tiny sphere. It was first Rutherford in 1911 to change this view [1.1]. According to him, an atom consisted of a minuscule, positively charged nucleus that is surrounded by a cloud of negatively charged electrons. In this model, electrons move around the nucleus like satellites around a planet (Fig. 1.1). In 1913, Bohr [1.2] announced a model that assumed that electrons move in circular orbits around the nucleus. These orbits correspond to certain energy levels that are now called quantum energy states of the electron. In 1916, Lewis and Langmuir [1.3] proposed a model in which protons and neutrons form a central dense nucleus. Each electron was envisioned as an occupant of a fixed position outside the nucleus. Two electrons surrounded the nucleus, and the remaining electrons occupied the corners of concentric cubes. It was proven in the 1930s that the nucleus consists of protons and neutrons, as predicted. The number of protons in the nucleus is the atomic number. In an atom with atomic number *Z,* the positive charge on the nucleus is equal to $+ Z{\cdot}e$, where e is the unit electron charge [1.60209×10^{-19} coulomb = 4.80×10^{-10} electrostatic units (esu)]. This charge is carried by protons, whose quantity is equal to Z. Each proton has the same mass as the unit hydrogen nucleus. Neutrons have almost the same mass as protons. The number of neutrons may vary from one atom to another, giving rise to different atomic weights of the same element that are called isotopes.

In 1929, Sommerfeld [1.4] extended Bohr's model, considering elliptical and circular electron orbits around the nucleus. Schrödinger [1.5] and Heisenberg [1.6] developed the wave mechanics model whose fundamental hypothesis is that matter is both a wave and a particle. Because of the wavelike properties, electrons cannot have well-defined orbits, but only the probability of existing in a certain orbit can be calculated. This probability is represented by a mathematical equation called the Schrödinger wave equation. The solution of the wave equation for electrons yields four parameters. These are n, l, m_l, and m_s, the principal quantum number, the angular (or subsidiary) quantum number, the magnetic quantum number, and the spin quantum number, respectively.

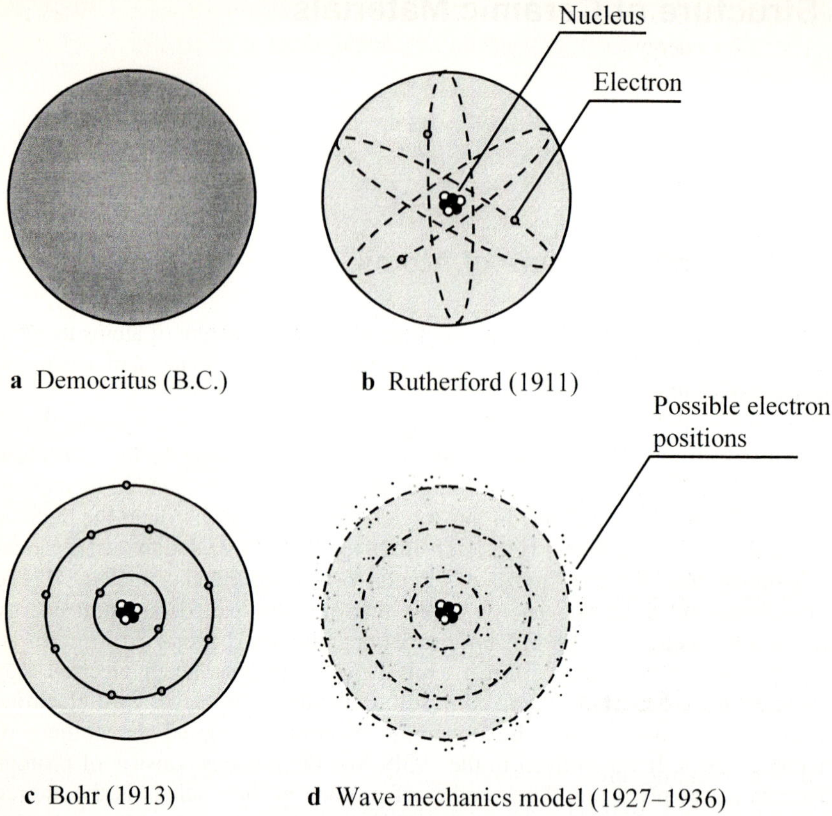

Fig. 1.1 a–d. Evolution of atomic structure models

The principal quantum number n is related to the total energy of an electron in a certain state and can equal a positive integer from 1 to 7. The subsidiary quantum number, l, may have values from 0 to n-1. The letters s, p, d, f, g, and h correspond to $l = 0, 1, 2, 3, 4, 5$ and represent subshells or orbitals. Thus, for example, the energy level for $n = 3$ and $l = 0$ is named the 3s level. The total number that is accepted in each shell is equal to $2n^2$. The first quantum shell (K shell) has two electrons both in the s orbital. The second shell (L) has $2{\cdot}(2^2) = 8$ electrons, two in the s orbitals and six in the p orbitals. Each orbital can contain a specific number of electrons, namely, 2, 6, 10, and 14 for s, p, d, and f orbitals, respectively. For example, Al has 13 electrons and its electron configuration is given by $1s^22s^22p^63s^23p^1$. As the atomic number and the number of electrons increase, the energy difference between electrons and between shells decreases, and overlapping between quantum shells occurs. Because of overlapping, there are groups of elements, such as that comprising Sc, Ti, V, Cr, Mn, Fe, Co, and Ni, in which the outer subshell is partially or fully filled before the shell below (3d in this case) is completely filled. For example, the 4s orbital of Fe is filled before the 3d orbital is full. This is shown in the electron notation by listing the order of filling energy levels in sequence from left to right:

$$Fe = 1s^2 2s^2 2p^6 3s^2 3p^6 3d^6 4s^2.$$

Such elements with incomplete d shells are called *transition elements* [1.7,1.8].

Modern electronic theories recognize elementary particles different from neutrons, protons, and electrons, as well as those, that aggregate to form neutrons and protons [1.9,1.10]. There are three particle families: leptons, hadrons (divided in two categories: barions and mesons), and bosons. Quarks are particles that can join together and form hadrons and mesons. Protons and neutrons belong to the hadron family. Pions and psi mesons belong to the meson family. Leptons are elementary particles that do not aggregate to form composite particles. Examples of leptons are electrons, electron neutrinos, muons, muon neutrinos, taus, and tau neutrinos. Gauge bosons, it is postulated, carry various physical forces. Gravitons are believed responsible for gravity, photons for electromagnetism, mesons for weak nuclear interactions, and gluons for strong nuclear interactions. Details of these particles are beyond the scope of this book and will not be discussed here.

1.2 Types of Bonding

Models of bonding have undergone significant changes throughout the last two centuries. In 1811, Berzelius, the inventor of currently used symbols for elements, proposed that atomic bonds are the result of electrostatic attraction between positive and negative atoms [1.11]. However, he could not explain how neutral atoms possess positive or negative charges. In 1916, Kossel developed a theory of ionic bonding, essentially a revival of Berzelius' electrostatic theory. In the same year, Lewis proposed a theory of covalent bonding based on shared electrons or electron pairs to achieve noble gas configurations [1.11]. From a quantum mechanical point of view, electron pairs are seen as the overlap of atomic orbitals or electron clouds. In 1935, Yukawa proposed a model to explain the extraordinary strong bonding between protons and neutrons in the nucleus. According to his model, bonding is achieved by particle exchange when protons and neutrons are close enough to each other. He called these particles mesons, whose existence was proven in 1947 by Powell [1.1].

There are four types of bonding which can dominate or coexist in solid materials [1.7,1.8,1.11–1.15]. Primary bond mechanisms are referred to as metallic, ionic, and covalent. A less effective type of bonding is secondary bonding, the van der Waals mechanism. Bonding occurs within atoms that have unfilled outer electron shells or excess electrons in the outermost shell. All noble gases (He, Ne, Ar, etc.) have full outer electron shells and therefore are very stable; they do not form primary bonds with other elements. Figure 1.2 shows the types of bonding schematically.

Primary (or chemical) bonds have strengths of about 20–100 kcal/mole. In contrast, secondary bonds have energies less than 10 kcal/mole. Ceramic materials generally have more than one type of bond in their structures leading to pronounced anisotropic properties.

1.2.1 Metallic Bonding

This type of bonding is seen in metals where an attractive force between an array of positive ions and a cloud of free electrons exists. Valence electrons are freely shared by all of the atoms in the structure. Mutual electrostatic repulsion of negatively charged electrons keeps them uniformly dispersed throughout the structure. The free electrons are not tied to any atom in particular, thus bonding is nondirectional. Electron mobility within the lattice of ions contributes to the material's thermal and electrical conductivity and explains the ductility, characteristic of most metals.

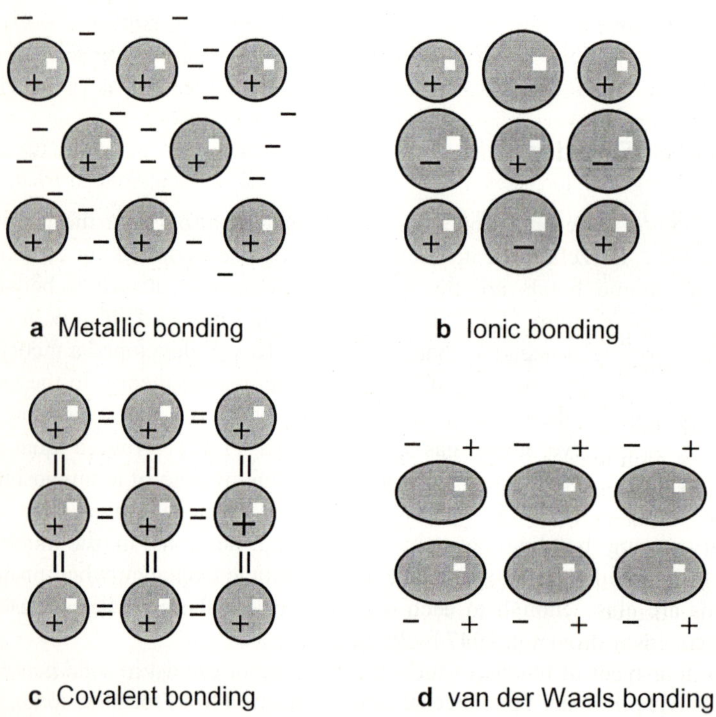

Fig. 1.2 a–d. Four types of atomic bonding

1.2.2 Ionic Bonding

The ionic bond is a result of electrostatic (or coulombic) attraction between oppositely charged ions. Coulomb derived a relationship describing the force between such ions [1.7].

$$F = -(k\, z_1\, z_2\, e^2)/r^2, \tag{1.1}$$

where z_1 and z_2 are the charges of particles, e is the unit charge, r is the distance between the centers of theoretical atomic spheres, and k is a proportionality constant. When the charges are given in esu, k becomes unity. When outer electron shells interact, the net coulombic forces of attraction are balanced by those of repulsion. For ionic bonding to occur, one atom needs to give up one or more outer shell electrons, and another atom needs to accept them. In this way, electrical neutrality is maintained, and each atom attains a stable, filled shell. For example, in the strongly ionic compound NaCl, Na atoms give one electron to Cl, producing Na ions with a net positive charge (Na^+) and Cl ions with negative charge (Cl^-). These opposite charges provide the coulombic attraction and hold the lattice together. In the NaCl crystal, each Na^+ ion is surrounded by six Cl^- ions and vice versa. In the MgS compound Mg^{2+} ions are attracted to S^{2-} ions, establishing neutrality.

In ionically bonded materials, the crystal structure is determined mainly by the size of positive and negative ions and by the number of each ion required to maintain charge neutrality. Figure 1.3 illustrates some examples of crystal structures common in ionic compounds. Notice that the bonds are *not* directional in these compounds. The relative size of ions determines the coordination number defined as the number of anions (negative ions) surrounded by cations (positive ions) in ionic compounds.

As a result of nondirectional bonding, most of the ionic compounds have close-packed structures. The close packing usually results in high coefficients of thermal expansion and high densities. Ceramic materials with high ionic character typically have low electrical conductivity at ambient temperature and ionic conductivity at high temperature. They are more easily sintered in the solid state than covalent ceramics.

The degree of ionic character of a compound can be estimated by using the electronegativity scale. Electronegativity is a measure of an atom's ability to attract electrons. The larger the electronegativity difference between atoms of a compound, the larger the degree of ionic character. Some characteristic features of ionic compounds are their high melting point, strength, and hardness, low electrical conductivity at ambient temperature, and conductivity by ion transfer in the molten state.

1.2.3 Covalent Bonding

This type of bonding involves mutual sharing of outer shell electrons among two or more atoms such that they fill their outer shells. Due to the geometry of the sharing action, covalent bonding is directional. Covalent bonding occurs between atoms of similar electronegativity or between two atoms of the same element. Nonmetals such as Si, Ge, and Se have partially covalent and partially metallic bonds, whereas nonmetallic atoms such as hydrogen, oxygen, and nitrogen form nearly pure covalent bonds. Except for elements in groups I–III in the periodic table, the number of bonds, N, is given by

$$N = 8 - V, \tag{1.2}$$

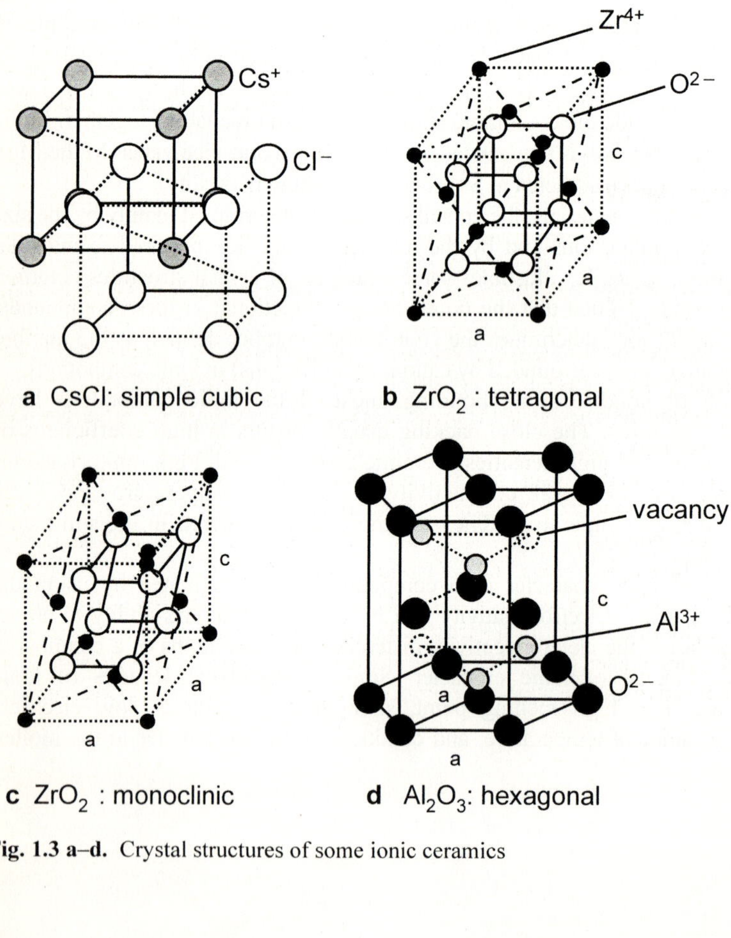

Fig. 1.3 a–d. Crystal structures of some ionic ceramics

where V is the number of valence electrons or the group number in the periodic table.

The directionality of covalent bonding prevents close-packed arrangements of atoms and results in open lattice structures. This has important consequences in many material properties such as density and deformability, as well as in atomic diffusion rates. Densities are lower due to the open structure. Coefficients of thermal expansion are usually lower than those of ionic ceramics since the thermal growth of atoms is partially absorbed by the open spaces in the structure. Nonoxide ceramics that are primarily covalently bonded have typically high hardness, strength, and melting points, and low electrical conductivity at both low and high temperatures. Since no ionic conduction is possible in the melt, covalently bonded materials are insulators even in their liquid state. Solid-state sintering is usually unlikely in covalent ceramics without sintering aids. Some examples of covalently bonded structures are shown in Fig. 1.4.

Covalent bonds are associated with semimetallic and nonmetallic atoms that are located in the upper right-hand corner of the periodic table. The elements can cluster and form large polyatomic units. If such groups form without enough electrons to fill the outer shell, they may receive additional electrons from external atoms upon which the unit becomes a polyatomic anion. If polyatomic groups link by periodic sharing of electrons, polymerization occurs. Polymers most commonly found in ceramics involve Si–O bonds, in contrast to organic polymers that usually contain C–C bonds.

1.2.4 The van der Waals Bond

This type of bonding originates from secondary forces of attraction caused by permanent electrical dipoles and temporary molecular polarization. As two atoms or molecules approach each other, the outer electron orbitals are mutually perturbed. If this perturbation is strong enough, these atoms form a primary bond. However, usually the perturbation is not too strong, and it results in the relative displacement of the centers of positive and negative charges, which in turn results in coulombic attractive forces and a temporary dipole moment.

Another type of van der Waals force is permanent molecular polarization where an electrical dipole is formed with asymmetric attraction forces of different atoms in molecules. Such molecules are called polar molecules. For example in hydrogen fluoride, when the single electron that is shared is orbiting in the fluorine nucleus, a dipole moment results, where the hydrogen side of the molecule has a net positive charge and the fluorine side has a net negative charge. This electron spends more time around the fluorine nucleus that is larger than the hydrogen nucleus. As a result, the centers of the positive and negative charges do not coincide, whereby a weak electric

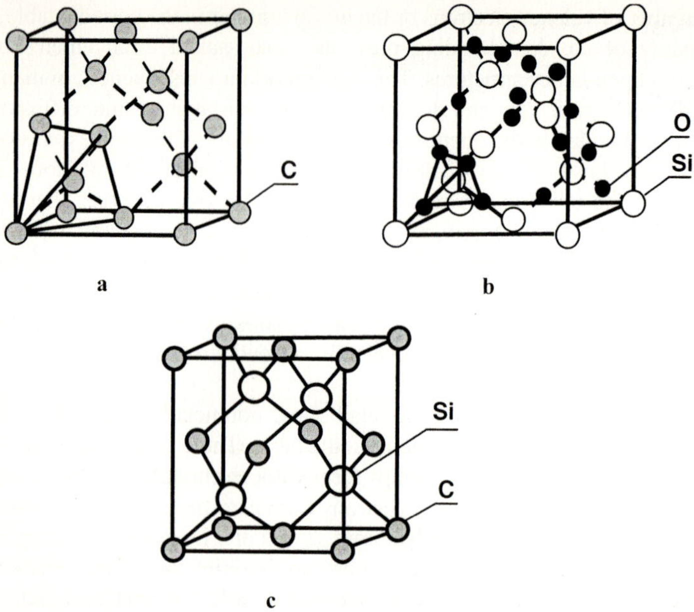

Fig. 1.4 a-c. Crystal structures of some covalent compounds: diamond with a periodic tetrahedral structure **a**, cubic silica with SiO_4 tetrahedra **b**, and cubic silicon carbide with the diamond cubic structure **c**

dipole occurs. Such a dipole can contribute to weak bonding of one molecule to another. NH_3, HCl, HBr, and H_2O are examples of polar molecules, and CO_2, CS_2, H_2, and N_2 are examples of nonpolar molecules.

van der Waals bonds assume a major role in structures of solidified inert and diatomic gases where there are no other binding forces operative. In ceramic materials, van der Waals forces are important in layered structures such as clays, graphite, and hexagonal BN. Within the layers, primary (covalent and/or ionic) bonds are effective, whereas between the layers only van der Waals bonds are active and hold these layers together. Due to the weak bonding, slip is easy between these layers, and plastic deformation is possible in the presence of a liquid. In the solid form, such ceramics can be machined with conventional cutting tools, and they can also serve as dry lubricants.

van der Waals forces promote physical adsorption of atoms, ions, or molecules onto solid and liquid surfaces (e.g., silica gels). They play an important role in dry and wet processing of ceramic particles such as dry or wet milling, slip casting, and cold pressing. The latter effect will be discussed in more detail in Sect. 3.1.

1.3. The Crystal Lattice

1.3.1 Crystalline Structures

In terms of atomic arrangement, solids can have either crystalline or short-range, ordered, noncrystalline structures. A crystalline structure can be defined as a periodic arrangement of atoms in three dimensions. Noncrystalline or amorphous structures, on the other hand, do not have a regular arrangement of atoms. They are built up of random clusters of ordered atoms, i.e., they have short-range order. Window glass is a typical example of a solid with an amorphous structure, whereas most solid metals and ceramics under equilibrium conditions have crystalline structures. Figure 1.5 illustrates crystalline and amorphous structures schematically.

In crystals, it is convenient to visualize the three-dimensional arrangement as an extension of one unit cell. The unit cell is the smallest three-dimensional lattice that can be extracted from the whole and has identical lattices around it. A unit cell can be defined by three vectors drawn from the corner of the cell taken as the origin (Fig. 1.6) or by the lengths and angles between these vectors. These lengths and angles are called the lattice parameters of a unit cell. By giving different values to these parameters, various unit cells can be produced. Bravais [1.7,1.12] demonstrated that there are only fourteen possible unit cells. These lattices are shown in Fig. 1.7. Some examples of materials with different unit cells are given in Table 1.1.

Metallic materials generally possess either body-centered cubic (BCC), face-centered cubic (FCC), or hexagonal close-packed (HCP) arrangements. Ceramic crystals are generally more complex than metals because of their bond structure. For example, Si_3N_4 can exist in three forms at ambient temperature: amorphous, covalent α, and covalent β phases. α and β are hexagonal structures composed of three-dimensional networks of SiN_4 tetrahedrons. The β-Si_3N_4 unit cell has six Si and eight N atoms whereas the α phase has 12 Si and 16 N atoms [1.16]. Some examples of ceramic crystal structures are illustrated in Figs. 1.3 and 1.4. Sometimes, it is convenient to view these structures as superpositions of simple cubic, FCC, and HCP structures and their interstitial positions.

1.3.2 Silicate Structures

Silica (SiO_2) has many polymorphs at various temperatures and pressures. The basic polymorphs are quartz, tridymite, cristobalite, and an amorphous form known as silica gel, diatomite, or vitreous quartz. Quartz has a low temperature (trigonal) and a high temperature (hexagonal) polymorph [1.17]. Natural or synthetic silicates have very complex structures. Classification of silicate and aluminosilicate structures based on structural units is the easiest approach in understanding the atomic arrangement. In SiO_4^{4-} ions, silicon atoms are surrounded tetrahedrally by four covalently bonded oxygen cations (Fig. 1.8).

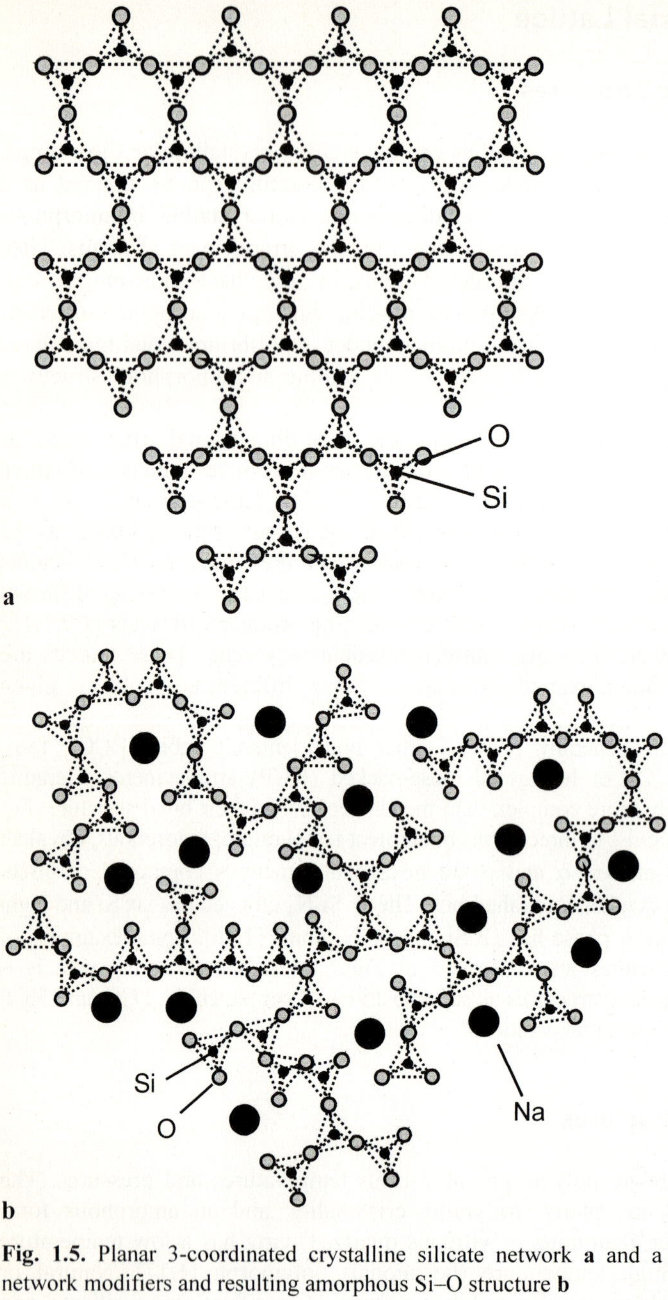

a

b

Fig. 1.5. Planar 3-coordinated crystalline silicate network **a** and a network broken by network modifiers and resulting amorphous Si–O structure **b**

Table 1.1. Crystal Structures of Some Ceramic Materials [a]

Material	Crystal structure	Structure name	Space group	Pearson	Strukturbericht
CsCl, CsBr, CsI	Simple cubic	Cesium chloride	Pm(−3)m	cP2	B2
UO_2, CeO_2, Li_2O, Na_2O, Rb_2O, ThO_2, ZrO_2	Simple cubic	Fluorite (CaF_2)	Fm(−3)m	cF12	C1
$CaTiO_3$, $BaTiO_3$, $PbTiO_3$, $PbZrO_3$	Simple cubic	Perovskite ($CaTiO_3$)	Pm(−3)m	cP5	$E2_1$
ZrC, HfC, TiC, NaCl, KCl, LiF, MgO, CaO, BaO, VO	Face-centered cubic	Rock salt (NaCl)	Fm(−3)m	cF8	B1
β-SiC, CdS, AlP, BN	Face-centered cubic	Zinc blende (ZnS)	F(−4)3m	cF8	B3
Diamond, Si, Ge, Sn	Face-centered cubic	Diamond	Fd(−3)m	cF8	A4
WC, WN	Hexagonal close-packed	Tungsten carbide (WC)	P(−6)m2	hP2	B_h
ZnS, ZnO, α-SiC, BeO, CdS, AlN	Hexagonal close-packed	Wurtzite (ZnS)	$P6_3mc$	hP4	B4
TiB_2, ZrB_2, TaB_2	Hexagonal close-packed	AlB_2			
Al_2O_3, Fe_2O_3, Cr_2O_3, V_2O_3, Ga_2O_3	Rhombohedral	Corundum (Al_2O_3)	R(−3)c	hR10	$D5_1$
B_4C	Rhombohedral	–	–	–	–
TiO_2	Orthorhombic	Brookite (TiO_2)	Pcab	–	–
AlF_3, WO_3	Triclinic	–	–	–	–
TiO_2, GeO_2, IrO_2, MoO_2, PbO_2, SnO_2, TaO_2, WO_2	Tetragonal	Rutile (TiO_2)	$P4_2/mnm$	tP6	C4
TiO_2	Tetragonal	Anatase (TiO_2)	–	–	–

[a] References

1. *Center for Computational Materials Science Database*: http://cst-www.nrl.navy.mil/lattice
2. *ART Handbook of Advanced Ceramic Materials.* (Advanced Refractory Technologies, 1991)
3. *Materials Handbook*. Cer. Ind. **136 [1],** 102 (1991)
4. L.H. Van Vlack: *Physical Ceramics for Engineers*. (Addison-Wesley Reading MA, 1964)
5. *CRC Materials Science and Engineering Handbook*. J.F. Shackelford, W. Alexander, J.S. Park (Eds.). (CRC Press Boca Raton FL, 1994)
6. D.W. Richerson: *Modern Ceramic Engineering*. (Marcel Dekker New York, 1992)

Single silicate tetrahedra act as ions in compounds that have a sufficient number of metal atoms to supply the necessary electrons. One example is forsterite (Mg_2SiO_4), where two Mg^{2+} ions correspond to each SiO_4^{4-} ion. Tetrahedral silicates are the universal building blocks of both crystalline and amorphous silicates [1.18]. Polymerization of tetrahedral silicates in two dimensions can form chain structures (Fig. 1.8c and d) such as those found in enstatite ($MgSiO_3$), spodumene ($LiAl(SiO_3)_2$), and wollastonite ($CaSiO_3$), ring structures such as those found in beryl ($Be_3Al_2Si_6O_{18}$), or sheet structures (Fig. 1.8e) which are bases of flaky minerals such as talc ($[Mg_3Si_4O_{10}(OH)_2]_2$), clays (for example, kaolinite ($[Al_2Si_2O_5(OH)_4]_2$) and pyrophyllite ($[Al_2Si_4O_{10}(OH)_2]_2$)), and micas (for example, muscovite ($KAl_3Si_3O_{10}(OH)_2]_2$) and phlogophite ($[KAlMg_3Si_3O_{10}(OH)_2]_2$).

When polymerization occurs in three dimensions, crystalline compounds such as feldspars, granites, synthetic zeolites, and quartz, or noncrystalline compounds, i.e., glasses are produced.

1.4 Noncrystalline Structures

When the long-range order of a crystalline structure is absent in a solid, the structure is referred to as noncrystalline or amorphous. Although these structures, mostly represented by glasses, gels, polymers, and rapidly solidified materials, lack the long-range order of crystalline materials, they do not lack a structure. The

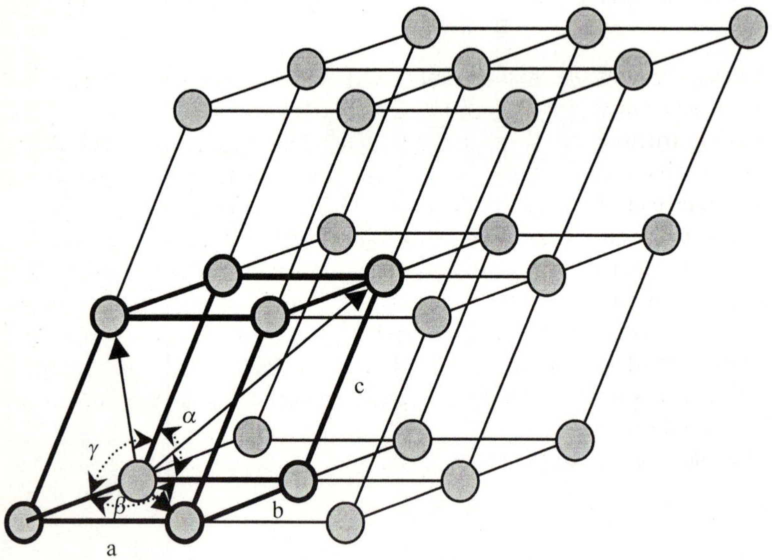

Fig. 1.6. Representation of a unit cell within a crystal network (triclinic structure-represented by the bold lines)

main characteristics of a glassy structure are first neighbor or short-range order and a continuous framework of strong primary (ionic or covalent) bonds. Silica glass is the most common material that has a glassy structure based on a single oxide. Its structure is shown in Fig. 1.5b. Tetrahedra are connected to each other at corners through an oxygen ion, thus producing a random, three-dimensional Si–O–Si network.

Zachariasen [1.19] developed the following rules to define glass structures based on interatomic bonds:

1. No oxygen atom can be linked to more than two cations.
2. The coordination number of oxygen atoms around any cation must be small (3 or 4).
3. Oxygen polyhedra must share corners, not edges or faces (Fig. 1.5a).
4. At least three corners of each polyhedra must be shared to achieve a three-dimensional network.

Zachariasen's rules are remarkably successful in predicting new glass-forming oxides and explaining common glass systems. However, to accommodate the vast range of more complex glasses, Zachariasen's rules had to be modified to describe systems that form glasses when a glass-forming oxide was mixed with a non-glass-forming oxide. These are the modified rules [1.20]:

1. An oxide glass is formed if a high percentage of cations is surrounded by oxygen tetrahedra or triangles.
2. Tetrahedra or triangles can share only corners.
3. Some oxygens are linked only to two network cations, i.e., there are bridging and nonbridging oxygens (Fig. 1.5).

From these rules, the terms *network forming oxides* and *network modifying oxides* arose. Network formers such as B_2O_3, SiO_2, GeO_2, P_2O_5, and As_2O_5 are oxides essential to glass structure. Network modifiers, including Na_2O, K_2O, CaO, BaO, MgO, and rare-earth oxides open up the network. The structure of B_2O_3 glass was analyzed by molecular dynamics simulation, where it was shown that Na_2O addition opens up the network, as experimental observations indicated [1.21]. Network modifiers contribute to the optical and thermal properties and also increase the plasticity of a glass by increasing its fluidity at high temperatures. These oxides cannot form a network by themselves, which, in fact, is in agreement with Zachariasen's third rule. A third group of oxides consists of the intermediates. These oxides can occasionally enter into the glass network by bonding to a limited extent. TiO_2 and PbO are such oxides.

When these rules are applied to oxides with specific formula types, it is found that [1.22]

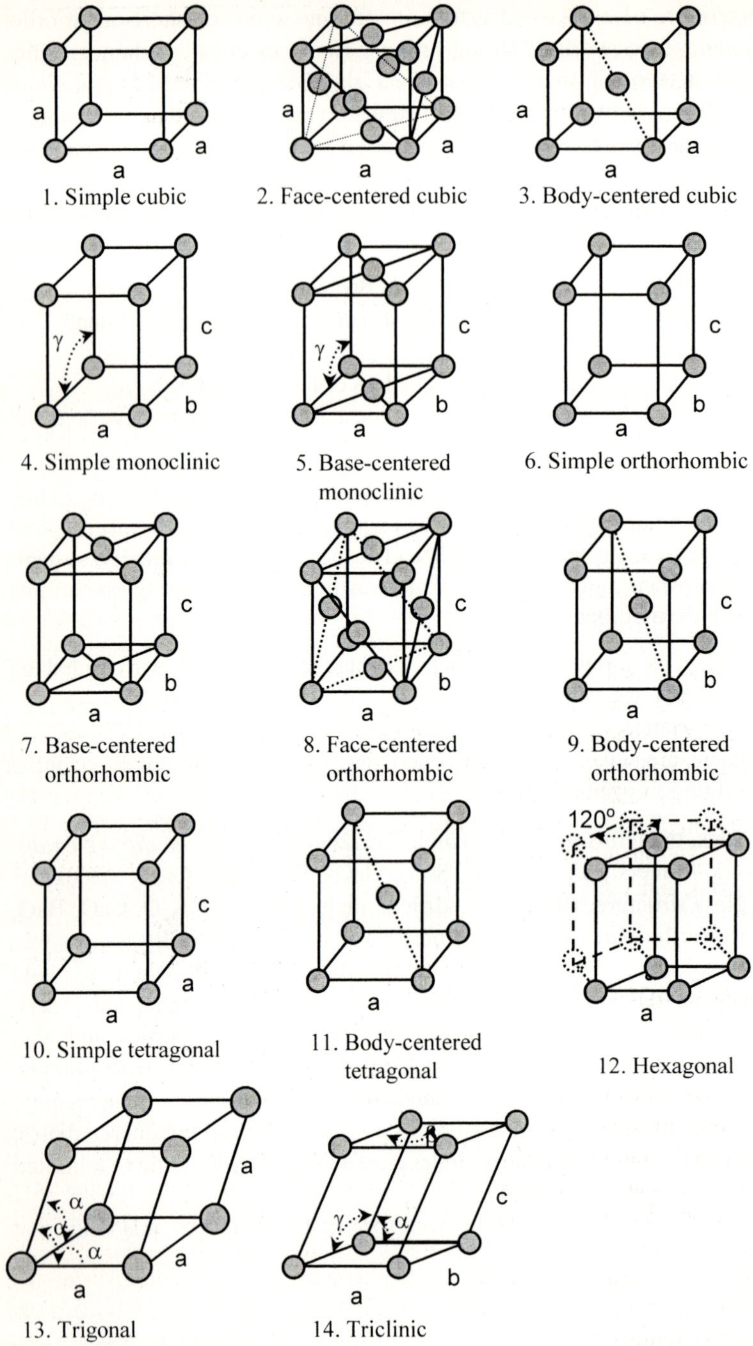

Fig. 1.7. Fourteen Bravais lattices

1. Metal oxides M_2O and MO cannot satisfy the rules and they cannot be used as network formers. For example, MgO has a structure containing MO_6 polyhedra sharing faces, which breaks rule 3. Similarly, PbO has a structure in which chains of PbO_4 share O–O edges, also contrary to rule 3.
2. M_2O_3 oxides can satisfy the rules if oxygens form planar triangles around the metal atom. B_2O_3 is a compound of this type.
3. MO_2 and M_2O_5 will satisfy the rules only when they form tetrahedra around the metal atom.
4. MO_3, M_2O_7, and MO_4 will satisfy the rules when oxygens form octahedra around the metal atom and when these octahedra are connected at the corners.

Factors that favor glass formation were summarized as follows [1.18,1.23]:

1. High viscosity at the nose of the related time–temperature–transformation temperature (TTT) curve (low transportation rates lead to impeded nucleation and crystallization).
2. Absence of heterogeneous nucleation centers.
3. A high crystal–liquid interfacial energy, since it reduces the nucleation rate.
4. A large concentration difference between liquid and crystal in multicomponent systems.
5. A low entropy of fusion.

Although various oxide glasses have been developed that do not contain silica, a key component, they usually do not have significant commercial application. B_2O_3 (borate)- and P_2O_5 (phosphate)-based glasses react with atmospheric water and degrade easily. Nevertheless, they have been studied to enhance our understanding of glass structure [1.18,1.22]. Among the nonsilica glasses that have some commercial use are boroaluminates (for example *Cabal* glasses with electrical resistivity higher than that of silica), alkaline earth aluminates as high temperature sealants and IR-transmitting glass, and V_2O_5-based glasses [1.22].

1.5 Phase Diagrams

Phase diagrams, also known as equilibrium diagrams, convey information about the structure and types of phases present at equilibrium or close to equilibrium conditions. They are essential for interpreting and predicting the distribution and evolution of phases and microstructure. There are three independent variables that may influence the state of a single- or multiphase ceramic: pressure, temperature, and composition. For most practical conditions, pressure is constant at atmospheric conditions. Thus, most industrially important phase diagrams are representations of structural changes that occur due to temperature and compositional variations [1.3]. Equilibrium conditions are achieved only under ideal conditions requiring extremely slow heating and cooling to allow phase changes. Some small discrepancies may occur between actual practice and available phase diagrams since heating or cooling faster than required for

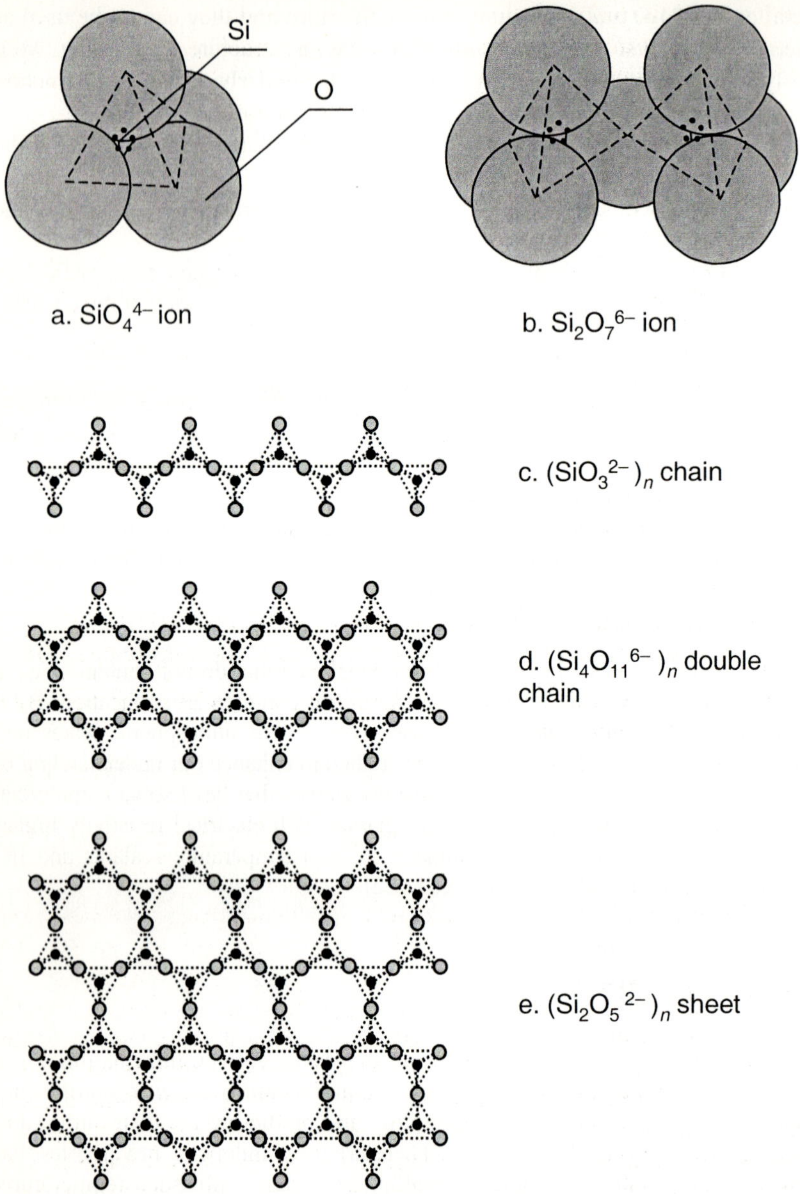

Fig. 1.8. Silicate structures: the SiO_4^{4-} monomer **a**, the $Si_2O_7^{6-}$ dimer with a bridging oxygen **b**, $(SiO_3^{2-})_n$ chain **c**, $(Si_4O_{11}^{6-})_n$ double chain **d**, and $(Si_2O_5^{2-})_n$ sheet **e**

equilibrium will take place.

Phase diagrams can be constructed from experimental data or thermodynamic calculations. Experimental methods include thermal analysis, microstructural analysis, and X–ray diffraction (XRD). Thermal analysis is the most widely used method. It involves constructing cooling curves at different compositions and combining this information to reconstruct the complete phase diagram. Microstructural analysis requires heating samples to certain temperatures, waiting for equilibrium conditions, and then fast cooling to freeze the high-temperature structure. XRD is a powerful, precise, and relatively simple method for studying structural variations during heating, under pressure, or from compositional changes. Phase diagrams can also be constructed by theoretical calculations based on minimizing the Gibbs free energy function and other thermodynamic models. Computer programs can be used to facilitate such calculations [1.24].

Phase diagrams provide useful information about a number of items [1.12,1.25]:

1. Phases present under equilibrium conditions
2. Chemical composition of phases at a given temperature
3. Relative amounts of each phase
4. Melting temperatures of each pure compound
5. The presence and range of solid solubility of one compound in the other
6. The presence of polymorphic transformations and corresponding temperatures

An important expression that describes the degrees of freedom within a system in equilibrium is the *Gibbs phase rule*:

$$F = C - P + 2. \quad (1.3)$$

Here, F represents the degrees of freedom, C is the number of components, and P is the number of phases in equilibrium. By the number of degrees, one understands the number of variables (temperature, pressure, and composition) that has to be specified to define an equilibrium system completely. A phase is a homogenous, physically distinct, and mechanically separable part of a system. Analytical techniques such as X-ray diffraction (XRD) or transmission electron microscopy (TEM) can be used to identify the types of phases in a system. The number of components, C, is the minimum number of independently variable constituents required to define the composition of each phase present. For example, in the $MgO–Al_2O_3$ binary system, there are two components: MgO and Al_2O_3.

1.5.1 Single-Component Phase Diagrams

These phase diagrams describe polymorphic phase transformations, if applicable, within a solid phase. The pressure and temperature ranges for solid, liquid, and gas phase occurrence can also be derived. According to the Gibbs phase rule, since $C = 1$ for a single-component system, $F = 1$ where two phases are present (for example, where solid and liquid meet), and $F = 0$ where three phases coexist (for example, solid, liquid, and gas). A classical example for $F = 0$ (zero degrees of freedom) is the triple point of the H_2O phase diagram. At 0.006 atm pressure and 0.0075°C, three phases (solid, liquid, and gas forms of water) meet at a triple point [1.26]. At this point, the degree of freedom becomes $F = 0$. Thus, any change in temperature or pressure will disturb this equilibrium.

Some examples of polymorphic transformations in technologically important single-phase ceramic components include transformations of ZrO_2, conversion of graphite to diamond under high static pressures, and the cubic → tetragonal transformation of $BaTiO_3$.

The stable form of ZrO_2 at room temperature has monoclinic symmetry, which upon heating transforms to a BCT phase at 1160–1190°C, and subsequently to a cubic phase at 2370°C. ZrO_2 melts at 2680°C. The monoclinic → tetragonal transformation is accompanied by a sudden volume increase (3–5%), which can be disruptive. The same volume expansion, however, can also be used for toughening a suitable ceramic matrix. The transformation is known to be of martensitic type.

$BaTiO_3$, an important electronic ceramic, has mainly four crystal structures. At T>132°C, it has a cubic Perovskite structure [1.27]. The distortion in this structure is believed to be the very reason for its high dielectric constant and other beneficial properties. Below this temperature, the crystal structure is tetragonal (Fig. 1.9). At about 0°C, an orthorhombic phase forms. All three phases exhibit ferroelectricity. Additionally, a trigonal (rhombohedral) phase exists at very low temperatures (below − 90°C).

A good example of a single-component phase diagram is that of carbon. Carbon has the form of graphite under ambient conditions and can be converted to diamond at high static or dynamic pressures. This phase diagram guided the historical synthesis of diamond in 1960 [1.28]. A recent version of the phase diagram reveals that there are two diamond phases, one cubic and the other hexagonal [1.29] (Fig. 1.10). The conversion of graphite to diamond requires the use of liquid metal catalysts such as Ni. At about 5000°C, a liquid phase forms.

1.5.2 Binary Phase Diagrams

Phase diagrams that involve two components, also called binary phase diagrams, are the most widely used type. This is due to the relative simplicity of their analysis and use. Actually, three variables may influence the phase equilibrium: temperature, composition, and pressure. This requires the use of a three-dimensional diagram. However, usually it can be assumed that the effect of pressure is minimal. Thus, a constant pressure (P =1 atm) system is employed mostly for simplicity. Most of the binary phase diagrams belong to one of the following categories or their combinations:

1. Isomorphous (complete solid solubility).

2. Eutectic diagram with partial solid solubility.
3. Peritectic diagram with partial solid solubility.

Two important factors that define these categories are the extent of solubility of one phase within the other and whether or not incongruent melting takes place. Incongruent melting is the dissociation of the compound into a liquid and another solid before melting completely.

1.5.2.1 Phase Diagrams of Isomorphous Systems

If two components have the same type of lattice structure, the same valence, if a small size difference between ions (typically less than 15%) and low chemical affinity exists, they may exhibit complete solid solubility. An isomorphous phase diagram is shown in Fig. 1.11. In such systems, there is only a single solid phase, thus the system is called *isomorphous*. Examples of ceramic systems with complete solid solubility are MgO–FeO, Al_2O_3–Cr_2O_3 [1.30] (Fig. 1.12) and NiO–MgO. The amount of each phase (liquid and solid) at a given composition and temperature can be found by the *lever rule*. According to this rule,

$$\% \text{ liquid} = (BC/AC) \times 100, \tag{1.4}$$

and

$$\% \text{ solid} = (AB/AC) \times 100, \tag{1.5}$$

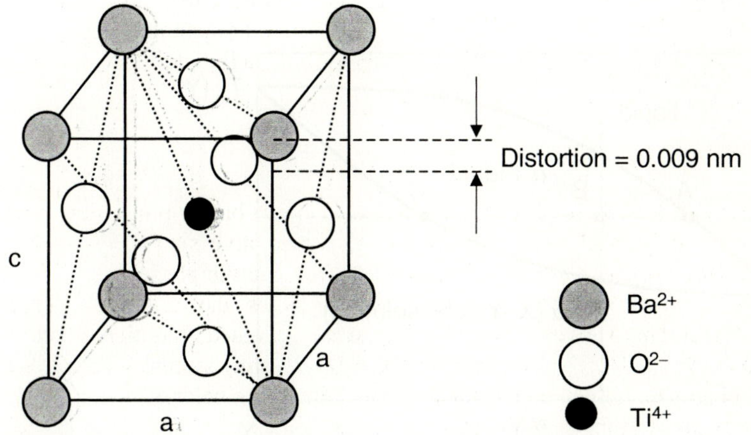

Fig. 1.9. The tetragonal unit cell of $BaTiO_3$ showing distortion of the cubic structure

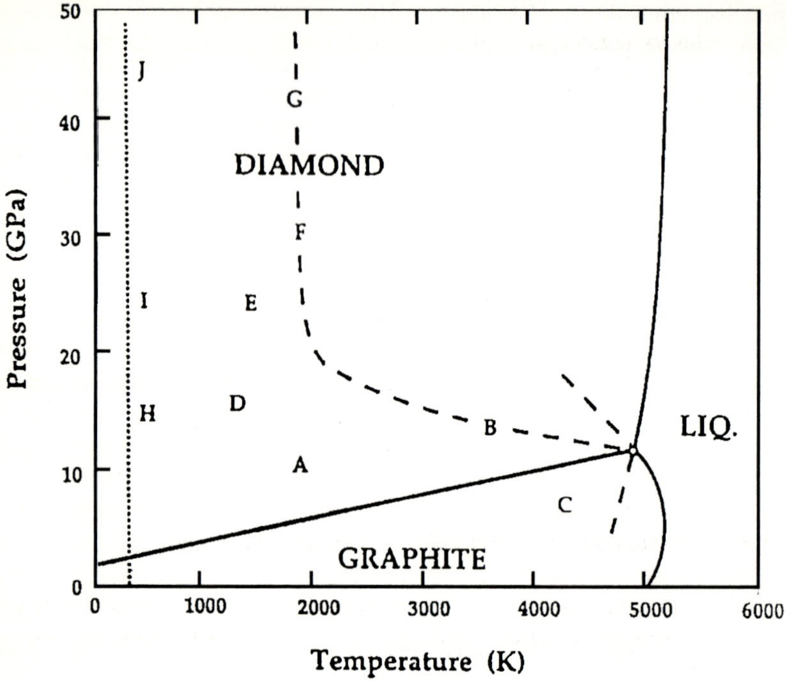

Fig. 1.10. Recent phase diagram of carbon. Solid lines represent equilibrium phase boundaries. Letters A to J represent various forms of carbon obtained by processes ranging from commercial synthesis of diamond from graphite by catalysis to shock compression/quench cycles (Reprinted from F.P. Bundy, W.A. Bassett, W.S. Weathers, R.J. Hemley, H.K. Mao, and A.F. Goncharov: *The pressure-temperature phase and transformation diagram for carbon, updated through 1994*. Carbon **34 [2]**, 141–153 (1996) Copyright 1996 Elsevier Science Oxford UK, with permission)

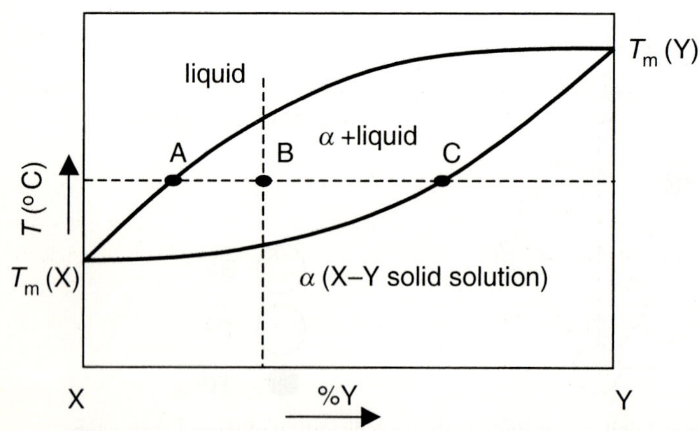

Fig. 1.11. Generic phase diagram exhibiting an isomorphous system

where B is the composition of the alloy in question, and A and C are intersection points of the relevant temperature (isotherm line) with liquidus and solidus curves, respectively, as indicated in Fig. 1.11.

1.5.2.2 Eutectic Phase Diagrams

This type of equilibrium diagram occurs between two compounds with partial solid solubility. If the addition of one compound to the other decreases the melting point compared to both compounds, a eutectic phase diagram will result (Fig. 1.13). The eutectic transformation occurs at the eutectic point during cooling. This is the point of intersection of liquidus curves. The eutectic transformation is described as

$$\text{liquid} \rightleftarrows \text{solid 1} + \text{solid 2}. \tag{1.6}$$

In other words, during cooling of a material with a eutectic composition, the liquid transforms into two discrete solid phases with different compositions. This phenomenon can be applied for the production of eutectic composites (see Sect. 3.4.8). The microstructure of eutectic ceramics will reveal two dissimilar phases

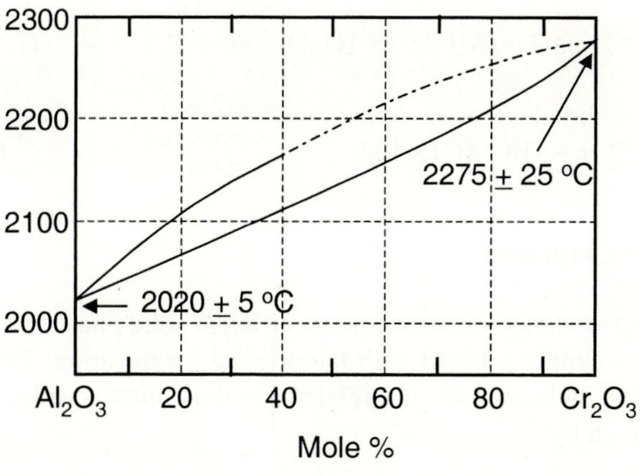

Fig. 1.12 Al_2O_3-Cr_2O_3 phase diagram exhibiting complete solid solubility (Reprinted from J. Res. Natl. Bur. Stand. **6 [6]**, 948 (1931), courtesy of the National Institute of Standards and Technology. Not copyrightable in the United States)

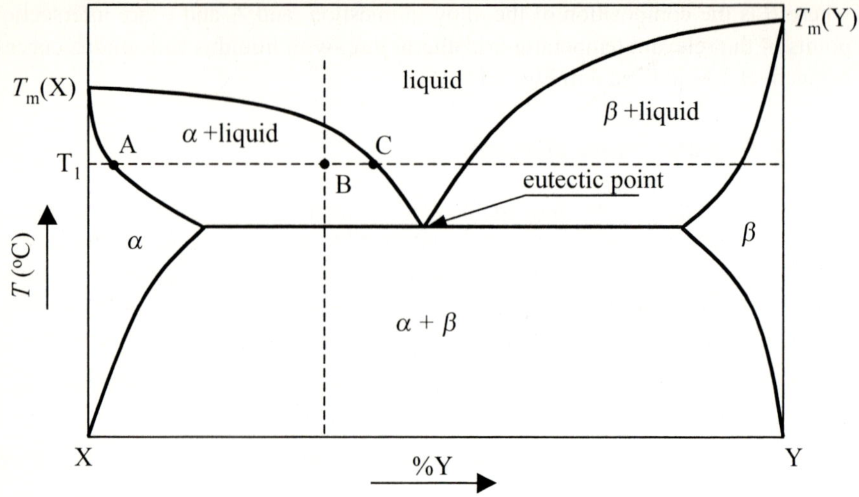

Fig. 1.13. Generic eutectic phase diagram

mixed with each other. Ceramic systems that exhibit simple eutectic phase diagrams include MgO–CaO (Fig. 1.14) and MgO–Al_2O_3 [1.24]. The lever rule for eutectic diagrams becomes (Fig 1.13),

$$\% \text{ liquid} = (\text{AB}/\text{AC}) \times 100, \tag{1.7}$$

and

$$\% \, \alpha = (\text{BC}/\text{AC}) \times 100. \tag{1.8}$$

1.5.2.3 Peritectic Phase Diagrams

During heating, a solid compound may dissociate into a different solid phase and a liquid instead of resulting in simply a liquid with the original composition. This behavior is called incongruent melting [1.8]. The related transformation is called a peritectic transformation, given by

$$\text{solid } 1 \rightleftarrows \text{ solid } 2 + \text{liquid}. \tag{1.9}$$

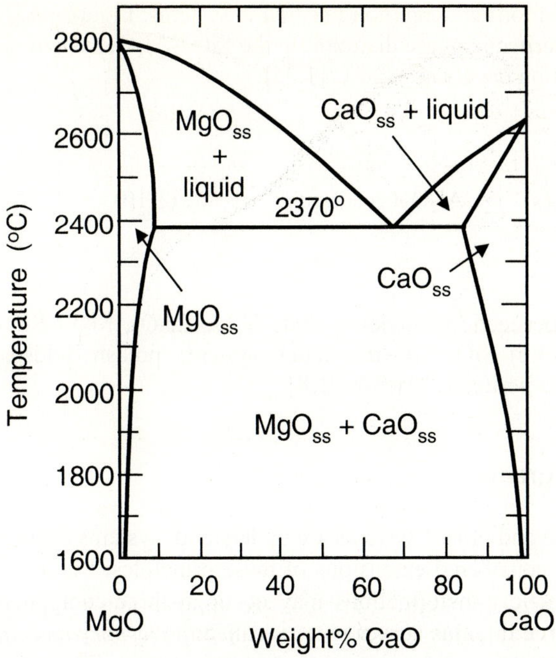

Fig. 1.14. The MgO–CaO eutectic phase diagram (From W.D. Kingery, H.K. Bowen, and D.R. Uhlmann: *Introduction to Ceramics.* Copyright 1976 John Wiley. Reprinted with permission of John Wiley & Sons, Inc.)

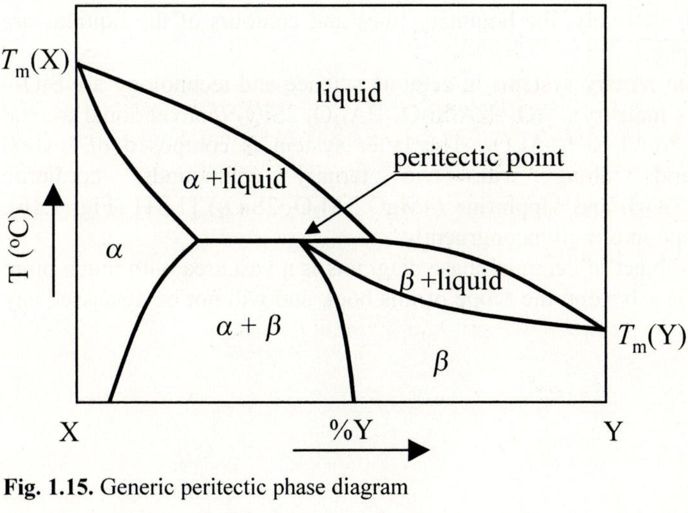

Fig. 1.15. Generic peritectic phase diagram

Peritectic phase transformations are common in ceramic systems. Figure 1.15 illustrates a simple generic peritectic phase diagram. In the SiO_2–Al_2O_3 system, the following peritectic reaction occurs at 1828°C [1.25]:

$$2SiO_2{\cdot}3Al_2O_3\,(s) \rightleftarrows Al_2O_3\,(s) + \text{liquid} \qquad (1.10)$$

Other systems that exhibit peritectic behavior include V_2O_5–Cr_2O_3, MgO–SiO_2, and $K_2O{\cdot}Al_2O_3{\cdot}4SiO_2$ (leucite)–SiO_2. In the latter system, potash feldspar ($K_2O{\cdot}Al_2O_3{\cdot}6SiO_2$) melts by a peritectic reaction [1.8].

1.5.3 Ternary Phase Diagrams

Once the principles for single and binary systems are understood, systems of three or more components can be considered extensions of these principles. Similar to binary phase diagrams, eutectic transformations may occur in three-component systems. The point where three liquidus curves meet is called the *ternary eutectic point*. Peritectic reactions are also common in many ternary systems. The major complexity of such systems lies in graphical representation since three-dimensional illustrations are difficult to exhibit in two dimensions. Other than the assumption of constant atmospheric pressure, a further simplification is required. One method is to cut a section from the three-dimensional phase diagram at a certain temperature and analyze the cross section, which has the form of a triangle. The composition of some point within the phase diagram is defined by drawing parallel lines to the three sides of the triangle. Another method is to cut a vertical cross section. Alternatively, the boundary lines and contours of the liquidus are projected onto a plane.

Some important ternary systems in ceramic science and technology are SiO_2–CaO–Na_2O (glass industry), SiO_2–$KAlSi_2O_6$–$3Al_2O_3{\cdot}2SiO_2$ (conventional triaxial whiteware), and MgO–SiO_2–Al_2O_3. The latter system is composed of several binary compounds along with two ternary compounds, cordierite ($2MgO{\cdot}2Al_2O_3{\cdot}5SiO_2$), and sapphirine ($4MgO{\cdot}5Al_2O_3{\cdot}2SiO_2$) [1.31] (Fig. 1.16). Both ternary compounds melt incongruently.

Although the subject of ceramic phase diagrams is a vast area with much more important detail, it is beyond the scope of this book and will not be discussed any further.

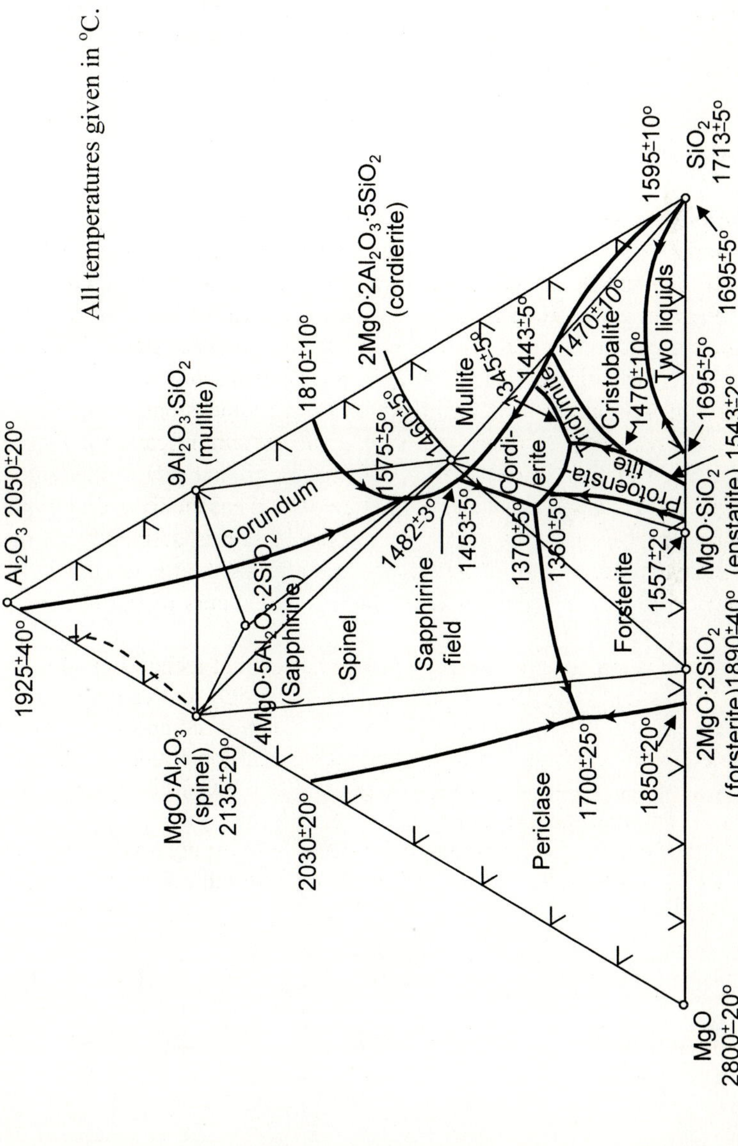

Fig. 1.16. The ternary system MgO–SiO_2–Al_2O_3 (From M.L. Keith and J.F. Schairer, *J. Geol.* **60,** 182 (1952). Copyright 1952 University of Chicago Press. Reprinted with permission of the University of Chicago)

2 Production and Properties of Ceramic Powders, Whiskers, Fibers, and Single Crystals

Typically, engineering ceramics have a controlled composition and higher purity compared to conventional ceramics. Most of the conventional ceramic products such as sanitary ware, tableware, wall tiles, and brick are made from raw materials found in nature. Raw materials for engineering ceramics, on the other hand, are normally not found in nature; they must be produced artificially.

Among the important engineering ceramics are metal oxides (e.g., Al_2O_3, ZrO_2, and MgO), nitrides (AlN, Si_3N_4, BN), carbides (WC, TiC, SiC), borides (TiB_2), magnetic ceramics ($Y_3Al_5O_{12}$, $BaFe_{12}O_{19}$, $PbFe_{12}O_{19}$), ferroelectric ceramics ($BaTiO_3$, $PbTiO_3$), nuclear fuels (UO_2, UN), and the new generation of high-temperature superconductors {$YBaCu_3O_{7-x}$, $Tl_2Ba_2Ca_2Cu_3O_{10}$, ($La_{1.85}$–$Sr_{0.15}$)CuO_4}. Ready-to-use ceramic powders are required for most of the processing methods available for producing these engineering ceramics. Ceramic whiskers and fibers are being used increasingly in the industry, mainly for reinforcement and toughening. Single crystals are used for optical, electronic, and magnetic applications and for research purposes. The following sections explain various production methods for powders, whiskers, fibers, and single crystals. Some of these methods are commercially applied, whereas some are still at the research scale.

2.1 Ceramic Powders

Today, an important fraction of ceramic powders is produced by conventional means. However, new methods are invented and existing ones are continuously improved to produce powders with higher purity as well as controlled grain size and distribution. High purity is essential in electroceramics since electronic properties depend critically on the level of purity (see Sect. 4.4). Smaller particle size, down to the level of nanometers, is required for the new nanoscale ceramics and composites. It is also generally accepted that a fine particle size with controlled size distribution is advantageous for ceramic consolidation.

2.1.1 Conventional Powder Production Techniques

2.1.1.1 Alumina (Al_2O_3)

Alumina powder is conventionally produced in large amounts by the *Bayer process* [2.1]. The process begins with bauxite ore that is a natural mixture of hydrated aluminum oxide, kaolin clay, hydrated iron oxide, and titanium dioxide. The steps to produce pure Al_2O_3 are as follows [2.2,2.3] (Fig. 2.1):

1. The ore is dried and calcined (i.e. heated to remove organic materials),
2. crushed and ground to –70 mesh to accelerate leaching,
3. leached in a NaOH solution (2-8 hours at 160°C and 400 KPa) to form soluble Na^+ and AlO_2^- ions by the following reaction:

$$Al(OH)_3 + NaOH \rightarrow AlO_2^- + Na^+ + 2\,H_2O \tag{2.1}$$

4. Insoluble residues of Fe, Si, and Ti are settled and filtered.
5. The solution is cooled to precipitate $Al(OH)_3$ with the aid of fresh $Al(OH)_3$ addition to nucleate precipitation.
6. The slurry is filtered and washed to remove $Al(OH)_3$ and the remaining NaOH.
7. $Al(OH)_3$ is calcined at 1100–1200°C to yield Al_2O_3.
8. The product is crushed and ground to the desired size range.

High-purity (up to 99.5%) alumina with a particle size ranging from several microns to submicron can be produced by this method [2.3].

High-purity Al_2O_3 has been produced by the *alum process* until recently [2.4]. Alum is the product of the following reaction:

$$Al_2(SO_4)_3 + (NH_4)_2SO_4 + 24H_2O \rightarrow 2NH_4Al(SO_4)_2{\cdot}12H_2O \tag{2.2}$$

Most of the Fe and Ti impurities are removed by maintaining the pH below 1 during reaction. Other impurities are removed during recrystallization. Alum is melted by heating to 80°C and yields a viscous liquid that foams due to volatile decomposition products. The remaining alumina is easily ground to a fine, low-density powder with few agglomerates and high purity (>99.99%).

2.1.1.2 Zirconia (ZrO_2)

Zirconia occurs in nature as the mineral baddeleyite, which always contains a small amount of hafnium oxide, and as zircon, a compound oxide with silica ($ZrO_2{\cdot}SiO_2$). Zirconia is usually produced from zircon, which is available in large quantities in nature [2.5]. Conventional methods of ZrO_2 powder production are

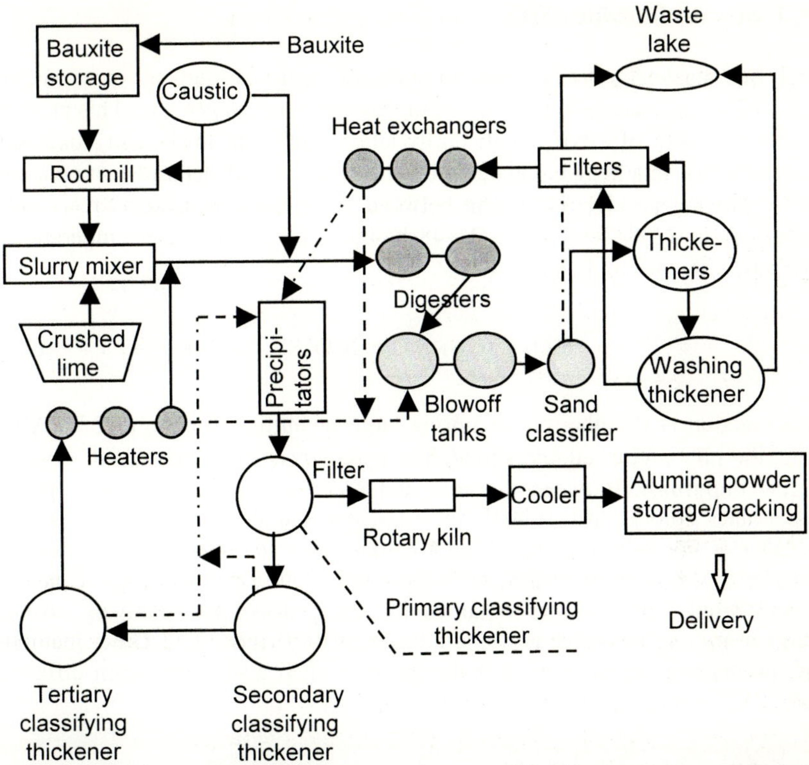

Fig. 2.1. Schematic flow diagram of the Bayer process for alumina powder production

chlorination-thermal decomposition, lime fusion, and alkali oxide decomposition [2.6]. Here we will describe the former method as an example.

In the chlorination-decomposition process, zircon is directly chlorinated in an electric arc furnace with carbon, usually after milling and pelletizing, by the following reaction:

$$ZrO_2{\cdot}SiO_2 + 4C + 4Cl_2 \rightarrow ZrCl_4 + SiCl_4 + 4CO \tag{2.3}$$

After heating at 800–1200°C, $ZrCl_4$ is distilled off and condensed at 150-180°C. $SiCl_4$ is condensed at −10°C. Next, hydrolysis with water is usually carried out to produce a solution of zirconium oxychloride. Crystals achieved by cooling the solution are separated and dried at 85°C. The oxide is then recovered by calcining these crystals.

2.1.1.3 Silicon Carbide (SiC)

Natural SiC has been found only in meteoric iron. Hexagonal or α-SiC is synthesized commercially in large quantities by the *Acheson process*. This process consists essentially of mixing high purity silica sand with low-sulfur coke (C), salt, and sawdust in a large elongated pile with large carbon electrodes at opposite ends. By passing an electric current between the electrodes, coke is resistively heated to about 2200–2500°C for 24–48 hours [2.1,2.3]. At these temperatures, SiO_2 reacts with coke and yields SiC:

$$SiO_2 + 3C \rightarrow SiC + 2CO \tag{2.4}$$

This reaction is endothermic, requiring large quantities of energy [2.7]. After cooling, the pile is crushed and sorted. The core consists of green α-SiC crystals with few impurities. Between the core and outer layer, SiC has lower purity (hence a black color) that is suitable for refractory and abrasive applications. The outer layer is not fully reacted. Therefore this layer is added to the following batch. The core is ground, treated with acids and alkalis, magnetically cleaned of iron, and dried. SiC powders obtained by this process are relatively coarse. Subsequent milling increases cost and introduces impurities [2.4]. Other methods of SiC production similarly involve the reaction of Si and C, but from different sources.

2.1.1.4 Silicon Nitride (Si_3N_4)

Commercially available Si_3N_4 is synthesized mostly by *direct nitridation* [2.8]. In this process, metallic silicon powder is refined, mixed with binders and catalysts, cast and dried, and reacted with nitrogen at 1200–1400°C [2.9]. The rate of nitrogen feed has to be controlled carefully since the reaction is highly exothermic. The product, predominantly α-Si_3N_4, is pulverized, classified, and refined to yield the final powder. The direct nitridation method is simple, but the particle shape and size are difficult to control [2.10]. The resulting purity is limited by the initial purity of the silicon powder.

The *solution* or *Ube* technique is used to produce higher purity Si_3N_4. For this, $SiCl_4$ solutions or silanes are reacted with ammonia:

$$SiCl_4\,(g) + 6NH_3\,(g) \rightarrow Si(NH)_2\,(s)\,(\text{silicon diimide}) + 4NH_4Cl\,(g), \tag{2.5}$$

$$3Si(NH)_2\,(s) \rightarrow Si_3N_4\,(s) + 2NH_3\,(g), \tag{2.6}$$

$$3SiH_4\,(g)\,(\text{silane}) + 4NH_3\,(g) \rightarrow Si_3N_4\,(s) + 12H_2\,(g). \tag{2.7}$$

The product is a very fine particle size amorphous Si_3N_4 which is crystallized by heat treatment at 1400°C [2.7].

2.1.2 Modern Powder Production Techniques

New powder production techniques are being developed for several reasons, including the need for higher purity, small and uniform particle size, controlled particle shape, and novel compounds. Submicron, monodispersed particles, which are ideal for sintering, have been produced by some of these new techniques. Similarly, the high purity of powders is a definite advantage in controlling sintering and mechanical, chemical, and physical properties. Important techniques and developments in powder production are discussed here.

2.1.2.1 Solution Techniques

This category consists of methods, which use solutions of desired ceramic cation components. In these techniques, the solution preparation step usually involves dissolving salts such as nitrates, sulfates, acetates, etc. in a solvent, typically water [2.11]. An alternative preparation method of solutions is to dissolve metals in acids.

2.1.2.1.1 Solvent Vaporization

Direct Evaporation

This technique simply involves heating a solution to evaporate the solvent. Segregation of chemical components occurs almost always during drying and largely limits the use of this method.

Spray Drying

In this method, solutions are dispersed into small droplets (10–100 μm in diameter) and injected into a hot air stream in which rapid drying takes place. Segregation is limited due to the small droplet size and rapid heating. High purity β-Al_2O_3 and reactive sodium rare-earth silicate powders for ion conductor applications have been produced by this method.

One recent adaptation of this process utilizes microwave heating during spray drying [2.12]. Submicrometer-sized ferrite, Al_2O_3, $YBa_2Cu_3O_7$, $BaTiO_3$, TiO_2, and mixed powders from aqueous inorganic and organic salts have been produced by this process. The novel technique also allows the decomposition of sol–gel combinations with subsequent recovery of the organic precursors and solvents.

Spray Roasting (Spray Decomposition, Aerosol Decomposition)

In this method, a carrier gas flows through an aerosol generator (ultrasonic atomizer, spray gun, nebulizer, etc.) where an aerosol or spray of the desired precursor solution is produced and directed into the furnace (Fig. 2.2). The related steps of powder formation were [2.13]

1. formation of the precursor aerosol,
2. evaporation of the solvent (drying) and solute precipitation, and
3. reaction/thermal decomposition of the solute to yield the ceramic powder product.

The process has been well adapted to the synthesis of oxide powders for ferrite production. Feed solutions are prepared from metallic chlorides, which are frequently by-products of metal pickling operations in the steel industry. In the case of ferrites, the resulting powders are hollow spheres (residues of solution droplets) 20–400 μm in diameter, with an average particle size of 0.1 to 0.25 μm. MgO, Al_2O_3–ZrO_2, ZrO_2, barium calcium titanate ($Ba_{0.86}Ca_{0.14}TiO_3$), BN, and sodium aluminosilicate powders have been produced by this process by various researchers. The advantages are precise control of the powder stoichiometry and avoidance of agglomeration. The disadvantage is poor control of particle morphology that usually results in porous or doughnut-shaped particles.

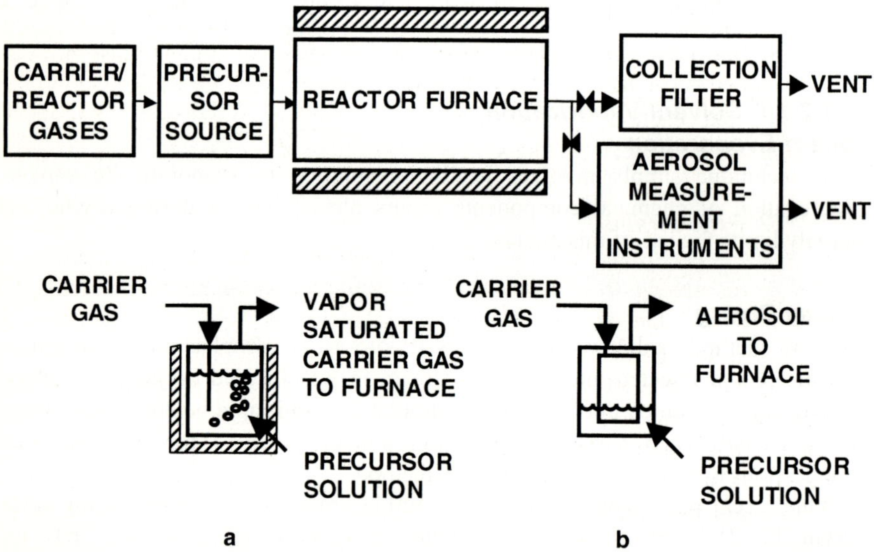

Fig. 2.2. Powder production by spray roasting (From C.L.J.Adkins, "Powder Preparation by Gas-Phase Techniques" in *Characterization of Ceramics*, edited by R.E.Loehman, Butterworth-Heinemann Stoneham, MA 1993. Reprinted by permission of Butterworth-Heinemann)

Emulsion Drying

This process minimizes component segregation by dividing the solution into small droplets. First, a solution is prepared in water. The solution is then mixed with preheated (~170°C) kerosene or an organic fluid that is liquid above 100°C such as toluene, and a liquid emulsifying agent. Water will evaporate immediately, and a salt mixture will be left dispersed in the organic liquid. The salts are screened from the liquid and calcined to form the oxide product.

Water can also be removed by using vacuum, freeze drying, or by combustion of the emulsion. If heated oils are used as the drying medium, the process is also referred to as the *hot kerosene process*. Although this method has not been widely used in the recent past, it is suitable for preparing both simple and complex compounds, including titanates, oxides, and superconductors [2.14].

Freeze Drying

Here, a salt solution is dispersed into small droplets (0.1 to 0.5 mm in diameter) and immediately frozen upon spraying into a cold liquid such as hexane to prevent segregation. The solvent is sublimated by slowly raising the pressure and temperature [2.3]. Vaporization can be quite slow since cations are relatively immobile in a solid. The final step is calcining, to decompose the crystallized salts and form very fine crystallites. Although freeze drying is a successful method for producing fine, homogeneous ceramic powders, it has been limited to research due to high capital costs and energy inefficiency. Other limitations are insolubility of some salts and reaction of others, leading to precipitation. Ferrites, β-Al_2O_3 ion conductors, and piezoelectrics have been produced by this method.

2.1.2.1.2 Precipitation

Precipitation, borrowed from techniques used for classical wet quantitative analysis, was among the first nonconventional powder preparation techniques developed. It is achieved either by adding a chemical precipitant to a solution or by changing the pressure or temperature. Thus, no special instrument is required for precipitation as a laboratory tool. Precipitation is a viable manufacturing tool and is being used to produce $BaTiO_3$ powders [2.4]. *Coprecipitation* is the precipitation of compounds with more than one cation. For example, a Ni–Fe solid solution oxalate can be precipitated from a solution of ammonium oxalate, iron sulfate, and nickel sulfate in H_2O (at 60°C) [2.3]:

$$NiSO_4 + 4FeSO_4 + 5(NH_4)_2C_2O_2{\cdot}H_2O \rightarrow$$

$$5Ni_{0.2}Fe_{0.8}C_2O_4{\cdot}2H_2O + 10NH_4^+ + 5SO_4^{2-} \qquad (2.8)$$

The solid solution is calcined below 500°C to yield a nickel ferrite compound:

$$3Ni_{0.2}Fe_{0.8}C_2O_4{\cdot}2H_2O + (2 + x/2)\, O_2\,(g) \rightarrow$$

$$Ni_{0.6}Fe_{2.4}O_4\,(s) + xCO\,(g) + (6 - x)CO_2\,(g) + 6H_2O\,(g). \qquad (2.9)$$

TiN precursors were precipitated by the reaction of tetrakisdimethylamidotitanium $\{Ti(N(CH_3)_2)_4\}$ with liquid ammonia (NH_3) at ambient temperature. Transamination reactions resulted in a $Ti_3(N(CH_3)_2)(NH_2)_2N_3$ polymer precipitate that was pyrolyzed in vacuum at 800°C to obtain TiN powder. ZrN and Nb(C,N) powders were also prepared by similar methods. In a similar effort, a dimethylaminoalane $\{H_2AlN(CH_3)_3\}$ solution in toluene reacted with excess NH_3,liq at –70°C. The precipitate was pyrolyzed in N_2 at 1000°C resulting in AlN powders with a surface area of 118 m^2/g and a particle size below 0.1 μm [2.15].

Important parameters of the precipitation technique are solution concentration, pH, mixing or stirring rate, and temperature. Precipitates are purified by digestion (coarsening of larger precipitates at the expense of finer ones in the solution), washing, and if required, reprecipitation before filtration. Other special precipitation methods are described here.

Air Oxidation

Some materials such as ferrites can be oxidized directly, whereas most oxide products can be achieved only after a thermal decomposition step. Air oxidation is therefore a specialized method and its use is limited. The method generally involves bubbling air or oxygen through a solution at 50–90°C that often yields spinels with unaggregated cubic particles. Examples include the preparation of Cr-substituted γ-Fe_2O_3 and transition-metal cobaltide spinels [2.11].

Hydrothermal Synthesis

This method is similar to air oxidation, but the solution temperatures are increased to 100–350°C using an autoclave at pressures up to 100 MPa. Due to the need for special equipment, this method is not widely used. Among many hydrothermally grown crystals, α-quartz single crystals are the only crystals successfully produced for commercial use [2.10]. One feature that makes hydrothermal synthesis advantageous in the production of electronic ceramic powders, particularly dielectrics, is the ability to produce solid solution particles of controlled size. Ceramic powders such as Y–TZP and dielectric PZT, Z5U, and Y5V, have been derived by this process [2.16]. Hydrothermal synthesis has also been used to produce fine oxide powders such as TiO_2, Al_2O_3, Nb_2O_5, and HfO_2 [2.10]. Hydrothermally grown ultrafine ferrite powders sinter to near theoretical density at 1000°C without sintering aids [2.17].

Molten Salt Precipitation

In this method, oxide precursors are mixed with low melting point salts such as NaCl, KCl, or their eutectics. Upon melting of the mixture, reactant oxides dissolve in the salt, and the desired compound precipitates due to its low solubility in the molten salt. The melt is cooled down, and the salt is dissolved to yield the desired powder product. Various ferrites, $BaTiO_3$, and $Bi_4Ti_3O_{12}$ powders have been produced by this method [2.11]. Particle sizes less than 0.1 μm have been achieved.

2.1.2.1.3 Sol–gel Processing

Ebelman and Graham performed the pioneering work in this domain on silica gels in the mid-1800s [2.18]. The work of Matijevich [2.19] and Barringer, Bowen, and co-workers [2.20] increased interest from the viewpoint of ceramic science and technology. Matijevich and co-workers demonstrated that a wide variety of shapes can be produced by colloidal techniques, as depicted in Fig. 2.3. They produced monodispersed submicron particles of a wide range of materials, including TiO_2, α-Fe_2O_3, $BaTiO_3$, CeO_2, zinc sulfide, hydroxides, carbonates, sulfides, mixed phases, and coated particles. Sol–gel routes involve converting a *sol* (a colloidal suspension of a solid in a liquid) to a *gel* (a semirigid colloidal dispersion of a solid in a liquid). *Sols* contain large molecules or colloidal particles and are intermediate between true solutions of low molecular weight species and coarse dispersions. *Gels* have a form of matter intermediate between liquids and solids. They are amorphous or nearly amorphous and generally look dry. A polymer gel is composed of a polymer network that is swollen by a solvent. A gel can be created by increasing the amount of the dispersed phase. A sol–gel transition occurs beyond which the viscosity increases drastically. Thermal dehydration of a gel produces shrinkage, resulting in an elastic solid called a *xerogel*.

In general, a sol–gel method uses one or several elements in the form of sol and/or gel to obtain a homogeneous, mainly amorphous solid. However, the term *sol–gel* is used for two different methods. One starts with a colloidal solution and the other with polymerized alkoxides or nitrates [2.18].

The first method uses raw materials in the form of colloidal particles such as clays and other natural colloids. The hydrolysis and precipitation of inorganic species in aqueous solutions to produce oxide colloids is a similar method. Compounds such as silicic acid, vanadic acid, and phosphates exhibit this reaction. Sol and gel formation are related to the acidic character of the framework or to the zeta potential. For these materials, the sol–gel transition can be reversible, depending on concentration and pH.

The second method uses chemical gels instead of physical gels. The essential reactions in this case are hydrolysis and polycondensation of metal/nonmetal alkoxides or metallo-organic reagents, usually in alcohol or hexane solutions, to yield a polymerized structure. A homogeneous solution is obtained by complexation reactions between alkoxides, except if preferential polycondensation occurs. A simple hydrolysis-polycondensation reaction is as follows [2.21]:

$$M(OR)_n + nH_2O \rightarrow M(OH)_n + nROH \text{ (hydrolysis)} \tag{2.10}$$

$$M(OH)_n \rightarrow MO_{n/2} + n/2\, H_2O \text{ (polycondensation)} \tag{2.11}$$

where $M(OR)_n$ represents a metal(M)alkoxide, $M(OH)_n$ is a monomer, and $MO_{n/2}$ is the polymer, the main substance of a chemical gel. Gelation depends on many

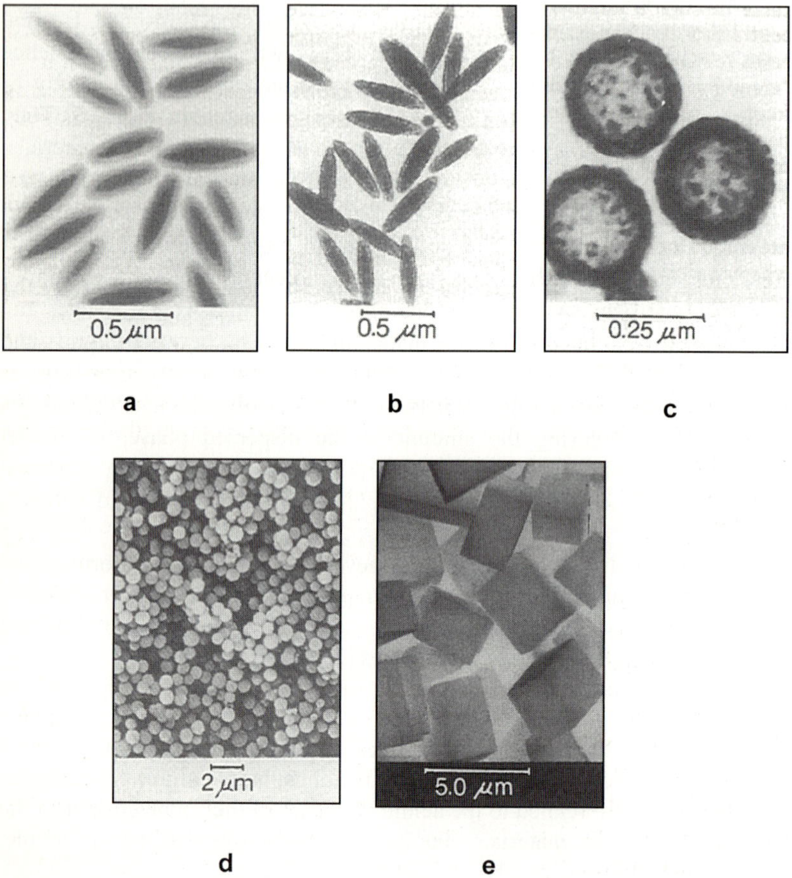

Fig. 2.3 a-e. Various forms of sol–gel derived powders: TEM of hematite (α-Fe_2O_3) particles coated with silica **a**, TEM of hematite particles coated with zirconium (hydrous) oxide **b**, TEM of hollow yttria particles **c**, SEM of nickel ferrite particles **d**, and SEM of lead sulfide particles **e** (Reprinted with permission from E. Matijevic: *Preparation and Properties of Uniform Size Colloids*. Chem. Mater. **5 [4]**, 412-426 (1993) Copyright 1993 American Chemical Society)

factors such as pH, water/alkoxide ratio, and the chemical properties of the alkoxides. It is believed that the gelation time is not an intrinsic property of the sol, since it depends on the size of the container [2.18].

A typical alkoxide route is to react a metal with an alcohol to form a metal alkoxide. The latter is then dissolved in a suitable alcohol. Next, the alkoxide is hydrolyzed upon the addition of water. After the pH of the solution is adjusted, the material polymerizes to form a gel. The gel consists of loosely bonded material and a water–alcohol mixture. This jelly-like material is heated to 200–500°C and is slowly dried to remove the liquid. Upon drying, powders with particle sizes in the range of 0.003 to 0.1 μm are obtained. This method can produce extremely homogeneous mixtures of two or more components because the mixing takes place at the atomic level in a liquid rather than at the particle level in the solid state. The uniform and fine particle size allows sintering to higher fractional densities than would be possible with conventional powders.

In the alkoxide method, the sol–gel transition (hydrolysis-polycondensation) is irreversible. A homogeneous distribution of various elements can be achieved by complexation in the liquid state.

Another method in the chemical gel category uses complex alkoxides or parent compositions in which different elements are present in the organic chains, thereby avoiding a complexation step. $(OBu)_2$–Al–O–$Si(OEt)_3$ and $(OiPr)_2$–Al–O–$SiMe_3$ are examples of such alkoxides, where Bu represents butyl, Et: ethyl, Pr: propyl, and Me: methyl groups, respectively.

Sol–gel methods are also classified according to the alkoxide/water ratio and hydrolysis time. The slow hydrolysis route involving a high alkoxide/water ratio (1–20 moles H_2O per one mole of oxygen in the final oxide composition) produces monolithic samples and is extensively studied for glass-forming formulations. Rapid hydrolysis with high water excess, on the other hand, leads to gel powders that can be pressed to high densities at ambient temperature.

Studies in sol–gel chemistry show that three different steps are usually involved [2.21]:

1. formation of solute precursors via polymerization of molecular species,
2. nucleation and growth of primary particles, and
3. aggregation of primary particles.

According to one theory, monodispersed particles form at a critical supersaturation of the solute precursors [2.15]. With a burst of nuclei, many particles start to form at once. The growth of colloidal particles from these nuclei proceeds by diffusion of solute species onto existing nuclei. However, this theory has been disputed on the basis of results of cryogenic transmission electron microscopy (TEM) studies. For the hydrolysis of $Si(OC_2H_5)_4$ solutions in ethanol and propanol with NH_4OH, it was hypothesized that the hydrolyzed monomer forms microgel polymers, which collapse above a critical size and cross-link density that renders them insoluble. The collapsed particles act as seeds that densify by condensation while growth occurs by adsorption of hydrolyzed monomers or polymers to their surface [2.15].

Ceramic powders for structural and electronic applications have been synthesized by sol–gel processing of alkoxides. As an example, $Pb(Fe_{2/3}W_{1/3})O_3$ powders are synthesized from mixed solutions of lead propoxide $\{Pb(OC_3H_7)_2\}$ and iron propoxide $\{(Fe(OC_3H_7)_3\}$ in methoxyethanol $\{(CH_3OCH_2CH_2OH)\}$ containing acetylacetone by

1. fluxing at 125°C,
2. adding a $W(OC_2H_5)_6$ solution in methoxyethanol,
3. refluxing the solution mixture at 80°C,
4. hydrolysis with a H_2O/ethylene glycol mixture in methoxyethanol,
5. gelation, drying, and comminution, and
6. calcination at 870°C.

These powders have applications in multilayer ceramic capacitors due to their low sintering temperatures and high dielectric constants [2.21].

The sol–gel route may also be used as a green forming method. This topic will be discussed in Sect. 3.2.8.

2.1.2.1.4 Electrochemical Synthesis

In the electrochemical synthesis of ceramics by a *redox reaction*, a metal ion or complex is oxidized at the electrode surface [2.23]. This method has been used to deposit films of metal oxides and hydroxides on substrates.

Another electrochemical method is the *electrogenerated base* method where cathodic processes are used to generate a base at an electrode surface. Metal ions or complexes are then hydrolyzed by the electrogenerated base. One example of base electrogeneration is given here:

$$2H_2O + 2e^- \rightarrow H_2 + 2OH^- \quad (2.12)$$

$$NO_3^- + H_2O + 2e^- \rightarrow NO_2^- + 2OH^- \quad (2.13)$$

$$\text{metal ion} \xrightarrow{\text{base}} \text{metal hydroxide} \rightarrow \text{metal oxide.} \quad (2.14)$$

CeO_2 powders with an average particle size of 1.8 μm were prepared by the cathodic electrogenerated base method [2.23]. Nanocrystalline CeO_2 powders were synthesized by the same method from cerium (III) nitrate hexahydrate and ammonium nitrate solution [2.24]. The average crystallite size depended on the solution temperature. It increased from 10 nm at 29°C to 14 nm at 80°C. An average synthesis rate of 1.5 g/h was achieved at a current density of 1 A/cm^2. A recent study [2.25] describes the preparation of TiO_2, ZrO_2, and $PbZrO_3$ films from *N*,*N*-dimethylformamide (DMF) solution. $TiCl_4$ solutions in hydrogen peroxide/DMF/water were used to form TiO_2 films. The presence of H_2O_2 or H_2O was necessary for film formation. It was proposed that these films are formed by the following sequence:

1. dissociation of $TiCl_4$ salt,
2. formation of a peroxo complex,
3. hydrolysis of the complex by the electrogenerated OH^-, and
4. decomposition of peroxytitanium hydrate.

Similarly, zirconia films were formed from $ZrO(NO_3)_2{\cdot}nH_2O$ solutions in DMF or DMF/H_2O_2 and $PbZrO_3$ films were prepared from $Pb(NO_3)_2$ and $ZrO(NO_3)_2{\cdot}nH_2O$ solutions in a molar ratio of Pb/Zr =1. All crystalline phases were obtained after annealing between 600 and 1000°C. It was also noted that high current densities and deposition times result in sedimentation of deposits within the electrolytic bath, which may be used to prepare ceramic powders of high purity.

An AlN precursor in the liquid state, $Al(NHR)_3$, was obtained by electrochemical synthesis. For this purpose, aluminum was anodically dissolved in acetonitrile containing a primary amine and a tetraalkylammonium salt to increase conductivity. The liquid precursor obtained by removing excess amine and solvent was polymerized to form a gel with the composition $Al_2(NR)_3$. Pyrolysis (fragmentation of a molecule at high temperature) of this gel at 800°C in an NH_3 atmosphere yielded AlN powders with an average particle size of 25 nm [2.15].

2.1.2.1.5 Reactive Electrode Submerged Arc Process

A relatively new technique called the *reactive electrode submerged arc (RESA) process* uses metal electrodes submerged in a dielectric fluid to produce metal nitrides, borides, and carbides of metal alloys [2.26]. An electric arc is struck across two electrodes inside the fluid. The spark provides a zone of extremely high temperature, resulting in the vaporization of the electrode metal and its reaction with the surrounding dielectric fluid, as shown in Fig. 2.4. A bubble is formed to contain the arc in which the reaction of metal vapor and dielectric vapor occurs. The product is quenched to submicrometer size spherical particles, dispersed as a colloid in the surrounding fluid. Kumar and Roy [2.26] synthesized fine particles of SiC and TiC by this method. In the case of SiC, silicone oil was used as the liquid and C electrodes were used as the C source. TiC was formed from Ti electrodes in various organic fluids such as mineral oil and aniline. Ultrafine AlN powders were fabricated by the RESA method using pure Al rods as electrodes and liquid nitrogen or liquid ammonia for the dielectric fluid. Spherical particles with sizes ranging from 5 to 100 nm were obtained [2.27].

2.1.2.2 Vapor-Phase Reaction Techniques

Vapor-phase methods play an important role in the commercial production of some ceramic powders such as TiO_2, C-black, ZnO, and SiO_2. The physical processes involved in vapor-phase synthesis are, as in synthesis from solutions, chemical reaction, mass transfer, nucleation, coagulation, and condensation [2.13]. The important advantages of vapor phase synthesis are listed here:

1. Due to the low density of gases, a low concentration of reactants can be achieved, whereby very fine particles are produced.
2. Direct formation of particles eliminates the need for an additional calcining step.
3. Most liquid-phase systems contain oxygen, but in vapor-phase synthesis, oxygen-containing gases may be avoided, facilitating high purity nonoxide powder production.
4. Intimate mixing of the components is achieved.
5. High purity and size distribution control is achieved.
6. Waste is minimized.

One important disadvantage of vapor-phase methods, however, is that high-temperature multiphase reactions are difficult to control due to the lack of a full understanding of the processes involved. Table 2.1 lists various powder products obtained by vapor-phase reactions and their corresponding reactants. Important vapor-phase techniques are briefly described here.

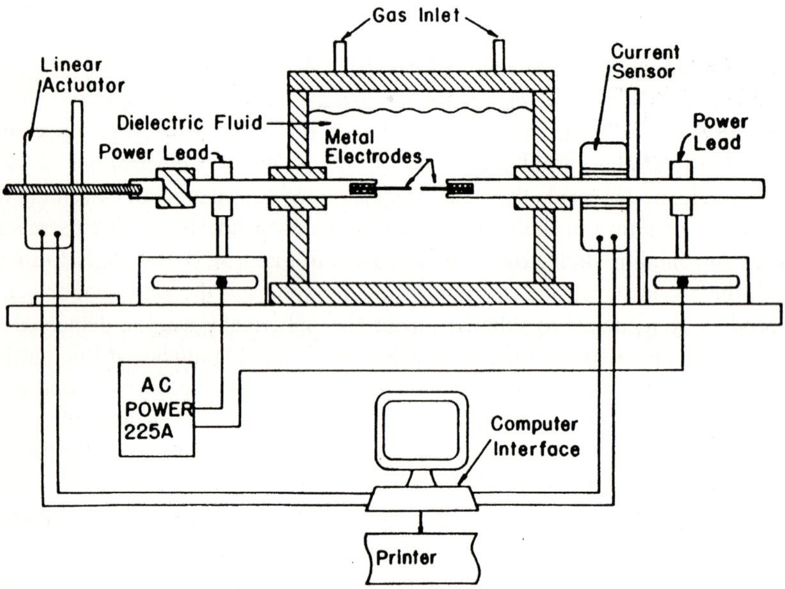

Fig. 2.4. Schematic of the reactive electrode submerged arc method (From A. Kumar and R. Roy: *Reactive-Electrode Submerged-Arc Process for Producing Fine Non-Oxide Powders*. J. Am. Ceram. Soc. **72 [2]**, 354-356 (1989). Copyright 1989 American Ceramic Society, reprinted with permission)

Table 2.1. Products Derived from Gas-Phase Reactions [a]

Product	Specific Surface Area (S) (m^2/g)	Primary Particle Mean Diameter (d) (nm)	Reaction Method	Reactants	Reaction Temp. (T) (°C)
AlN	160–260	20–110	Plasma	$Al + N_2 + NH_3$	3500
AlN	–	17	Thermal (CVD)	$(Me_2CHCH_2)_3Al + NH_3$	800–1400
B_4C [b]	–	20–30	Plasma	$BCl_3 + CH_4 + H_2$	–
B_4C	55	34	Laser	$BCl_3 + CH_4$ or $C_2H_4 + H_2$	–
B_4Si	–	80	Plasma	$B_2H_6 + SiH_4$	–
Mo_2C	–	10–40	Thermal	$MoCl_5 + CH_4 + H_2$	>1200
Mo_2C	–	10–40	Thermal	$MoO_3 + CH_4 + H_2$	>1200
MoO_xC_y	20–59	10	Thermal	$Mo(CO)_6 + CO + H_2$	1100–1800
SiC	60–164	4–60	Plasma	$SiH_6 + CH_4$ [+ H_2]	–
SiC	–	10–20	Plasma	$SiCl_4 + CH_4$	–
SiC	61	10–30	Plasma	SiO_2 or $SiO + CH_4$	–
SiC	40–80	8–50	Plasma	Chloromethyl–silane (e.g. Me2SiCl2 or MeSiCl3) [+ H2]	–
SiC	–	10–60	Plasma	$SiMe_4 + H_2$	–
SiC	22–240	10–90	Laser	$SiH_4 + CH_4$ or C_2H_2 or C_2H_4	1650–1830
SiC [c]	8–150	–	Laser	$SiH_2Cl_2 + C_2H_4$	–
SiC [c]	73–233	5–200	Gas Evaporation	Si + C [+ CH_4]	>2200
SiC	42–69	5–150	Thermal	$SiMe_4$ or polysilane [+ H_2]	700–1500
SiC [c]	–	10–150	Thermal	$SiH_4 + CH_4 + H_2$	1000–1400

(Continued)

Table 2.1. (continued)

Product	Specific Surface Area (S) (m^2/g)	Primary Particle Mean Diameter (d) (nm)	Reaction Method	Reactants	Reaction Temp. (T) (°C)
$SiC–Si_3N_4$	–	20–70	Thermal	$SiMe_4 + NH_3$ [+ H_2]	1200
Si_3N_4 [c]	40–163	>8	Laser	$SiH_4 + NH_3$	–
Si_3N_4	–	10–60	Laser	$SiCl_4$ or SiH_2Cl_2 + NH_3 (+SF_6 sensitizer) [+ H_2]	–
Si_3N_4	26	–	Thermal	$SiH_4 + NH_3$	820
Si_3N_4	35	–	Thermal	$SiCl_4 + NH_3 + H_2$	900–1500
Si_3N_4	30–70	5–40	Plasma	$SiCl_4 + NH_3$ + [H_2]	–
Si_3N_4	160	110	Plasma	$Si + NH_3$	–
Si_3N_4	145–250	5–60	Plasma	$Si + N_2$	–
Si_3N_4	35–152	–	Plasma	$SiH_4 + NH_3$	–
SiAlON	186	110	Plasma	$Si + Al + NH_3 + O_2$	–
TaC	72	5–100	Plasma	$Ta + CH_4$	–
TiB_2	30–50	>20	Laser	$TiCl_4 + B_2H_6$	–
TiB_2	3–25	80–600	Plasma	$TiCl_4 + BCl_3 + H_2$	–
TiB_2	24	>50	Plasma	$TiCl_4 + BCl_3 + C_2HCl_3 + HCl$	–
TiC	–	50	Gas Evaporation	Ti + C	–
TiC [d]	80–120	10–40	Plasma	$TiCl_4 + CH_4$ [+ H_2]	–
TiN	28–31	10	Plasma	$Ti + N_2$ [+ H_2]	–
TiN	–	10–400	Thermal	$TiCl_4 + H_2 + NH_3 + N_2$	250–700
TiN	–	5–50	Plasma	$TiCl_4 + NH_3 + H_2$	–
$VN_{1.55–1.87}$	70	–	Thermal	$VCl_4 + NH_3 + N_2 + H_2$	700–1100

(Continued)

Table 2.1. (continued)

Product	Specific Surface Area (S) (m^2/g)	Primary Particle Mean Diameter (d) (nm)	Reaction Method	Reactants	Reaction Temp. (T) (°C)
WC	–	2–16	Plasma	$W + CH_4$	–
WC	–	5–8	Plasma	$WCl_6 + CH_4$ [+ H_2]	–
WC, W_2C	–	20	Thermal	$WCl_6 + CH_4 + H_2$	1000–1400
ZrN (<80%)	–	<50	Thermal	$ZrCl_4 + NH_3 + N_2 + H_2$	1000–1500

[a] Adapted From: R.W. Chorley and P.W. Lednor, *Synthetic Routes to High Surface Area Non-Oxide Materials.* Adv. Mater. **3 [10],** 474-485 (1991) (used with permission of Wiley – VCH)
[b] Various stoichiometries
[c] Contains Si and C impurities
[d] Contains Ti and C impurities

2.1.2.2.1 Vaporization-Condensation

Most of the ceramics that are technologically important have very high melting points and are difficult to vaporize. In addition, many important ceramics such as SiC and Si_3N_4 sublime instead of evaporating. Therefore, this method is not attractive at present.

2.1.2.2.2. Vapor Decomposition

This is a method similar to spray (aerosol) decomposition. Instead of the spray, the precursor material is introduced into the reactor as a vapor. The vapor is obtained by bubbling a carrier gas through the precursor solution, or, in the case of a solid precursor, the carrier gas is passed through a heated, packed bed of the material [2.13]. The resulting vapor-saturated carrier gas is introduced into a furnace or plasma, where thermal decomposition takes place. If plasma is used, nucleation and growth of ceramic powders occur outside the plasma region. Al_2O_3, ZrO_2, and Fe_2O_3 powders at a size range of 10–20 nm were prepared from nitrate solutions by the plasma decomposition technique. ZrO_2–SiO_2 powders were also prepared where small ZrO_2 powders were trapped in the latter. Such a unique morphology may be useful for certain applications, for example, in promoting the advantageous effect of ZrO_2-toughening in ceramic matrices.

Zirconia powders were produced by the pyrolysis of zirconium *tert*-butoxide [2.28]. Thermal decomposition of tetramethylsilane, $(CH_3)_4Si$, and trimethylchlorosilane, $(CH_3)_3SiCl$, at 600–1200°C, can yield controlled size SiC powders in the particle size range of 0.1 to 4 μm [2.29]. Similarly, high-purity α-Si_3N_4 powders can be produced by the decomposition of silicon imide ($Si(NH)_2$ or $Si(NH_2)_4$) [2.10]. Titania (TiO_2) pigment is commercially produced from a titanium tetrachloride ($TiCl_4$) precursor by reacting the vapor at around 1000°C [2.13]. Alternative precursors such as titanium isopropoxide [$Ti(OC_3H_7)_4$] were also studied to lower the reaction temperature. A particle size range of 0.01–0.06 μm was achieved at 400°C, and amorphous particles were produced below 400°C. Similar results were observed for the thermal decomposition of silicon tetraethoxide [($Si(OC_2H_5)_4$] to silica and aluminum tri-*sec*-butoxide [$Al(OC_4H_9)_3$] to alumina.

2.1.2.2.3 Vapor–Vapor Reactions

Ceramic powders can be produced by the reaction of two vapor species with the aid of a suitable heat source such as laser, plasma, or simply a furnace. These processes are commonly employed for high-purity commercial powder production, especially due to the ease of vaporizing metal halides and hydrides. Table 2.2 compares the pros and cons of three processes involving different heat sources.

Conventional Heating

In this method, reactant gases are passed through a hot tube, or in flame oxidation, through a high-temperature flame. The product is collected afterward. Chlorides are generally preferred as reactants for the formation of oxide as well as nonoxide powders due to their generally low vaporization temperatures. TiO_2 powders were formed by the oxidation of $TiCl_4$ in the presence of additives such as $SiCl_4$, $ZrCl_4$, and $AlCl_4$. Forty nm powders of Si_3N_4 were prepared from SiH_4 and NH_3 as reactants. Other nitrides and carbides were also prepared using a similar heating method. Oxides were produced by the oxidation of vaporized metallo-organics. For example, Al_2O_3 was prepared by oxidizing aluminum acetylacetone with O_2. The aluminum compound was exposed to a N_2 carrier stream and then mixed with oxygen in a heated reaction tube [2.11]. Commercial applications of flame oxidation include SiO_2 powder production from $SiCl_4$, Al_2O_3 from $AlCl_3$, and TiO_2 from $TiCl_4$ in H_2/O_2 flames [2.30].

Plasma Heating

Plasma synthesis is usually a single-step, continuous process. Powders are formed from reactant gases in a high-temperature plasma, generated by radio frequency (RF), inductive coupling, or a plasma generated between two electrodes [2.31] (Fig. 2.5). Ultrafine (<100 μm), ultrapure powders, often in a metastable high-temperature phase, are produced by rapidly quenching the hot gases. The process is chemically nonspecific since the high temperatures within the plasma (7000–10000 K) lead to the decomposition of near-arbitrary precursors [2.32]. No milling or grinding is required. For example, Si_3N_4 has been produced with a 5- to 30-nm

Table 2.2. Comparison of Vapor-Phase Synthesis Techniques [a]

Heat Source	Advantages	Disadvantages	Examples
Conventional heating	Simple, economical, high purity	Low yields, reactant gases expensive, limited product range	Oxides, nitrides, carbides
Plasma heating	Ultrahigh purity, ultrafine size range, metastable phases possible, highly efficient	Hazardous by-products, nonspecific product, requires high power, large capital and operating costs, some agglomeration	Oxides, carbides, nitrides, mixed phases, metastable phases
Laser heating	High efficiency, ultrahigh purity, improved process control, uniformity, limited agglomeration, reproducibility	Requires large capital and operating costs	Si, nitrides, carbides, borides, oxides

[a] Adapted with permission from: L.M. Sheppard: *Vapor-Phase Synthesis of Ceramics*. Adv. Mater. Proc. **131 [4]**, 53-58 (1987). (©1987 ASM International® Materials Park, OH)

particle size, in $\alpha + \beta$ or in the amorphous form. Ultrafine SiC powders have been produced from silane and methane gases by this method. SiC powders were also prepared by the plasma reaction of SiO(g) and carbon. One disadvantage of the process is that by-products including HCl are produced, leading to problems such as reactor corrosion and contamination [2.33]. Some powder types produced by plasma synthesis in the industry or on a laboratory scale and their corresponding reactant gases are listed in Table 2.1.

Laser Heating

Direct coupling of laser radiation to gas molecules can overcome problems faced in other vapor-phase methods such as low power yield, high energy requirements, low reproducibility, and formation of hard aggregates. The laser synthesis method is typically clean due to a wall-free environment and a well-defined reaction zone that has high temperatures (>1000°C), steep temperature gradients, and short reaction times. For laser heating to take place, one of the reactants must absorb the laser radiation strongly. Otherwise, a sensitizer is added to the gas mixture. Powders such as Si, Si_3N_4, and SiC have been synthesized by using a CO_2 laser.

$SiCl_4$ or $SiH_2Cl_2 + NH_3$ mixtures were used to derive Si_3N_4 powders. $SiH_4 + CH_4$ or C_2H_2 were employed for SiC powders and $TiCl_4 + B_2H_6$ for TiB_2 powders [2.34]. Haggerty [2.35] produced Si_3N_4 by the reaction of silane (SiH_4) and ammonia (NH_3) as well as SiC by the reaction of silane and ethylene (C_2H_4), yielding nonporous particles with a mean size of 0.02–0.2 μm. Other powders

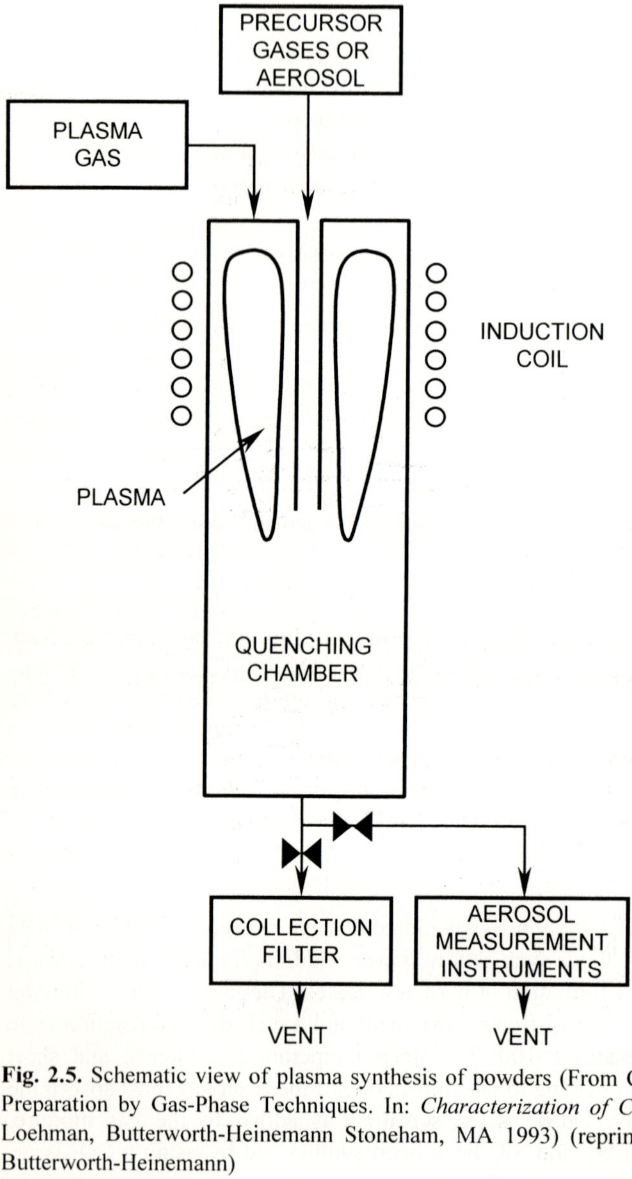

Fig. 2.5. Schematic view of plasma synthesis of powders (From C. L. J. Adkins: Powder Preparation by Gas-Phase Techniques. In: *Characterization of Ceramics* edited by R.E. Loehman, Butterworth-Heinemann Stoneham, MA 1993) (reprinted with permission of Butterworth-Heinemann)

produced industrially include B_4C and TiO_2, which are loosely agglomerated and have a particle size of 10–100 nm.

An alternative approach in laser synthesis is to inject the precursor material as fine aerosol droplets into the laser beam. This approach was employed to synthesize ultrafine Si(C,N), B, TiB_2, B_4C, and BN powders. BN particles with a particle size range of 11–31 nm were produced at a laser power of 400–600 W by the overall reaction,

$$BCl_3 + NH_3 \rightarrow BN + 3HCl. \tag{2.15}$$

The resulting turbostratic structure was converted to hexagonal BN upon annealing [2.36]. For other examples, see also Table 2.1.

2.1.2.3 Powders from the Solid State

2.1.2.3.1 Salt Decomposition

Thermal decomposition or *calcination* is often the final step in conventional powder production. Many of the solution methods require a final decomposition step. For single-phase oxides, a simple decomposition method such as thermal decomposition of carbonates can be used. High-purity powders can be obtained since carbonates do not leave any residues behind. γ-Fe_2O_3, used in magnetic recorders, was prepared by decomposing $N_2H_5Fe(N_2H_3COO)_3{\cdot}H_2O$ or $Fe(N_2H_3COO)_2(N_2H_4)_2$ salts at 250°C in a single step. Multicomponent metal oxides were prepared by calcining carbonates of the calcite structure. Uranium doped thoria was prepared by decomposing mixed nitrates in a microwave oven. Ground powders were sinterable to high densities [2.11]. Another example is the decomposition of salts such as $(NH_4)_2MoS_4$, $(NH_4)_2WS_4$, and $(NH_4)WO_2S_2$ under H_2–H_2S atmospheres, yielding sulfides having 40–85 m^2/g surface areas [2.34].

2.1.2.3.2 Self-Propagating High-Temperature Synthesis

Combustion synthesis or *self-propagating high-Temperature synthesis (SHS)* involves the ignition of powder mixtures that exhibit an exothermic reaction and produce temperatures in the range of 1000–3000°C under adiabatic conditions [2.37]. The SHS process is also used for direct production of ceramic bodies, which will be discussed in Sect. 3.4.6. Many single or multiphase ceramic compositions can be achieved by the SHS process. A simple laboratory setup for SHS studies is shown in Fig. 2.6.

Some typical examples describing the chemical reactions are given here:

- Simple binary compounds

$$Ti\ (s) + C\ (s) \rightarrow TiC\ (s), \tag{2.16}$$

$$2Ti\ (s) + N_2\ (l,g) \rightarrow 2TiN\ (s), \tag{2.17}$$

$$Si\ (s) + C\ (s) \rightarrow SiC\ (s). \tag{2.18}$$

- Complex compositions,

$$Ti\ (s) + C\ (s) + N\ (l) \rightarrow Ti(CN)\ (s). \tag{2.19}$$

- Cermet compositions

$$wTi\ (s) + xB\ (s) + yC\ (s) + zCu\ (s) \rightarrow\ lTiB_2\ (s) + mTiC\ (s) + zCu\ (s). \tag{2.20}$$

- Reactions with chemical activators

$$3Ti\ (s) + NaN_3\ (s) \rightarrow 3TiN\ (s) + Na\ (g). \tag{2.21}$$

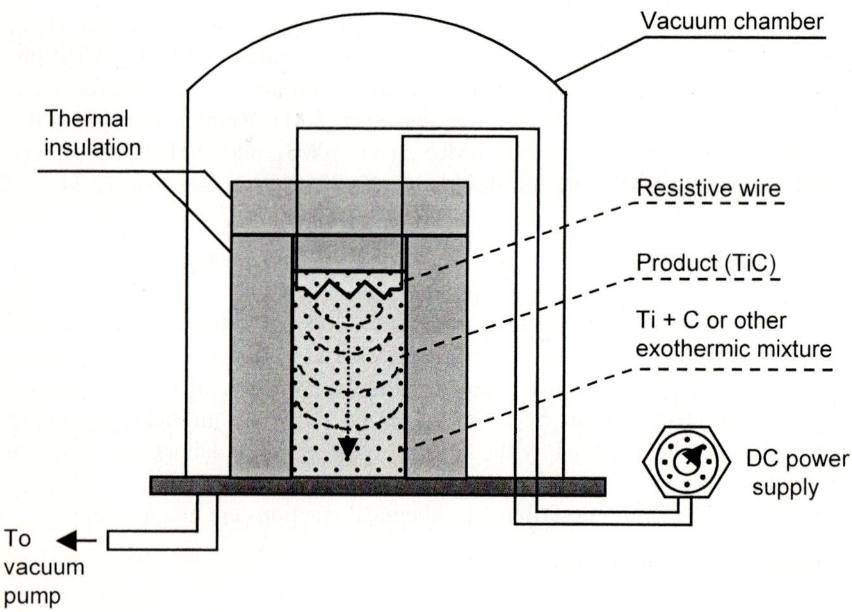

Fig. 2.6. Experimental setup for SHS studies

Yamada et al. [2.38] obtained β-SiC powders by igniting of Si and C mixtures initially at room temperature, by directly passing an electric current through the reactive mixture. The heat generated by the resistivity of the powders heated the samples to 720°C and initiated a rapid chemical reaction. Similarly, Pampuch et al. [2.37] synthesized sinterable β-SiC powders by heating pellets of powder mixtures to 1200°C in graphite containers. Maximum temperatures of about 2050°C were attained during the reaction. Submicron, high-purity powders were produced by this method.

In another example, α-Al_2O_3 and related oxide mixtures were synthesized by the combustion of stoichiometric compositions of metal nitrates and urea. Heating the mixtures in a furnace at 500°C in a Pyrex dish caused reaction and temperature increases to over 1600°C. Al_2O_3 was recovered as foam. Well-mixed Al_2O_3–ZrO_2 powders were also produced by a similar method [2.39].

A new method was developed to prepare ultrafine (~0.1 μm) powders from cakes produced by SHS. In this method, instead of mechanically disintegrating the cake, chemical disintegration is used. The cake is placed in a liquid solvent at a specified temperature, dissolving the defect layers at crystallite boundaries and leaving completely disintegrated powders. High surface area powders of AlN, BN, SiC, and B_4C were produced by this approach [2.40].

The SHS process has found some industrial applications worldwide. In Russia, United States, and Japan, different types of ceramic powders are produced on an industrial scale. At least three types of ferrite are produced industrially in Russia by SHS. An economic efficiency study showed that various nitride, carbide, and boride powders could be produced at a cost of 13 to 45% of the present production costs in the industry. Other advantages of the SHS process include simplicity and the possibility of producing high-purity composite powders, solid solutions, and nonstoichiometric phases [2.40].

2.1.2.3.3 Shock-Induced Synthesis

Some novel materials and metastable phases that cannot be produced by conventional methods can be formed under dynamic pressures achieved by explosives, gas guns, or other means. One of the striking examples of this process is the transformation of graphite to diamond powder. Early efforts to synthesize diamond from graphite started in the 1960s [2.41]. An explosion was used to accelerate a metal plate into another where the carbon in between was converted to diamond. Subsequent developments improved the productivity, lowered the cost of diamond, and commercialized the process. Du Pont's patented process used two concentric tubes and an explosive charge around them. The innermost pipe is filled with graphite and metal powders. The driver tube is accelerated toward the inner tube with tons of explosive around it in a large underground limestone mine. The explosion generates a pressure between 10 and 30 GPa inside the tubes and synthesizes microcrystalline diamond powder in the shock process. Each microcrystalline particle contains 10 to 10000 microcrystallites. Because of trace amounts of metal in the crystal lattice, the diamond powder is black. The density is 3.45 g/cc. Such powders are useful as abrasives and in fine polishing operations.

The production of cubic BN under high dynamic pressures was first achieved by Wentorf in 1957 [2.42]. Cubic BN is the second hardest material (c-BN H_v = 50–55 GPa, diamond H_v = 130 GPa), and its thermal conductivity is also very high (theoretical thermal conductivity = 13 W/cm K). Recently, direct shock synthesis of cubic BN has been realized from ammonia borane (NH_3BH_3) under high pressures (90 GPa) and temperatures (>1000°C) [2.42]. Aggregates with diameters of about 200 μm were recovered.

2.1.2.3.4 Mechanosynthesis of Carbides and Silicides

A novel approach to synthesizing carbides and silicides was recently developed by LeCaer and Matteazzi et al. [2.43,2.44] using a process involving comminution of elemental powders. In this process, elemental mixtures are milled with suitable milling media. No external heating is used. To synthesize various carbides, a powder to ball weight ratio of 1:10 to 1:5 was used. WC or hard steel balls were used as milling media. Transition-metal powders with a mean size of 45 μm were mixed with graphite and milled with a vibratory or planetary ball mill for 24 h. In this method, the temperature rise of the mill due to a fraction of the impact energy converted to heat usually remains below 100°C. Among compounds that can be produced are TiC, Ti–Al–V–C, V_2C, VC, HfC, Fe–C, and ZrC. Although oxidation is not observed, contamination from milling media is unavoidable unless a fluid energy mill is used or the milling media are from the same material as the product. The main advantages of the process are the possibility of forming nonequilibrium phases, high defect concentrations, and new compounds. The method is fairly simple and requires no external heating.

2.1.2.3.5 Polymer Pyrolysis

The principles of polymer pyrolysis are discussed in Sect. 3.4.10. Polymer pyrolysis is a rather new method that is under development for producing powders, fibers, and bulk ceramics. The method has the advantages of high purity, intimate mixing on a molecular or atomic level, and resulting particle sizes in the nanometer range. Polymer pyrolysis is usually one of the steps involved in sol–gel processing (see Sect. 2.1.2.1.3).

Alcohol exchange reactions have been employed to synthesize polymeric precursors. Metal alkoxides were reacted with polyhydric and monohydric alcohols at 100–200°C in diethylene glycol methyl ether. Removal of the by-product alcohol left an insoluble polymer containing M–O–C bridges that was collected as fine powder. Calcining of this polymer resulted in a coke. For example, AlN powders of 0.2–1.0 nm particle size were produced by

1. reaction of $Al(OC_3H_7)_3$ with a mixture of glycerol and furfuryl alcohol,
2. pyrolysis at 800°C in N_2, and
3. carbothermal reduction at 1000°C [2.15].

Si_3N_4 powders can be obtained by the pyrolysis of silicon diimide $\{[Si(NH)_2]n\}$ at about 1000°C. This method has attracted much interest because the system is free of C, thus offering the possibility of yielding powders free of C contamination [2.34].

Recently, the reactions of various polymers were investigated to obtain nitrides and oxynitrides [2.34]. It was reported that pyrolyzing a vinylic polysilane (containing no nitrogen) in a nitrogen atmosphere yields partially crystalline, SiO_2-containing nanosized crystals. Si_3N_4 has been synthesized via decomposition of preceramic polymers [2.9]. An amorphous polymeric polysilazane precursor, $Si(CH_3)_3)_2NH$, manufactured by the Mitsubishi Corporation has been used to obtain nanocrystalline SiC/Si_3N_4 ceramics that exhibit superplasticity [2.45].

2.1.2.4 Gas–Solid Reactions

The advantage of this process over solid–solid reactions is improved contact between reactant phases. It is used to produce fine powders with high surface area [2.34]. *Carbothermal reduction* and nitridation of silica to produce Si_3N_4 is a well-known process of this type. In this process, silica is reduced by carbon in a nitrogen or ammonia atmosphere,

$$3SiO_2 + 6C + 2N_2 \rightarrow Si_3N_4 + 6CO \tag{2.22}$$

The Toshiba company currently uses this method on a commercial scale [2.9]. Weimer et al. [2.46] developed the *direct nitridation process* for manufacturing AlN powders at pilot scale. The process involves heating atomized aluminum metal and nitrogen gas in a transport flow reactor where rapid reaction takes place and fine AlN powder is obtained. Additional milling and heat treatment are necessary for full conversion to AlN. Although fully dense AlN parts could be produced by pressureless sintering of these powders with 3 wt% Y_2O_3 additions, the overall sinterability was reportedly lower than commercially available powders produced by a carbothermal reaction.

The reaction of silica with flowing ammonia at 1200°C or less results in silicon oxynitride of various stoichiometries. For example, Si_2N_2O powder having a 150 m^2/g surface area was produced by this method. Similarly, oxynitrides of other metals such as Al and Ti can be produced by the reacting metal powders with flowing ammonia [2.34].

2.2 Ceramic Whiskers

Whiskers are fibrous single crystals whose diameters are in the micrometer range. Strengths extending to theoretical limits have been achieved in whiskers because of the absence of grain boundaries and their very small dimensions that limit the number of imperfections such as dislocations, vacancy clusters, porosity, or twins. High-strength, low-density ceramic whiskers are excellent strengthening agents for both metal and ceramic matrices. Recent developments reveal that whiskers can also be used as effective toughening agents in most ceramics. The role played in toughening ceramic bodies is far more important than strengthening them since most ceramic materials already have relatively high strength but intrinsically low toughness values. Whiskers or large single crystals have also found optical, magnetic, and electronic applications due to their superior properties, which results from the absence of imperfections.

The physical and mechanical properties of whiskers, which are directly related to the production method, are very important in many strengthening and toughening applications. Table 2.3 lists the properties of various experimental and commercial whiskers. The properties of whisker-reinforced ceramic matrix composites (CMCs) will be discussed in Sect. 4.1.

2.2.1 Growth of Whiskers by Vapor-Phase Reactions

2.2.1.1 Evaporation–Condensation

Although widely used for growing metallic whiskers, this method is limited to the growth of a few types of whiskers, including Al_2O_3, ZnO, ZnS, W–O, and CdS. For example, CdS whiskers were grown in $<0001>$ and $<000\bar{1}>$ directions by sublimation from an atmosphere containing excess Cd or S vapor respectively [2.47]. The method involves evaporating or sublimating of the whisker source material, mass transport through the vapor phase, and condensation at the growth site under low supersaturation. The growth chamber consists of a tube with a controlled thermal gradient between the source and the growth site. Although the setup is simple, the whiskers obtained are typically small and few. Working at melting temperatures complicates the method, and the growth rate is low [2.48].

2.2.1.2 Chemical Reduction

This method involves reacting the source material with a gas phase to produce a volatile material, vapor transport to the growth site, and condensation under low supersaturation. For oxide whiskers, wet hydrogen or reductive materials have been used. The experimental setup used for whisker growth by chemical reduction is shown in Fig. 2.7. Whiskers grow on the sides and top of the crucible. Growth times range from 0.5 to 20 hrs. Whisker diameters range from submicron to 1 mm, and lengths up to 10 mm can be achieved.

Al_2O_3 whiskers were produced by passing a stream of moist hydrogen over molten aluminum or $TiAl_3$. Nucleation and growth occurred in the cooler portion of the growth chamber. Whiskers 1–30 mm long and 3–50 μm in diameter were obtained [2.48]. A large-scale production technique was presented for Al_2O_3 whiskers. In this method, H_2 gas was passed through a humidifier, and the wet gas was passed over an Al_2O_3 boat filled with molten Al metal. Whisker lengths up to 4 cm and diameters about 0.5 μm were obtained. Growth times ranged from 0.5 to 10 hrs at a constant furnace pressure of 3 cm Hg [2.48].

Dry hydrogen has also been used for growing Al_2O_3 whiskers. When silica-rich refractory boats and chambers are used, the following reactions occur:

$$H_2\,(g) + SiO_2\,(s) \rightarrow H_2O\,(g) + SiO\,(g) \tag{2.23}$$

The advantage of this method is improved production control capability. SiC whiskers were grown by the reduction of complex silicates in graphite crucibles, specially designed to restrict the escape of gaseous products. A similar system was used to produce MgO whiskers by W and C reduction as well as by wet H_2 reduction [2.48].

Wada and Wang [2.49] synthesized β-Si_3N_4, β-SiC, and β'-sialon whiskers by a single carbothermal reduction process. Their results show that SiC, Si_3N_4, and sialon whiskers can be produced by the same process by varying the initial reactant phases or by controlling the activities of C and N. Vapor–solid reactions with graphite were found responsible for whisker formation. SiO, one of the reactive gases, was formed from SiO_2 by the following reaction:

$$SiO_2\,(s) + C\,(s) \rightarrow SiO\,(g) + CO\,(g) \tag{2.25}$$

Whisker-forming reactions were

$$SiO\,(g) + 3CO\,(g) \rightarrow SiC\,(s) + 2CO_2\,(g) \tag{2.26}$$

for β-SiC whiskers,

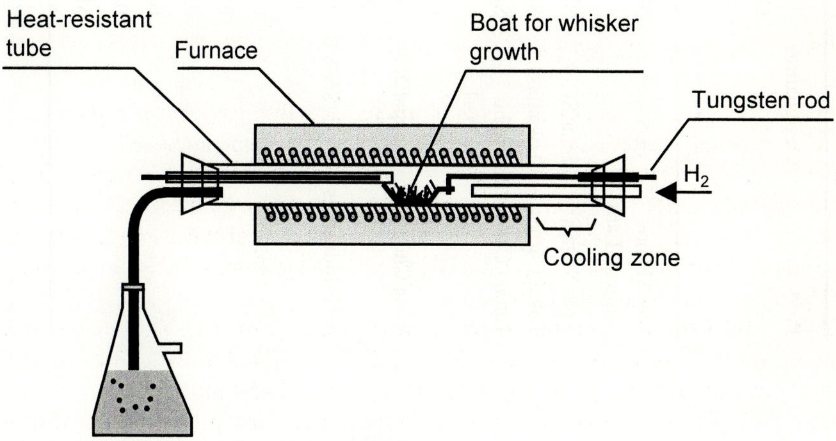

Fig. 2.7. Whisker growth by chemical reduction

Table 2.3. Properties of Commercial and Experimental Whiskers

Producer	Designation	Typical Composition (wt%)	Density (g/cm^3)	Average Diameter (µm)	Average Length (µm)	Elastic Modulus (GPa)	Tensile Strength (GPa)	Melting or Sublimation Point (°C)	CTE (10^{-6}/°C)	Ref. [a]
Los Alamos National Labs	VLS	SiC	3,2	3–10	500–10000	580	8	2687	4.5	[1]
Arco Metals	Silar SC-9	SiC	3,2	0.5–0.6	10–80	600	–	2687	4.5	[2]
Tokai Carbon	Tokamax	SiC	3,2	0.1–1.0	50–200	600	–	2687	4.5	[3]
Tateho Chemical Ind.	Tateho SCW–1	SiC	3,2	0.05–0.2	10–40	600	–	2687	4.5	[4]
		α - SiC	3,2	–	–	485	7–35	2316	–	[5]
Vista Chemical	Catapal XW	Al_2O_3	4	4–7	40–100	400	–	2037	7	[6]
		Al_2O_3	3,9	–	–	550	14–28	2082	–	[5]
UBE Industries	UBE–SNWB	Si_3N_4	3,2	0.05–0.5	5–100	370	–	1899	2.5	[7]
		Si_3N_4	3,2	–	–	380	3–11	1899	–	[5]
		Graphite	2,2	–	–	980	21	3593	–	[5]
		B_4C	2,5	–	–	480	14	2426	–	[8]
		AlN	3,3	–	–	335	14–21	2198	–	[5]
		BeO	1,8	–	–	700	14–21	2549	–	[5]
		B_2O_3	2,5	–	–	450	7	2449	–	[5]
		MgO	3,6	–	–	310	7–14	2799	–	[5]

(Continued)

Table 2.3. *(continued)*

Producer	Designation	Typical Composition (wt%)	Density (g/cm^3)	Average Diameter (μm)	Average Length (μm)	Elastic Modulus (GPa)	Tensile Strength (GPa)	Melting or Sublimation Point (°C)	CTE (10^{-6}/°C)	Ref. [a]
–	–	Quartz	–	–	–	80	4.1	–	–	[9]
–	–	ZrO_2	–	–	–	430	4.1	–	–	[9]
–	–	Si	–	–	–	160	3.8	–	–	[9]
–	–	Iron	7,8	–	–	190	13	–	–	[10]

[a] References

1. Ceramic Matrix Composites. R. Warren (ed). (Blackie and Son London England, 1992)
2. Arco Metals Co., Manufacturer's Data
3. Tokai Carbon Co., Manufacturer's Data
4. Tateho Chemical Ind. Co., Manufacturer's Data
5. J.V.Milevski. In: *Concise Encyclopedia of Composite Materials*. A. Kelly (ed). (Pergamon Press Oxford England, 1989)
6. Vista Chemical Co., Manufacturer's Data
7. UBE Industries, Manufacturer's Data
8. *Whisker Technology.* A.P. Levitt (ed). (Wiley New York, 1970)
9. *ASM Engineered Materials Reference Book*. (ASM International Metals Park OH, 1989)
10. Z.D. Jastrzebski: *The Nature and Properties of Engineering Materials*. (Wiley New York, 1977)

$$3SiO\,(g) + 2N_2\,(g) + 3CO\,(g) \rightarrow Si_3N_4\,(s) + 3CO_2\,(g) \tag{2.27}$$

for β-Si_3N_4 whiskers, and,

$$(6-x)SiO\,(g) + (4-x/2)N_2\,(g) + (6-2x)CO\,(g) + xAl\,(g) \rightarrow$$

$$Si_{(6-x)}Al_xO_xN_{(8-x)}\,(s) + (6-2x)CO_2\,(g) \tag{2.28}$$

for β'-sialon whiskers, where x ranges from 0.8 to 1.1. The average whisker diameter from this process was 0.75 μm.

Hashimoto and Yamaguchi [2.50] reported the formation of $MgAl_2O_4$ whiskers in an Al_2O_3 boat. Al (g) and Mg (g) were produced from powder mixtures of Al/C and MgO/C/Al at 1500°C under P_{CO} = 0.1 MPa. The addition of CO_2 increased the oxygen partial pressure, causing condensation of Al (g) and Mg (g) as $MgAl_2O_4$ whiskers. Under optimal conditions at 1500°C and 8h, whiskers with average diameters of 3.1 μm and lengths in the range of 4 mm were grown in the [2.111] direction.

2.2.1.3 Vapor–Vapor Reactions (Chemical Vapor Deposition–CVD)

In this method, the whisker-forming compound is formed by the reaction of gaseous species and deposition of this compound on a substrate. Important process parameters are partial pressures of reactant gases, temperature, flow rate, and the particle size of the nuclei. Low supersaturation favors whisker growth, whereas intermediate supersaturation favors the formation of dendrites or platelets. At high supersaturation, powders are generally formed. Al_2O_3 whiskers were grown by the oxidation of $AlCl_3$ in an atmosphere of $CO/CO_2/H_2/Cl_2/AlCl_3$ at temperatures ranging from 1275 to 1700°C and total pressures ranging from 5 to 17 mmHg [2.48]. For nuclei other than Al or Al_2O_3, no whisker growth occurred. Whiskers of Si_3N_4, cristobalite, and boron were grown by the vapor-phase reaction method. Futamoto et al. [2.51] reported the growth of HfC and HfN whiskers by a CVD process. Similarly, Lackey et al. [2.52] studied the growth of HfC whiskers by CVD. They observed that most of the whiskers grow by the vapor–solid mechanism.

Gorny and Pampuch [2.53] studied the growth of SiC whiskers by using CVD from methyltrichlorosilane (CH_3SiCl_3) and $SiCl_4$ + CCl_4, respectively, at various temperatures. H_2 was used as a carrier gas. SiC whiskers were grown on inductively heated graphite substrates covered by a SiC layer. The whiskers were grown for 2 hours at 1400, 1450, and 1500°C. It was found by X-ray and TEM analysis that, depending on the temperature and reactant gases, 2H, 3C, 15R, and 6H polytypes of SiC can be grown on certain growth axes.

2.2.2 Vapor–Liquid–Solid (VLS) Method

One important feature of the VLS method that distinguishes it from other whisker growth processes is the presence of a catalyst metal in liquid form [2.54]. Initially, the catalyst is distributed on a compatible substrate. The substrate is then placed in a suitable enclosure and heated to a temperature high enough to melt the catalyst. Then, process gases are allowed inside. Once the liquid becomes supersaturated with elements supplied from the vapor, precipitation occurs at the liquid/solid interface. Whisker growth occurs by further precipitation from the nucleus as shown in Fig. 2.8.

To grow SiC whiskers, gaseous SiO and CH_4 as reactants and Fe- or Mn-based alloy particles as catalysts were used [2.55]. The most important parameter of the process was the partial pressures of the reactant gases. An average tensile strength of 7 GPa was reported for SiC whiskers derived by the VLS process. Some whiskers exhibited strengths higher than 16 GPa. The average elastic modulus in <111> directions was 580 GPa.

SiC whiskers were also grown using stainless steel catalysts on graphite plates. The growth atmosphere was 1% methane, 80%H_2, 10%CO, and 9%N_2. The graphite reactor also contained porous insulation brick impregnated with a fine powder mixture of SiO_2 and C, which react at 1400°C and form SiO. Resulting whiskers had 4–6 μm diameters and ~10-mm lengths with average tensile strengths of 8.4 GPa and Young's moduli of 581 GPa [2.56]. Figure 2.9 shows SEM micrographs of such whiskers obtained by the VLS process.

In other studies using the VLS method, MgO whiskers were grown using Al, Ca, or Si as the catalyst, and α-Al_2O_3 whiskers were grown from Pb catalysts [2.47]. B whiskers were prepared by reduction of BBr_3 and BCl_3 with hydrogen, using Ag, Pt, Cu, and Au as liquid metal catalysts [2.57].

The VLS process offers controlled whisker length and diameter, high purity, good high temperature stability, the possibility of in-situ coating, and successful scale-up. Therefore, this process is one of the preferred methods of whisker production on a commercial scale.

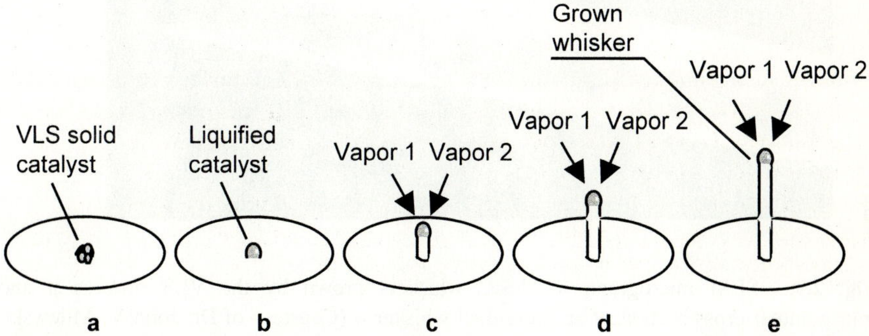

Fig. 2.8 a–e. Whisker growth by the VLS process

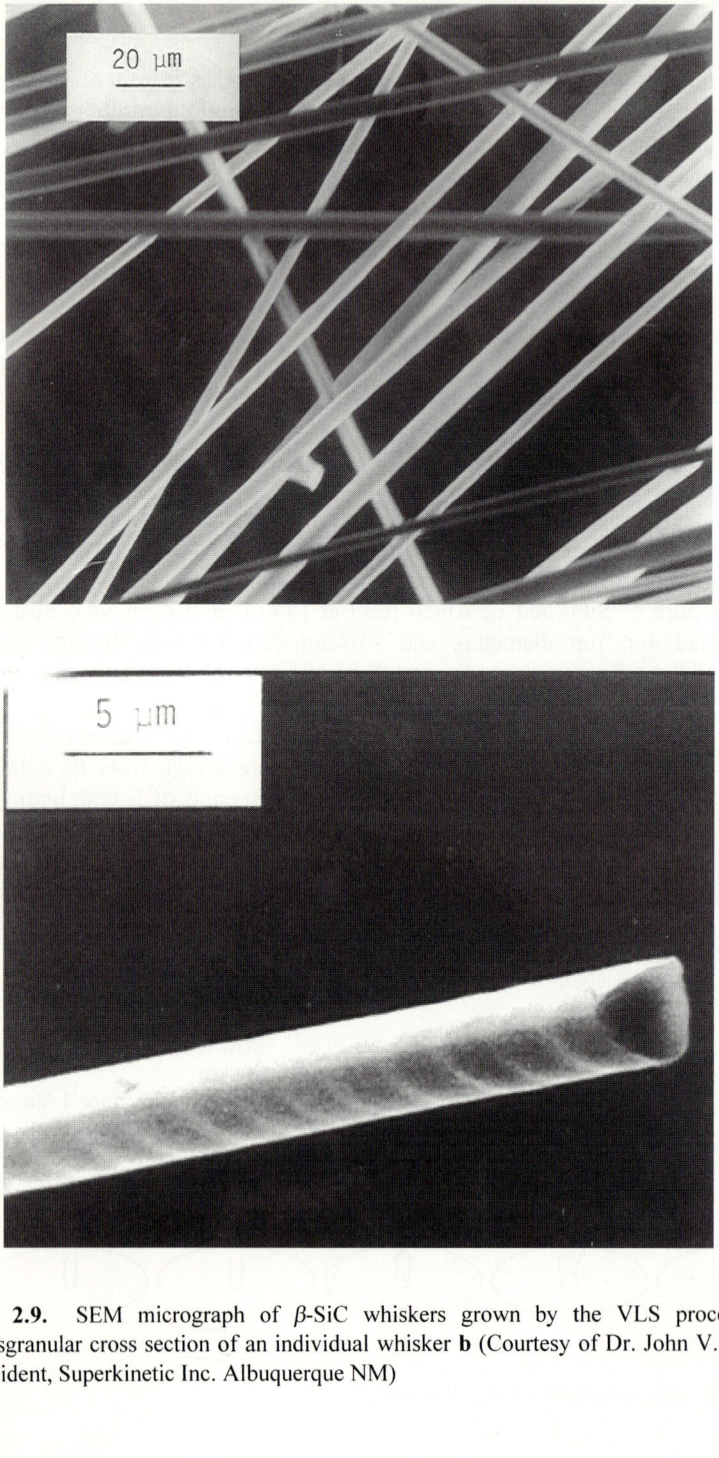

a

b

Fig. 2.9. SEM micrograph of β-SiC whiskers grown by the VLS process **a** and transgranular cross section of an individual whisker **b** (Courtesy of Dr. John V. Milewski, President, Superkinetic Inc. Albuquerque NM)

2.2.3 SiC Whiskers from Rice Hulls

SiC whiskers are currently produced commercially by several companies using the *rice hull process*. Originally, SiC whiskers were produced from rice hulls by Lee and Cutler [2.58]. Rice hulls contain about 20 wt% silica, 38% cellulose, 22% lignin, and 18% pentosan. The remainder consists of some other inorganic material and organic material concentrated in the inner epidermis of the rice husks.

The first two are the raw materials necessary for SiC synthesis. The inorganic matter is concentrated in the outer epidermis (outer face) of the rice husks and silica occurs in the amorphous form and mainly in the horns (Fig. 2.10). The process involves two steps. In the first step, rice husks are coked in an inert atmosphere at 400–700°C. In the second step, coked rice husks are pyrolyzed in an inert atmosphere at 1400–1750°C. Additional steps such as removal of excess carbon and silica may also be necessary [2.59]. Direct pyrolysis instead of the two step process has also been attempted with positive results [2.60]. Cubic SiC whiskers with diameters ranging from 0.4 to 1 μm and lengths of up to 50 μm are produced. TEM micrographs prove that single-crystal SiC whiskers can be produced (Fig. 2.10 b and c). Generally, the purity of these whiskers is not as high as that of VLS whiskers. Small SiC particles coexist with rice hull-derived whiskers, which may reduce the amount of attainable strengthening or toughening with whiskers (Fig. 2.11).

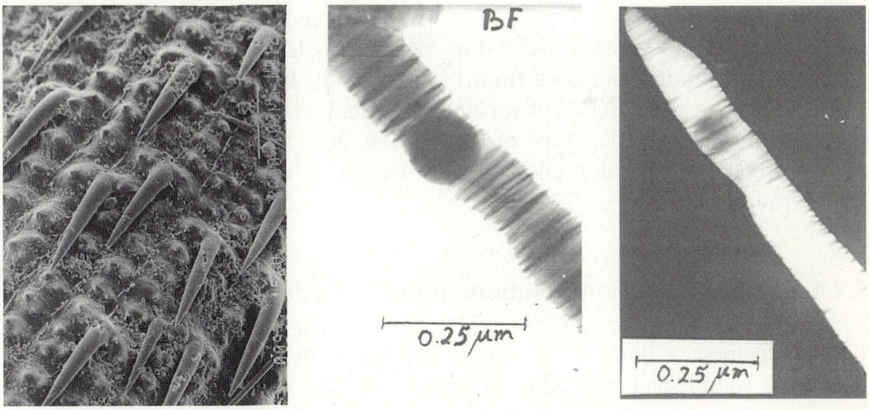

a b c

Fig. 2.10. SEM micrograph of the outer face of rice husks **a**, bright field TEM micrograph of *β*-SiC whisker grown from rice hulls **b**, and dark field TEM micrograph of the same whisker **c** (Courtesy of Dr. Ajoy Kumar Ray, National Metallurgical Laboratory, MTP Division, Jamshedpur, India)

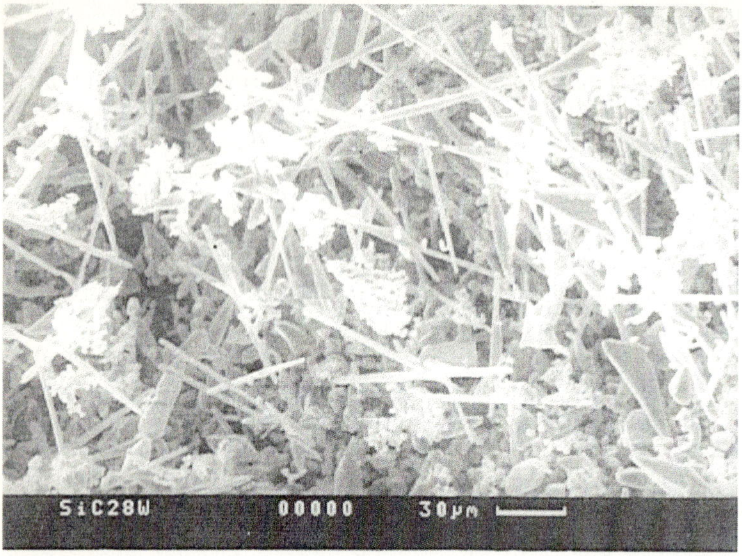

Fig. 2.11. General view of SiC whiskers grown from rice hulls (Courtesy of Dr. Raghavarapu V. Krishnarao, Defense Metallurgical Research Laboratory Hyderabad India)

Divecha et al. [2.61] quote tensile strengths of 13 GPa and elastic moduli of 700 GPa for these whiskers. Ray et al. [2.59] studied the effect of catalyst and temperature on the whisker yield from rice husks. It was observed that as the pyrolysis temperature increased from 1400 and 1750°C, the yield increased from 1.4 to 13.2% in the absence of a catalyst. A similar trend was observed in the presence of a catalyst. The type of catalyst (Fe, Ni, Pt, or Pd) did not affect the yield but better whisker morphology was obtained in the presence of Pt, Ni, and Pd catalysts.

2.2.4 Growth from Melt Solutions (Flux Growth)

For this method, a flux, i.e., a solvent is required in which the whisker source material is soluble and from which it crystallizes. The flux with is selected from a low melting point salt low volatility, low viscosity, low toxicity, and low reactivity with the crucible, but high purity and a wide liquid range. On the other hand, the resulting whisker phase should not form a solid solution with the flux, and its initial liquid solubility should be ~25 wt% [2.62]. Usually a homogeneous solution is prepared above the melting temperature of the flux, and slow cooling at rates between 1–10°C/h results in spontaneous nucleation and crystal growth. After the completion of crystal growth, the flux is cooled rapidly and cleaned by a solvent or by other means. CaF_2 crystals were grown by this technique from alkali halides as a flux. TiO_2 whiskers were grown from a solution of titanium compounds with a melt of an oxyacid salt (Li_2SO_4, Na_2SO_4, etc.) at 600°C. β-Si_3N_4 whiskers were grown from a silicon melt in a Si_3N_4 crucible [2.47,2.57]. Ferrite whiskers were grown from fluxes of sodium ferrite, $BaO–B_2O_3$, or PbO. Ni–Zn single crystal ferrites were prepared using $PbO–PbF_2$ flux under an $Ar–O_2$ atmosphere [2.63]. Whiskers of NiO, MgO, and their solid solutions were produced by reacting a heated mixture of Ni and/or $MgCl_2$ and a suitable auxiliary salt with water vapor [2.47].

Recent studies by Bamberger et al. [2.63] revealed the formation of TiN whiskers by reacting molten alkali metal cyanide with several sodium and potassium titanates. Initially, TiN powder was contacted with NaCN at 1000°C, and some whisker formation was observed in graphite or Al_2O_3 crucibles. Subsequent studies showed that contamination of NaCN melts with oxide ions was necessary for whisker growth. TiN whiskers were produced from mixtures of alkali metal cyanides with oxides of alkali metals and of Ti. It is speculated that the growth mechanism involves the formation of alkali metal titanates and alkali metal titanium bronze. TiN whiskers were obtained along with particles. Whiskers had well-developed faces and diameters about 5–10 µm (Fig. 2.12a). Depending on the reactants, different morphologies such as needles (Fig. 2.12b) and fibrils were also obtained.

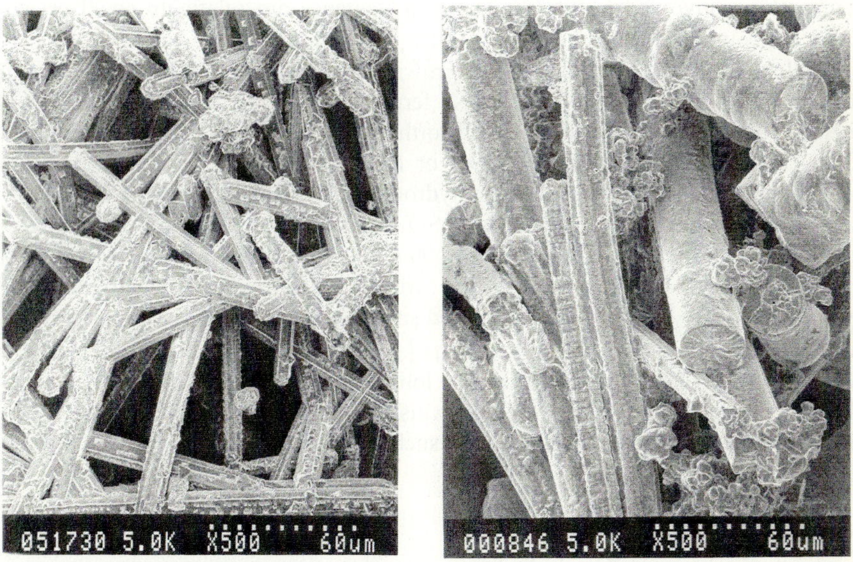

a b

Fig. 2.12. SEM micrographs of TiN whiskers grown from oxide-containing cyanide melts by reacting 1/5 $Na_8Ti_5O_{1}4$ with 25 NaCN **a** and ½ $K_2Ti_2O_5$ with 34 NaCN **b** (Courtesy of Dr. Carlos E. Bamberger, Oak Ridge National Laboratory Oak Ridge Tennessee)

2.2.5 Growth from Gels

Some whiskers that cannot be obtained by other techniques can be prepared by growth from a gel. The gel acts as an inert growth medium, prevents convection currents or turbulence, and allows smooth growth from a crystal nucleus that is firmly held in position. Elements of the whisker compound are fed into the gel (usually silica gel) and allowed to diffuse through it. Whiskers of $Ca(IO_3)_2$ were produced by injecting $NaIO_3$ solution into a SiO_2 gel and feeding $CaCl_2$ solution above it. TlI whiskers were grown by diffusion of NaI and $TlNO_3$ solution through SiO_2 hydrogels. Similarly, SnI_2 and SnI_4 whiskers were grown by diffusion and controlled reaction of $SnCl_4$ and KI in a SiO_2 gel [2.47,2.57].

2.2.6 Hydrothermal Growth

Whiskers can be grown in aqueous or acid solutions by hydrothermal synthesis. Many materials, which are insoluble in these liquids below the boiling points of the liquids, become soluble at higher temperatures, for example, water at ~400°C. To reach these temperatures, pressure is applied by an autoclave or by other means. Crystallization is achieved in the solution. The main advantages of hydrothermal synthesis are the high level of crystal perfection and low temperatures involved (<500°C). The disadvantages are relatively small crystal sizes and complexity, thus high cost, of the process [2.62].

Willemite (Zn_2SiO_4) crystals were grown from HCl solution by hydrothermal synthesis [2.47]. Single crystals of Ni ferrite and Zn ferrite were grown from aqueous solutions. YIG and rare-earth orthoferrite single crystals were obtained from concentrated solutions of NaOH or KOH at 400°C and 500 bar pressure [2.62]. m-ZrO_2 whiskers were grown hydrothermally from YSZ single crystals by selective dissolution of yttrium in an H_2SO_4 solution. Many whiskers were grown simultaneously by treatment for 3 h at 600°C and 100 MPa in an aqeous 30 wt% H_2SO_4 solution [2.64]. In a recent work, the growth of hydroxyapatite [$Ca_{10}(PO_4)_6(OH)_2$] whiskers of controlled size and morphology was reported from a beta-tricalcium phosphate [β-$Ca_3(PO_4)_2$] solution in citric acid at 200°C and 2 MPa [2.65]. Whiskers were 20–30 μm long and 0.1–1 μm in diameter. Due to their biocompatibility, hydroxyapatite whiskers are proposed as a substitute for biohazardous fibrous-reinforcing agents such as asbestos.

2.3 Ceramic Single Crystals

Single crystals are required for special optical, electronic, and magnetic applications and for research in material science, chemistry, and physics. Whiskers discussed in Sect. 2.2. are a family of single crystals with diameters of a few micrometers and high aspect ratios. They are usually aimed at improving the mechanical properties of a given metal or ceramic matrix material. Single crystals other than whiskers are usually rod shaped but have larger diameters and lengths; typically the diameters are in the millimeter to centimeter range, and lengths can even reach a couple of meters. Single crystals in other forms such as platelets and fibers can also be produced. Examples of applications are quartz crystals for oscillators and timing circuits, silicon and germanium crystals for transistors and other semiconductor products, and magnetic single crystals for magnetic memories. Crystallographic analysis of single crystals, for example, by X-ray crystallography or electron diffraction, is very important for precise measurements. Studies conducted with single crystals have contributed immensely to the understanding of mechanical, physical, chemical, and other properties of materials since they are nearly perfect and they represent the building blocks of polycrystalline materials [2.66]. Important single-crystal growth techniques are described later. Some of the whisker growth techniques such as *hydrothermal growth* are also applicable to single crystals, as discussed in Sect. 2.2.6.

2.3.1 Verneuil (Flame Fusion) Method

In this method, powders of the source material are fed from the top of a tube and during their free fall, are melted by an oxyhydrogen burner that produces a flame, pointing downward [2.62] (Fig. 2.13a). The molten material falls onto a crystal holder and grows upward. An important advantage of the Verneuil method is the

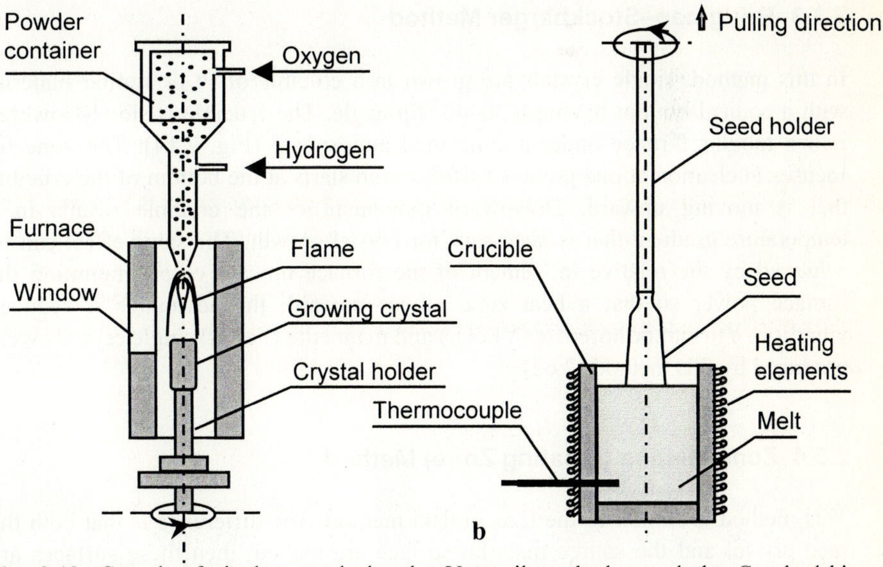

Fig. 2.13. Growth of single crystals by the Verneuil method **a** and the Czochralski method **b**

absence of a crucible, hence no possibility of contamination by either the crucible material or a liquid–crucible interface to cause spurious nucleation. However, the high-temperature flame results in a steep temperature gradient that, in turn, produces residual stress in the single crystals produced. Single crystals of spinel ferrite were grown by this method, but their quality was not of the desired level.

2.3.2 Czochralski Method

Among single-crystal growth methods from melts, this is the most successful in producing high-quality, single crystals with large dimensions, free of residual stress or dislocations. Initially, the source material that should only melt congruently is heated to slightly above the melting point. A seed crystal is immersed in the melt, where the immersed part starts to melt. Cooling from the melt temperature slowly results in crystal growth from the seed. The seed holder is slowly rotated and pulled out of the melt to accompany crystal growth (Fig. 2.13b). The aim of rotation is to maintain the crystal symmetry and decrease the solute-rich layer at the solid–liquid interface [2.67]. No contact exists between the crucible and crystal/melt interface that can cause spurious nucleation, whereas care should be taken to prevent contamination by the crucible material. Single crystals of Fe_3O_4, Zn, Co, Mn, and Li ferrites were grown by the Czochralski method [2.62]. Single silicon crystals 10 cm in diameter and 1.5 m long were also grown by this method [2.68].

2.3.3 Bridgman–Stockbarger Method

In this method, single crystals are grown in a crucible of Pt or similar material with a conical bottom having a 30–60° tip angle. The crucible is slowly lowered into a tubular furnace under a controlled atmosphere (Fig. 2.14). The cone tip focuses nucleation at one point. Crystallization starts at the bottom of the crucible that is moving upward. Downward movement of the crucible results in a temperature gradient that is necessary for crystal growth. The same effect can be achieved by the relative movement of the furnace upward or programming the furnace power so that a heat zone moves upward; the advantage is reduced vibration. Yttrium orthoferrite ($YFeO_3$) and magnetite (Fe_3O_4) single crystals were produced by this method [2.62].

2.3.4 Zone Melting (Floating Zone) Method

This method is similar to the Czochralski method. The difference is that both the seed crystal and the source material surface are melted, then these surfaces are joined and pulled away from each other [2.47,2.62]. The molten zone is held in place by surface tension, sometimes aided by a magnetic field [2.67]. Melting can be achieved by various techniques, including radio frequency, arc, arc image furnace, electron bombardment, laser, or gas flame heating. As is the case with the Czochralski method, a single crystal is grown from its own melt, and the material must melt congruently. Monocrystalline filaments of Al_2O_3, ZrO_2, ThO_2, B, and Si were grown by the molten zone technique [2.47]. Magnetic single crystals such as YIG, $LaCoO_3$, Ni ferrites, and Ni Zn ferrites were grown by various heating techniques [2.62]. Large magnetite and Ni–Fe ferrite single crystals were grown by the modified floating zone technique with a drawing speed of 3 cm/h and a rotational speed of 50 rpm [2.69].

2.3.5 Edge Defined Film Fed Growth (EFG) Method

This method was developed by LaBelle and Tyco Laboratories [2.47,2.57]. Single-crystal filaments are grown individually in self-feeding capillary tubes that

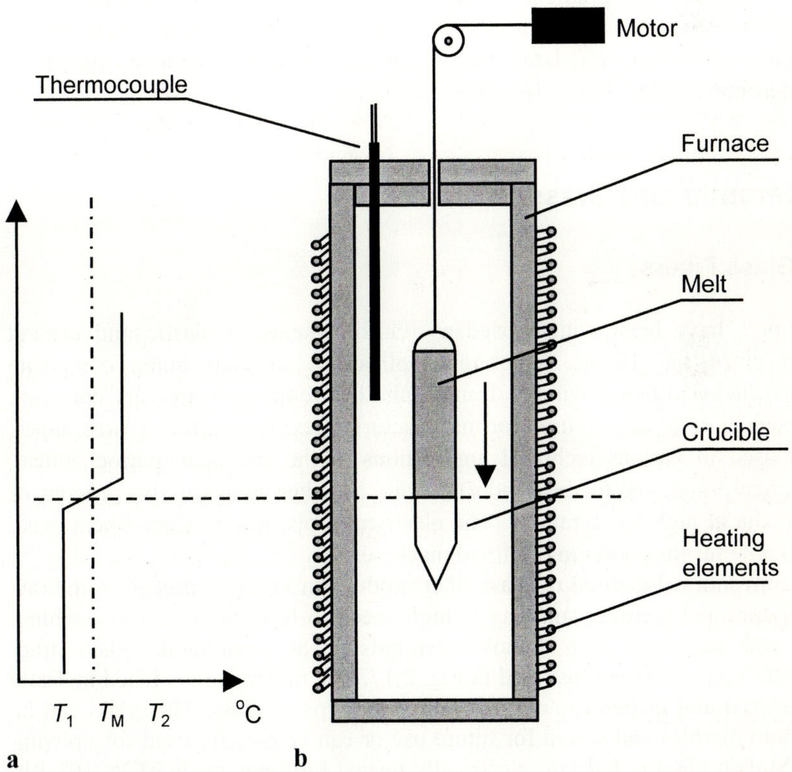

Fig. 2.14 Temperature profile **a** and growth of single crystals **b** by the Bridgman–Stockbarger method

hold the molten material and become heated in a crucible. In one effort, 25 Al_2O_3 filaments 0.25 mm in diameter were grown simultaneously at the rate of 60 m/h using 25 capillary tubes in a single crucible. Tensile strengths of 3 GPa were reported.

2.3.6 Heat Exchanger Method (HEM)

The HEM crystal growth furnace has an isothermal heat zone in which there are no intrinsic temperature gradients. A heat exchanger is placed at the bottom of the isothermal zone [2.70] (Fig. 2.15). This is a special heat exchanger where both the heat input and extraction are independently controlled. This unique feature allows dynamic control of the solidification process and removal of the temperature gradient of the solidified crystal for in-situ annealing. A seed crystal is placed at the center of the crucible. Helium flows through the heat exchanger to control the heat extraction, and the furnace controls the heat input. This allows independent control of temperature gradients in the solid and liquid regions during crystal growth. No crucible, heat zone, or crystal movement takes place during solidification. High-quality, large, single crystals can be produced by the HEM process, as indicated by Fig. 2.16.

2.4 Ceramic and Glass Fibers

2.4.1 Glass Fibers

Glass fibers have been widely used as reinforcements in plastic and cement matrices since the 1940s. Structural applications in boat hulls, composite airplanes, chemical tanks, window frames, and bathroom units are very common. Continuous glass fibers are used for manufacturing textile materials (yarns, tapes, fabrics) used in various technical applications in the chemical, petrochemical, metallurgical, food, and medical industries and for filtering aggressive gaseous or liquid media at high temperatures. The electrical properties of glass fibers make them suitable for insulation in electrical machines.

The conventional method of glass fiber production involves preparing the raw glass mixture and melting, spinning at high speed to form the fibers, quenching, sizing, and gathering into bundles (strands). The continuous glass fiber manufacturing process is illustrated in Fig. 2.17. The raw materials listed in Table 2.4 are mixed and melted to achieve the desired type of glass. This glass can be shaped into marbles and stored for future use or can be directly used for drawing [2.71]. Molten glass is fed into electrically heated bushings made of Pt–10%Rh, Pt–20%Rh, or SiC [2.72]. It is forced through about 200 orifices in the bushing and is drawn away by a spinneret. Individual fibers are quenched by a water spray and coated by *sizing*, which is a protective and adhesive layer, before they are gathered by a spinneret. The sizing lubricates fiber surfaces to prevent damage of fibers from surface flaws and binds them together. Different sizing materials can be used for different purposes. For example, a size based on poly(vinyl acetate) and a resin-coupling agent is used for reinforcement in polyester, epoxy, and phenolic matrix resins. The final fiber diameter can be between 3 and 11 μm. Typical properties of different glass fibers are given in Table 2.4.

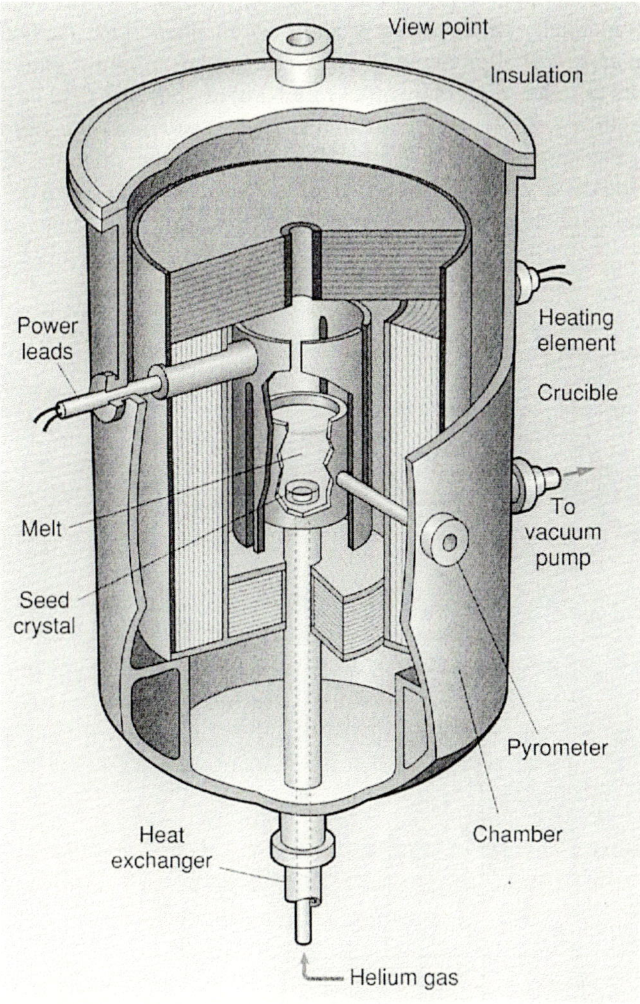

Fig. 2.15. The HEM crystal growth furnace (From F. Schmid, C.P. Khattak, and D.M. Felt: *Saphire Sparkles in Many Optical Elements.* Laser Focus World, June 1996 Copyright 1996 PennWell Publishing Company, used with permission)

Glass fibers have also been produced by the sol–gel technique in recent years. A homogeneous solution is prepared from metal alkoxides, converted to a sol, gelled to form a viscous suspension, and heat treated to form the glass. Temperatures to form glass by this process are much lower than for conventional glass processing [2.71].

2.4.2 Boron Fibers

Two methods are commercially employed to produce boron fibers. One method involves *thermal decomposition* of diborane (B_2H_6) from gaseous mixtures onto a core. Deposition occurs at reduced pressures (0.3 to 3 KPa) at 450–650°C [2.73]. The low temperatures allow using low melting point substrates and make it a more efficient method than the reduction of boron halides. However, the strengths of filaments are inferior to B fibers from halides (2 to 2.5 GPa compared to 3.4 to 3.9 GPa, respectively), and an increased possibility of explosion hazards is involved in the former.

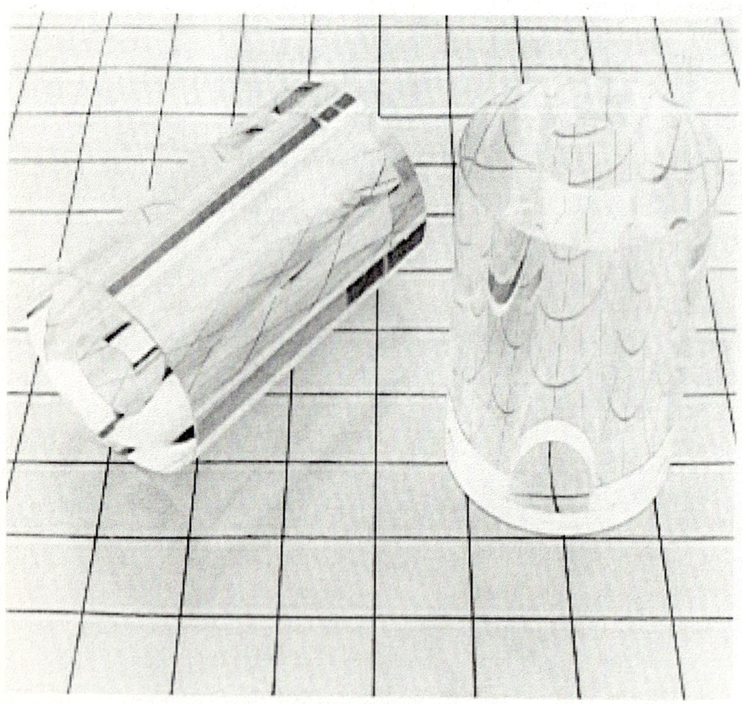

Fig. 2.16. Sapphire tube with scratch-free polish for high-pressure applications produced by the HEM process (courtesy of Crystal Systems Salem MA)

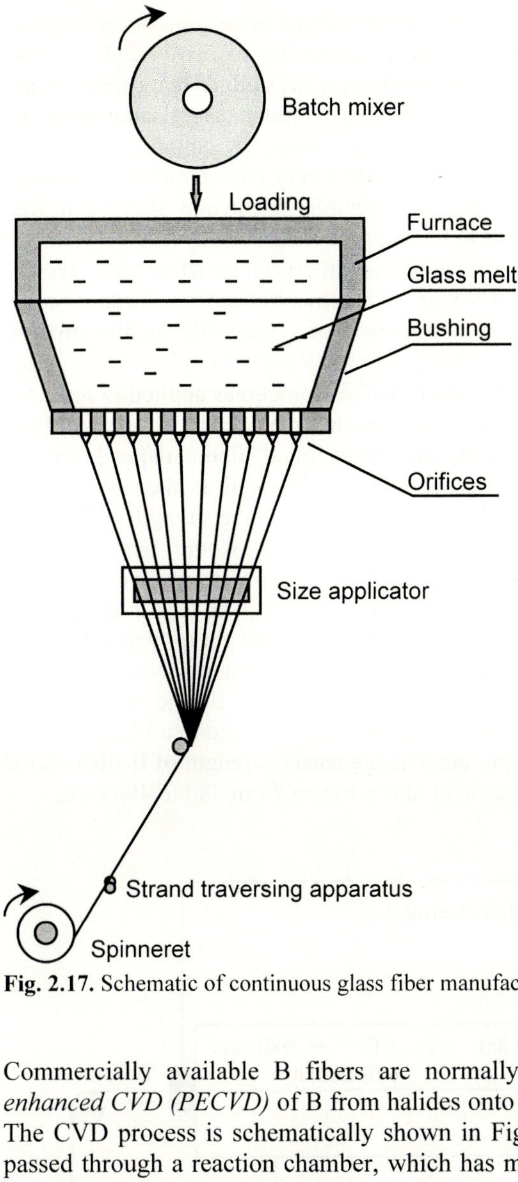

Fig. 2.17. Schematic of continuous glass fiber manufacturing process

Commercially available B fibers are normally produced by *CVD* or *plasma enhanced CVD (PECVD)* of B from halides onto heated W wires or carbon fibers. The CVD process is schematically shown in Fig. 2.18. The W or C substrate is passed through a reaction chamber, which has mercury seals at both ends. These seals also act as electrical contacts to heat the substrate resistively [2.73].

Hydrogen is used to reduce boron halides:

$$2BX_3\ (g) + 3H_2\ (g) \rightarrow 2B\ (s) + 6HX\ (g) \qquad (X = Cl, Br) \tag{2.29}$$

Deposition is carried out for 30–60 s at temperatures ranging from 1100 to 1200°C. Below 1000°C, deposition rates are very low. Above 1200–1300°C (depending on the wire speed), deposition of large crystalline B masses occurs, which is undesirable. The desired structure of B filaments is amorphous or microcrystalline, which yields the highest tensile strengths among such fibers. This structure is built up in successive layers with a very large number of stacking faults between them. Based on X-ray and electron diffraction studies, it was concluded that what appeared earlier to be amorphous, is actually microcrystalline B with very fine grains (~2nm) and a rhombohedral crystal structure. The typical macrostructure of B fibers is like the texture of a corncob. Each nodule starts as an individual nucleus on the substrate and then grows conically as the process continues.

A thin coating (0.5–8 μm) of SiC, BN, or B_4C is sometimes applied to improve the oxidation resistance of the fibers, although this may result in up to a 10% decrease in strength. The mechanical properties of B fibers depend on the fabrication temperature, fiber length, and diameter, as well as the amount and distribution of residual stress. Since B is a brittle material, the strength of B fibers can easily be influenced by structural flaws, which would act as stress concentration sites. On the other hand, B fibers are relatively insensitive to surface flaws due to the presence of residual compressive stresses. The tensile strength of B/W fibers decreases with increasing temperature and drops abruptly around 500–600°C due to grain growth of B and chemical reaction between the core and B. During coating of W, several tungsten borides including W_2B_5 and WB_4 may be formed. In C fiber substrates, however, the formation of boron carbides has not been observed. At room temperature, the average tensile strength of B fibers with W substrates is 3-4 GPa, and the elastic modulus ranges from 380 to 400 GPa.

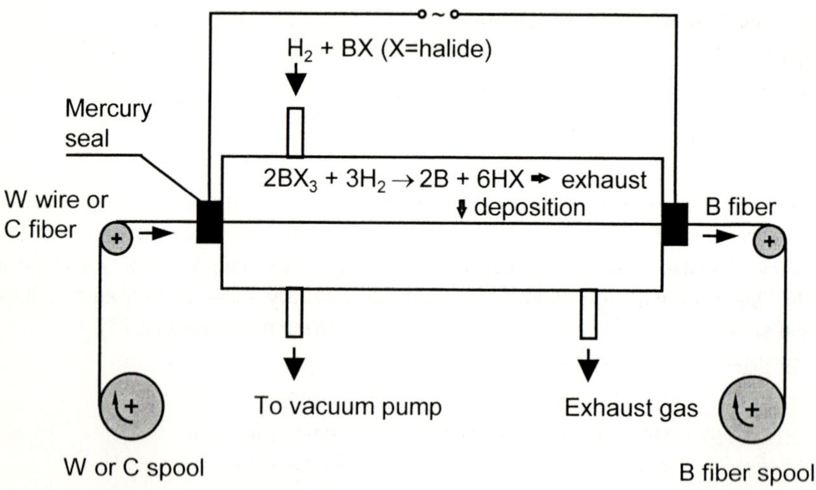

Fig. 2.18. CVD process for continuous B fiber production

Table 2.4. Properties of Commercial Glass Fibers [a]

	Constituent	A Glass (wt%)	C Glass (wt%)	E Glass (wt%)	S Glass (wt%)	Fused Silica
	SiO_2	72.4	66.9	52–56	64.2	99.95
	Al_2O_3	1.6	4.1	12–16	24.8	–
	B_2O_3	–	4.0	5–10	0.01	–
	MgO	4.5	2.7	0–5	10.27	–
	CaO	7.5	6.5	16–25	0.01	–
	Na_2O	–	–	0–2	0.27	–
	K_2O	–	–	0–2	–	–
	Fe_2O_3	0.1	0.2	0–0.8	0.21	–
	F_2	–	–	0–1	–	–
	R_2O	13.2	11.4	–	–	–
	Li_2O	–	0.6	–	–	–
	ZnO	–	3.5	–	–	–
	BaO	–	–	–	0.2	–
Tensile strength of single fiber (MPa)						
at $-190^{\circ}C$	–	–	5400	5300	8700	–
at $22^{\circ}C$	–	3000	3300	3800	4500	5800
at $370^{\circ}C$	–	–	–	2600	4400	–
at $540^{\circ}C$	–	–	–	1700	2400	–
Tensile strength of strand (MPa)	–	–	–	2400	3100	–
Young's modulus at $25^{\circ}C$ (GPa)	–	69	–	75	85	72
Density (g/cm^3)	–	2.50	2,56	2.52–2.62	2.45–2.49	2.19
Refractive index	–	–	–	1.546–1.560	1.524–1.528	1.458
CTE ($10^{-6}/^{\circ}C$)	–	8.6 (1/K)	–	5	3	–

[a] References:

1.Commercial Glasses. In: Advances in Ceramics Vol.18. D.C. Boyd and J.F. MacDowell (eds). (The American Ceramic Society Columbus OH, 1986), pp. 57–59

2. B.D. Agarwal in: *Handbook of Ceramics and Composites,* Vol.1. N.P. Cheremisinoff (ed.). (Marcel Dekker New York, 1990)

3. *Introduction to Fine Ceramics.* N.Ichinose (ed.). (Wiley New York, 1987)

4. *ASM Engineered Materials Handbook.* (ASM International Metals Park OH, 1989)

Boron fibers are used for reinforcing polymer and metal matrices. A major part of composites containing B fibers are aluminum-based. Another important application is in the horizontal stabilizers of F-14 airplanes. Components made of B-reinforced plastics and Al are used in military and civil airplanes and helicopters [2.71]. Although B fibers have high strength and elastic moduli, one major limiting factor in their application is their high price. Their low fracture toughness also makes them unsuitable for specific applications where some elastic deformation is required.

2.4.3 Carbon Fibers

Carbon fibers have attracted considerable attention due to their extremely low density and high specific mechanical properties. Although they were initially developed for the aerospace industry, C fibers are suitable for a wide variety of applications where low weight, high strength, elastic modulus, and fatigue resistance are critical requirements. Carbon fibers are relatively cheap, and they can be produced from many sources. A major drawback of C fibers is their rapid oxidation in oxidizing environments above ~300°C. Many precursors have been used in the fabrication of C fibers, but only the most important will be mentioned here. Commercially important precursors include polyacrylonitrile (PAN), cellulose, and pitch. Other sources investigated include coal; polymerized resins such as polybutadiene, polyacetylene, vinyl, polyethylene, and polystyrene; and polycondensation resins such as polyamides, polyaramides, polyurethanes, and polybenzimidazoles [2.47].

The precursor material is selected on the basis of the potential for a high carbon yield, low weight loss during carbonization, and a high degree of molecular orientation. Heat treatments prior to the carbonization process aim to stabilize the fiber to prevent melting or degradation during carbonization. In some cases, stretching is also employed to maintain or improve molecular orientation. The process parameters of carbonization/graphitization steps have to be selected carefully since these play a determining role in the strength and elastic modulus of the product fiber.

Various surface treatments may be used for specific applications:

1. Coating with organic materials to improve wetting characteristics aimed at carbon fiber-resin composite processing.
2. Coating with inorganic materials to impart oxidation or corrosion resistance.
3. Surface activation by etching or roughening for improved adhesion to metal matrices [2.47].

2.4.3.1 Fibers from PAN Precursors

In recent years, polyacrylonitriles have become the most important commercial precursors for C fiber production. Commercially available PAN-based fibers are listed in Table 2.5. The advantages of PAN fibers are the high degree of molecular orientation, higher melting point, and the high yield of the process. Original fibers used for C fiber production are usually acrylic fibers containing more than 95% polyacrylonitrile.

Four basic steps are involved in C fiber production from PAN:

1. Cyclization by heat treatment under inert conditions, which converts triple C/N bonds to double bonds (C=N) and adds the missing N–C bonds between N and C atoms in the chain structure.
2. Oxidation in air at 200–400°C. This step results in a cross-linking reaction between molecular chains and converts thermoplastic PAN into a nonplastic ladder compound capable of enduring high temperatures. Shrinkage during this step results in poor mechanical properties, hence fibers are kept under tension.
3. Carbonization of stabilized PAN at temperatures up to 1700°C in an inert or reducing atmosphere. At this stage, about 50 wt% of the fibers is lost as gases. PAN fibers, which are essentially textile fibers of low strength and modulus, are converted to brittle ceramic fibers with high strength and modulus. During carbonization, open organic polymer chains shrink longitudinally to high hexagonal ring layers of C atoms.
4. Graphitization at about 2500°C where stretching is again applied to improve mechanical properties. Fibers do not lose any significant weight but ordering and orientation of crystallites in the fiber axis help the formation of a more graphitic structure [2.47,2.74].

Mechanical properties of PAN-based C fibers can vary extensively, depending on the particular manufacturing process. The final heat treatment temperature has a direct effect on strength and other mechanical properties [2.75]. The carbonization temperature and stretching during various stages of heat treatment are also important parameters that determine mechanical properties.

2.4.3.2 Fibers From Cellulosic Precursors

Among cellulosic precursors, rayon fibers made from wood pulp (Fig. 2.19) are most suitable because they can be continuously drawn by wet spinning. Other natural sources such as cellulose fibers of cotton and ramie are discontinuous, have low degree of orientation, and contain more impurities. Therefore, they are not suitable for C fiber production.

Processing steps are similar to those of PAN-based fibers. During preliminary heat treatment, water is lost from cellulose at 25–250°C. Thermal cleavage of glycosidic links and C–O bonds occurs at 240–400°C, and aromatization occurs above 400°C. Carbonization and graphitization are the next steps. Fiber stretching during graphitization is necessary to increase the elastic moduli of fibers. Elastic moduli up to 620 GPa can be achieved by fiber elongation of 400% at 2750°C. This elongation is much higher than that involved in PAN fibers. The C yield of

Table 2.5. Properties of Selected C Fibers

Manu-facturer	Designation	Precur-sor	Filament Count	Den-sity (g/cm^3)	Tensile Strength (MPa)	Elastic Mo-dulus (GPa)	Strain to Failu-re (%)	Ref. [a]
AMOCO Union Carbide (USA)	Thornel 75	Rayon	10K	1.9	2500	517	1.50	[1]
	Thornel T300	PAN	1, 3, 15K	1.75	3300	228	1.40	
	Thornel P55	Pitch	1, 2, 4K	2.0	1700	379	0.50	
	Thornel P75	Pitch	0.5, 1, 2K	2.0	2100	517	0.40	
	Thornel P100	Pitch	0.5, 1, 2K	2.15	2200	724	0.31	
	Thornel T500	–	–	–	3600	240	1.51	
	Thornel T700	–	–	–	4500	248	1.83	
Hercules (USA)	AS-4	PAN	6, 12K	1.78	4000	235	1.60	[2]
	IM-6	PAN	6, 12K	1.74	4900	296	1.73	
	IM-7	PAN	12K	1.77	5300	276	1.81	
	UHMS	PAN	3, 6, 12K	1.87	3400	441	0.81	
Toray Indus-tries (Japan)	Torayca T400H	PAN	–	–	4500	250	1.80	[2]
	T800H	PAN	6, 12K	1.81	5500	294	1.90	[1]
	T300	PAN	1, 3, 6, 12K	1.76	3500	230	1.50	[1]
	T1000	PAN	12K	1.82	7100	294	2.40	[1]
	T1000G	PAN	12K	1.80	6400	294	2.10	[1]
	M46J	PAN	6, 12K	1.84	4200	436	1.00	[1]
	M40	PAN	1, 3, 6, 12K	1.81	2700	392	0.60	[1]
	M55J	PAN	6K	1.93	3900	540	0.70	[1]
	M60J	PAN	3, 6K	1.94	3900	588	0.70	[1]
Celanese (USA)	Celion G-40	–	–	–	4700	275	1.70	[2]
	Celion G-50	–	–	–	–	358	–	
Mitsubi-shi Rayon (Japan)	Pyrofil T-2	Rayon	–	–	4000	240	1.67	[1]
	K135	Pitch	2, 4K	2.10	2600	540	0.50	
	K139	Pitch	1, 2, 4K	2.12	2800	740	0.40	
Enka AG (Ger-many)	Tenax IM 500X	–	–	–	4700	295	1.60	[2]

(Continued)

Table 2.5. *(continued)*

Manufacturer	Designation	Precursor	Filament Count	Density (g/cm^3)	Tensile Strength (MPa)	Elastic Modulus (GPa)	Strain to Failure (%)	Ref. [a]
Afikim Carbon Fibers (Israel)	ACIF XHT	–	–	–	3600	230	1.55	[2]
Tonen (Japan)	FT500	Pitch	2, 4K	2.14	3000	490	0.61	[1]
	FT700	Pitch	1, 2K	2.17	3200	690	0.47	
BASF (Germany)	Carbon Yarn	–	–	1.7–2.0	1400	200–500	–	[3]

[a] References:

1. G. Savage: Carbon–Carbon Composites. (Chapman & Hall London, 1993)
2. D.R. Lowell: In: *Carbon Fibers*. The Plastics and Rubber Inst. (ed). (Noyes Park Ridge NJ, 1986)
3. T.I. Mah, M.H. Mendiratta, I. A.P. Katz, and K.S. Mazdiyasni: Am. Ceram. Soc. Bull. **66 [2]**, 304–308 (1987)

the process is also low (15–30 wt%) compared to about 50% for PAN precursors. Due to their disadvantages in processing, cellulosic precursors are not preferred for high-strength fiber production. Rather, carbon textile fibers are produced from cellulose for industrial applications such as activated fibers and special refractory clothes [2.47].

Tensile strengths of stress-graphitized cellulose-based fibers are proportional to their Young's moduli at low strength and stiffness values, but as the modulus is increased, this relationship ceases. At a modulus of 140 GPa, the fracture strain is about 1%, whereas when the modulus is increased to 690 GPa, fibers fail at a strain of 0.5%. Strength values of cellulose-based fibers change from 1 to 4 GPa, whereas elastic moduli can vary from 100 to 800 GPa, depending on the graphitization temperature and stretching load.

a

b

Fig. 2.19 a, b. SEM micrograph of rayon-based C fiber at two magnifications: 1000X **a** and 5000X **b**. (Courtesy of Tenax Fibers GmbH&Co. Wuppertal Germany.)

2.4.3.3 Fibers From Pitch Precursors

Pitch is an attractive precursor for C fiber production because it has a high C yield (~80%) and is inexpensive. Important types of pitch include poly(vinyl chloride) (PVC), petroleum bitumen, and coal tar. Two alternative ways of orienting C crystals and increasing elastic moduli are stretching during graphitization and spinning at low temperatures. Ordinary commercial pitches are mixtures of aromatic compounds. When these pitches are heated for a sufficient time above 350°C, the molecules undergo dehydrogenative condensation reactions and form large molecules that aggregate into a liquid crystalline phase. This phase, called the *mesophase*, forms small liquid spheres that later grow and combine. When the mesophase content reaches about 50%, an interconnected network of mesophase and isotropic pitch develops that forms a mosaic structure. A high mesophase content is preferred since it eases the attaining of high-strength graphite fibers. Thus, some studies aimed to raise the mesophase content of pitches. A special treatment resulted in 75% mesophase from coal or ethylene for pitch heated above 400°C [2.47].

The mesophase can be spun at appropriate temperatures, that yield fibers with a high degree of preferential alignment of planar molecules along the fiber axis. These fibers can be conveniently converted into high-strength, high-modulus C fibers. The conversion process is analogous to that of the PAN fiber process. Tensile strengths up to 3 GPa and elastic moduli up to 800 GPa can be obtained. The density of C fibers fabricated from mesophase pitch is greater than that of fibers produced from other precursors.

2.4.4 Carbide and Nitride Fibers

The conventional method of producing SiC fibers is similar to B fiber production. Using the *CVD process*, thick deposits of SiC are formed on heated, thin, W or C substrates. Usually a mixture of alkylsilanes and H_2 is used to obtain SiC. This is an example of the chemical reaction involving methylchlorosilane [2.57]:

$$CH_3SiCl_3\,(g) \rightarrow SiC\,(s) + 3HCl\,(g) \tag{2.30}$$

Another method is called *host conversion*. In this process developed by Carborundum, the host fiber is converted to the product fiber by reaction with a gas. B_4C or SiC fibers can be fabricated from a rayon-based C fiber by reaction with BCl or SiCl. Strengths of 1200 MPa and elastic moduli of 400 GPa have been achieved in B_4C fibers. BN fibers were also produced by this method. The initially inferior strength values of 340 MPa and elastic moduli of 27 GPa were improved to 2100 MPa and 310 GPa, respectively, by strain annealing [2.57].

Yajima and co-workers [2.76] developed the *polymer pyrolysis* method for producing continuous SiC fibers. This method consists of melt spinning polycarbosilane into fibers and converting them to SiC in three steps: condensation of polymers, thermal decomposition of side chains such as hydrogen and methyl, and crystallization of β-SiC. Polycarbosilane is cured in air where the polymers are cross-linked by oxidation and hence the condensation of polycarbosilane in the first stage is restrained. Polycarbosilane is initially synthesized by thermal decomposition of polydimethylsilane obtained by dechlorinating dichlorodimethyl silane. Diameters of SiC fibers produced by this method vary between 7 and 10 μm. Tensile strength and elastic modulus values vary from 2000–3400 MPa and 147–196 GPa, respectively.

A similar process has been used to produce Si_3N_4 and silicon oxynitride fibers. In this case, polycarbosilane fibers are cured by electron irradiation or oxidation. These fibers are then heat treated in flowing NH_3 gas. Fibers cured by electron irradiation yield Si_3N_4 eventually, whereas oxidation curing yields silicon oxynitride fibers. Fibers 11–13 μm in diameter are formed by this process. Reported tensile strengths are between 1000 and 1500 MPa and elastic moduli between 100 and 150 GPa for nitride and oxynitride fibers, respectively. These properties depend on the oxygen content introduced and on the nitridation temperature. A higher nitridation temperature results in higher moduli with reduced tensile strengths.

The production of Nicalon fibers in Japan involves basically the same process developed by Yajima and co-workers. This fiber has a composition of Si–C–O and is formed by the conversting polycarbosilane into melt-spun fibers, which are cured and heated at high temperatures in inert atmospheres. Other polymeric precursors used for Si–C–N–O fiber production are methylpolydisilylazane (MPDZ) and hydridopolysilazane (HPZ). MPDA- and HPZ-derived fibers have average tensile strengths of 2 GPa [2.77]. Classical brittle failure was observed in such fibers. Characterization of these fibers demonstrated the presence of an amorphous matrix, which is likely to be the source of glass-like mechanical behavior.

Tyranno fibers are produced by Ube Chemicals by a process similar to that used for Nicalon fibers. The manufacturing steps are production of polydimethylsilane $(Si(CH_3)_2)_n$, mixing with titanium alkoxide in a ratio of about 1:10, and polymerization at 340°C in N_2 for 10 h [2.78]. The properties of Nicalon, Tyranno, and other fibers are listed in Table 2.6.

Langley et al. [2.79] studied the microstructure, chemical structure, and properties thoroughly. The structure of polycarbosilane-derived fibers such as Nicalon and Tyranno contain 10–20 wt% oxygen. X-ray line broadening and TEM analysis demonstrated that the structure of β-SiC in these fibers is nanocrystalline, and the average grain size is 1–2 nm. Nearly all fibers obtained from organosilicon polymers contained over 10 vol% porosity. These pores, determined by various techniques, are of nanoscale and of the closed type that arises during the pyrolysis process. Amorphous structures were observed as the major phase when pyrolysis temperatures were not sufficiently high. Above 1400°C, crystallization of SiC or Si_3N_4 occurred, depending on composition. Heat treatment of Nicalon and Tyranno fibers revealed that over 1300°C the tensile strength decreases rapidly. Among many possible reasons, oxidation, reaction of SiO_2 and excess C in the structure, crystallization, grain growth, and evaporation of CO are some of those more likely to cause such deterioration.

2.4.5 Oxide Fibers

Oxide fibers, in general, have higher densities compared to C fibers, but they may be preferred over C due to their ultimate oxidation resistance and superior corrosion properties, especially at high temperatures. The general process of oxide fiber production is the *slurry process*. This involves preparation of a slurry from oxide powders, extrusion of the slurry through an orifice, heat treatment to devolatilize and calcine, and sintering. Various components such as dispersants and rheology aids are also added to control the viscosity. The diameters of these fibers are usually in the region of 20 μm. Fiber FP (α-Al_2O_3 fibers) of DuPont is produced by the slurry process [2.57]. These fibers are continuous, thus suitable for combining in yarns. They have low strain to failure values that limit handling capability, but they retain their strength well up to 1000°C, and no strength loss occurs even after 300h at this temperature [2.78]. Another product developed by DuPont and identified as PRD-166 is a polycrystalline composite fiber of Al_2O_3–20wt% ZrO_2. These fibers have superior mechanical properties compared to Fiber FP [2.80]. The tensile strength of PRD-166 and its retention during heating is significantly better than that of Fiber FP.
Heating to 1400°C results in a 35% drop in strength almost immediately, but no further drop is observed well after 100h.

The *relic process* is a different route. In this process, an organic filament is soaked in an inorganic salt solution, dried, and pyrolyzed to produce a precursor to the final fiber or cloth. Union Carbide developed a ZrO_2 fiber using this method. Zircar Products Inc. produces such fibers in the form of fiber, felt, fabric, board, paper, etc. with the trade name Zircar. These products are suitable for refractory insulation but due to their relatively poor mechanical properties, they are not fit for use as structural reinforcements [2.81].

The *solution process* involves spinning viscous solutions. In this process, an aqueous solution is prepared from soluble precursors to yield the metal oxide. For example, salts such as aluminum oxychloride or formate can be used to produce Al_2O_3. The viscous solution is extruded as filaments usually in air, dried, and heated to evaporate volatile species. Safimax and Saffil fibers of ICI are also produced by a solution spinning process. Saffil fibers are a shorter form of Safimax. The initial solution consists of an aluminum salt, e.g., aluminum chlorohydroxide, $AlCl_x(OH)_y$. A viscous solution is formed by eliminating water and polymerization. By suitable heat treatment, organic species are burnt off, and the alumina salt is converted to alumina.

A variation of the solution process, the *sol–gel process*, involves an aqueous phase containing an oxide sol, together with an organic polymer. The sol is extruded into a precipitating medium where gelation occurs. This step preserves the fiber shape. Oxide fibers are obtained by a final heat treatment. Sumika (γ-alumina) fibers of Sumitomo are produced from an organoaluminum compound such as trialkyl aluminum or trialkoxy aluminum. This compound is polymerized by the addition of water to yield polyaluminixane, an organometallic powder. A viscous gel is prepared from the polymer by dissolving it in an organic solvent,

Table 2.6. Properties of Some Commercial Fibers

Manufacturer	Designation	Typical Composition (wt%)	Density (g/cm^3)	Average Diameter (μm)	Typical Composition (wt%)	Tensile Strength (MPa)	Strain to Failure (%)	Production Method	Max. Use Temp (°C)	Ref. [a]
AVCO	SCS–6	SiC on C core	3.0	140	SiC on C core	3900	0.8	CVD mono-filament	1300	[1]
AVCO	–	SiC	2.7–3.3	6–10	SiC	2800	–	–	–	[1]
Dow Corning/ Celanese	MPDZ precursor fiber	47 Si, 30 C, 15 N, 8 O	2,3	10–15	47 Si, 30 C, 15 N, 8 O	1750–2100	1.1	–	–	[1]
Dow Corning/ Celanese	HPZ precursor fiber	59 Si, 10 C, 28 N, 3 O	2,4	10	59 Si, 10 C, 28 N, 3 O	2100–2450	–	–	–	[1]
Dow Corning/ Celanese	MPS	69 Si, 30 C, 1 O	2.6–2.7	10–15	69 Si, 30 C, 1 O	1050–1400	–	–	–	[1]
Nippon Carbon	Nicalon	65 SiC, 15 C, 20 SiO_2	2,5	10–20	65 SiC, 15 C, 20 SiO_2	2700	1.4	Polymer-derived	1200	[1]
Ube Industries	Tyranno	Si,C,O,Ti	2,4	9	Si,C,O,Ti	3000	1.5	Polymer-derived	1300	[1]

(Continued)

Table 2.6. *(continued)*

Manufacturer	Designation	Typical Composition (wt%)	Density (g/cm^3)	Average Diameter (μm)	Typical Composition (wt%)	Tensile Strength (MPa)	Strain to Failure (%)	Production Method	Max. Use Temp (°C)	Ref. [a]
Toa Nenvyo Kogyo	TNSN	Si_3N_4	2,5	10	Si_3N_4	3300	–	–	1200	[1]
Berghof	Sigma	SiC	3,4	100	SiC	3450	–	Mono-filament	1250	[1]
3M	Nextel	Alumino-borosilicate	2,5	10	Alumino-borosilicate	1700	1.1	–	–	[1]
Dupont	Kevlar 49	Aramid	1,4	–	Aramid	3600-4100	2.6	–	–	[1]
J.B. Stevens	Astroquartz	Pure fused silica	2,2	9	Pure fused silica	3450	–	–	1800	[1]
–	–	B on W wire	2,3	–	B on W wire	6900	–	CVD mono-filament	–	[2]
–	–	E Glass	2,5	–	E Glass	3700	–	–	–	[2]

[a] References

1. Manufacturer's Data
2. Z.D. Jastrzebski: *The Nature and Properties of Engineering Materials.* (Wiley New York, 1977)

which contains alkylsilicate. Spinning of the gel and heat treatment at 760°C and 970°C yield γ-alumina/15wt% amorphous silica fibers with lower Young's modulus than that of α-Al_2O_3 fibers but with higher strain to failure values. The fiber diameters exhibit large variations.

High-purity SiO_2 fibers are made by spinning a polymer of silicon alkoxide from a sol–gel process. Heating the initial fibers to 1000°C produces quartz fibers with heat resistance up to 1000°C [2.57]. A similar process is employed to produce alumina, lead zirconate titanate (PZT), yttrium iron garnet (YIG), and yttrium aluminum garnet (YAG) fibers. The yield of alumina sols reaches 30 wt% and that of PZT more than 70 wt%. Initially, the fibers are in the amorphous state. Crystalline fibers are developed by suitable heat treatment. Dense fibers are obtained after 1 hour of heat treatment at 1300°C with crystallite sizes of ~300 nm. The fiber diameter lies in the range of 10–15 μm [2.82].

Nextel ™ fibers of the 3M company (Fig. 2.20) are produced by the solution process [2.57] (see Table 2.7 for properties and composition). These fibers have found many applications, such as high-temperature fabrics in flame curtains, seals, furnace belts, filter bags, thermal protection of the Space Shuttle, and flame shielding composites for jet engine compartments [2.81].

Fine-grained polycrystalline ZrO_2 fibers have been formed from acetate precursors. Strengths of 1.5–2.6 GPa have been achieved. Addition of Y_2O_3 allows tetragonal ZrO_2 retention at room temperature. Fibers are obtained by drawing suitable mixtures of oxide and water.

Haggerty et al. [2.83] developed the *laser heated floating zone directional solidification process* to produce single-crystal ceramic oxide fibers. In this process, a ceramic melt is supported by a feed rod, and single crystals that have a composition that is in equilibrium with the melt are developed by directional solidification. A laser beam is employed to produce and maintain the ceramic melt. Operational limitations establish a minimum diameter of 10 μm for fibers grown from CO_2 laser-heated melts. Al_2O_3 fibers doped with Cr_2O_3, TiO_2, or MgO have been produced by this process. Flexural strengths up to 9.6 GPa, tensile strengths up to 3.2 GPa, and better creep rates in undoped fibers were achieved.

The *inviscid melt spinning process* is a fast and low-cost process similar to melt spinning described for glass fibers. $CaO \cdot Al_2O_3$ fibers were first produced by this technique [2.84]. Other studies followed, including the production of $BaO \cdot TiO_2$ short fibers and $CaO \cdot Al_2O_3 \cdot MgO$ (CAM) fibers. CAM fibers were continuous, ~200 μm in diameter, with a thin C coating and nonequilibrium crystalline phases. Upon heat treatment, these phases transformed into equilibrium phases. Surface roughness and porosity were noted on these fibers.

Properties of commercially available oxide fibers are listed in Table 2.7.

Table 2.7. Properties of Commercial Oxide Fibers [a]

Manufacturer	Designation	Typical Composition (wt%)	Density (g/cm^3)	Average Diameter (μm)	Elastic Modulus (GPa)	Tensile Strength (MPa)	Strain to Failure (%)	Max. Use Temp (°C)	Melting Point (°C)	CTE (10^{-6}/°C)	C or D [b]
Du Pont	Fiber FP	>99 α-Al_2O_3	3,95	20	380	1400	0.4	1320	2040	–	C
Du Pont	PRD-166	80 Al_2O_3; 20 ZrO_2	4,2	20	380	2070	0.6	1430	1830	9.0	C
ICI	Saffil	95 Al_2O_3; 5 SiO_2	3,3	3	300	2000	-	1400	1980	–	D
ICI	Safimax	96 Al_2O_3; 4 SiO_2	3,3	3	300	2000	0.7	–	–	–	D
Sumitomo	Sumika	85 Al_2O_3; 15 SiO_2	3,2	17	200	1500	1.1	1250	–	8.8	C
Carborundum	Fiberfrax	50 Al_2O_3; 50 SiO_2	2,7	4	105	1000	–	–	–	–	D
3M	Nextel 312	62 Al_2O_3; 24 SiO_2; 14 B_2O_3	2,7	11	155	1720	1.1	1200	1800	3.5	C
3M	Nextel 440	70 Al_2O_3; 28 SiO_2; 2 B_2O_3	3,1	11	189	1720	1.1	1430	1890	5.0	C
3M	Nextel 480	70 Al_2O_3; 28 SiO_2; 2 B_2O_3	3,1	3	224	2300	1.0	1430	>1800	–	C

[a] Reference: Manufacturer's data

[b] Continuous (C) or discontinous (D)

Fig. 2.20. General view of 3M ™ Nextel ™ Ceramic Fibers (courtesy of 3M Ceramic Textiles and Composites Department St. Paul MN)

3 Production of Ceramic Bodies

3.1 Preconsolidation Processing

The term *consolidation* can be defined as producing a dense body to attain ultimate material properties. A majority of methods used for fabricating ceramic bodies, be they monoliths or composites, require various preparatory steps before consolidation. Depending on the densification method as well as the size and shape of the component, a number of preconsolidation methods may be required. In the recent years, research efforts have shown that these preparatory steps can be very important in determining the properties of the final product. In this section, an overview of important preconsolidation processes will be given, and their effects on the product will be discussed.

3.1.1 Milling and Sizing

For most production techniques, ceramic powders need to have a small average particle size and a narrow particle size distribution. Industrial methods to achieve such powders include milling, screening, separation, and centrifugation.

Milling is the process for reducing coarse particles in size by removing material from particle surfaces and by particle fracture. Various energy consuming phenomena associated with milling are [3.1]

1. increased surface energy,
2. plastic deformation, and
3. elastic deformation of particles.

Usually, when a particle is repeatedly fractured, the resultant particles will be stronger and particle agglomeration will occur more easily. Thus, there is a limit to particle size attainable in milling. Common techniques used for particles below 1 mm are ball milling, vibratory milling, attrition milling, fluid energy milling, and roller milling [3.2].

In *ball milling*, ceramic powders are mixed with suitable cylindrical or spherical milling media. Common milling media are TZP, Al_2O_3–SiC, and WC–Co. Milling can be done dry or in a liquid. Plastic bottles, ceramic jars, or steel shells with ceramic linings are used as mills. Rotation of the mill provides a cascading action where the tumbling media grind by impacting and shearing the particles (Fig. 3.1). Many parameters can affect the milling performance. Important ones are the size and angular speed of the mill, the relative amounts of milling media and powders, the relative amount and viscosity of the liquid medium in wet milling, the degree of powder agglomeration, and the size of milling media. Smaller milling media are suitable for finer powders due to increased contact area. If particles agglomerate during dry milling, wet milling or the addition of dry lubricants is necessary. The liquid to solid volume ratio should not be very high during wet milling to minimize drying costs. This ratio should not be very low either otherwise movement will be restricted by high viscosity.

Besides size reduction, ball milling is frequently used for mixing different types of powders. If the densities or settling rates of the powders differ significantly from each other, the viscosity of the slurry should be increased. Keeping the solid load high relative to the liquid or additions of thickeners to the liquid can increase the viscosity and prevent separation during drying. Thickening or gelation can be achieved in water-based systems by using methylcellulose or hydroxypropyl methylcellulose to form a thermally reversible gel [3.3]. Chemical gelation can be achieved by using sodium alginate in the presence of alkaline earth ions such as Ca^{2+} [3.1]. Chemical gelation is mostly irreversible. Many types of polysaccharides (agar, alginate, and carrageen extracted from seaweed, gum from trees, pectin from fruits, starch from vegetables and cellulose from wood fibers) are suitable for use as thickening agents [3.4].

Vibratory ball milling uses a vibratory action transmitted into the mill containing the powders, liquid, and grinding media (Fig. 3.2). Vibration causes a cascading action as well as local impact and friction and reducies the particle size effectively. The mill chamber is usually lined with rubber or polyurethane. This method yields finer particles compared to those achieved by ball milling. In *fluid energy milling*, powders are impacted with each other in a high velocity stream, as depicted in Fig. 3.3. Compressed air, nitrogen, CO_2, water, or other types of gases and liquids may be used as the energy transfer medium. Particle to particle impact at sonic velocities

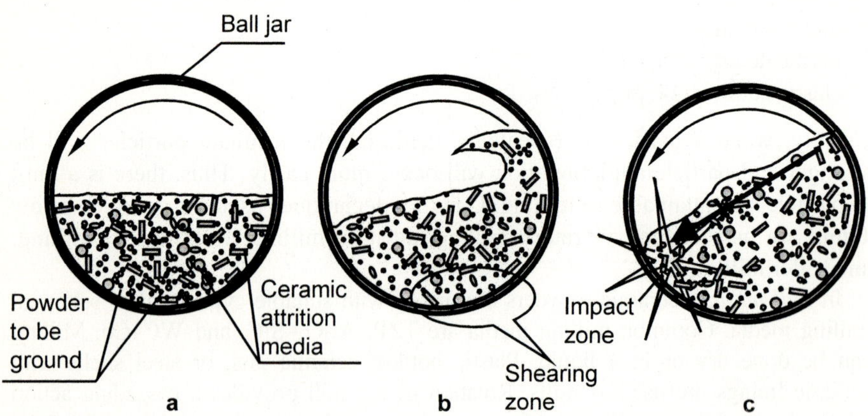

Fig. 3.1 a–c. Ball milling in a tumbling mill

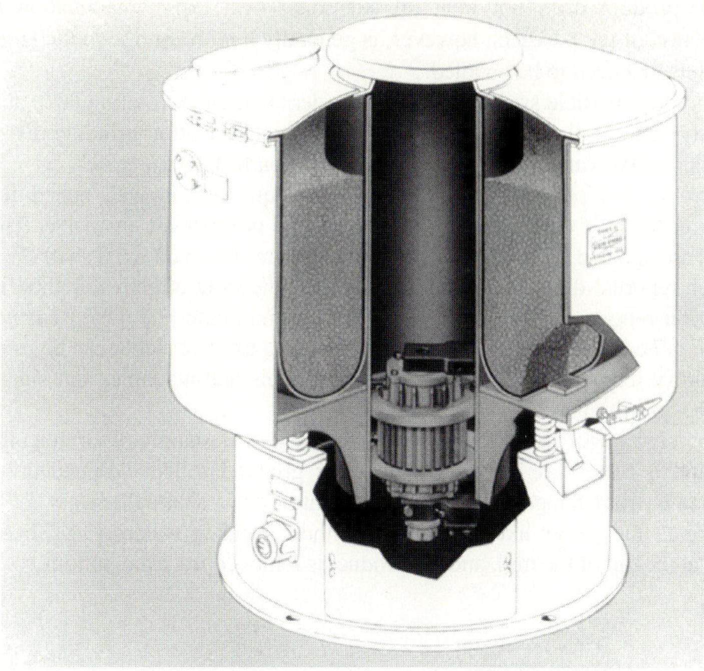

Fig. 3.2. Vibratory ball milling (courtesy of Sweco Inc. Florence Kentucky)

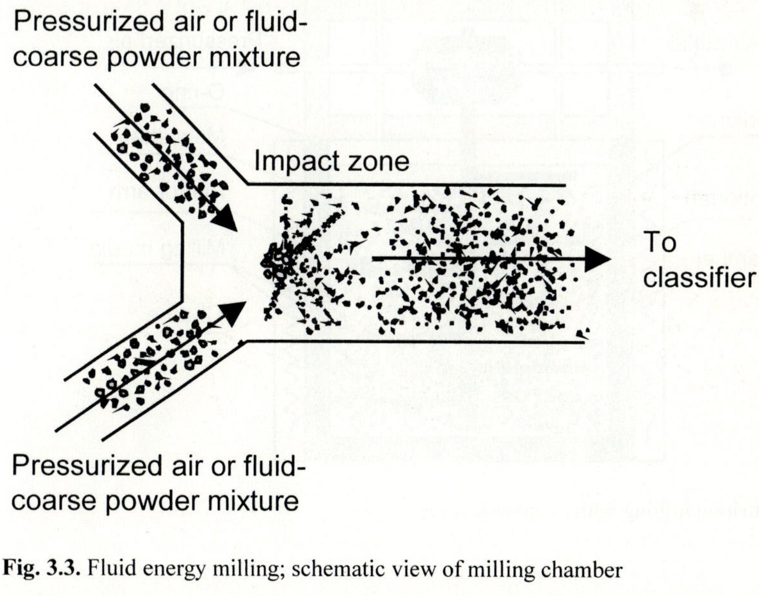

Fig. 3.3. Fluid energy milling; schematic view of milling chamber

reduces their size effectively. Because grinding is introduced by high velocity air or liquids, the product does not pick up attritional heat and contamination is minimal. Collection of the powders, however, is generally a problem due to the large volume of carriers that need to be handled.

Roller mills reduce particle size by grinding powders between a rotating table and arm supported rollers. Similarly, *disk mills* use stationary and rotating grinding disks. Differential roller speeds can be employed to control the particle size precisely.

Attrition milling is a powerful comminution technique that is well suited for submicrometer grinding if beads harder than the charge material are available. The grinding media are agitated by a series of string arms connected to an axial shaft (Fig. 3.4). It has been reported that attrition milling is by far the most effective method in the submicrometer range and keeps contamination at a minimum [3.5,3.6]. Reactive powders can be milled under controlled atmosphere, and a thermal jacket can be used to adjust the slurry temperature [3.7]. Continuous models are available, and slurry capacities can be quite high.

Planetary or *centrifugal mills (Szego mills)* consist of a stationary or rotating support disk and a composite rotor with helically grooved rollers suspended by flexible steel shafts from flanges connected to a central drive shaft. Planetary mills owe their name to the planet-like movement of their grinding rollers [3.8]. Feed materials enter at the top of the mill, and the product is removed from the bottom.

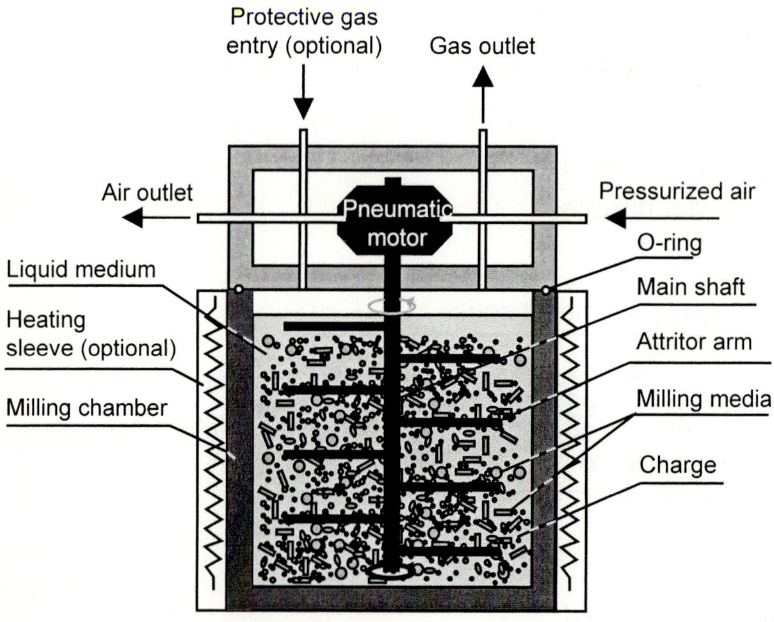

Fig. 3.4. Attrition milling with pneumatic drive

Fig. 3.5. Equipment used for vibratory screening (courtesy of Brinkmann Instruments Inc. Westbury NY)

The crushing force is created by radial acceleration of the rotors and the high velocity gradients, due to the opposing rotational movement of the rollers and the support disk, that produce a large shear force. Operational product sizes range from 50 to 150 μm in this process. Depending on the feed material, average particle sizes of 1 μm can also be achieved.

Other specialized methods include ultrasonic systems, electrohydraulic crushers, thermal decrepitation, explosives, and chemical comminution.

Screening is the selective separation of powders using sieves with standard openings. Continuous screening, for example, by a disk-screener, is used to separate relatively coarse particles. Finer particles are separated by differences in their precipitation rates, i.e., by classification. Vibratory screening is a simple method for low-volume operations. Figure 3.5 shows a commercial apparatus for this purpose. Table 3.1 lists standard mesh sizes and corresponding particle size ranges.

Vibratory wet screening is a faster method than dry screening since clogging of the sieves is less likely (Fig. 3.6). The passage of particles through sieve openings is aided by the vibration and movement of the liquid carrier. Dry forced-air screening and sonic sieving methods are used in the range of 37 to 850 μm in industry. Finer sieve sizes are generally not suitable for industrial use, but they are available for research purposes. Sieving in this particle size regime is called *microsieving* and usually necessitates a dispersing medium. Successful microsieving requires that the velocity of the dispersing medium through the sieve is high, the sieve surface is immersed in the dispersing medium to reduce surface tension effects, and the particles are fluidized by vibration to reduce resistance to passage through the sieve [3.9]. Cyclone and centrifugation methods are used for sizing in the subsieve range of particle size in industry.

Table 3.1. ASTM Standard Screen Sizes [a]

"Mesh" Sieve	Sieve Opening	
Designation	(mm)	(inch)
4	4.76	0.187
6	3.36	0.132
10	2.00	0.0787
12	1.68	0.0661
16	1.19	0.0469
20	0.84	0.0331
40	0.42	0.0165
80	0.177	0.0070
120	0.125	0.0049
170	0.088	0.0035
200	0.074	0.0029
230	0.063	0.0025
270	0.053	0.0021
325	0.044	0.0017
400	0.037	0.0015

[a] Adapted from ASTM E11, Annual Book of ASTM Standards.
Copyright 1970 ASTM West Conshohocken PA

Cyclones have no moving parts and particle separation is achieved by centrifugal force. The feed suspension enters the column tangentially under pressure and follows a spiral path downward. The centrifugal force on a particle is opposed by the inward drag force of the carrier fluid or air (in air classification) directed inward. Coarse particles move downward, and finer ones migrate into a secondary vortex moving upward and are discharged with most of the carrier fluid.

Centrifuges use the differential settling of particles described by Stokes' law, which relates the settling time to particle diameter by [3.2]

$$v = a^2 (D_P - D_L)\, g / 18\eta_L, \tag{3.1}$$

where v is the constant velocity of a particle released in a fluid of viscosity η_L, D_P is the particle density, D_L is the fluid density, a is the particle diameter, and g is the acceleration due to gravity or centrifugal action.

Typically, a centrifuge consists of a drive system, usually an electric motor, a rotor, and centrifuge buckets or tubes [3.10] (Fig. 3.7). When the rotor revolves, a centrifugal force is exerted on the slurry in the centrifuge buckets. Larger particles will sediment first. Smaller particles will remain in the suspension. This technique is called *differential pelleting*. The yield of such a procedure can be low. Another method is *rate-zonal centrifugation*. Upon obtaining a density gradient in the centrifuge tube through a suitable solution or a gradient maker, particles move at

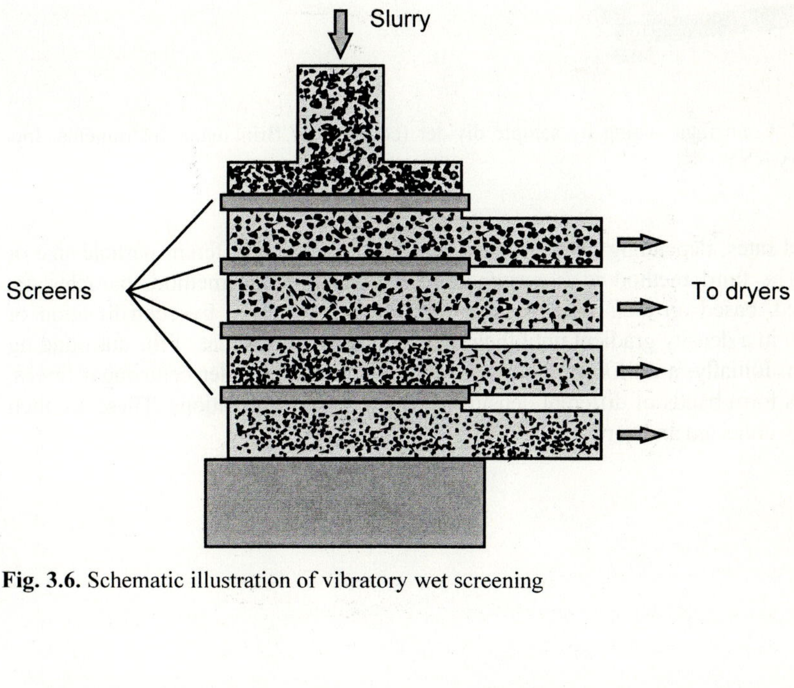

Fig. 3.6. Schematic illustration of vibratory wet screening

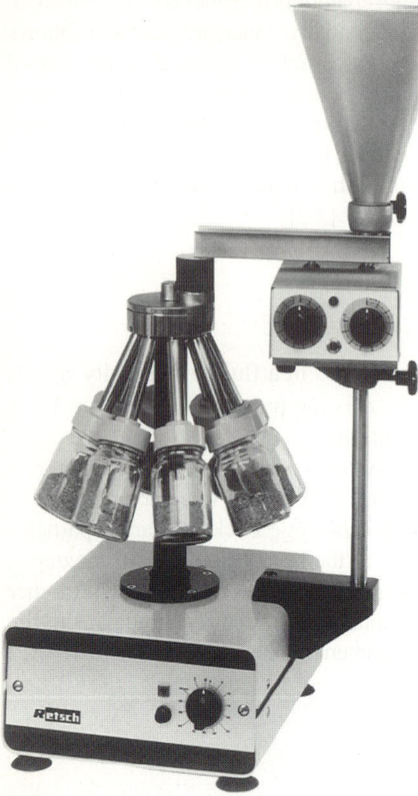

Fig. 3.7. Centrifugal sizing by sample divider (courtesy of Brinkmann Instruments, Inc. Westbury NY)

different rates, depending on their mass, and form layers of different particle size or mass. The third method is *isopycnic centrifugation*. In this method, particles are separated, based on their density. This separation is achieved by centrifugation of particles in a density gradient until their density is the same as that of the surrounding medium. Initially, a uniform mixture of sample is formed. Under centrifugal forces, particles form bands of different densities at their isopycnic positions. These are then carefully collected and sorted.

3.1.2 Dispersion and Rheology of Slurries

3.1.2.1 Rheology

Rheology is the science of deformation and flow. The rheology of ceramic slurries and pastes plays an important role in many aspects of ceramic processing. Among these processes are preconsolidation operations (e.g., mixing and milling), green forming operations (e.g., slip casting, tape casting, fiber drawing), coating processes (e.g., dipping, spinning, spraying), as well as consolidation and drying (e.g., liquid-phase sintering) [3.11].

Liquids are classified as Newtonian, pseudoplastic, and dilatant, according to their flow behavior. Typical shear rate ($d\varepsilon/dt$)–shear stress (τ) curves for different flow characteristics are illustrated in Fig. 3.8. The rheological properties of a slurry can be determined from experimental steady-state shear curves that display flow stress–shear rate relationships.

In a *Newtonian liquid*, the shear stress required to initiate and maintain steady flow is proportional to the velocity gradient,

$$\tau = \eta\,(dv/dr), \tag{3.2}$$

where η is the proportionality constant better known as viscosity, v is the fluid velocity, and r is the distance from the container wall. The magnitude of viscosity is an indication of the resistance of the liquid to flow due to internal friction between liquid molecules. In suspensions with moderate solids loading and liquids containing large molecules, the whole system will be connected by weak bonds. Such a gel-like structure resists flow at low shear rates, but the bonds are broken at higher shear rates, enabling easier flow. This liquid behavior is called *pseudoplasticity* or *shear- thinning* and can be described by an empirical power law [3.2]:

$$\tau = K(\dot{\gamma})^n \qquad (\text{where } n < 1), \tag{3.3}$$

where $\dot{\gamma}$ is the shear rate, K is the consistency index, and n is the shear-thinning constant. The apparent viscosity of such a liquid is given by

$$\eta_a = K\,(\dot{\gamma})^{n-1}. \tag{3.4}$$

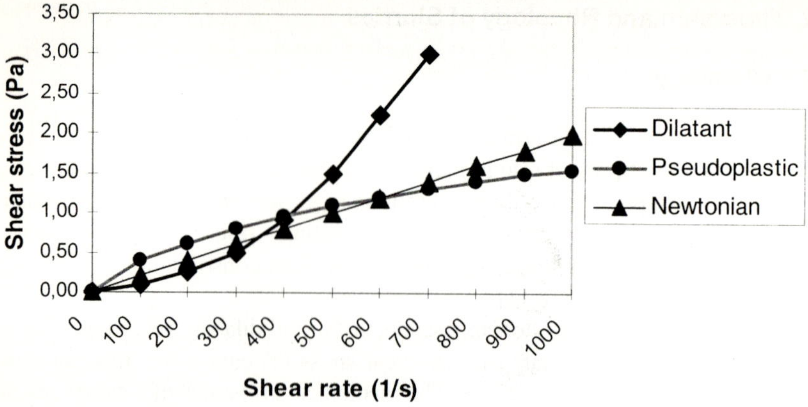

Fig. 3.8. Typical shear rate–shear stress curves for different flow characteristics.

In some systems, particles tend to collide and stick to each other at higher shear rates. This is called *shear-thickening* or *dilatancy*. In this case, Eq. (3.4) applies with $n > 1$.

If the suspension resists flow until a certain shear stress is exceeded and then exhibits Newtonian flow, such a suspension is called a *Bingham plastic*. Similarly, some systems may exhibit pseudoplastic or dilatant flow with a yield point (Fig. 3.8). The flow behavior of a Bingham plastic is described by

$$\tau - \tau_y = \eta_p \dot{\gamma} . \tag{3.5}$$

For some suspensions, viscosity and shear stress decrease with time under a given shear rate. This behavior is called *thixotropy* and is commonly observed in pseudoplastic or Bingham suspensions when the orientation and bonding of molecules or particles change with time during shear. In most ceramic systems, thixotropy is reversible, i.e., when flow stops, the viscosity increases to its original value after some time. The reverse behavior, that is, a viscosity increase with time, under shear, is also possible, for example, in systems where breakup of agglomerates is a time-dependent phenomenon. This type of suspension is called *antithixotropic* or *rheopexic* [3.4].

Particles in a liquid are under attractive and repulsive forces. It can generally be assumed that the net force can be found simply by adding up all the contributions to surface forces [3.12]. However, these forces are not always strictly independent of each other, so that their additivity should be treated with caution [3.11].

In almost all green forming operations that deal with a dispersion, such as slip casting, injection molding, extrusion, or colloidal/sol–gel processing, a homogeneous dispersion of the components is essential to achieve desired properties. A stable dispersion is one in which interparticle repulsive forces are larger than attractive forces. Otherwise, particles tend to form undesirable flocs or agglomerates. The different contributions to interparticle forces that may occur are outlined here.

3.1.2.2 van der Waals Forces

In some materials van der Waals forces form weak bonds as discussed in Sect. 1.2. When two solid particles in a certain medium are considered, these forces are always attractive if the particles are similar. van der Waals forces are caused by interactions between electron orbitals of two neighboring particles and the electron orbitals in the surrounding liquid. In other words, they arise from the interaction of atomic and molecular electric dipoles whose orientations are such that they attract each other. The types of such interactions are

1. *Keesom interaction*: A permanent molecular dipole results in an electric field, and this field orients other permanent dipoles. These dipoles are attracted to each other.
2. *Debye interaction*: Induced dipoles in polarizable atoms, molecules, or media are oriented so that they are attracted to the inducing dipole.
3. *London* or *dispersion force*: A temporary dipole induces other dipoles in surrounding atoms or molecules, which are then attracted to the first one.

Generally, London and van der Waals forces are used in the same sense. If there are no permanent dipoles in a system, then the first two contributions will not occur, which leaves only the London attractive force effective among all van der Waals forces. If particles are not far apart, the interparticle force is given by [3.13]

$$F_L = -(A R_s) / (12D^2), \tag{3.6}$$

where R_s is the particle radius assuming a spherical shape, D is the distance between two particle surfaces, and A is the *Hamaker constant*. The Hamaker constant is characteristic of a system and depends on such factors as polarizability, composition, and the structure of the particle and liquid.

The potential energy of attraction is roughly given by [3.14]

$$U_A = -(A R_s) / (12D). \tag{3.7}$$

For powders dispersed in liquids, the *Hamaker constant* is given by [3.4]

$$A = (\sqrt{A_2 - A_1}\,)^2, \tag{3.8}$$

where A_2 and A_1 are Hamaker constants of the particle and the liquid, respectively.

3.1.2.3 Electrostatic or Ionic Forces in Nonpolar Media

Particle surfaces can become charged by adsorbing or desorbing electrons and ions, which may cause coulombic interactions among them. Particles of the same type are likely to acquire the same charge and thus repel each other, but they may be attracted to a different type of particle. A charged particle is also attracted to a neutral material whose dielectric constant is higher than that of the surrounding medium. Direct electrostatic forces are generally stronger and have a longer range than other surface forces; therefore, if they are present, they will dominate in the system.

3.1.2.4 Electrostatic Forces in Polar Media

In polar media such as water, electrostatic charging occurs by dissociation of ionizable species in the liquid, followed by preferential adsorption of the anion or cation onto particle surfaces [3.15]. These ions are known as counterions. For entropic reasons, counterions remain in solution and form a diffuse layer of charge next to the surface. The surface charge and the diffuse layer of opposite charge form an electrical double layer. The thickness of this layer depends on the concentration of ions in the medium. More ions available result in a thinner double layer.

Similar particles acquire the same charge and repel each other, although the magnitude and range of this repulsion are dramatically reduced due to counterions. So far, the best formulation for the ionic repulsion in polar systems is given by the *DLVO theory*, developed by Derjaguin and Landau [3.16] and Vervey and Overbeek [3.17]. According to this theory, the net force between particles immersed in a polar liquid is given by the sum of electrical double-layer repulsion and van der Waals attraction. For two spherical particles of radius R_s, the potential energy of repulsion is given by [3.18]

$$U_R = [\varepsilon R_s \psi_o^2/2] \ln[1 + \exp(-\kappa D)] \tag{3.9}$$

where ε is the permittivity of the medium, κ is the *Debye constant*, which is the inverse of the electrical double layer thickness, and ψ_o is the electrical potential at the particle surface. The electrical double layer thickness is given by the *Debye length* [3.13],

$$\Lambda = \kappa^{-1} = [(e^2 \Sigma \rho_i z_i^2)/(\varepsilon_o \varepsilon_r k T)]^{-1/2} = [(2000\, e^2 N_A L)/(\varepsilon_o \varepsilon_r k T)]^{-1/2}, \tag{3.10}$$

where e is the electronic charge, ρ_i is the number density, z_i is the valence of ionic species i, ε_o is the dielectric constant of vacuum, ε_r is the dielectric constant of the solvent, N_A is the *Avogadro number*, k is *Boltzmann's constant*, T is the absolute temperature, and L is the ionic strength. Equations for various particle geometries and surface potentials can be found in the literature [3.19].

If no other force exists, the energy barrier resulting from the repulsive force prevents particle agglomeration. If the particles collide with sufficient kinetic energy to overcome that barrier, attractive forces will cause them to adhere strongly to each other (Fig. 3.9a). Where the total energy is negative, attraction is possible at a finite separation in concentrated electrolyte solutions (Fig. 3.9b). This yields a much weaker and potentially reversible adhesion between particles, unlike the irreversible attraction that occurs at very small separations.

The amount of ions in a solution determines the degree of electrostatic repulsion. With many oxide powders in water, the surface charge is determined by proton association or dissociation with surface hydroxyl groups, and hence, is pH dependent. For each suspension, a critical *zeta potential*, ξ, and a range of *double- layer thickness* exist. In a suitable suspension, the double-layer repulsion energy, U_R, is larger than van der Waals attraction, U_A, and produces a barrier to flocculation [3.2]. By changing ξ, flocculation of particles can be prevented. ξ and ψ_o (the potential at the

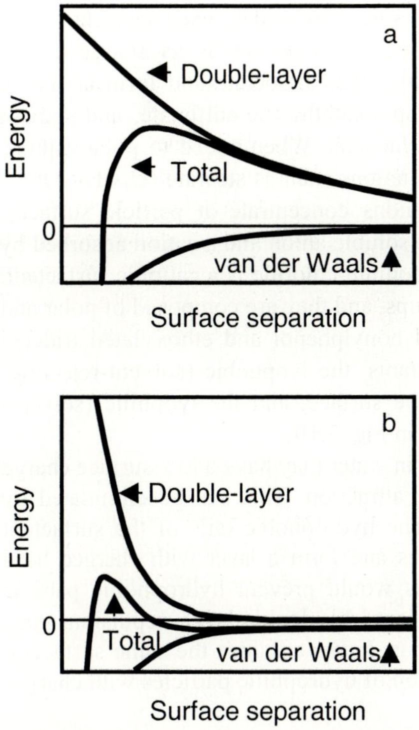

Fig. 3.9. Energy of interaction between two particle surfaces across a polar liquid at low electrolyte concentration **a** and high electrolyte concentration **b** (From R.G. Horn: *Surface Forces and Their Action in Ceramic Materials*. J. Am. Ceram. Soc. **73 [5]**, 1117–1135 (1990). Copyright 1990 American Ceramic Society, Columbus OH, reprinted with permission)

particle surface, used for practical purposes) are related to the pH of the solution. At a pH < pH_{isoel} (pH at the isoelectric point), the surface reacts with H^+ ions and becomes positively charged. At pH > pH_{isoel} the surface reacts with OH^- ions and acquires a negative charge. When ξ is between +30 and –30 mV, ionic repulsion becomes so low that thermal vibrations can push particles toward each other, in which case adhesion and flocculation are likely [3.4]. Beyond these limits, which can be adjusted by pH control, suspension stability, i.e., deflocculation, may be achieved. The *isoelectric points* of some oxide ceramics are 1.5 for δ-MnO_2, 2.2 for SiO_2, 4.5 for SnO_2, 6.0 for TiO_2, 7.0 for Cr_2O_3, 7.3 for β-MnO_2, 8.6 for Fe_2O_3, 9.0 for Al_2O_3 and ZnO, 10.3 for PbO, and 12 for MgO [3.4].

3.1.2.5 Stabilization via Surfactants

The word surfactant stands for surface-active agent and describes the tendency of such molecules to adhere to surfaces. Surfactants are compounds that have a strong tendency to concentrate near surfaces. Soaps, detergents, wetting agents, foaming agents, antifoam agents, and emulsifiers are widely used surfactants. An important property of a surfactant for use as a dispersant is its charge on the interfacial ion. *Anionic surfactants* are those that dissociate and form anions at surfaces. Sodium stearate, sodium disopropylnaphthalene sulfonate, and sodium dodecyl benzene sulfonate are anionic surfactants. When added to polar liquids, they dissociate to soluble Na^+ cations and anions such as stearate ($C_{17}H_{35}COO^-$) in the case of sodium stearate [3.2]. Anions concentrate at particle surfaces. Similarly, a *cationic surfactant* produces a soluble anion and a cation adsorbed by the particle surface. Dodecyltrimethylammonium chloride is a cationic surfactant. *Nonionic surfactants* have no ionizing groups, and they are composed of polar and nonpolar ends. Examples are ethoxylated nonylphenol and ethoxylated tridecyl alcohol [3.2]. In all three types of surfactants, the lyophobic (solvent-rejecting) ends of the molecules are adsorbed by the surface, and the lyophilic (solvent-liking) ends extend to the liquid, as shown in Fig. 3.10.

Hydrophobic (water-rejecting) particles in water may have a low surface charge and be agglomerated due to hydrophobic attraction. This can be eliminated by using a suitable ionic surfactant [3.13]. The hydrophobic tails of the surfactant molecules will be adsorbed by the particles and form a layer with charged head groups exposed to water (Fig. 3.10c). This would prevent hydrophobic particle interaction and, in addition, provides electrical double-layer repulsion, thus stabilizing the system. One must be cautious, however, since the same surfactant would have the reverse effect in a suspension of hydrophilic particles with charges of opposite polarity.

Other forces can also be affected by surfactant adsorption, in particular van der Waals attraction. If the refractive index of the surfactant is intermediate between that of the particle and that of the liquid, as is often the case, attraction will be weakened by the surfactant. Furthermore, adsorbed monolayers prevent close contact of particles, where attractive forces are dominant. This effect called steric stabilization, is discussed in more detail in the next section.

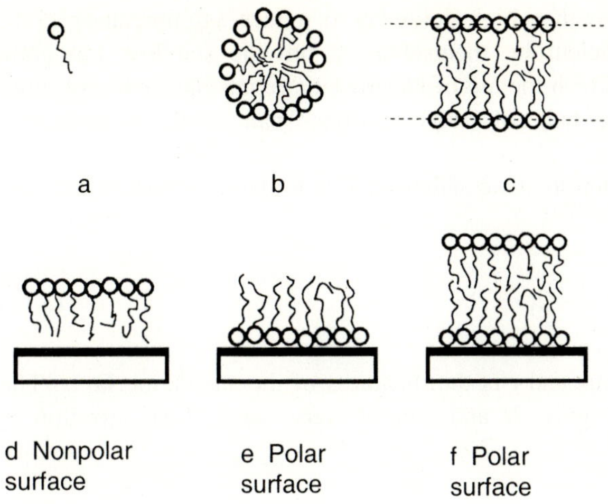

Fig. 3.10. A single surfactant represented by a circle for the hydrophilic head group and by a squiggly line for the hydrophobic tail **a,** aggregation of surfactants into spherical micelles **b** and extended bilayers **c,** adsorption of surfactant monolayers on nonpolar and polar surfaces, **d** and **e,** respectively, and adsorption of a double layer on a polar surface **f** (From R.G. Horn: *Surface Forces and Their Action in Ceramic Materials*. J. Am. Ceram. Soc. **73 [5],** 1117-1135 (1990). Copyright 1990 American Ceramic Society Columbus OH, reprinted with permission)

With all of these possibilities, the addition of surfactants to a suspension is one of the simplest and most powerful methods of controlling stability. Nevertheless, it has to be exercised carefully due to the possibility of adverse effects.

3.1.2.6 Steric Stabilization

Steric stabilization is a useful alternative to stabilization via surfactants in colloidal suspensions with high solids concentrations or high electrolyte concentrations. Steric stabilization involves coating particles with a suitable polymer, as illustrated in Fig. 3.11. Homopolymers and random copolymers are usually not suitable for stabilization [3.20]. In low concentrations of high molecular weight polymers in a poor solvent, it is possible for one polymer molecule to adhere to more than one particle. This phenomenon is called bridging (Fig. 3.11). This problem can be overcome by using block copolymers. These polymers contain two different types of homopolymer blocks attached to each other, in a way similar to surfactants. Block copolymers provide a coating of nonadsorbing and nonbridging polymers on particles with one end extending to the liquid, resulting in effective stabilization.

In most commercial practice, semisteric stabilization is preferred to steric stabilization [3.20]. *Semisteric stabilization* is defined as the use of adsorbed short-chain molecules rather than high polymers to reduce agglomeration [3.4]. Such polymers are insufficient by themselves to provide stability, but good dispersion can be achieved with the aid of electrostatic repulsion. Low molecular weight polymers with the help of polyelectrolytes can result in semisteric stabilization.

The simplest approximation to steric stabilization is to define a particle collision diameter as [3.14]

$$d = 2(a + t), \tag{3.11}$$

where a is the particle radius and t is the thickness of the stabilizing layer. The range of interaction is roughly $2t$ and can be very steep if stabilization is effectively used.

3.1.2.7 Capillary Forces

It is common for a vapor to condense in a narrow gap or capillar. This phenomenon can also occur between solid surfaces in close proximity. The most familiar example of capillary forces occurs in powders, which can flow freely in a dry

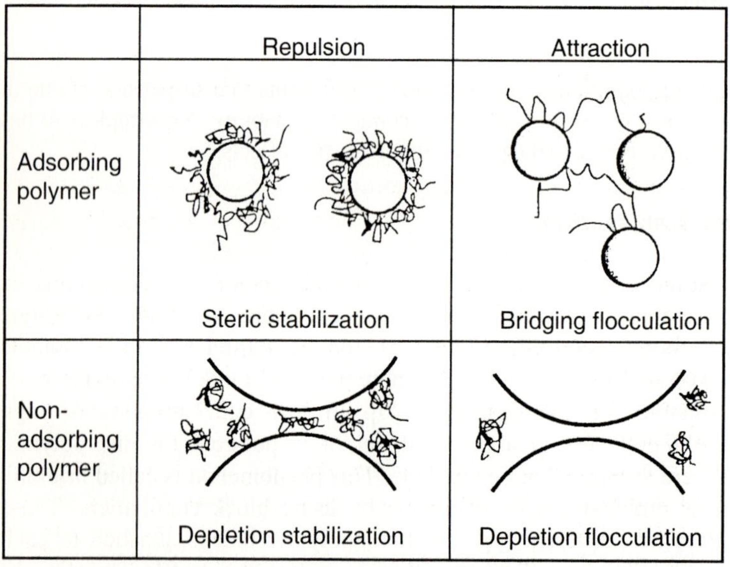

Fig. 3.11. Important mechanisms of stabilization and flocculation by uncharged homopolymers (From R.G. Horn: *Surface Forces and Their Action in Ceramic Materials*. J. Am. Ceram. Soc. **73 [5]**, 1117-1135 (1990). Copyright 1990 American Ceramic Society Columbus OH, reprinted with permission)

atmosphere, but tend to stick together when exposed to a humid environment. It is generally not realized, however, that the same effect can occur in particles suspended in a liquid when trace amounts of a second, immiscible liquid are present. In the general form, the critical distance above which, a capillary bridge will form is given by the Kelvin radius [3.13],

$$r_K = (\gamma_L V) / (RT \log [c/c_s]), \tag{3.12}$$

where γ_L is the liquid–liquid or other type of interfacial tension, V is the molar volume, R is the gas constant, T is temperature, c is the concentration of the second liquid, and c_s is the saturation concentration. The resulting force can be given by

$$F = 2\pi R_s \gamma_L \cos\Theta, \tag{3.13}$$

where R_s is the particle radius and Θ is the contact angle.

To give an example of capillary forces, consider hydrophilic particles that are suspended in a nonpolar oil, which has been exposed to air at 50% relative humidity. The oil will also be 50% saturated with water, which means about 100 ppm water is dissolved in oil. Water will tend to condense between nearby particles, i.e., segregate from the oil and form bridges across particles. Therefore, the oil can act as a reservoir of water molecules, and capillary forces can exceed any other active surface forces present.

3.1.2.8 Other Important Forces

Various other types of forces exist in different systems. Charge correlation effects, structural forces, solvation forces, and hydrophobic forces are among them. *Charge correlations* are significant when surfaces are highly charged and counterions are polyvalent. Nonuniform charge distributions in interacting double layers can result in an attractive contribution to surface forces. *Structural forces* occur as a consequence of the arrangement of the dispersion medium. The force between two smooth solid surfaces separated by a simple liquid shows alternating maxima and minima as the gap between them changes. The spacing between successive maxima or minima is approximately equal to the molecular diameter, and the amplitudes are highest next to particle surfaces.

Forces between particle surfaces and the surrounding liquid are often affected by the interaction of that liquid with the surface. The surface may be solvated, for example, a polar surface may orient the molecules of a polar liquid such as water, or the liquid molecules may form hydrogen bonds with the surface. The magnitude of such bonding falls off rapidly with distance from the surface. A surface will thereby modify the structure of the adjacent liquid, and the liquid may modify the structure of the solid. Generally, an inhomogeneous liquid structure is caused. When two solvated surfaces meet, their solvation layers overlap, producing a net force. There is, however, no general theory of *solvation forces* because they depend on specific features of the surface and solvent.

Hydrophobic forces are strong, long-range attractive forces that occur between macroscopic hydrophobic surfaces immersed in water. They are significantly stronger than van der Waals attraction and their range extends as far as 70 nm. The source of this force is not very clear, and different explanations exist, although the force is experimentally shown to be present.

3.1.3 Drying and Granulation of Powders

Drying of powder/liquid mixtures to achieve powders or granules is necessary for dry forming methods such as uniaxial pressing and isostatic pressing. Granulated material is used for hot-pressing operations, as feed material for calcining and melting processes, and as catalytic supports.

The simplest way of eliminating the liquid from a slurry is *pan drying*. This may be an acceptable practice for single-phase ceramics, but even in that case, powders tend to form agglomerates or a cake during drying. To produce powders from the cake, an additional dry milling operation is required. Multiphase ceramics pose the problem of differential settling during drying. The time, t, for settling a height, H, is [3.2]

$$t = (18H\,\eta_L) / [a^2\,(D_p - D_L)\,g], \tag{3.14}$$

where η_L is the liquid viscosity, a is the diameter of a spherical powder, D_p and D_L are the powder and liquid densities, respectively, and g is the acceleration of gravity. In a multicomponent system, powders of different sizes and/or densities will sediment at different rates, which will cause inhomogeneities in the dried body. If there is a binder in the system, it will rise to the top and dry first. These problems can be overcome by stirring continuously during drying. For small volumes, this can be achieved with a magnetic stirrer on top of a hot plate.

Granules may be formed directly by *compaction, extrusion*, or *spray granulation.* All of these methods involve dry powders and a liquid or a binder. These granules generally have diameters in the millimeter range. *Spray drying* is the most suitable method for uniform drying of powder dispersions and production of very fine granules at the same time. This process is traditionally used within the ceramic industries for powder production aimed at dry pressing. During spray drying, the slurry or solution is atomized and injected into a hot gas stream, usually air. Upon contact with the hot gas, the liquid evaporates and the spherical shape of the droplet is preserved [3.21]. Binders, dispersants, soluble salts, and even small amounts of residual liquid hold these spheres together. Dry powders are then separated from the gas stream and collected.

A typical spray dryer consists of three major parts: an atomizer, a drying chamber, and a collection chamber (Fig. 3.12). Spraying can be done in a fountain configuration or along the gas flow. If the powders being produced have a mean diameter larger than about 75 μm, it is possible to collect the powder directly at the bottom of the drying chamber. Finer particles, however, need to be separated by a cyclone attached to the drying chamber. There are various means of atomization. These include centrifugal (rotary), pressure nozzle, pneumatic, and ultrasonic atomization.

The slurry should have the maximum possible solids content to increase productivity, yet it should contain just enough liquid to make sure that nozzles do not clog during spraying. A high slurry viscosity due to large binder content can also be a hindrance to spraying. The viscosity must be controlled over a wide range of shear rates to provide reproducibility. Water is frequently used as the liquid medium, but organic liquids, e.g., alcohol, have the advantage of decreased drying times. The type of liquid also depends on whether it provides enough stabilization.

In the *rotary atomization* technique, a high-speed rotating wheel is used to mechanically divide a liquid stream into fine droplets. The wheel is hollow and perforated around its circumference. The liquid stream is fed into the center of the wheel and the centrifugal force projects the liquid through perforations. After the droplets form, they are dried before touching the chamber walls. Rotary atomizers can handle a variety of feeds, including abrasive materials and coarse particles. Atomizers with capacities ranging from a few to 150000 kg/h are available commercially. The main disadvantage of this method is that the liquid is directed horizontally toward the drying chamber walls and that restricts the chamber dimensions.

In the *hydraulic nozzle* technique, atomization is provided by forced liquid feed through an orifice. This requires a high-pressure pump, which can be a costly item for abrasive slips due to frequent maintenance requirements. Plugging of nozzles is

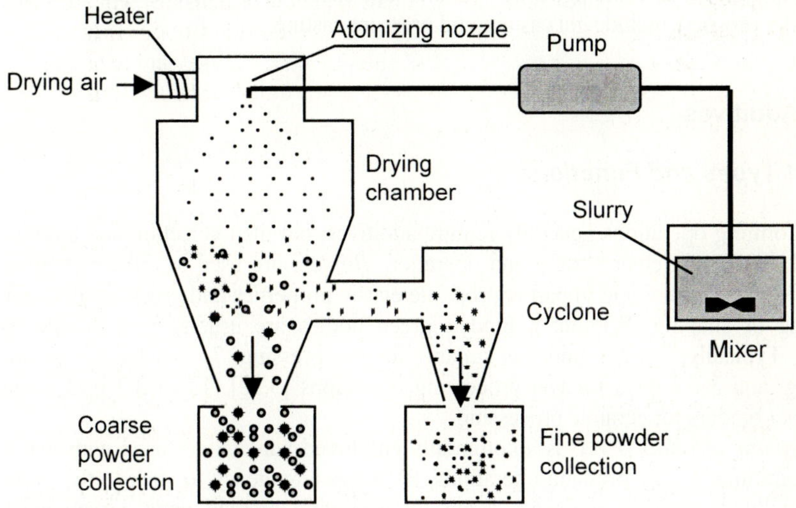

Fig. 3.12. Schematic view of a spray dryer

common. The advantage comes from the fact that the spray is projected along the vertical axis of the drying chamber. As a result, relatively large agglomerate sizes can be produced (up to around 300 μm diameter) if the drying chamber is sufficiently tall.

Two-fluid (pneumatic) atomization is another method. Here, compressed air is used for atomization. The liquid passes through an orifice with only enough differential pressure (175–350 KPa) to create a hollow, cone-shaped, liquid mist. The mass flow ratio of air to liquid determines the particle size. Disadvantages include a broad particle size distribution and low dryer efficiency due to the heat used up by pressurized air.

One of the more recent techniques uses *ultrasonic nozzles* and force-feeding of the slurry through the nozzle. The ultrasonic tip sprays the slurry into very fine droplets. Abrasive powders may harm relatively soft nozzles, but special nozzles with hard coatings can eliminate this problem.

Regardless of the atomization technique, very effective mixing of powders can be achieved since there is too little time for segregation and very short distances for different phases to travel. The particle shape depends on the solid concentration, film-forming solutes, evaporation rate, and feed particle shape and size. Spherical or donut-shaped agglomerates are commonly obtained by spray drying. Ceramic whiskers can be effectively mixed with ceramic particles. Examples of spray-dried powders and powder–whisker mixtures are shown in Fig. 3.13.

3.2 Green Forming

Green forming is the stage where ceramic powders are given the desired near net shape before densification. Important green forming operations include dry pressing, slip casting, injection molding, extrusion, and tape casting. Newly developed methods still under progress include gel casting and pressure casting.

3.2.1 Additives

3.2.1.1 Types and Functions

Green forming operations generally require additives, but their selection and amount depend on the particular system and operation. *Binders* are used to adhere ceramic particles to each other and impart appropriate elastic properties and green strength for handling or shaping. Without a binder, green bodies are usually very fragile to handle. Typically, 3 wt% binder is suitable for dry pressing, 7–20 wt% for plastic forming, and 3–17 wt% for wet processing operations [3.22]. Table 3.2 lists some important binders for ceramic processing.

Poly(vinyl alcohol) (PVA) is a commonly employed hard binder used with many oxide ceramics. In dry pressing operations, PVA has a tendency to stick to the mold and usually necessitates a release agent for removing pressed parts. Green parts pressed with PVA are strong but brittle.

Poly(vinyl butyral) (PVB) imparts medium strength to green bodies. It can be used in extrusion or dry pressing. Thin films or filaments can be prepared with the aid of this binder. PVB is soluble only in organic solvents. Parts pressed from this binder system can withstand green machining.

Cellulose ethers are water-soluble polymers, which can serve as lubricants and binders. They impart high plasticity to ceramics, and thus, they are suitable for

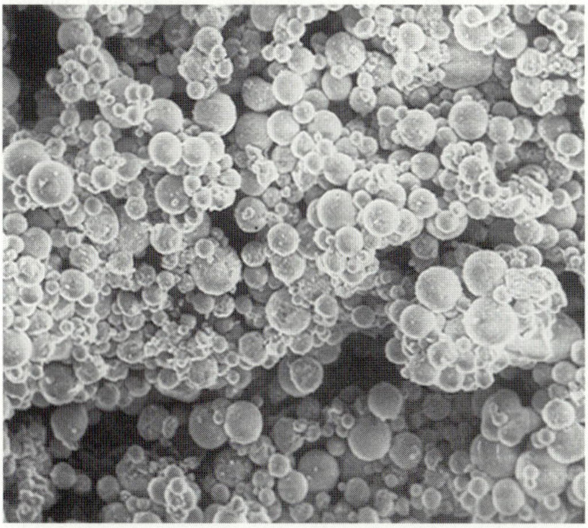

a

b

Fig. 3.13. Scanning electron micrographs of spray-dried alumina–zirconia **a** and alumina–silicon carbide whisker mixtures **b** (From M. Bengisu, Ph.D. Thesis, New Mexico Institute of Mining and Technology Socorro NM, 1992)

extrusion techniques. One important characteristic of these binders is thermal gelation. Thermal gelation is primarily caused by hydrophobic reaction between molecules containing methoxyl substitution. At low temperatures, molecules are hydrated, and there is little polymer–polymer interaction except chain entanglement. At higher temperatures, molecules dehydrate. Sufficient dehydration causes the system to form an infinite network structure, which increases the viscosity abruptly. These network structures or gels are completely reversible; they return to their original consistency when cooled. The gel structure prevents binder migration, which in turn eliminates inconsistent drying and surface defects. In oxidizing atmospheres, 99–100% burnout is achieved at around 250–300°C. Other binders specific to various green forming operations will be discussed in the relevant sections.

Resistance to particle movement and packing occurs under pressure when powders are coated with a binder. Some binders such as PVA and acrylic resins are referred to as hard, i.e., they form agglomerates and green parts that are hard and brittle. Such binders are generally not self-lubricating and require small additions of lubricants to obtain maximum green density. Other binders such as waxes and gums are soft and sensitive to temperature. They often do not require lubricants and are preferred if high purity is required. They may cause sticking to die walls during dry pressing, and the burnout ranges of some of these binders are narrow, making binder burnout a critical process.

Paraffin wax, stearates, stearic acid, oleic acid, polyglycols, talc (in the platelet form), graphite, and hexagonal boron nitride are commonly used *lubricants*. In addition, additives such as silicone provide a water-repellent coating and reduce the tendency to sticking. Lubricants facilitate movement under pressure and reduce die friction. Oleic acid and Neofat possess very good lubricating qualities. Generally around 1% lubricant suffices. These lubricants have a narrow burnout temperature range, and burnout occurs rapidly, posing the threat of crack formation. In contrast, graphite has a wide burnout range and slow burnoff qualities, making related crack formation less likely. Graphite has demonstrated good lubrication capabilities in powder or emulsion form. Butyl stearate is also a good lubricant but must be used with a wetting agent, i.e., a surfactant. It does not burn off rapidly and has a moderate burnout range; 0.1% butyl stearate is usually sufficient for green forming operations. This additive has also been used as a mold release agent.

Dispersants are another important group of additives. They are used mainly to prevent agglomeration in liquid–solid mixtures, as discussed in Sect. 3.1.2. A commonly used dispersant category is *polyelectrolytes* with trade names such as Darvan #7, Darvan C, Tamol 731, Triton X-100, and Aerosol OT-75%. Among inorganic polyelectrolytes are sodium carbonate, sodium silicate, sodium borate, and tetrasodium pyrophosphate. Examples of organic polyelectrolytes are sodium polyacrylate, ammonium polyacrylate, sodium citrate, sodium tartrate, ammonium citrate, and sodium polysulfonate [3.23].

A lubricant must be miscible in the medium used, and therefore, compatible with the binder. However, some lubricants are not miscible in certain media. For example, if water is used as a binder and an oil-based lubricant is to be added to the system, immiscibility occurs. This problem can be solved by the aid of a *surfactant* such as

Table 3.2. Important Binders for Ceramic Processing [a]

Trade name	Composition	Materials	Amount (wt%)	Process	Pressure Range (MPa)	Burnout Range (°C)
PVA	Poly(vinyl alcohol)	Al_2O_3, ZrO_2, $MgAl_2O_4$	0.1–2	Slip casting, extrusion, dry pressing, tape casting	7–70	250–400
PVB (B-76)	Poly(vinyl butyral)	Al_2O_3, ZrO_2	2–15	Tape casting, dry pressing, extrusion	70–400	250–300
Cimarec	Tertiary amide polymer	Oxides	–	Dry pressing	–	–
Pyrofine PV	Polysilazane	Advanced ceramics	–	Dry pressing	–	–
Methocel	Methylcellulose	Al_2O_3, ZrO_2	1.5–3	Extrusion, dry pressing	35	250–450
Polyox	Resin	Al_2O_3	1.5	Dry pressing	35	200–300
Carbowax 20M	Poly(ethylene glycol)	Oxides	1–3	Dry pressing	20–70	250–350
Darvan No.7	Sodium polyelectrolyte	Oxides	–	Spray drying	–	–
	Cellulose ethers	Oxides	7–20	Extrusion	–	250–300
SR350/SR355	Silicone	Oxides	–	Injection molding	–	900
Ross wax	Paraffin wax	Oxides	3	Dry pressing, injection molding	20–35	200–400

[a] References:
1. Manufacturer's data
2. L.M. Sheppard: Am. Ceram. Soc. Bull. **69 [5],** 802 (1990)
3. *Engineered Materials Handbook* Vol.4: *Ceramics and Glasses*. (ASM International Materials Park OH, 1991)
4. J. S. Reed: *Introduction to the Principles of Ceramic Processing*. (Wiley New York, 1988)

silicone that reduces the surface tension of water and oil and permits an emulsion to be formed.

Dispersants frequently cause foaming, which is undesirable in operations such as milling and spray drying. To prevent foaming, *defoamers* are required. However, defoamers may also act as lubricants, which would produce an excessive amount of lubricant. The binder quality may be destroyed under such circumstances. In such cases, the slurry formulation should consider the total amount of lubricants.

Plasticizers are added to modify the viscoelastic properties of the binder film on particles and to impose formability on the system. Water is a common plasticizer in clays. A viscosity modifier that increases the viscosity of the liquid can increase plastic strength and formability. In other instances, a substance such as bentonite may be added to provide plasticity, for example, in clay-based ceramics [3.24]. Figure 3.14 shows the effect of plasticizers in plastic forming operations.

Less common additive types are *preservatives*. They are used if enzymatic degradation of a binder is possible. Important characteristics of lubricants, dispersants, surfactants, foaming and defoaming agents, and preservatives are listed in Table 3.3.

3.2.1.2 Removal of Organic Additives

Thermal debinding or removal of organic additives, is a critical process in systems containing large amounts of these additives. Such is the case in injection molding, extrusion, and tape casting where 10–50 vol% organic compounds are employed. Careless heat treatment for binder burnout may result in defects such as warps and cracks that may be healed by only hot isostatic pressing (HIP), which is an expensive consolidation method. There is also the possibility that C residues remain that may be detrimental to material properties. Large green bodies with high amounts of organic additives may have to be heat treated for days to prevent such defects.

Three mechanisms of thermally activated debinding have been identified [3.25]:

1. capillary migration in the liquid phase,
2. permeation in the gas phase, and
3. diffusion in the gas phase.

Capillary migration often occurs in the initial period of heat treatment. Capillary forces cause the liquid binder phase to move progressively from larger to smaller pores. Capillary action can be intensified by the wicking technique where the ceramic body is placed in a powder bed of fine-grained alumina or graphite. The powder bed also acts as a support to prevent sagging and reduces temperature gradients [3.26]. Capillary action distributes the liquid binder within the body, and if the vapor pressure is larger than the external pressure, evaporation takes place.

Diffusion and *permeation* are gas transport mechanisms that occur in the later stages of heat treatment. In these stages, the liquid binder continues to evaporate, or it decomposes into volatile species. Polymers decompose into gaseous species within a narrow temperature range. This is the most critical stage since the internal pressure

Table 3.3. Important Characteristics of Lubricants, Flocculants, and Defoaming Agents [a]

Function	Trade Name	Composition	Materials	Amount (wt%)	Process	Burnout Range (°C)	Company
Lubricants	–	Oleic acid	All	0.5–1	Dry pressing	350–375	–
	Neofat	Tall oil	All	0.5–1	Dry pressing	350–375	–
	Dry lubricants	Graphite, BN, talc (plate form)	Al_2O_3, ZrO_2, SiC	1	High temp. processes	550 and up (BN and talc remain)	–
	–	Butyl stearate	All	0.1–0.5	Dry pressing	275–500	–
	–	Polyglycols	All	1	–	–	
Flocculants	Glucoplus 110	Pre gelled, cationic corn starch	For colloidal silica and alumina binder	–	–	–	Wesbond
Defoaming agents	–	Tributyl phosphate	All	–	–	250–500	–
	DB–31	Silicone-based	All	–	–	250–500	–
	DB–33	Silicone-based	All	–	–	250–500	
	Foamaster		All	–	–		Nopco

[a] References:

1. Manufacturer's data
2. L. M. Sheppard, Am. Ceram. Soc. Bull. **69 [5],** 802 (1990)
3. *Engineered Materials Handbook,* Vol.4: *Ceramics and Glasses*. (ASM International Materials Park OH, 1991)
4. J. S. Reed: *Introduction to the Principles of Ceramic Processing*. (Wiley New York, 1988)

Fig. 3.14. Effect of plasticizers in green forming operations. (Deairing pug mill, courtesy of NETZCH Incorporated Exton PA)

rises within narrow limits. Rapid heating may cause swelling and cracking. Thus, extremely slow heating (down to 1–2°C/h) is common practice at this stage. Debinding with diffusion means gaseous molecule transport through a gaseous environment. Permeation means gaseous molecule transport through a porous material. In other words, when diffusion is the rate-controlling mechanism, debinding will be a function of the frequency of molecular impact, which in turn is a function of temperature. On the other hand, if permeation is the rate-controlling mechanism, the debinding rate will be a function of the frequency of gas molecule to pore wall impact during transport. This frequency is not a function of temperature. Diffusion-controlled debinding usually takes place at low vapor pressures and small pore sizes and vice versa for permeation.

The debinding temperature range varies from one organic compound to another. Usually debinding starts at 20°C and continues up to 600°C. Higher temperatures are needed to eliminate organic matter fully and thus avoid C contamination [3.2]. For fine organic matter, full oxidation is achieved at ~700°C, but coarse particulate carbon may require higher temperatures. Low amounts of decomposed carbon (<1000 ppm) may bond at interparticle regions above 1000°C. In some cases, multiphase binders are employed to spread the decomposition over a wide temperature range and thus decrease the risk of processing-related defects [3.2,3.26].

Thermal degradation of organic compounds usually occurs by one or more of bond scission, depolymerization, and side radical elimination mechanisms [3.25,3.26]. In oxidizing atmospheres, the characteristics of decomposition change since in addition to thermal reactions, oxidation reactions will also take place.

Oxidation may lower the decomposition temperature and increase the debinding rate [3.26].

3.2.2 Dry Pressing

3.2.2.1 Uniaxial Dry Pressing

This method is one of the basic processes for producing simply shaped ceramic parts such as electronic ceramics, ceramic tile and porcelain products, engineering ceramics (cutting tools, sensors, etc.), grinding and cutting wheels, and structural clay products. Figure 3.15 shows a selection of typical dry-pressed parts.

Uniaxial dry pressing is a simple and cheap process. The process has been successfully automated. Powders, preferably granulated, are loaded into a die and pressed from one or both ends. The stages of dry pressing are sliding and rearrangement of particles, particle fragmentation, and elimination of pores (Fig. 3.16). In the majority of ceramic systems without binders, plastic flow is completely absent. Therefore, pores and interstices will persist into stage III. Suitable binders can impart some plasticity to powders, and lubricants are used to improve particle sliding. Binders also provide green strength to the pressed parts and makes handling possible. Common binders for dry pressing are lignosulfates, which are by-products of the sulfite paper process, dextrines, starches, celluloses, and PVA [3.27]. Other additives used in dry pressing are lubricants (such as paraffin and poly(ethylene glycol)) to impart plasticity and mold release agents (including stearic acid and colloidal graphite) to reduce frictional effects during pressing and for easy removal of the pressed products [3.27].

Dry pressing is performed in the pressure range of 10–100 MPa. Vibration can be used for better rearrangement and packing. Initially, the rate of densification is high, but it rapidly decreases above 10 MPa. Plastic deformation ceases at high

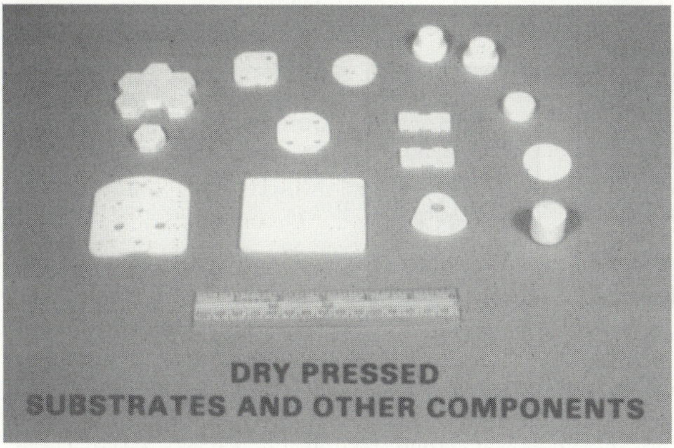

Fig. 3.15. A selection of typical dry pressed parts (courtesy of Fred L. Kennard, AC DELCO Systems Division, General Motors Corporation Flint MI)

pressures, and elastic energy is stored above the deformation limit. Upon the release of pressure, the part expands to release the elastic energy. This is called springback. Excessive springback causes defects such as lamination and cracking during ejection from the die. Laminations can also be caused where the applied stress varies across the component cross section, at changes in diameter, and due to die wall friction [3.22]. The possibility of lamination can be reduced by increasing the green strength of the system and reducing the compaction pressure. Green strength can be enhanced by improved powder characteristics, fill density, lubrication of the die, and the use of a suitable binder [3.28]. It was shown [3.29] that environmental humidity is also important in dry pressing. The bulk density of pressed bodies reaches a peak value for a certain level of humidity named optimum humidity. The optimum humidity decreases with applied pressure in semi-plastic ceramics such as clays or very fine brittle ceramics [3.30].

Among many models that attempt to correlate dry pressing parameters and the pressed density, those developed by Lukasiewicz and Reed [3.31] and Cooper and Eaton [3.32] are noteworthy [3.33]. The semiquantitative model of Lukasiewicz and Reed [3.31] yields the following equation:

$$D_g = D_f + m \ln (P_a / P_y), \tag{3.15}$$

where D_g is the green compact density, D_f is the fill density of powders, m is an empirical constant, P_a is the applied pressure, and P_y is the yield pressure of powders/granules.

The probabilistic model proposed by Cooper and Eaton [3.32] has the following form,

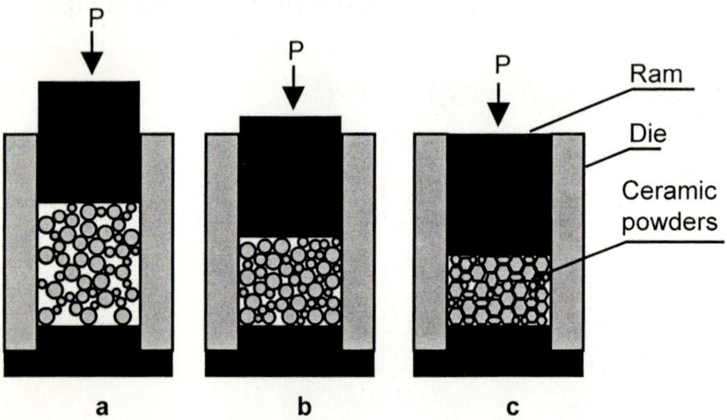

Fig. 3.16. Stages involved in dry pressing: sliding and rearrangement **a**, particle fragmentation **b**, and elimination of pores **c** (From J.S. Reed: *Introduction to the Principles of Ceramic Processing*. Copyright 1988 John Wiley &Sons. Adapted with permission of John Wiley & Sons, New York)

$$V^* = (V_o - V) / (V_o - V_\infty) = A_1 \exp(-k_1 / P_a) + A_2 \exp(-k_2 / P_a), \quad (3.16)$$

where V^* is the fractional volume compaction, V is the volume of the compact at P_a, V_o is the initial volume, V_∞ is the volume at infinite pressure, i.e., the volume at full density, and A_i and k_i are experimental constants. The final density can be calculated from the compact mass and the calculated volume V at pressure P_a.

3.2.2.2 Isostatic Pressing

Problems such as density variations, lamination, and restricted mold shapes associated with uniaxial dry pressing can be eliminated by the isostatic dry pressing process. In the *wet bag technique*, a deformable plastic bag or mold is filled with the ceramic powder and water or an oil-based fluid surrounding the mold is pressurized at 30 to 700 MPa (Fig. 3.17). The mold material must have good resilience or "memory" and resist deformation during pressing. Important process parameters include peak pressure, pressure increase–release rates, dwell time, setup, and alignment [3.34].

Dry bag isopressing is similar to the wet bag technique except that pressure is applied radially by a pressure transmitting liquid between a flexible mold and a rigid shell (Fig. 3.18). Wet bag isopressing has the advantage of low tooling costs compared to dry bag isopressing since the bags are inexpensive. Dry bag isopressing offers higher production rates and automation [3.35].

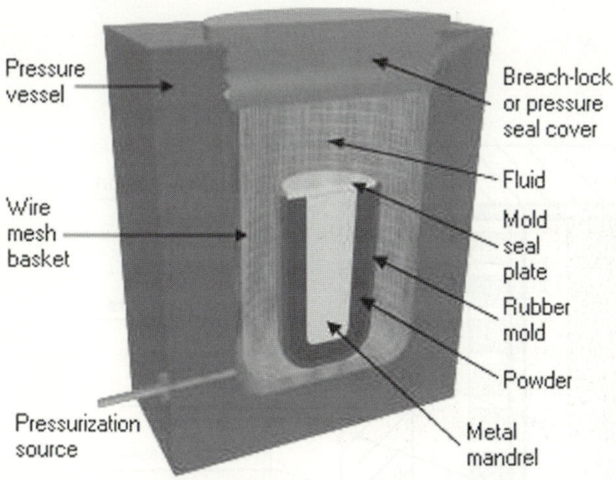

Fig. 3.17. Schematic representation of the wet bag isostatic pressing process (illustration reprinted with permission from www.designinsite.dk)

Large parts can be pressed by isopressing due to the hydrostatic nature of the pressure. It is mostly a batch process, which makes it rather slow and expensive. In spite of its disadvantages, isopressing is commonly used to form hollow shapes, large tubes, multiple heterogeneous articles, and composites with high whisker or fiber content. Large ceramic heating elements, electrical insulators for overhead power lines, spark plugs, large wire-drawing cones, reaction chamber liners, crucibles, grinding balls, large capacitors, radomes, and long closed-end tubes are important application areas of cold isostatic pressing [3.36]. Examples of wet and dry bag isopressed components are shown in Figs. 3.19 and 3.20, respectively [3.37].

3.2.3 Slip Casting

This process, one of the oldest green forming techniques, is used mostly for conventional ceramics such as sanitary ware, tableware, and porcelain. A slurry made from fine powders and a suitable liquid is poured into a mold with micropores. Partial drying occurs by capillary withdrawal of the liquid through the mold walls, leaving a fragile shell or body. The drying rate decreases with time due to saturation of pores with liquid and decreased permeability of the shell upon thickening. Typical slip casting methods are shown in Fig. 3.21. The slip may be connected to a feeding line or sprue, a method similar to metal casting, to produce bulk articles. Shell or hollow structures are produced by partially drying the slip and draining the remaining liquid.

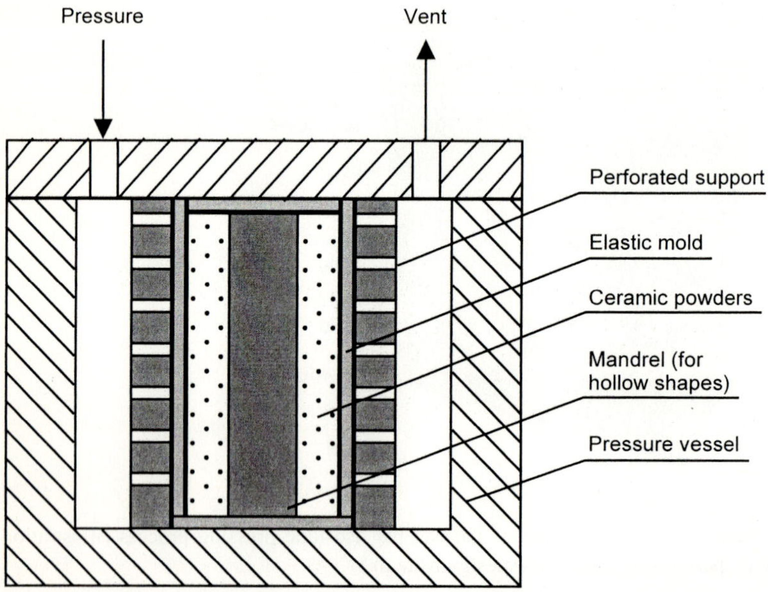

Fig. 3.18. Schematic representation of the dry bag isostatic pressing process

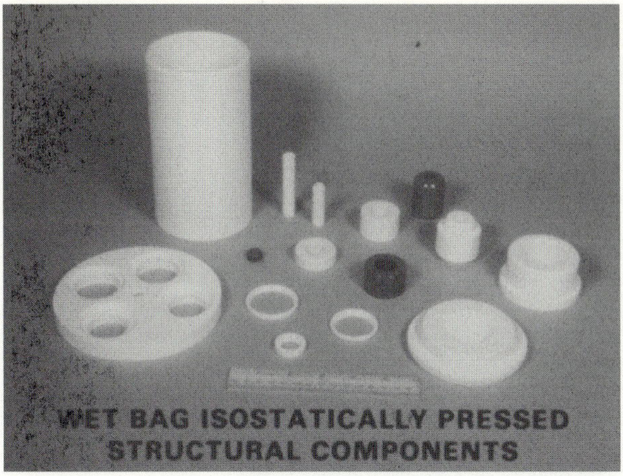

Fig. 3.19. A selection of wet bag isostatic pressed parts (courtesy of Fred L. Kennard, AC DELCO Systems Division General Motors Corporation Flint MI)

Gypsum ($CaSO_4 \cdot 2H_2O$) is the most common mold material formed by the reaction of plaster of Paris and water. The advantages of this material are its affordability, relatively smooth surface, and small pore size. Water is a common liquid used for slip casting. Whiteware and refractory slips are water-based. Typical compositions of these systems are given in Table 3.4. Some ceramic powders, however, react with water. MgO and CaO quickly hydrate in water. Nitride ceramics such as Si_3N_4 and AlN react with water and liberate ammonia [3.22]:

$$Si_3N_4 + 6H_2O \rightleftarrows 3SiO_2 + 4NH_3 \tag{3.17}$$

$$2AlN + 3H_2O \rightleftarrows Al_2O_3 + 2NH_3 \tag{3.18}$$

In slip casting such powders, nonaqueous liquids such as alcohols, trichloroethylene, and methyl ethyl ketone (MEK) are used [3.38]. Methanol and ethanol are suitable for slip casting MgO-based systems [3.39]. High drying rates associated with such solvents cause cracking. This can be eliminated by controlled drying for 36 hours at high alcohol partial pressures.

The cake thickness is conventionally given by [3.40]

$$L^2 = (2P_m\, g\, E^3\, t) / [5\, S^2\, \eta\, (y-1)\, (1-E)^2], \tag{3.19}$$

where P_m is the pressure drop across the mold, E is the porosity of the cake, t is time, S is the surface area of suspended particles, η is the viscosity of the slip, y is the

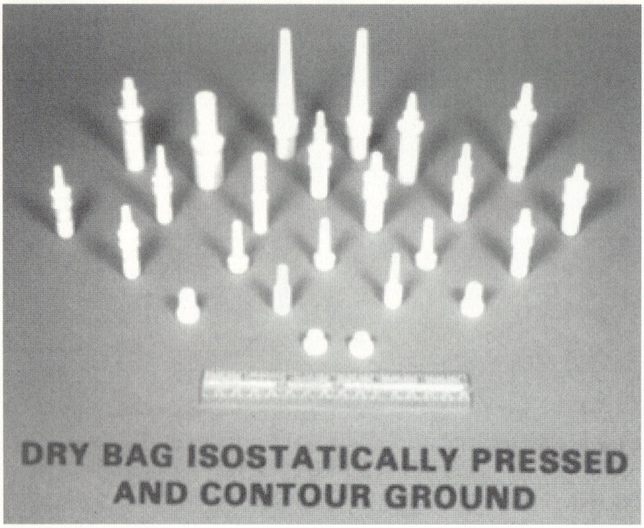

Fig. 3.20. A selection of dry bag isostatic pressed parts (courtesy of Fred L. Kennard, AC DELCO Systems Division, General Motors Corporation Flint MI)

volume of the slip containing a volume equal to 1–E of solids, and g is gravitational acceleration. P_m is given by the *Laplace equation* [3.41],

$$P_m = S_m \sigma \cos\gamma, \tag{3.20}$$

where S_m is the specific surface area of the mold, σ is the surface tension of the liquid, and γ is the contact angle ($\cos\gamma = 1$ for plaster of Paris in contact with water). P_m varies between 0.03 and 0.1 MPa in water–plaster systems [3.41].

Recently, Hampton et al. [3.42] modeled the rate of shell thickening during slip casting. Their model takes into account two distinct regions. The lower region contains a high concentration of smaller particles, whereas the ratio of small to large particles in the upper region is the same as for the slip (Fig. 3.22). This model shows that there is a parabolic relationship for the increase in the thickness of region 1 and region 2 with time, but each has a different rate.

Slip casting is to some extent still regarded as an art because a large number of interacting parameters affect the process. Achieving a desirable and reproducible slip necessitates close control of many parameters such as solid and additive fractions, particle size, pH, viscosity, mold properties, time, temperature, and drying rate. Combining these parameters is critical for obtaining the optimum green density and strength in the cast body. Rheological properties, discussed in Sect. 3.1.2, are especially important in slip casting.

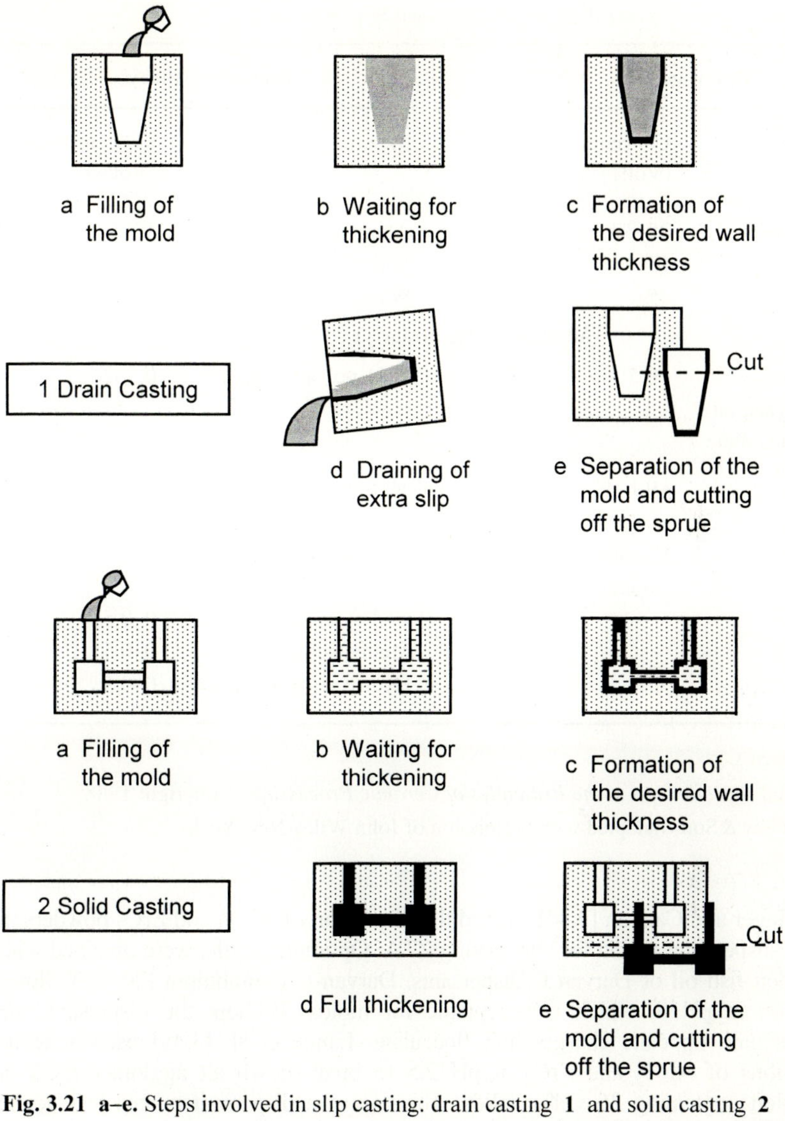

Fig. 3.21 a–e. Steps involved in slip casting: drain casting **1** and solid casting **2**

Torre and Bigay [3.43] reported that slip casting is a suitable method for the manufacturing of complex parts from ultrafine Si_3N_4 powders. Their study shows that good deflocculation can be achieved at very high pH values. However, damage of plaster molds and corrosion of mixing and handling equipment may occur at such high pH conditions. These problems were solved by using a suitable deflocculant at pH levels compatible with plaster molds.

Table 3.4. Typical Compositions of Water-Based Slip Systems [a]

Whiteware Slip			Refractory Slurry	
Material	Concentration (Vol%)		Material	Concentration (Vol%)
Nonplastics	25–30		Alumina (<45 μm)	40–50
Clay	15–25		Ball Clay	0–10
Water	45–60		Water	50–60
		Additives (wt% in slurry)		
Na_2SiO_3, Na–polyacrylate, Na–lignosulfate	<0.5	Deflocculant	NH_4–polyacrylate	0.5–2
$CaCO_3$	<0.1	Coagulant	$MgSO_4$	0–0.1
$BaCO_3$	<0.1			
Clay <1 μm	Variable amount	Binder	NH_4–alginate	0–0.5
			Carboxymethyl cellulose	0–0.5
			Methyl cellulose	0–0.5
			Hydroxyethyl cellulose	0–0.5

[a] Reference

J.S. Reed: *Introduction to the Principles of Ceramic Processing.* Copyright 1988 John Wiley &Sons (adapted with permission of John Wiley New York)

Wilfinger and Cannon [3.44] studied the dispersion of Al_2O_3 and ZrO_2 powders in various suspending media for slip casting ZTA. Optimum results were obtained with Menhaden fish oil or Darvan C dispersants. Darvan C (ammonium PMAA) allows dispersion at pH 8, which is acceptable for molds. Without the dispersant, the powders gain opposite charges and flocculate. Lange et al. [3.45] used aqueous suspensions of Al_2O_3 and ZrO_2 at pH 2.5 to break down all agglomerates in a suspension containing 25 vol% solids. The use of a dispersant makes less acidic conditions possible and provides good slip casting characteristics. The names and suppliers of various dispersants used in water-, ethanol-, and MEK-based suspensions are listed in Table 3.5. Typical additives used in slip casting include organic polymers as rheology modifiers, plasticizers, binders, acids and bases for pH control, sodium salts, phosphate esters, lignite, and gum acacia as dispersants (deflocculants), and octyl alcohol or tributyl phosphate as antifoaming agents [3.40]. It was shown that the use of ultrasonic dispersion before or during slip casting is beneficial for high green densities and microstructural uniformity [3.46]. Mangels [3.47] noted that broad

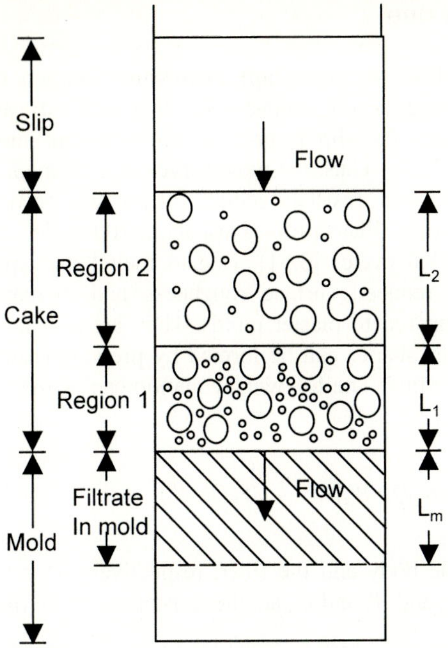

Fig. 3.22. Schematic representation of the cake model in slip casting (From J.H.D. Hampton, S.B. Savage, and R.A.L. Drew: *Experimental Analysis and Modeling of Slip Casting*. J. Am. Ceram. Soc. **71 [12],** 1040–1045 (1988). Copyright 1990 American Ceramic Society Columbus OH, reprinted with permission)

particle size distributions are advantageous in slip casting and yield higher green densities.

Slip casting is a batch process and relatively slow due to the many steps involved. The advantage of the process is the rather complicated shapes formed by this method, such as the gas turbine rotor shown in Fig. 3.23. It is economically feasible for limited production requirements. Industrial applications include traditional ceramic production such as kitchenware, porcelain products such as bathroom fixtures, refractories, highly porous thermal insulators, and thin-walled products, including alumina combustion and thermocouple tubes. Slip casting has also been used to produce aluminum titanate exhaust port liners for the Porsche engine [3.48]. Recent technical developments have resulted in nontraditional casting processes, which offer the intricate shape capabilities of slip casting while approaching faster production rates, higher fill rates, and uniformity. These new processes are explained in the following sections.

3.2.4 Pressure Casting (Wet Pressing)

This process is similar to pressure filtration (filter pressing). In pressure filtration, the liquid is removed from the slip by pressing the slip against a flat filter with a die and punches [3.49,3.50]. In pressure casting, the slip is pressed into a porous mold consisting of one or more pieces (Fig. 3.24). Thus, the mold serves as a filter [3.2]. Plaster molds are usually not strong and wear resistant enough for pressure casting. Therefore, special polymeric molds are used. Plaster molds last about 100 cycles, but polymeric molds can last at least 10,000 cycles [3.51]. Due to their better wear resistance, polymeric molds provide reduced surface roughness and increased consistency of edges and corners, compared to plaster molds. They are also more resistant to chemical reactions. Complex shapes can be formed by pressure casting with higher production rates than slip casting. The following equation can be used to calculate the body thickness on a filter [3.52]:

$$D_s = \{ (Pt/W_s) + [(W_f D_f)/W_s]^2 \}^{1/2} - (W_f D_f / W_s), \tag{3.21}$$

where D_s and D_f are thicknesses of the body and the filter, respectively, P is the filtration pressure, t is the filtration time, and W_f and W_s are the resistances of the filter and the body to fluid flow, respectively.

Slip resistance is related to the rheology of the slip. For special mold materials with high porosity, the two factors related to W_f are negligible, which reduces Eq. (3.21) to

$$D_s \approx (P\,t\,/\,W_s)^{1/2}. \tag{3.22}$$

The practical implication of this equation is that casting time can be rapidly improved for a given thickness, as the slip pressure is increased. For example, to form a body thickness of 6 mm from a porcelain slip, the filtration time of 120 seconds at 2 MPa can be reduced to about 60 seconds at 5 MPa. When the slip thixotropy is reduced, casting time can also be decreased. Higher pressures (3–4 MPa) are suitable for smaller parts, whereas intermediate pressures (0.3–0.4 MPa) are necessary for increased production rates.

Highly stabilized slips with fine particles pose an important risk of filling up the very large pores (typically 30 to 40 times larger than the particle size) of polymeric molds during pressure casting [3.51]. This risk is minimized by optimizing the slip properties in solids loading, particle size distribution, and degree of stabilization. Lyckfeldt et al. [3.51] determined that a slip should have pseudoplastic rheology to be suitable for pressure casting and that the initial pressure increase should be slow. Once a layer of cake is developed, the risk of pore plugging is minimized, and the casting behavior is governed by the pore properties of this cake.

Unlike slip casting, in which filtration occurs by capillary action (0.1–0.2 MPa), pressure casting is achieved by applying an external pressure (up to 4 MPa). After compaction is achieved and the pressure is released, large elastic strains are

Table 3.5. Various Dispersants (Deflocculants) Used in Ceramic Processing [a]

Suspension System	Trade Name of Dispersant	Supplier	Generic Name	Notes
Oxide ceramic–ethanol	1. Hypol WD-4000	Grace Chem.	Water-borne urethane	
	2. Emphos PS-21A	Witco Chem.	Phosphate ester	
	3. Mono Fax 785	Mona Ind.	Phosphate ester	
	4. HCl (reagent grade)			Hazardous and corrosive
	5. Zonyl A	Dupont	Ethoxylate	
	6. Zonyl FSK	Dupont	Fluorinated surfactant	
	7. Mono Fax 786	Mona Ind.	Phosphate ester	
	8. Z3 Menhaden fish oil	Spencer Kellog	Fatty acid	Insoluble
Oxide ceramic–methyl ethyl ketone	1. Z3 Menhaden fish oil	Spencer Kellog	Fatty acid	
	2. Emphos PS–21A	Witco Chem.	Phosphate ester	
	3. Zonyl A	Dupont	Ethoxylate	
	4. Monazoline T	Mona Ind.	Substituted imidazoline (1–hydroethyl *z*–alkyl imidazolines) from tall oil	
	5. Aerosol C–61	Amer. Cyan.	Ethanolated alkylguanidine amine complex	
	6. Alkazine 0	Alkaril Chem.	Oleic hydroxyethyl imidazoline	

(Continued)

Table 3.5. *(continued)*

Suspension System	Trade Name of Dispersant	Supplier	Generic Name	Notes
Oxide ceramic $-H_2O$	1. Darvan C and Darvan #7	R.T.Van-derbilt	Ammonium polyelectrolyte	All aqueous slips (0.1–0.5 wt% added)
	2. Daxad 21	W.R. Grace	Calcium salt of polymerized alkyl substituted benzoid alkyl sulfonic acid	
	3. Daxad 27	W.R. Grace	Sodium salt of polymerized alkyl and substituted benzoid alkyl–sulfonic acids	
	4. Daxad 32	W.R. Grace	Ammonium salt of a carboxylated polyelectrolyte	
	5. Emphos CS–1361	Witco Chem.	Organic phosphate ester	
	6. Tamol–731	Rohm &Haas	N/A	(0.1–1 wt% added.) A powerful dispersant but forms a tough foam. A large amount of antifoam required to break it
	7. Triton X–100	Rohm &Haas	N/A	Works as dispersant or surfactant (1–3 wt% added)
	8. Spinks 211	H.C. Spinks Clay Co.	Sodium polyacrylate	For tile and sanitary ware slurries
	9. Carbam 50	AAA Paint Additives	Salt of polyamine amide and polycarboxylic acids	Under certain conditions, it contributes to positive effect on corrosion resistance of water-borne systems

[a] References

1. K. Wilfinger and W.R. Cannon: Ceram. Eng. Sci. Proc. **7 [9–10]**, 1169–1181 (1990)
2. L.M. Sheppard: Am. Ceram. Soc. Bull. **69 [5]**, 802 (1990)
3. T. Morse: *Handbook of Organic Additives*, study notes (unpublished work)
4. Manufacturer's data

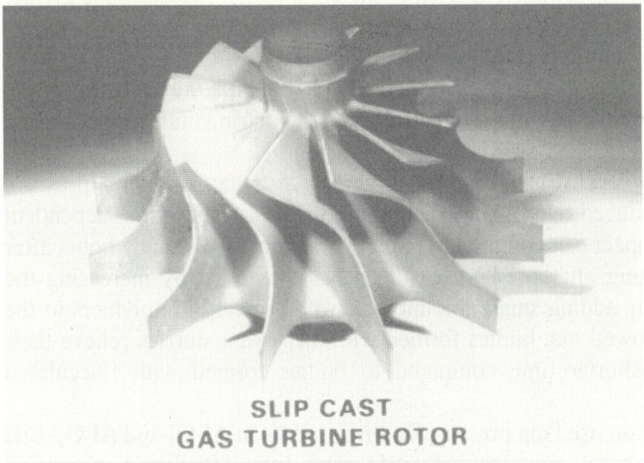

Fig. 3.23. Example of complex shape capability of slip casting: slip cast gas turbine rotor (courtesy of Fred L. Kennard, AC DELCO Systems Division, General Motors Corporation Flint MI)

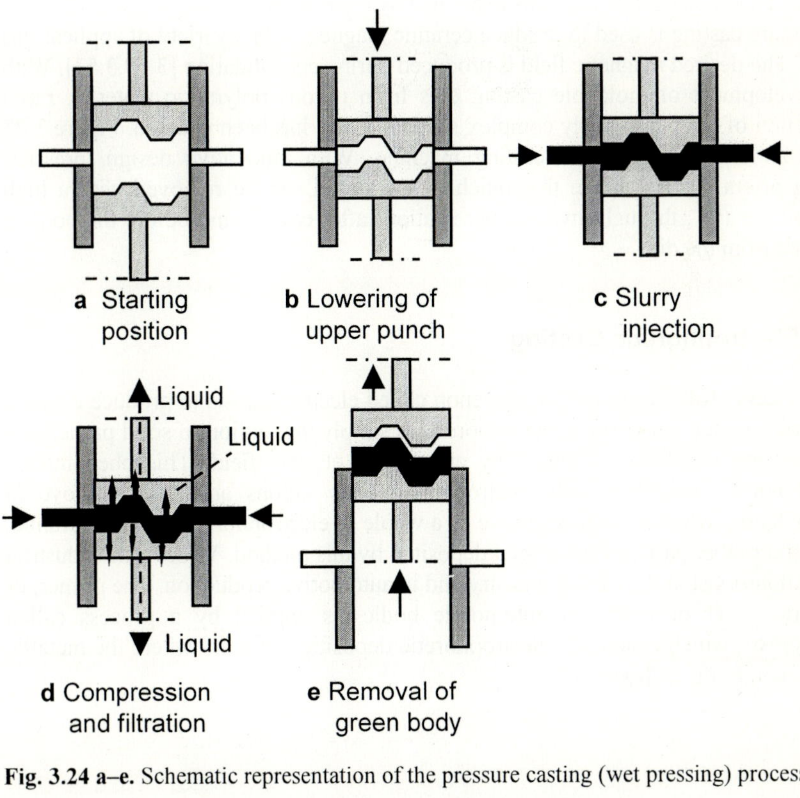

Fig. 3.24 a–e. Schematic representation of the pressure casting (wet pressing) process

recovered, and the compact grows. The greater the compaction pressure, the greater the recoverable strain. Inclusions or inhomogeneities within the powder compact which are stiffer (e.g., agglomerates, whiskers, fibers) or more compliant (e.g., organic inclusions) will store a different stress relative to the bulk during compaction. The differential strain relieved by inclusions may produce detrimental stresses during strain recovery.

For dry-pressed powders, strain recovery is almost instantaneous after pressure release. In ompacts produced by pressure casting, on the other hand, time-dependent recovery occurs. A compact can continue to release strain and grow many hours after pressure release. Cracking after pressure casting can be avoided by increasing the green toughness, e.g., by adding small amounts (<2 wt%) of certain polymers to the slurry. Lange [3.53] showed that bodies formed with dispersant slurries relieve their strain within a much shorter time compared to bodies formed with flocculated slurries.

Grazzini and Wilkinson used the pressure casting process in Al_2O_3 and Al_2O_3–SiC systems [3.54]. In some cases, cracking occurred during drying. Preferred orientation of SiC particles was observed in the composites. Well-dispersed slips did not respond to pressure in the expected manner, but coagulated slips resulted in higher relative green densities at higher pressures. These ranged from 61.5% theoretical density (TD) at 0.1 MPa to 66.5% TD at 0.17 MPa. The difference in the casting behavior of dispersed and coagulated slips was explained by their different viscosities.

Pressure casting is used to produce ceramic magnets with a variety of applications [3.50]. The desired magnetic field is produced during consolidation [3.50,3.53]. With the development of moldable casting dies from porous polymeric material, rapid production of large, relatively complex shaped bodies has been enabled. Figure 3.25 shows an industrial pressure casting machine. With innovative design, pressure casting offers the advantage that much of the liquid can be removed with a high pressure gas (i.e., through invasion percolation) after casting and before the body is removed from the die.

3.2.5 Electrophoretic Casting

This process profits from the phenomenon called electrophoresis to produce ceramic parts with limited thickness. Electrophoresis is simply the motion of solid particles in a suspension electrically charged by an applied electric field. This phenomenon should not be confused with electroplating. Only anions and cations move in electroplating, whereas particles move as a whole in electrophoresis. Metal, ceramic, glass, and rubber particles have been deposited by this method. Wide-scale industrial applications exist in the rubber industry and in automotive production. The primer, or first gray layer of paint, on automobile bodies is applied by a process called cataphoresis, which is actually electrophoretic deposition of paint where the metallic body becomes the cathode.

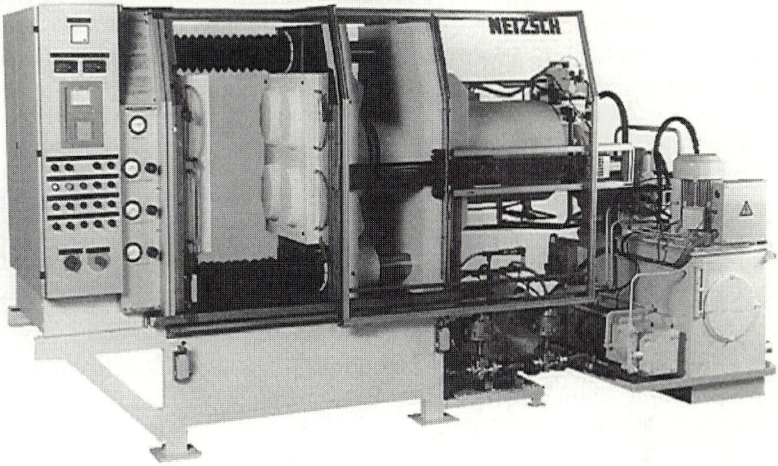

Fig. 3.25. Industrial pressure casting machine (courtesy of NETZCH Incorporated Exton PA)

Large-scale applications of ceramic deposition have been limited until now. Two important applications are continuous clay strip forming and production of β''-alumina tubes for the sodium/sulfur battery [3.55]. Other smaller scale and potential applications include green forming of porcelain cups, sanitary ware, and alumina radomes; electronic thin or thick film coatings; vitreous enameling; and corrosion-resistant, wear-resistant, and heat-resistant coatings for various purposes.

Ceramic particles can be charged both positively or negatively. The most common method for charging solid particles is by adsorption, usually of a proton or another cation. Positive charging of β-Al_2O_3 occurs when an acid is added to the solution [3.56],

$$\beta\text{-}Al_2O_3 + HA \rightarrow \beta\text{-}Al_2O_3 / H^+ \text{ (adsorbed on surface)} + A^- , \tag{3.23}$$

where HA represents an acid. Wilfinger and Cannon [3.44] showed that ammonium PMMA reverses the charge on alumina. The thickness of an electrophoretic deposit is given by *Hamaker's equation* [3.55,3.57]:

$$Y = \int_0^t aACF\mu \, \mathrm{d}t \ , \tag{3.24}$$

where Y is the mass of the deposit produced in time t (s), A is the mandrel area (cm^2), C is the particle concentration (g/cc), F is the electric field (V/cm), μ is the electrophoretic mobility (cm^2/s-V), and a is a coefficient based on the fraction of particles that become deposited. The electrophoretic mobility is given by [3.58]

$$\mu = \frac{K\varepsilon\zeta}{\pi\eta}, \tag{3.25}$$

where K is a constant that ranges from 1/6 to 1/4, ε is the dielectric constant, ζ is the *zeta potential*, and η is the viscosity. The velocity of particles is given by

$$v = \mu F. \tag{3.26}$$

The viscosity and field strength can be varied within a limited range. The zeta potential is important for both mass transport and the deposition process. Similarly, a high dielectric constant leads to high particle velocity, thus, high mass transport. Therefore, the high dielectric constant of water is advantageous.

There are basically three steps in electrophoretic casting. First, a suitable suspension is produced from the powders to be deposited and an organic liquid. Powers [3.56] showed that electrophoretic deposits are obtained only from liquids with dielectric constants in the range of about 12 to 25. The conductivity also increases rather sharply with the dielectric constant of the liquid. Acetone, 1-hexanol, 1-propanol, 2-propanol, pentanone, and ethanol are suitable liquids for depositing β -alumina, whereas benzene, methyl acetate, methanol, and ethylene glycol do not yield any deposition. Water can be used as the liquid medium but has some disadvantages. The main disadvantages are gas evolution at electrodes at high field strengths (necessary for reasonably short deposit times) and Joule heating of the suspension due to high current densities. Joule heating leads to electrochemical attack on metallic electrodes that results in contaminating the deposit. Thus special procedures are required to prevent bubble formation and contamination of the deposit.

The second step is deposition by the application of a dc voltage between the mandrel and the counterelectrode (Fig. 3.26). Electric fields of 280–1400 V/cm usually make deposition times of about 1 min possible with no need to stir the suspension. The last step is removing green parts from the mandrel. Mandrels usually need to be tapered for easy removal.

A number of shapes can be formed using simple tooling with the limitation that deposit thicknesses cannot be greater than ~3 mm when using a mandrel. In the continuous clay strip forming process, strips of thickness ranging from 2 to 20 mm can be produced [3.55]. Another disadvantage of the process is the need for an organic liquid or other special procedures to reduce degassing at mandrels during the application of high voltage. Amyl alcohol, the preferred liquid for alumina casting,

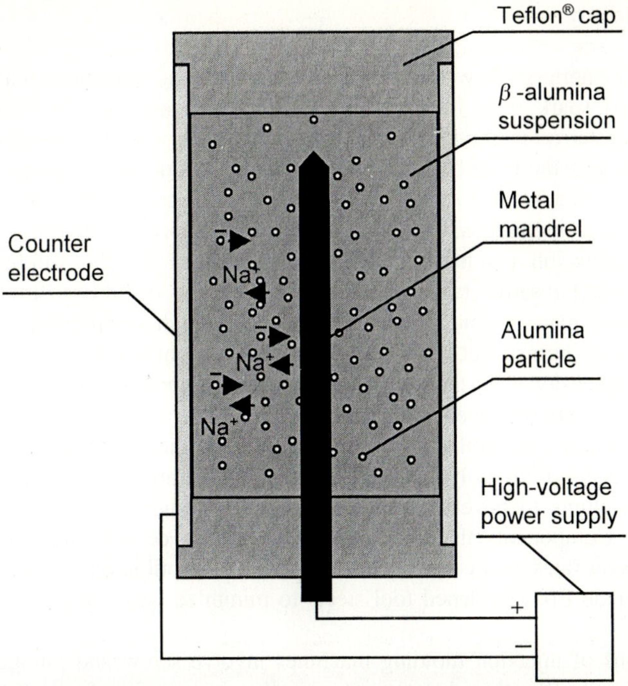

Fig. 3.26. Schematic of electrophoretic deposition of alumina

has an unpleasant smell and is flammable, and thus requires special safety precautions. Its advantages are short forming times (1–5 min), excellent dimensional control of electrophoretically cast articles, and a fine distribution of micropores that makes sintering easier.

Various solutions to problems encountered in electrophoretic casting were presented in a recent article [3.58]:

1. Use a high volume slurry reservoir to prevent reduction of concentration.
2. Mix the slurry to eliminate the possibility of sedimentation.
3. Use poled special electrodes to fight corrosion.
4. Use an auxiliary electrode system to linearize the electric field.

It was also stated that a combination of pressure casting and electrophoresis makes cast layers thicker than 30 mm possible. This combination also allows using water-based suspensions and accelerates the process.

3.2.6 Injection Molding

Injection molding of ceramic parts was initiated in the 1930s and became a mass production method for spark plug insulators in 1937 [3.59], although isostatic pressing became the preferred method later. Recently, interest in injection molding has renewed due to the possibility of high-volume, net shape production of complex ceramic parts such as turbine rotor blades [3.37] (Fig. 3.27), integral stators, thread guides, bushings, wire-drawing guides, electronic parts, ceramic cores for investment casting of alloys, combustion prechamber inserts, and nozzles for the welding industry [3.60]. The major disadvantages of this process are the inability to mold large components without macroscopic internal defects and the high-volume fraction of binders that need to be burned out before sintering. Other problems include low reproducibility and dimensional instability. Nevertheless, components with thin (<10 mm) sections are already produced industrially [3.61].

Ceramic injection molding is similar to plastic injection molding. Basically, the same equipment that is used for polymers can be used for ceramic components. However, there are attempts to develop machines specially designed for ceramic materials [3.62,3.63]. Components that require replacement in ceramic injection molders, due to high wear rates are pistons, liners, and dies. Susceptible regions such as these are usually made from hardened tool steels to minimize wear by abrasive ceramic particles.

Two basic variations of injection molding machines involve screw and plunger type molders. Plunger units offer the advantage of a smaller chance of contamination due to wear [3.64]. In a plunger unit, significant pressure losses can occur in the cylinder, affecting the nozzle pressure, whereas in a screw unit, the ram pressure and nozzle pressure are about the same [3.65]. A typical plunger unit is illustrated in Fig. 3.28.

Three steps are involved in injection molding of ceramic components:

1. *Blending and compounding:* Ceramic powders are mixed with suitable binders and additives, and the rheology is adjusted to give an injectable blend.
2. *Molding:* The molding mix is pressed by a ram or plunger with hydraulic, pneumatic, or screw mechanisms into a cold die cavity with the desired net shape.
3. *Ejection and binder removal:* After the part is removed from the mold, the binder is slowly removed by heat treatment.

Major variables in injection molding are injection speed and pressure, barrel and mold temperature, and hold time. Other variables are solids load, binder properties, and particle size, all of which affect slurry rheology.

Binder systems used in injection molding are mostly composed of a major binder, a minor binder, and a surfactant. The major binder determines the range of binder properties. The binder system should provide

1. a slurry that does not separate but flows freely during molding,
2. ease of separation from the mold without sticking, and
3. burnout characteristics that do not harm the molded body.

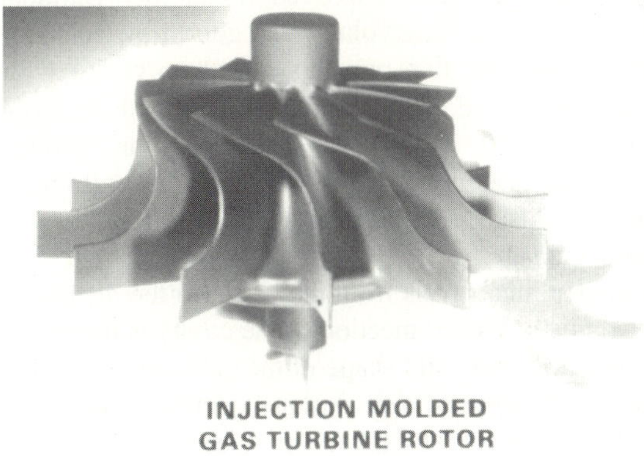

Fig. 3.27. Injection molded gas turbine rotor. (Courtesy of Fred L. Kennard, AC DELCO Systems Division, General Motors Corporation, Flint MI.)

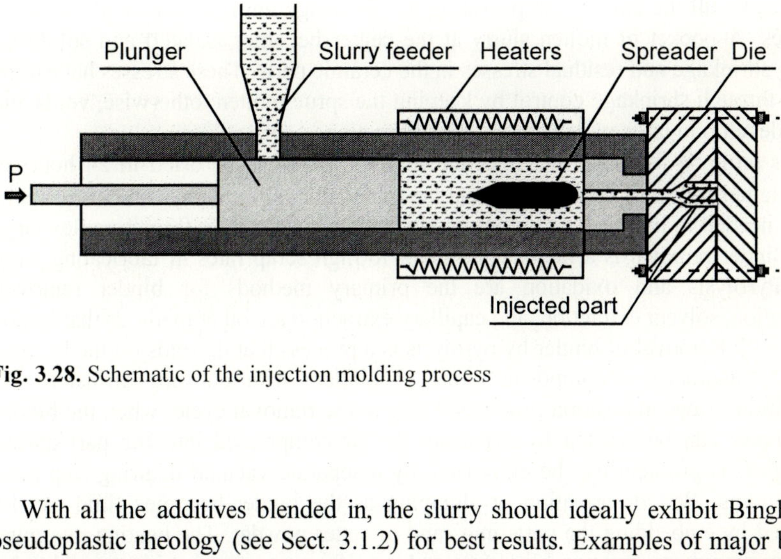

Fig. 3.28. Schematic of the injection molding process

With all the additives blended in, the slurry should ideally exhibit Bingham or pseudoplastic rheology (see Sect. 3.1.2) for best results. Examples of major binders are thermoplastic polymers (polyethylene, polystyrene, polypropylene), thermosetting polymers (silicone, phenolic resin), and paraffin wax. Minor binders are added to increase the fluidity of the system and facilitate binder removal. Generally an oil or low molecular weight thermoplastic that can be removed early during binder burnout is used. Surfactants improve wetting between the ceramic powder and the binder during blending. In many cases stearic acid is used as a surfactant.

A broad particle size distribution is preferred for improved moldability. Bimodal particle size distribution provides improved packing volume and good flow behavior of highly filled slurries. In recent years, very fine powders started to be used for injection molding. SiC, Si_3N_4, and Al_2O_3 powders with particle size less than 1 μm and specific surface areas between 5–20 m^2/g have been injection molded successfully, although such powders tend to agglomerate. Some success has also been achieved with injection molding of whisker-reinforced ceramics in spite of the difficulties encountered with whisker damage.

During injection molding, the molten mix travels from the barrel to the die cavity through passages called sprues or runners. The connection to the cavity is through a small orifice called the gate. The location, size, and shape of the gate are critical for good results. Quackenbush et al. [3.64] reported that during injection molding of Si_3N_4 with the gates located at one end of the mold, the material initially shot to the far end of the cavity. Then it folded back and forth against itself during filling. This effect called *jetting,* is detrimental to the strength of injection molded SiC [3.66]. When the gates were enlarged and altered to inject across the shortest dimension of the cavity, jetting was eliminated.

Another problem encountered during injection molding is void and crack formation during solidification. A solid layer forms at the cavity surface immediately after injection and its thickness increases during shrinkage. The static hold pressure continues to fill the cavity, compensating for shrinkage until the center of the sprue solidifies. A pocket of molten slurry at the center becomes isolated and solidifies, causing shrinkage and residual stresses in the ceramic parts. These stresses have to be limited through shrinkage control by keeping the sprue molten; otherwise, voids and cracks develop in large parts with thick cross sections.

After removal from the die, injection molded parts can be dried in 24 hours at moderate temperatures up to 100°C. On the other hand, several days and even weeks may be needed to remove certain polymeric binders successfully from complex/large parts. Binder removal is a major contributor to high scrap rates in fabricating large parts. Pyrolysis and oxidation are the primary methods for binder removal. Evaporation, solvent extraction, and capillary extraction are other methods that can be used [3.49]. Removal of binder by pyrolysis is a process that depends on the balance of two phenomena: decomposition rate of binders with outward diffusion and evaporation of decomposition products. Early in the removal cycle, when the binder melts, flaws can be created by expansion of air compressed into the part during molding. This problem can be eliminated by a separate vacuum deairing step after compounding. Part deformation or slumping as the binder becomes fluid can be prevented by embedding the parts in a sand or setter powder. The heating rate must be slow enough to prevent blistering and cracking associated with rapid binder evolution. Typical heating/cooling rates are 2–10°C/h. Internal cracking or delamination is another characteristic flaw that can be generated during binder removal. This cannot be prevented by slow binder removal. The cause of such a flaw is rearrangement of ceramic particles in the interior during binder burnout. These flaws may be controlled by increasing the solids loading level and changing the surfactant that has been used. Computer simulation of binder removal is attempted to establish critical properties of the organic vehicle and critical process control parameters.

Aqueous injection and extrusion molding processes have been developed by Rivers [3.67] and Schuetz [3.68], based on the thermogelation properties of methylcellulose polymers, to avoid some problems involved with conventional polymeric binders. Methylcellulose polymers are soluble in cold water, and they gel in hot water due to hydrophobic, self-aggregation processes. Fanelli, Silvers, and co-workers [3.69,3.70], on the other hand, used agarose (a derivative of a red algae class of seaweed produced by a series of extraction and bleaching operations) as the major binder in aqueous injection molding. Agarose is soluble in hot water and gels on cooling, contrary to methylcellulose polymers. The rigid framework structure of agarose keeps pore channels open, allowing water to evaporate from the part easily. Agar also provides exceptionally high strength to molded parts in the green state, which makes machining easier. Molded parts can be fired without special debinding precautions. The debinding time is reduced from several days to several hours.

Another alternative to conventional binders is metallo-organic polymeric precursors such as carbosilane for SiC bodies. During thermal treatment, such binders transform to a reactive ceramic that surrounds the original ceramic particles, producing a dense green body. Also in this case, the debinding time is reduced from several days to several hours [3.60].

3.2.7 Extrusion

Extrusion is the process of forcing a plastic mass through an orifice to produce green parts with the desired cross section. The main difference from injection molding is that there is no die cavity in extrusion, which allows forming parts with unlimited length. Traditional ceramics such as brick, tile, and electrical porcelain insulators have been produced extensively by extrusion for many years, whereas extrusion of engineering ceramics is less developed. Refractory furnace tubes, honeycomb catalyst supports used in heat exchangers and catalytic converters for emission control applications [3.37] (Fig. 3.29), transparent Al_2O_3 tubes for lamps, very small electronic and magnetic ceramic parts, as well as graphite electrodes up to 1 ton, are current applications of the extrusion method. The two types of extruder units are the *plunger (piston)* type and the more common *auger (screw) extruder*. Figure 3.30 illustrates an auger extrusion unit schematically. Plunger units have the advantage of high-pressure capability. However, they are batch machines that hold only a limited amount of material and increase production cost. Incremental reloading can also interfere with the flow and result in trapped air. The auger extruder has the advantage of continuous, uninterrupted extrusion. The disadvantage is a higher wear rate of revolving parts that leads to high maintenance costs and sometimes contamination.

Generally, the process involves the following steps:

1. *Blending:* Ceramic compositions are uniformly mixed and distributed in a liquid medium on a macroscopic scale.

2. *Pugging:* The plastic mix is placed in an evacuated cylinder or pug mill where it is kneaded to remove air and form a uniform liquid layer around particles. The deaired mix is transported to the extruder that is connected to the pug mill.
3. *Extrusion:* The part is ejected through a die by application of pressure or by the screw movement, and the desired profile is achieved.
4. *Cutting and Drying:* After the desired length is extruded, the green parts are cut, typically with a wire, and then dried before sintering.

A typical extrusion mix is similar in composition to an injection mix. Usually it is a mixture of liquid, a suitable binder, surfactant, deflocculant, coagulant, lubricant, plasticizer, and preservative [3.71]. Organic polymers such as cellulose ethers are usually used as the primary binder in the extrusion of engineering ceramics. PVA, polyacrylimides, and polysaccharides are other binders employed. Clay is used as the binder for some alumina and silica compositions in an aqueous mix. BeO parts have been extruded using additions of 10–18% Mogul starch and 15–18% water–glycerol.

Similarly, MgO parts have been extruded using 30–40% flour paste additions [3.49]. $CaCl_2$, $MgCl_2$, $MgSO_4$, $AlCl_3$, and $CaCO_3$ can be used as coagulants (additives used to produce a coagulated or flocculated slurry). Among typical lubricants for extrusion are various stearates, silicones, petroleum oil, colloidal talc, and colloidal graphite.

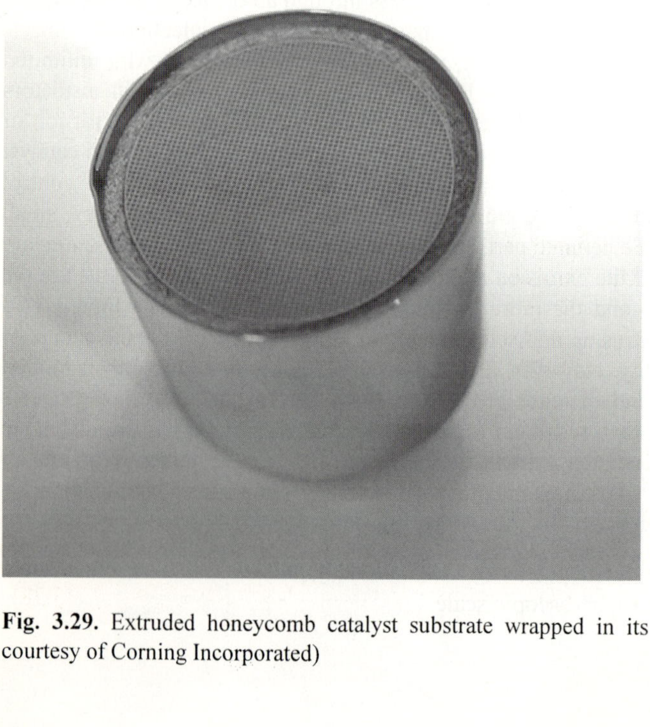

Fig. 3.29. Extruded honeycomb catalyst substrate wrapped in its housing (photograph courtesy of Corning Incorporated)

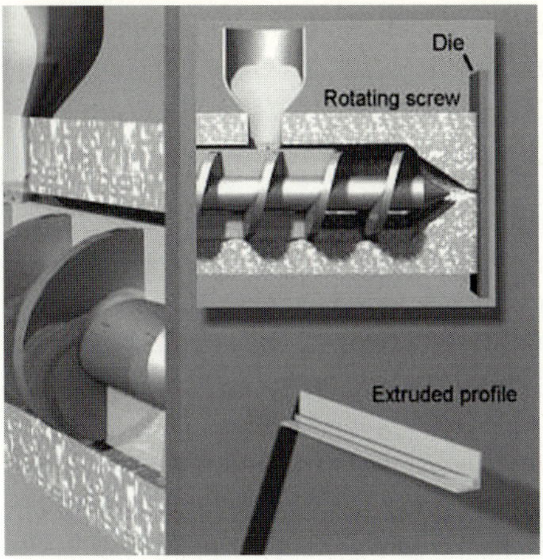

Fig. 3.30. Schematic view of an auger extrusion unit (illustration reprinted with permission from www.designinsite.dk)

During extrusion, differential shear flow under compression occurs as the material flows through the throat or the die. According to Benbow et al. [3.72], flow can occur by two mechanisms. One is plastic deformation at the die entry, and the second one is plug flow along the die-land where no cross-sectional change occurs. When pore saturation occurs and the properties of the plastic body are constant with time and position, the flow behavior can be approximated by the Bingham equation, as given before in Eq. (3.5) [3.2]:

$$\tau - \tau_y = \eta_p \dot{\gamma} , \tag{3.5}$$

where τ is the applied shear stress, τ_y is the yield stress to initiate flow, and η_p is the plastic viscosity, indicating that a stress larger than τ_y will cause a certain shear rate, which can be derived from the following equation;

$$\eta_p = (\tau - \tau_y) / \dot{\gamma} . \tag{3.27}$$

The steady-state flow of a Bingham plastic material in a linear tube can be described by the Buckingham–Reiner equation [3.2],

$$\bar{v} = \frac{R^2 \Delta P}{8L\eta_p}\left[1-(4/3)\frac{2L\tau_y}{R\Delta P}+(1/3)\left\{\frac{2L\tau_y}{R\Delta P}\right\}^4\right]+v_{\text{slippage}} , \tag{3.28}$$

where $\bar{v}$ is the mean flow velocity, R is the tube radius, $\Delta P/L$ is the differential axial pressure for the length L, and $v_{slippage}$ is given by,

$$v_{slippage} = (\varepsilon\, R\, \Delta P) / (\eta_b\, 2L) = \varepsilon\, \tau_{wall} / \eta_b \tag{3.29}$$

where η_b is the viscosity of a lubricating boundary film of thickness ε. Plug flow with slippage at the wall occurs when τ_y is larger than $\tau_{wall} = (R\, \Delta P) / (2L)$. When $\tau_y < \tau_{wall}$, differential laminar flow occurs. The shear stress developed increases radially and reaches τ_y at a critical radius, r_c (Fig. 3.31). Uniform flow without shear occurs at r_c and differential shear flow occurs when $r > r_c$.

When the cross section varies, the material is subjected to compressive (σ) and shear (τ) stresses. The ratio σ/τ decreases as the entry angle α of the die (at cross-sectional variation) increases, whereas the differential flow velocity between the center and the wall increases when α increases. The applied pressure for a circular cross section is given by [3.72]

$$P = 2(\sigma_y + \alpha_1\, V)\ \ln(D_o / D) + (4L / D)(\tau_o + \beta\, V), \tag{3.30}$$

where V is the mean paste velocity in a die-land of length L. σ_y, α_1, τ_o, and β are the extrusion parameters. They depend on the rheological properties of the paste under extrusion conditions. D_o and D are the initial and final diameters, respectively, if the cross section varies. For intermediate extrusion rates, the extrusion pressure varies directly with the shear strength of the paste (τ_y). As an example, the minimum pressure for a circular cross section occurs near an entry angle $\alpha = 60$–$70°$ for Al_2O_3–3% clay + 15.5% glucose syrup + 5% water mixtures.

These are common defects in extruded materials:

1. Cracks or laminations caused by differential springback and/or differential drying shrinkage.
2. Surface craters and blisters caused by insufficient deairing.
3. Periodic surface laminations due to slip-stick wall friction.
4. Laminations caused by divided paste flow.
5. Curling on drying or nonuniform flow through the die.
6. Gradients in extrudate stiffness caused by liquid migration down the column (bleeding) during piston extrusion.

Such defects can usually be eliminated by carefully controlling process parameters. For example, to minimize shrinkage, the density of the mix can be adjusted; friction on drying fixtures can be reduced, the part can be hung vertically, moved, or rotated during drying, and longer, slower drying cycles can be used [3.71].

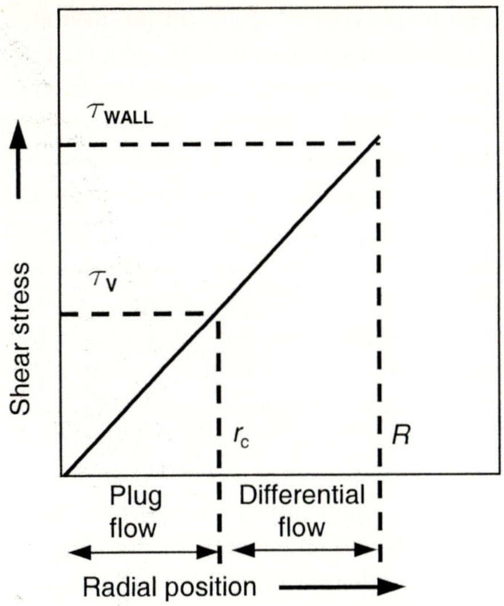

Fig. 3.31. Variation of shear stress with radial position for flow in a tube (From J.S. Reed: *Introduction to the Principles of Ceramic Processing*. Copyright 1988 John Wiley &Sons. Adapted by permission of Wiley New York)

3.2.8 Sol–Gel Processing and Gel Casting

3.2.8.1 Sol–Gel Processing

The principles of sol–gel processing of powders were described in Sect. 2.1.2.1.3. The same principles also apply to sol–gel processing of ceramic bodies. Here, we will discuss the sol–gel routes applied to produce semirigid bodies instead of powders.

The first dense ceramics obtained by sol–gel routes were nuclear reactor fuels [3.73], cubic ZrO_2 [3.74], and PLZT [3.75]. The demonstration by Yoldas [3.76,3.77] and Yamane et al. [3.78] that monolithic ceramics can be produced by sol–gel routes is pinpointed as one of the major driving forces for the enormous research effort in sol–gel science today [3.79]. The monolithic sol–gel route has been used to produce oxide glasses. Sakka and Kamiya [3.80] outlined the following requirements for glass making by this route:

1. a high alkoxide/water ratio, as pointed out by Yoldas [3.81],
2. use of containers made from non hydrophilic materials, and
3. slow heating and drying rates.

Cracking during drying and aging of monolithic gels is a major problem, especially for large components. Chemical agents to control drying are used to optimize gelation, aging, and drying by reducing the magnitude of capillary stresses induced during drying. Another method used to prevent cracking during drying is hypercritical evacuation of the solvents. Hypercritical evacuation is obtained by heating the gel in an autoclave into which a sufficient amount of solvent is added to achieve hypercritical conditions. Gels prepared by this method are called *aerogels*. Aerogels can have porosities up to 98% and densities as low as 0.08 g/cc [3.82].

Aksay and his co-workers [3.83,3.84] prepared monolithic gels of mullite by the colloidal route. These monoliths could be sintered to 98% theoretical density at 1250°C in 4 h. Ultrafine (~15 nm) particles of AlOOH were used as the alumina source, and controlled hydrolysis of tetraethoxysilane (TEOS) around AlOOH particles acted as the SiO_2 source. An important requirement for good results was control of the degree of chemical segregation at the nanometer scale. The monolithic route appears complex and limited in composition. Although rapid processing without cracking and achieving reproducible shrinkage during gelation remain critical issues to be solved, the monolithic route can be used where complex shapes are needed or low temperature processing is required. Applications are generally those that demand high purity, such as optical and electronic components, or porosity that leads to novel properties. Porous aerogels are used as transparent or translucent superinsulators to improve energy efficiency [3.79].

The advantages of sol–gel processing are extremely high purity, which is required in the electronic and optical industry, a high degree of uniformity, very fine and controlled particle size, and mixing of constituents at the atomic level. Most sol–gel derived materials can also be sintered at lower temperatures, compared to components prepared by other preconsolidation methods. Sol–gel processing allows the preparation of glasses with novel compositions and properties by avoiding limitations imposed by the critical cooling rate for glass transformation. Novel microstructures and phase distributions and controlled porosity are also among the advantages of sol–gel processing [3.85,3.86]. For example, Fig. 3.32 shows nanoscale $AlTi_3O_9$ precipitates in a mullite matrix, obtained by a sol–gel process.

The following disadvantages of this process, however, should be realized:

1. Porosity, which leads to loss of mechanical and other properties.
2. Long hydrolysis and drying times. For instance, some compositions of the Na_2O–ZrO_2–SiO_2–P_2O_5 system require one month to solidify.
3. Carbon traces that remain in thermally treated material, which may cause discoloration in glasses and other problems in applications where high purity is required.
4. Special treatments required for drying thick ceramic parts.

3.2.8.2. Gel Casting

Gel casting begins with a suspension of fine ceramic powders in a solution containing an organic binder and a dispersant as depicted in Fig. 3.33. Unlike sol–gel processes, which depend on particular inorganic chemistry, the gel casting gel system is totally organic. This feature allows using gel casting to cast parts from any ceramic powder [3.87]. Although organic solvent systems have also been developed, water is the preferred medium in gel casting. The suspension is poured into a mold at room temperature and allowed to polymerize or gel [3.88]. After the resulting gelatin-like material is removed from the mold and allowed to dry, it can be machined if desired. Following binder burnout, the part is ready for sintering. The polymer binder content remaining in the dried body is about 2–4 wt%.

In the original work depicting gel casting, the essential components of the process are reactive organic monomers that can be polymerized. The monomer solution is composed of a solvent (typically water), a monomer that forms a polymer chain, a cross-linking monomer, and a free-radical initiator.

Chain formers: Methacrylamide (MAM) monomer is most suitable for this process. Other commonly used monomers are hydroxymethylacrylamide (HMAM), *N*-vinyl pyrrolidone (NVP), and methoxy poly(ethylene glycol) monomethacrylate (MPEGMA). Combinations of these monomers can also be used.

Cross-linking agents: Methylene bisacrylamide (MBAM) and poly(ethylene glycol dimethacrylate) (PEGDMA) are preferred cross-linkers.

Free-radical initiators: Ammonium persulfate (APS) is the preferred radical initiator. Frequently, tetramethylethylene diamine (TEMED) is added to catalyze the breakdown of APS, which in turn accelerates polymerization and cross-linking.

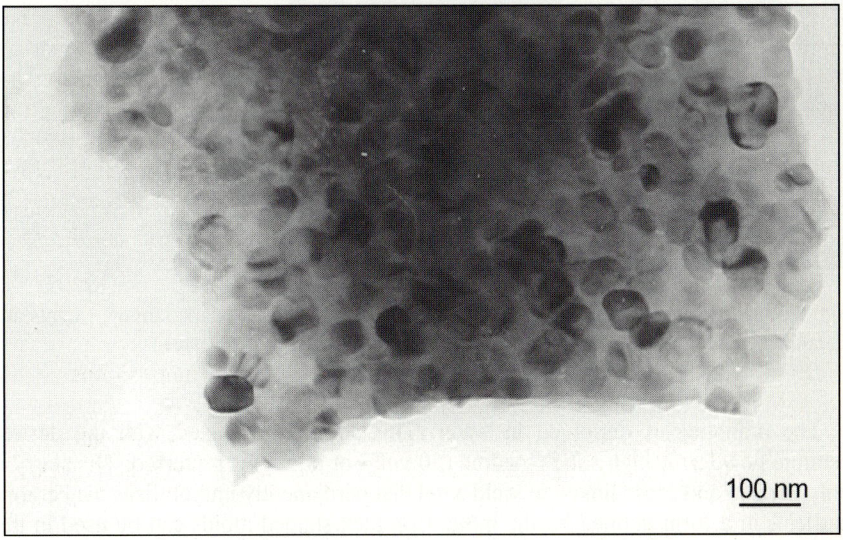

Fig. 3.32. Nanoscale $AlTi_3O_9$ precipitates in optically clear mullite matrix prepared by sol–gel processing (courtesy of Dr. Ph. Colomban, Centre National de la Recherche Scientifique, Laboratoire Spectrochimie Infrarouge et Raman Thiais France)

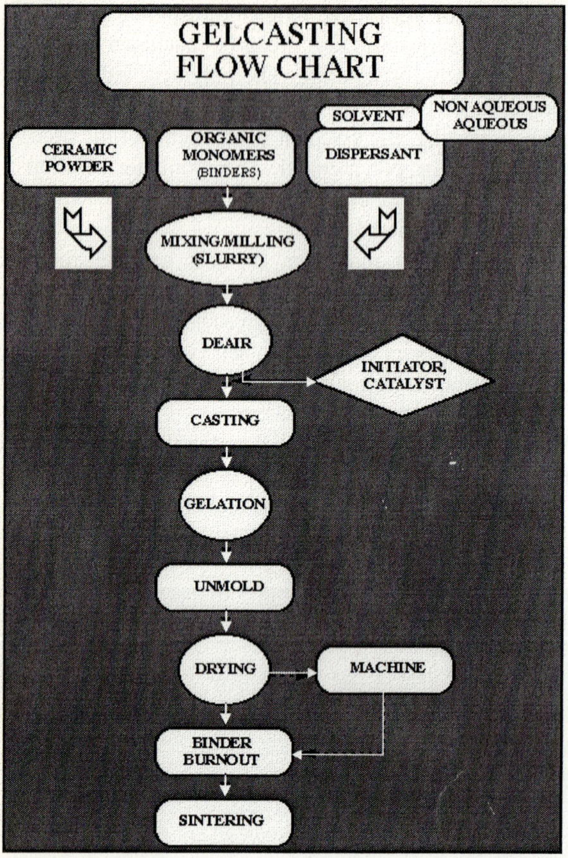

Fig. 3.33. Flow chart of the gel casting process (courtesy of Dr. Mark Janney, Metals and Ceramics Division, Oak Ridge National Laboratory Oak Ridge Tennessee)

The monomer is dissolved in water. This solution is mixed with the desired ceramic powder. A high solids loading (50 vol% or above) is preferred. The slurry is polymerized and cross-linked to yield a gel that permanently immobilizes the ceramic particles in a form defined by the mold. Complex shaped molds can be used in this process.

Gel casting equipment is not significantly different from that used for more common ceramic processes such as slip casting and spray drying. No special molding machines or presses are required. Thus, gel casting requires a minimal departure from existing industrial practice.

Air bubbles entrapped during casting can be a problem and have to be removed by deairing before casting. As for similar processes such as die pressing and isopressing, large components prepared by gel casting require slow heating rates for binder burnout. Carbonaceous residues may cause some problems in the final part.

Procedures for binder burnout are similar to those used for die pressed or isopressed bodies. Some chemicals used in gel casting are irritants, and a few of them are potentially hazardous materials. The acrylamide monomer used initially in the development of gel casting is a neurotoxin and has been largely discontinued because of health concerns [3.89]. Thus, various alternative gelling systems are being investigated.

Recent studies in gel casting obtained gelation by polymeric species in the suspension instead of polymerization of monomers. Some systems with satisfactory results include cross-linking of poly(vinyl alcohol) by titanium ion complexation in aqueous solutions [3.90,3.91], thermoreversible gelation with biopolymers such as agarose [3.89] and carrageenan [3.92], thermal cross-linking of chitosan with dimethoxy-dihydrofuran (DHF) [3.93], and cross-linking of chitosan at ambient temperature with glutaraldehyde [3.94]. Most of these studies employed alumina as the ceramic phase, while the gel casting of silicon carbide, silicon nitride, and zirconia have also been evaluated.

The advantages of gel casting over other green forming processes are as follows:

1. Near-net complex shapes can be produced which are also significantly strong, allowing machining operations after drying. Complex shapes such as turbine rotors and thin sections as small as 0.2 mm have been obtained.
2. The slurry can be processed in an entirely closed system. This prevents the introduction of contaminants and foreign material during mixing, filtration, deairing, and mold filling steps.
3. The time available for casting the slurry (idle time) can be controlled by process parameters such as temperature, acidity, and concentration of monomers/polymers and cross-linking agents.
4. Fine surface detail can be obtained.
5. It is a versatile process. Parts lighter than 1 g and heavier than 6 kg have been produced. Large parts with simple shapes can be produced. For example, large rods up to 10 cm in diameter have been made.
6. Almost any ceramic powder can be used in gel casting.
7. The process exhibits high reproducibility.
8. Many types of metal or polymer mold materials can be used.

3.2.9 Tape Casting

Tape casting or doctor blade processing is used for manufacturing thin ceramic plates. Ceramic capacitors, substrates for microelectronic circuits, multilayer circuitry, piezoelectric devices, heat exchangers, and sensors are commonly produced by this method [3.95,3.96]. Recent applications include heat exchangers and complex three-dimensional geometries by green forming of tapes [3.97].

Tape casting involves a slurry based on ceramic powders, solvents, dispersants, binders, and plasticizers. The degassed, well-dispersed slurry is fed onto a moving carrier and is flattened by a thin blade (Fig. 3.34). After a highly uniform sheet is formed, the solvent is evaporated, resulting in a dry tape strong enough to be cut and fired. Green sheets can be laminated to produce multilayer composites. Tape casting also offers the possibility of tailoring functionally gradient materials for special applications. These materials can be produced by laminating tapes with different compositions. Composite materials with improved properties can be produced by orienting whiskers or platelets by tape casting. This effect was demonstrated in tape casting $Bi_4Ti_3O_{12}$ powders with platelet morphology and SiC whiskers or platelets in an Al_2O_3 matrix [3.97]. Various constituents can be used in tape casting slurry formulations. Two typical systems are given in Table 3.6.

Tape casting offers all of the advantages common to colloidal processing methods. Furthermore, the maximum defect size that may occur in the green tape is determined by the tape thickness (generally 100 μm). This can be further improved by thermocompression to laminate individual tapes.

Ceramic powders should be free of hard agglomerates, which interfere with dense particle packing and full densification during sintering. Since high forming pressures do not exist in tape casting, full deagglomeration is essential to prepare very thin tapes. Sol–gel processing can also be used for better surface smoothness, homogeneity, and purity. Sol–gel processing also makes milling and deairing steps unnecessary. The disadvantages of using sol–gel methods in tape casting include higher cost and problems associated with drying due to high shrinkage rates.

Plucknett et al. [3.97] studied the effect of employing nanometer size powders on the tape casting behavior of a ceramic slurry. It was shown that finer powders (250–300 nm average diameters) than those used conventionally increased the required amounts of dispersant, plasticizer, and binder for optimum results.

Solvents are used to dissolve the binder, plasticizer, and dispersant. They should not react with the ceramic. Aqueous or nonaqueous solvents can be used for tape casting Al_2O_3 or ZrO_2. Aqueous solvents are not suitable for tape casting nonoxides such as SiC, AlN, and Si_3N_4, since they react with water and generate hydrogen. This changes the pH of the system and causes bubble formation. In the case of SiC, a trichloroethylene/ethyl alcohol system can be used. For AlN, a mixture of toluene/ethyl alcohol with PVB as the binder and dibutyl phthalate as the plasticizer has been used [3.95]. Recently, water-resistant AlN powders have been developed with the aid of particle conditioning and formation of hydrophobic surfaces [3.98]. These powders make tape casting and roll compaction of aqueous slurries possible. Tapes as thin as 100 μm have been formed using this system. In the case of roll compaction, AlN tapes have been formed with thicknesses ranging from 0.05 to 1 mm. *Roll compaction* is a process similar to tape casting where the feed material is compressed between two rollers to form flexible green tapes. Other common solvents used in tape casting are azeotropic mixtures such as 63/34 vol% 2-butane/ethanol or 72/28 vol% trichloroethylene/ethanol [3.99]. Organic solvents evaporate more easily than water, which results in shorter drying time, less heat flux, and lower drying temperatures. However, they can be an explosion hazard and may be toxic. Water, on the other hand, is cheaper and safer than organic solvents but can hydrolyze the surface of ceramic powders in the slurry.

a

b

Fig. 3.34. Laboratory tape casting bench **a** and partial view of casting station in operation **b** (courtesy of NETZCH Incorporated Exton PA)

Dispersants are added in concentrations of 0.5–2 wt%. To judge the effectiveness of a dispersant and to find the optimum concentration, the sedimentation behavior and viscosity of the slurry can be determined as a function of concentration. Alternatively, an absorption isotherm, i.e., a functional relationship between the amount of adsorbed dispersant and concentration, can be used. By using a dispersant concentration at which the adsorption reaches a constant value, a maximum degree of deagglomeration is achievable.

To get the best effect from a dispersant, the solvent, dispersant, and ceramic powders are milled before the binder and plasticizer are added [3.100]. This saturates surfaces with the dispersant. Menhaden fish oil is one of the most commonly used dispersant for Al_2O_3 in trichloroethylene/ethanol systems. It has variable properties dependent on location and season since it is derived from natural resources. Therefore, a synthetic dispersant may be advantageous for industrial applications.

Table 3.6. Examples of Nonaqueous Tape Casting Slip Formulae [a]

Constituent	Function	Batch Weight (g)
Example 1		
Al_2O_3	Ceramic	100.0
Menhaden fish oil	Dispersant	1.8
Xylene	Solvent	21.0
Anhydrous ethyl alcohol	Solvent	13.7
Mixed phthalates	Plasticizer	3.6
Poly(alkylene glycol)	Plasticizer	4.3
Poly(vinyl butyral) (PVB)	Binder	4.0
Example 2		
Al_2O_3	Ceramic	100.0
Phosphate ester (Beycostat C213)	Dispersant	8.0
Trichloroethylene ethanol	Solvent	38.8
Dibutyl phthalate	Plasticizer	0 – 21.0
Poly(vinyl butyral) (PVB)	Binder	2.0 – 7.0

[a] References:

1. *Engineered Materials Handbook* Vol.4, *Ceramics and Glasses.* ASM International Metals Park OH, 1991), p.163
2. M. Descamps, G. Ringuet, D. Leger, and B. Thierry: J. Eur. Ceram. Soc. **15 [4],** 357–362 (1995)

The *binder* gives the green tape flexibility and strength to enable handling, storage, and processing, e.g., cutting and punching. It should be compatible with the dispersant and should provide lubrication between particles. It should not interact during solvent evaporation. A high molecular weight is desired in a binder for high toughness and strength and a low glass transition temperature. A plasticizer is usually required to soften the binder and improve tape flexibility. Another important feature required in tape casting is a green tape that does not stick to the carrier. Binders and plasticizers that do not strongly adhere to the carrier should be selected. Release agents can also be mixed into a slurry for easy removal of the tape. Typical binders for nonaqueous systems are PVB, PVA, polyethylene, and PMMA. Commonly used plasticizers are dioctyl phthalate, dibutyl phthalate, butyl benzyl phthalate, and poly(ethylene glycol). In aqueous systems, acrylic polymers, methylcellulose, and PVA are often used as binders. Glycerin, ethylene glycol, and dibutyl phthalate are used as plasticizers. Water-based polymer emulsions can also be used. In such emulsions, organic substances are dispersed with the aid of surfactants. An example of such a system is given in Table 3.7 [3.95].

Binder burnout can be a problem in tape casting and other green forming operations. One approach to eliminating the polymeric binder is polymerization of the solvent and dispersant after casting. Landham et al. [3.96] produced a barium titanate slip from a monomer/solvent, organotitanate dispersant, cross-linking agent, plasticizer, initiator, and accelerator. Another advantage of eliminating the binder was a lower slip viscosity that allowed higher powder loading in the slip. An alternative to this method is to use colloidal gels without organic binders. Table 3.8 shows examples of slip formulations used for both of these approaches.

Three types of *carriers* are usually used for tape casting: rigid glass plate, continuous steel belt, and flexible plastic film. Ordinary glass plate with a thickness of 6 mm or more is satisfactory for most tape casting operations [3.101]. However, handling glass plates represents a potential hazard. Continuous steel belts form a good surface for casting but they wear out during tape casting. Flexible films, such as 0.05–0.2 mm polyester, overcome most of the problems and are mostly used for thick tape production. Cheaper films such as polyethylene and acetates are not as strong but may also be used. Flexible films have the added advantage of easy removal of the tape with less damage. If adhesion is strong, silicone, waxes, or soybean derivatives can be sprayed or applied to the carrier as temporary release agents.

3.2.10 Centrifugal Casting

Two approaches can be used for centrifugal casting. In the first, a hollow cylindrical plaster mold is rotated about its axis with the slip in the interior. This approach yields tubular and other axisymmetrical green shapes. Alternatively, a mold is placed at the bottom of a centrifugal bucket whose rotation produces the centrifugal action in the direction of the mold [3.102]. Accelerations of a few tens to thousands of "*g*" and casting times of 1–2 h are employed.

Centrifugal casting has the advantage of much shorter casting times compared to slip casting, adaptability to both small and large complex shaped components, and reduced risk of stress gradients in the green part. During centrifugation, particles within the slip move through the fluid in a direction perpendicular to the gravitational force, at a rate dependent on this force and particle drag. The liquid, on the other hand, moves in the opposite direction. Thus, the total volume of liquid in the slurry does not need to be transported through the cake, which may otherwise leave defects such as filtration channels (as in the case of slip casting and pressure filtration) [3.102,3.103].

One drawback of this process is segregation due to differential settling rates of suspension constituents by density and size. Segregation may result in particle size separation in single-phase suspensions and phase separation in multiphase suspensions. Segregation may be prevented by various approaches. These include flocculation or coagulation of particles in the slip or using suspensions with repulsive, non touching networks that contain high solids volume fractions (40–50 vol%). Flocculation and coagulation result in open structures and lower green densities whereas high solids content suspensions result in bodies with up to 68% TD [3.103]. The seeming disadvantage of segregation may also be used to advantage to produce

Table 3.7. Water-Based Slurry Formulation for Tape Casting [a]

Material	Function	Amount (wt%)
Al_2O_3	Matrix	66.5
MgO	Grain growth inhibitor	0.1
Water	Solvent	16.3
Condensed acrylsulfonic acid	Dispersant	2.5
Acrylic polymer emulsion	Binder	7.0
Poly(ethylene glycol)	Plasticizer	4.2
Butyl benzyl phthalate	Plasticizer	3.1
Nonionic octylphenooxyethanol	Wetting agent	0.2
Wax-based emulsion	Defoamer	0.1

[a] Reference:

J.C. Williams. In: *Treatise on Materials Science and Technology. F.F. Yang (ed.).* (Academic Press Orlando FL, 1976), pp.173–198 (adapted with permission of Academic Press)

layered or graded components for special applications. For example, a gear with gradient layers parallel to the outer surfaces was produced by Steinlage et al. [3.102] to demonstrate this effect.

Microstructures with minimal surface and volume defects are achieved by combining ultrasonic treatment and filtration to eliminate agglomerates, deflocculation with the addition of deflocculants, and centrifugal casting. The microstructure, postconsolidation mechanical properties, and reliability were equal or superior to those obtained by isostatic pressing [3.103,3.104].

Table 3.8. Examples of Slip Formulation Without the Need for Binder Burnout [a]

Constituent	Function	Content (wt%)
1. Monomeric solvents and dispersants that polymerize after casting		
$BaTiO_3$	Ceramic	67–77
Butyl methacrylate	Solvent/binder	21–25
Organotitanate	Dispersant/coupling agent	1–7
Diethylene glycol dimethacrylate	Cross-linking agent	1.5–2.0
Benzoyl peroxide	Initiator	0.6
Dioctyl phthalate	Plasticizer	4.5–5.2
N, *N*-dimethyl-*p*-toluidine	Accelerator	0.6
2. Whiteware slips with colloidal inorganic binder		
Nonplastics	Ceramic	25–30
Clay	Ceramic	15–25
Water	Liquid medium	45–60
Na silicate, Na polyacrylate, or Na lignosulfate	Deflocculant	< 0.5
$CaCO_3$	Coagulant	< 0.1
$BaCO_3$	–	< 0.1
Clay < 1 μm	Binder	Variable

[a] References:

1. R.R. Landham, P. Nahass, D.K. Leung, M. Ungureit, W.E. Rhine, H.K. Bowen, and P. Calvert: Am. Ceram. Soc. Bull. **66 [10]**, 1513–1516 (1987)

2. J.S. Reed: *Introduction to the Principles of Ceramic Processing.* (Wiley New York, 1988)

3.2.11 Other Forming Methods

3.2.11.1 Vacuum Casting

The driving force for suction in a permeable mold can be increased up to five times by applying reverse pressure, i.e., vacuum created around the mold. This can increase the casting rate and uniformity, and improve the economics. This process is generally used for producing very porous refractory insulation with complex shapes.

3.2.11.2 Freeze Casting

This novel casting method employs an impermeable mold filled with slurry. After freezing the slip at a suitable temperature, liquid is removed by freeze drying. This method allows casting thick slips and novel compositions, provided that freezing is rapid before any settling occurs.

3.2.11.3 Vacuum Forming/Blow Molding

These are processes that shape green ceramic preforms that are easily formable with a mold and vacuum in vacuum forming or with pressurized gases in blow molding [3.105]. Blow molding has been used to shape clay bodies and alumina suspensions containing water-soluble polymers for producing alumina gas discharge lamp tubes. Vacuum forming has been applied to ceramic suspensions based on thermoplastic polymers with ceramic loadings at the 50–60 vol% level to prepare thin-walled ceramic shapes. A thin sheet of ceramic is produced by extrusion or another process. Uniform mixing is essential for defect-free products. The sheet is then cut, placed on the forming device (Fig. 3.35), heated to the melting temperature of the polymer, and drawn under vacuum. The green shape is placed on a ceramic powder bed as a support and subsequently heated to debind.

3.3 Conventional Consolidation Methods

3.3.1 Solid-State Sintering

Sintering can be defined as increasing the relative density of a body that has initial porosity by diffusional processes [3.49]. The mechanical and physical properties of a porous body are generally improved drastically upon sintering. Unlike metals, polymers, and intermetallics for which casting and hot or cold deformation are feasible forming methods, sintering is the major route for forming dense ceramics. Alternative processes exist, however, that will be discussed in Sect. 3.4.

The driving force for solid-state sintering is the tendency of particles to reduce their surface energies. There are, however, two alternatives that lead to surface energy reduction. These are grain growth (coarsening) without pore removal and

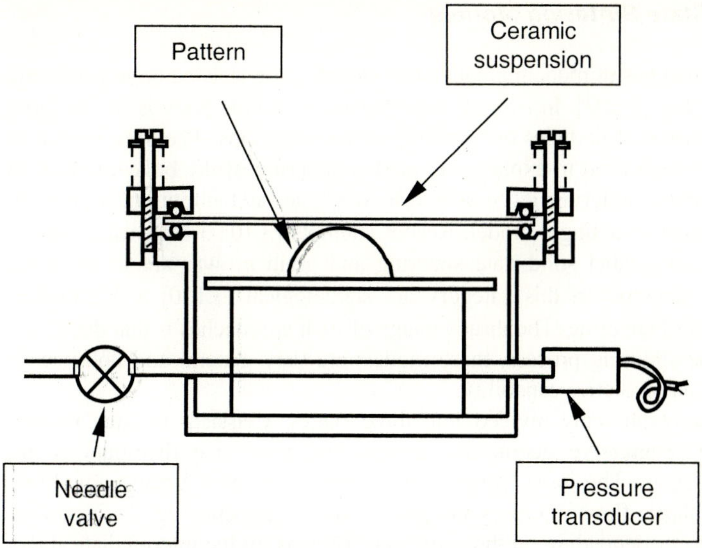

Fig. 3.35. Schematic view of vacuum forming (From K.M. Haunton, J.K. Wright, and J.R.G. Evans: *The Vacuum Forming of Ceramics*. Br. Ceram. Trans. J. **89 [2]**, 53–56 (1990) (adapted with permission of The Institute of Materials London UK)

densification with pore shrinkage. These two phenomena are competitive and interactive. Another important factor in sintering is densification kinetics. Even if densification is thermodynamically favored, sintering kinetics may be too slow for practical purposes. There are many factors that can affect sintering rates. This makes the sintering phenomenon extremely difficult to model quantitatively. Some of the better identified factors are temperature, pressure, atmosphere, impurity concentration, second phases, pore size distribution, particle size, particle shape distribution, and other initial microstructural features. Other contributions are attributed to heating/cooling rates, sintering schedules, as well as defect density and distribution. It is obvious that explaining all sintering phenomena with one big universal equation is not an easy task, yet many attempts successfully take the main features of sintering into account.

Sintering involves transport of atoms. Figure 3.36 displays various paths that atoms can take. Among these transport mechanisms, only lattice diffusion and grain boundary diffusion can provide densification, i.e., a decrease in center-to-center distances of adjacent particles. Evaporation of atoms from grain boundary surfaces and condensation at necks formed between particles can cause only coarsening. From this point of view, one should sinter under conditions that enhance lattice and grain boundary diffusion and suppress coarsening. However, recently it was suggested that grain growth could be helpful during the final stages of sintering to avoid pore/grain boundary separation [3.106]. This theory will be discussed later in this chapter. Various aspects of solid-state sintering are discussed in the following sections.

3.3.1.1 Solid-State Sintering Models

Two general approaches in modeling solid-state sintering are isolated approaches and universal approaches [3.107]. In isolated approaches, the whole body is divided into periodic arrangements of particles or particle/pore combinations. Theories applied to these portions are then used to explain and predict sintering of the bulk material. In universal approaches, sintering is visualized as a whole, and important aspects of sintering are applied to a single model. Recent attempts [3.108,3.109] using Monte Carlo simulation to predict solid-state sintering and grain growth are in the latter category. Earlier attempts in this category are stereological [3.110] and statistical [3.111] treatments of sintering. The disadvantage of such approaches is that due to the complexity of the sintering process, these models are very general and nonspecific and result in limited predictive capability.

Sintering is conventionally divided into three stages. Classical isolated models assume that the progressive geometric changes and pore size distributions are constant in each stage. The initial stage is characterized by neck growth involving diffusion and/or plastic flow processes and particle rearrangement (Fig. 3.37a and b). Shrinkage may occur depending on the active mechanisms. In the intermediate stage, pores remain interconnected and necks are relatively large (Fig. 3.37c). A major portion of densification occurs in the intermediate stage. The final stage begins when pores are closed and separated from grain boundaries (Fig. 3.37d and e).

At this stage, pores are typically spherical. The densification rate decreases rapidly and eventually ceases in practical terms.

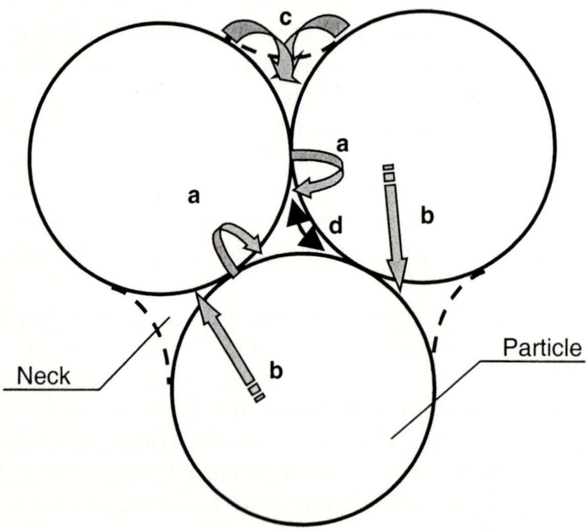

Fig. 3.36. Transport paths for atoms during solid-state sintering: grain boundary diffusion **a**, lattice diffusion **b**, evaporation-condensation **c**, and surface diffusion **d**.

An important notion in sintering is the *dihedral angle* (Fig. 3.38). The grain boundary energy created upon neck formation determines the shape of pores. From the balance of forces, the dihedral angle can be calculated from [3.79]

$$\gamma_{gb} = 2\gamma_{sv} \cos (\psi /2), \tag{3.31}$$

$$\psi = 2 \arccos (\gamma_{gb} / 2\gamma_{sv}), \tag{3.32}$$

where γ_{gb} is the grain boundary specific interfacial energy, γ_{sv} is the specific solid–vapor interfacial energy, and ψ is the dihedral angle. The tension created by the curvature in Fig. 3.38a produces a higher vacancy concentration; therefore atoms will diffuse from the grain boundary to the pore surface and provide shrinkage. On the other hand, in multigrain situations (Fig. 3.38b), grains have a positive curvature, and ψ is small. This arrangement reduces the vacancy concentration at the pore surface, induces diffusion away from the pore and causes pore growth. Therefore, pore shrinkage or growth depends on the character of ψ and the number of grains surrounding a pore. A smaller number of surrounding grains increase the possibility of pore shrinkage. Agglomerates similar to those in Fig. 3.38b, however, provide locations for pore growth.

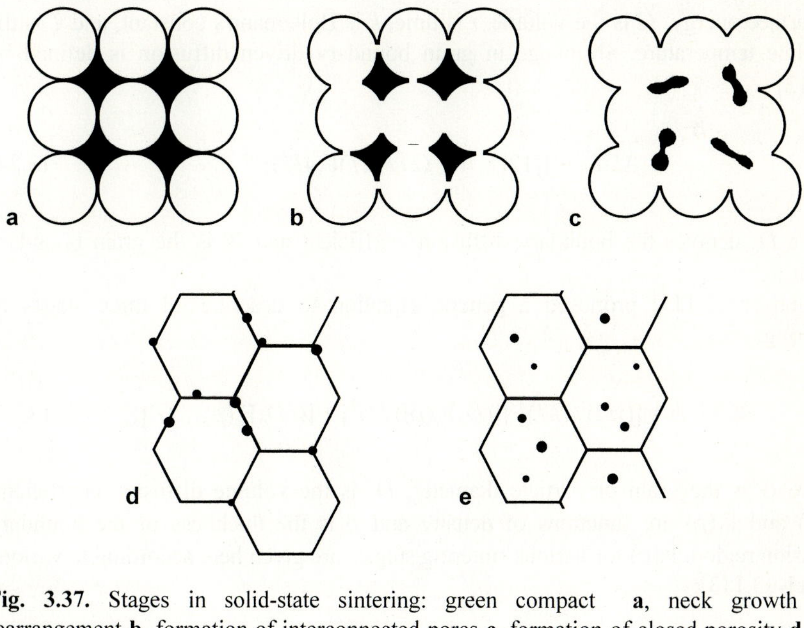

Fig. 3.37. Stages in solid-state sintering: green compact **a**, neck growth and rearrangement **b**, formation of interconnected pores **c**, formation of closed porosity **d**, and pore separation **e**

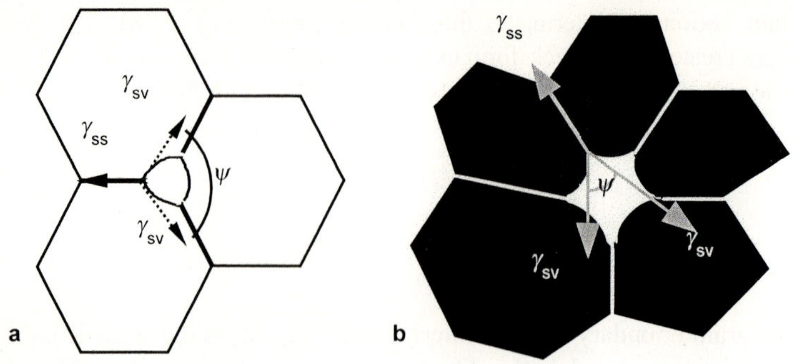

Fig. 3.38. Shrinkage of pores at three grain junctions **a** and coarsening of pores surrounded by many grains **b**

Coble [3.112] derived two equations that describe initial-stage sintering by lattice diffusion or grain boundary diffusion for spherical particles with average radii, R. In lattice diffusion dominated sintering, shrinkage is defined by

$$\Delta L/L_o = [(4D_L \gamma \Omega t) / (R^3 kT)]^{1/2}, \tag{3.33}$$

where $\Delta L/L_o$ represents linear shrinkage, D_L is the lattice diffusion coefficient, γ is thesurface energy, Ω is the volume, t is time, k is Boltzmann's constant, and T is the absolute temperature. Shrinkage in grain boundary driven diffusion is defined by [3.112]

$$\Delta L/L_o = \{[12D_b W \gamma \Omega t / (kT)]/(4R^4)\}^{1/3}, \tag{3.34}$$

where D_b denotes the boundary diffusion coefficient and W is the grain boundary width.

Johnson [3.113] proposed a general equation to describe all three stages of sintering:

$$-dL / L\, dt = [(\gamma\Omega) / (kT)]\ \{[(D_v F_v(\rho)) / G^3] + [(\delta D_b F_b(\rho)) / G^4]\}, \tag{3.35}$$

where G is the grain or particle diameter, D_v is the volume diffusion coefficient, $F_v(\rho)$ and $F_b(\rho)$ are functions of density and δ is the thickness of the boundary diffusion region. $F_i(\rho)$ for various sintering stages are given here according to various models [3.113]:

1. For initial-stage sintering ($\Delta L/L_o < 0.05$), $F_v(\rho) = 20 / (\Delta L/L_o)$ and $F_b(\rho) = 12.5 / (\Delta L/L_o)^2$.

2. For intermediate- stage sintering, $F_v(\rho) = 250$, and $F_b(\rho) = 330/(1-\rho)^{1/2}$ or $F_v(\rho) = 240/\rho$, and $F_b(\rho) = 27 / [\rho(1-\rho)^{1/2}]$.
3. For final-stage sintering, $F_v(\rho) = 288\,(1-(1-\rho)^{2/3}) / \{3(1-\rho)^{2/3} - [1+(1-\rho)^{2/3}]\ln(1-\rho) - 3\}/\rho$, and $F_b(\rho) = 1047\,f(\rho) / [\rho\,(1-\rho)^{1/3}]$.

Kuczynski's Statistical Sintering Theory

Kuczynski [3.111,3.114] developed a statistical sintering theory to explain intermediate and final-stage sintering. The model recognizes that pore shrinkage is always accompanied by Ostwald ripening, i.e., grain growth. This fact was neglected in Coble's theory for intermediate-stage sintering. Kuczynski related pore size, grain size, and porosity using *Zener's relationship*,

$$\bar{a} = K\bar{r} / P\ , \tag{3.36}$$

where $\bar{a}$ is the average grain size, $\bar{r}$ is the average pore radius, K is a constant, and P is porosity.

In the intermediate stage, it is considered the body contains cylindrical pores of total length per unit volume, L_v, with variable radii, r, along that length. This pore radius is a statistical variable, whereas L_v is a well-determined function dependent only on time. In the final stage, the body contains N_v discrete spherical pores of variable radius per unit volume. Another function introduced by Kuczynski is the pore size distribution function, $f_i(r,t)$ where $i = 1$ for the intermediate stage and $i = 2$ for the final stage. Therefore, during the intermediate stage of sintering,

$$L_V = \int_0^\infty f_1(r,t)\,dr\ , \tag{3.37}$$

and during final-stage sintering,

$$N_V = \int_0^\infty f_2(r,t)\,dr\ . \tag{3.38}$$

The volume fraction of pores to the total volume, i.e., porosity, in the intermediate stage is defined as [3.115]

$$P = \pi\,L_V\,\bar{r}^2\ , \tag{3.39}$$

and in the final stage,

$$P = \frac{4\pi N_V \bar{r}^3}{3} , \tag{3.40}$$

where

$$\overline{r^n} = \frac{\int_0^\infty r^n \, f_i(r,t) \, dr}{\int_0^\infty f_i(r,t) \, dr} \qquad i = 1,2 . \tag{3.41}$$

Equations (3.37) and (3.38) can be represented by the differential equations,

$$\frac{d \ln L_V}{d \ln P} + \frac{d \ln \bar{r}^2}{d \ln P} = 1 \tag{3.42}$$

and

$$\frac{d \ln N_V}{d \ln P} + \frac{d \ln \bar{r}^3}{d \ln P} = 1 , \tag{3.43}$$

whose simplest solutions are

$$L_V = L_V^0 \, (P/P_0)^{x_1} \;\; ; \;\; \bar{r} = \bar{r}_0 \, (P/P_0)^{(1-x_1)/2} , \tag{3.44}$$

$$N_V = N_V^0 \, (P/P_0)^{x_2} \;\; ; \;\; \bar{r} = \bar{r}_0 \, (P/P_0)^{(1-x_2)/3} , \tag{3.45}$$

where $0 < x_1 < 1$ and the subscript "0" represents the initial conditions of intermediate- or final-stage sintering. Porosity is expressed as a function of time in this model as follows [3.111]:

$$\frac{\mathrm{d}P}{\mathrm{d}t} = 2K_0\, L_V^{1/2}\, \alpha \left(\frac{\bar{r}}{\bar{a}^2}\right) \overline{(1/r^* - 1/r)}\ , \tag{3.46}$$

for the intermediate stage, and

$$\frac{\mathrm{d}P}{\mathrm{d}t} = 3K_0'\, N_V^{2/3}\, \alpha \left(\frac{\bar{r}}{\bar{a}^2}\right)\left(\frac{\bar{r}}{r^*} - 1\right)\ , \tag{3.47}$$

for the final stage. In equations (3.46) and (3.47), K_o's are proportionality constants, $\bar{a}$ is the average grain size given by Eq. (3.41), and α is given by

$$\alpha = (C\,\gamma\,\Omega\,D)/kT\,, \tag{3.48}$$

where $C = 3$ for tubular pores and $C = 6$ for spherical pores, γ is the isotropic surface tension, Ω is the atomic volume, D is the diffusion coefficient, and k and T have their usual meanings. r^* is the particle radius whose magnitude remains constant with time:

$$r^* = \beta\,\bar{r}\ ;\ \beta > 1\ . \tag{3.49}$$

Finally, a more generalized equation is obtained,

$$\frac{1}{P^n} - \frac{1}{P_0^n} = K_P\,\alpha\,t\,, \tag{3.50}$$

where K_P is a proportionality constant, $n = (3/2)x_1 + s_1$ for intermediate-stage sintering, and $n = (x_2 + 2/3 + s_2)$ for final-stage sintering, where s_i are positive numbers less than one.

Computer Simulation of Sintering

Hare [3.116] used computer simulation to estimate coordinational and surface area changes during initial stage sintering, i.e., local densification-rearrangement. The model uses uniform spheres which actually do not represent real particles that are usually rough, non-spherical during densification, and nonuniform. This study, however, is useful in describing the green state where packing of particles is generally random unless special techniques such as sol–gel processing are used.

One approach to modeling sintering is to combine statistical approaches and Monte Carlo simulation. Recently, based on studies of grain growth [3.108,3.117]

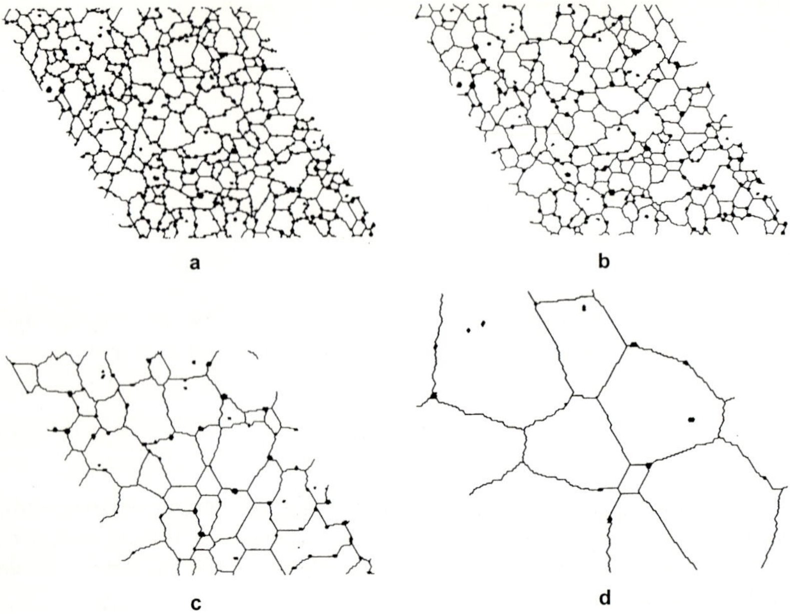

Fig. 3.39. Microstructural evolution of a porous polycrystalline solid undergoing final-stage sintering where $D_{gb}/D_m = 0.1$ and $D_s/D_m = 1$. Initial porosity, pore size, and grain size were 9.3%, 2.2, and 14.2, respectively. The microstructure calculated after 1 **a**, 10^4 **b**, 10^5 **c**, and 10^6 **d** Monte-Carlo steps (MCS) is depicted (From I.W. Chen, G.N. Hassold, and D.J. Srolovitz: *Computer Simulation of Final-Stage Sintering: II, Influence of Initial Pore Size.* J. Am. Ceram. Soc. **73 [10],** 2865-2872 (1990) (copyright 1990 the American Ceramic Society Columbus OH, reprinted with permission)

and recrystallization [3.118,3.119], Hassold, Chen, and Srolovitz developed a computer code using Monte Carlo statistics to predict the final-stage sintering behavior of well-defined porous bodies [3.109,3.120]. The microstructural evolution derived by this technique successfully resembles observations (Fig. 3.39). One important contribution of these studies is the ability to predict the effect of initial pore size on the final microstructure and kinetics of densification. It was found that pores effectively pin grain boundaries at a volumetric fraction of 3–9%. Below 3%, the remaining pores, as predicted, are ineffective in grain boundary pinning, and allow grain growth. Smaller pores were also found less effective in retarding grain growth.

3.3.1.2 Effect of Additives, Impurities, and Dispersed Phases on Sintering

It has been long recognized that very small amounts of MgO additions can lead to densification of Al_2O_3 to theoretical levels [3.121]. The actual mechanism, by which MgO improves densification, however, still remains a matter of discussion. One theory that is agreed upon more frequently states that grain growth is inhibited by the reduction of grain boundary mobility upon forming a solid solution, i.e., by the mechanism of solid drag [3.107,3.122,3.123]. Other explanations for accelerated densification of Al_2O_3 by MgO additions are increased lattice diffusion coefficients [3.124] and decreased surface diffusion rates [3.107]. One of the more recent hypotheses is based on the assumption that grain boundary movement in Al_2O_3 occurs by the addition and subtraction of atoms to and from kinks in steps on adjoining grains. It is suggested that Mg^{2+} ions can "poison" these kinks and retard atom attachment or detachment and thus reduce grain boundary mobility [3.125]. Experimental evidence suggests that the role of MgO can be attributed mainly to its ability to lower the grain boundary mobility [3.126–3.128]. A similar role has been assigned to ThO_2 in Y_2O_3, where pore and grain boundary separation are inhibited [3.127].

The effect of impurities on the sintering of oxides was analyzed theoretically by Johnson [3.129]. According to his study, two types of structures are possible in the presence of an impurity: a solid solution or a second phase. Solid solutions can improve shrinkage rates by increasing the chemical potential gradient and the number of defects in the structure. Second-phase particles, if small enough, can segregate to grain boundaries and act as grain growth inhibitors. This is demonstrated in Fig. 3.40 in zirconia-reinforced alumina (ZTA). Grain growth inhibition is mutual in Al_2O_3–ZrO_2 composites. Just as ZrO_2 particles reduce grain growth in Al_2O_3, Al_2O_3 particles can also act as grain boundary pinning agents in ZrO_2. Hori et al. [3.130] demonstrated that the grain growth of Al_2O_3 can be suppressed effectively during sintering by the addition of >1 wt% ZrO_2 with a particle size of 0.25 μm. It was found that ZrO_2 particles exhibit sufficient self-diffusion to move with Al_2O_3 four-grain junctions during grain growth [3.131]. This exerts a dragging force at the junctions, limiting grain growth. However, ZrO_2 particles are also known to limit densification in alumina [3.132]. In the presence of large (~5 μm) pores, it was observed that grain growth inhibition can provide slightly higher final densities, compared to the matrix alone [3.107]. This can be related to the prevention of pore and grain boundary separation by inclusions. Once pores are separated from grain boundaries, their shrinkage practically stops; this phenomenon will be discussed further in the following section. Lange and Hirlinger [3.133] studied the effect of Al_2O_3 inclusions on the sintering of ZrO_2. The inclusions acted as grain growth inhibitors when the ZrO_2 grain size increased to about 1 μm. The degree of grain growth inhibition increased with the volume fraction of Al_2O_3 inclusions.

Similar to the effect of ZrO_2, most nonreacting inclusions limit the densification of Al_2O_3. The addition of 5 vol% B_4C or SiC to alumina drastically reduces the densities attainable [3.134]. However, up to 98% theoretical density

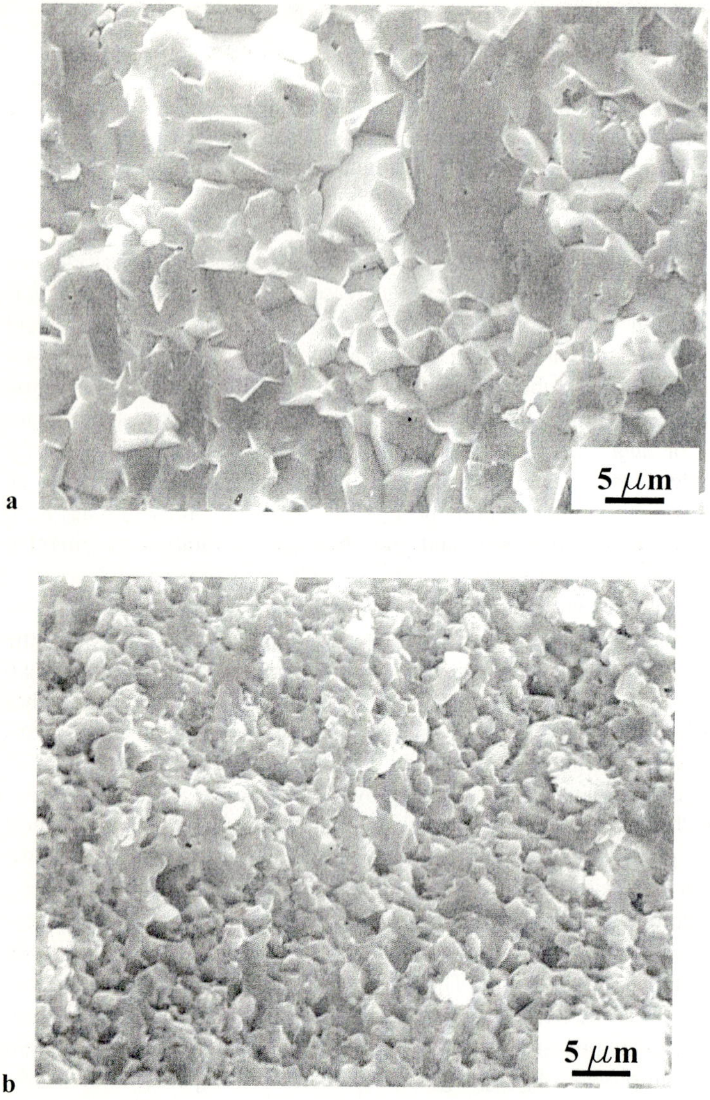

Fig. 3.40. The role of ZrO_2 particles as a grain growth inhibitor. Al_2O_3 powders **a** and Al_2O_3–5 vol%ZrO_2 powders **b** sintered at 1600°C for 1 h showing the reduced grain size in the sintered body upon the addition of zirconia particles (From M. Bengisu, Ph.D. Thesis, New Mexico Institute of Mining and Technology Socorro NM, 1992)

can be attained when 5 vol% TiC is incorporated (Fig. 3.41). Note that nondensifying inclusions, including various metal nitrides and carbides, are more detrimental to sintering than sinterable particles such as oxides. The difference seems to lie in the rates of densification. Differential sintering becomes more pronounced when nonsinterable inclusions are introduced into a readily densifying matrix, causing rearrangement (local densification) and pore growth between fast shrinking islands. On the other hand, if differential shrinkage is less acute, overall densification becomes easier.

Impurity segregation to grain boundaries can affect grain boundary diffusion and mobility as in the case of MgO in Al_2O_3. Many types of impurities can form a small amount of liquid phase at grain boundaries, which, in turn, can improve sintering. Usually, however, such amorphous phases are detrimental to mechanical properties. Examples of such systems are 5–10% Y_2O_3 in AlN, producing AlN–Al_2O_3–Y_2O_3 glass at the sintering temperature or similarly 1% CaO or SrO, which aids densification by forming liquid phases at grain boundaries [3.135]. The mechanism to which densification can be attributed is liquid-phase sintering, discussed in more detail in Sect. 3.3.2.

Some impurities can retard densification and cause pore growth by evaporation-condensation mechanisms. For example, it has been demonstrated that chlorine ions can cause grain growth and cuboidization in MgO at lower sintering temperatures (900–1000°C) that result in pore growth [3.136]. Another example is CuO in SnO_2. Two mol% CuO can increase the number of lattice defects and improve sintering, and the evaporation-condensation mechanism increases the rate of grain and pore growth, and retards densification [3.137]. In systems with such competitive phenomena, ideal sintering conditions must be found experimentally.

Second-phase additives have been successfully implemented in the pressureless densification of some covalently bonded ceramics. One method is to produce a thin liquid layer between particles by careful compositional design. The effect of the second phase is to react with the matrix and other phases to form a new phase with a relatively low melting point. This method is widely practiced for liquid-phase sintering of Si_3N_4. SiO_2 and other oxides, e.g., Y_2O_3, are added to Si_3N_4 upon which a liquid silicate forms around Si_3N_4 particles and facilitates liquid-phase sintering [3.138].

Another group of additives is found in B/C-SiC systems. Prochazka [3.139] first showed that B and C additions to SiC can provide pressureless sintering of SiC close to theoretical densities. Various roles have been attributed to B and C as sintering aids. Microanalytical examinations of sintered SiC showed that no grain boundary

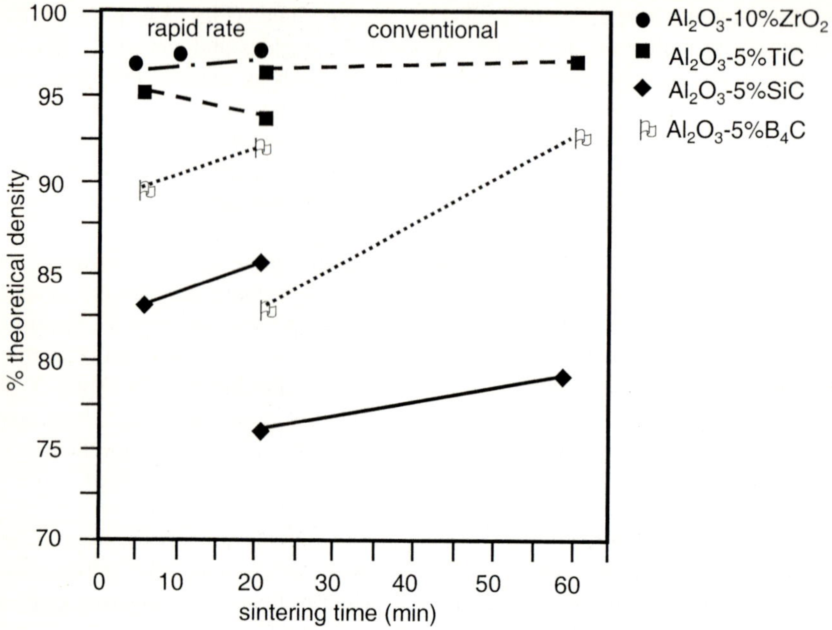

Fig. 3.41. Effect of second-phase particles on the attainable density of alumina (Reprinted from M. Bengisu and O.T. Inal: *Rapid-Rate Sintering of Particulate Ceramic Matrix Composites.* Ceram. Int. **17 [3],** 187–198 (1991) Copyright 1991 Elsevier Science Oxford UK, with permission)

enrichment of impurities or sintering aids occurs with B/C compositions; this indicates that B exists as a solid solution in the crystal [3.140,3.141]. Indeed, the optimum B concentration for sintering, 0.36 to 0.5, relates well to the solubility limit of B or B_4C in SiC (0.5%B_4C) at typical sintering temperatures (2100–2200°C) [3.142]. On the other hand, it is believed that C removes oxygen impurities and binds excess Si [3.142], reduces grain growth, or forms a liquid phase [3.143]. C was detected as graphite-like inclusions of 1–5 μm diameter in sintered SiC with C additive [3.140, 3.141, 3.144]. Rühle and Petzow [3.145] observed precipitates of B and C between SiC grains but no amorphous layer at the grain boundaries. Similar results were observed by others [3.146]. One explanation of improved densification is that B doping increases both C and Si diffusion coefficients, as suggested by Birnie [3.147]. Densities of 99% TD have been achieved using ultrafine SiC powders with optimal additions of 0.4 %B (or the corresponding amount of B_4C) and 3–4%C [3.142] at 2150°C. It was also observed that a 3% C addition to submicron, stoichiometric B_4C yields densities up to 98% TD at 2250°C [3.148]. Dole and Prochazka made similar observations [3.143]. According to Dole and Prochazka, C additions to B_4C promote densification by impeding surface diffusion and, thus, retarding the coarsening process. It was also observed that additives such as SiC,

Be_2C [3.143], and AlF_3 [3.149] activate the sintering of B_4C, resulting in densities of up to 94% TD. Lange et al. [3.149] conclude that the addition of AlF_3 activates sintering by altering the chemical nature of the grain boundary and by increasing grain boundary diffusivity.

Ceramics prepared under typical laboratory conditions contain significant amounts of impurities (at least several thousand ppm) from the starting powders and from contamination during handling and sintering. It is important to realize that many real ceramics which are nominally single phase, contain a residual glass phase at their grain boundaries due to such impurities [3.150]. Ünal et al. [3.151] demonstrated that such a glassy phase can even be encountered in a Si_3N_4 monolith prepared from high-purity Si_3N_4 powder, as seen in Fig. 3.42. Such phases can affect the sintering behavior of compacts. Morris [3.152] showed that in Al_2O_3, regions that contain large concentrations of specific impurities such as Si and Ca experience broad dihedral angle distribution, enhanced pore-boundary separation, and discontinuous grain growth. The presence of MgO may reduce the effect of chemical inhomogeneities by increasing the solubility of impurities or by reducing the segregated impurity

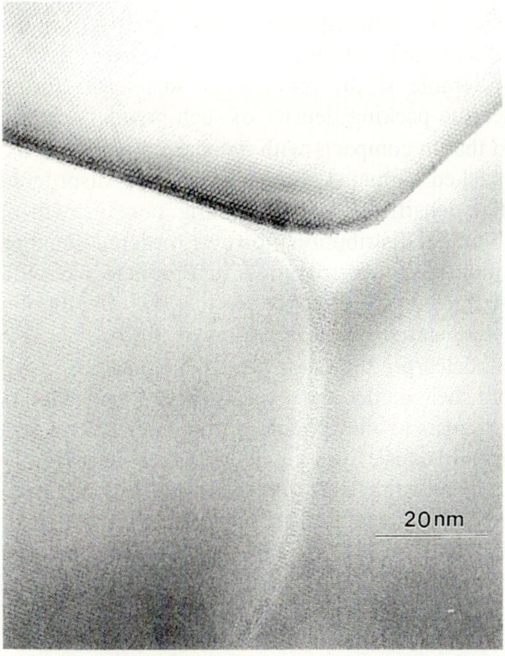

Fig. 3.42. HREM image of single-phase Si_3N_4 showing the continuous amorphous phase along the grain boundaries and at a triple-point pocket (From O. Ünal, J.J. Petrovic, and T.E. Mitchell: *Mechanical Properties of Hot Isostatically Pressed Si_3N_4 and Si_3N_4/SiC Composites.* J. Mater. Res. **8 [3],** 626-634 (1993), reprinted with permission from the Materials Research Society Pittsburgh PA)

concentration and eliminating the liquid phase. It was observed that MgO eliminated pore-boundary separation and discontinuous grain growth [3.152], supporting this hypothesis.

3.3.1.3 Effect of Green Microstructure

The initial microstructure can have a profound effect on sintering. Among important microstructural parameters that influence sintering are green density, average particle size, particle size distribution, and whether agglomerates, elongated grains, whiskers, fibers, or second-phase particles are present. To achieve high densities by pressureless sintering, it is advisable to use a preconsolidation process that yields the maximum green density possible. A fine particle size ($\leq$1 μm) is typically desired due to increased specific surface area, i.e., a larger area for diffusion. There are mainly two approaches that are used for particle size distribution. Barringer and Bowen [3.153] experimentally showed that equiaxed, monosized oxide particles produced by special techniques can be sintered almost to full density. On the other hand, Han et al. [3.154] and Yeh and Sacks [3.155, 3.156] used a particle size distribution with two average sizes. In such a distribution, small particles fit in the voids between larger particles, and they attained 99 to 100% TD upon sintering. Two-size distribution may be preferable if the amount of shrinkage during sintering should be minimum, since the packing density of such powders can be higher. Milne et al. [3.157] showed that in compacts with monosized particles, the structure is composed of close-packed domains interspersed with disordered packing. This promotes preferential densification and, in turn, pore growth at domain boundaries. In the case of bimodal distribution, however, ordered domains disappear, which allows attaining more uniform structures. In a particle size with normal distribution, it is possible to analyze the effect of particle size distribution directly in terms of median grain size and standard deviation of the particle size distribution. Ting and Lin [3.158] modeled the shrinkage behavior in terms of these characteristics. They showed in both their model and experimentally that alumina with a larger initial median grain size sinters to a lower density and that a narrow size distribution (keeping the median grain size equal) results in a higher final density.

Agglomerates are undesirable in the green microstructure since they promote local densification and reduce sinterability. They can remain in powders if sufficient milling is not achieved. Figures 3.43 and 3.44 show the effect of hard agglomerates produced during shock processing of powders [3.159]. These agglomerates remain as individual islands where sintering is faster than the rest of the compact; this is detrimental to overall densification.

Sintering theories predict that densification rates may be enhanced by using a fine particle size due to the larger driving force and shorter diffusion distances. However, the advantage can be obscured due to the increased tendency of finer particles to agglomerate. Another problem is the increased role of surface diffusion that can lead to coarsening. Nevertheless, if agglomeration can be prevented, the advantages of fine particles can be realized. Recent studies on sintering nanocrystalline powders reveal many clues in this direction. Hahn et al. [3.160] studied the sintering behavior of nanocrystalline TiO_2. Between 600–900°C, sintering of particles with an average diameter of 14 nm led to densities up to 99% TD, but rapid grain growth took also place. Application of an isostatic

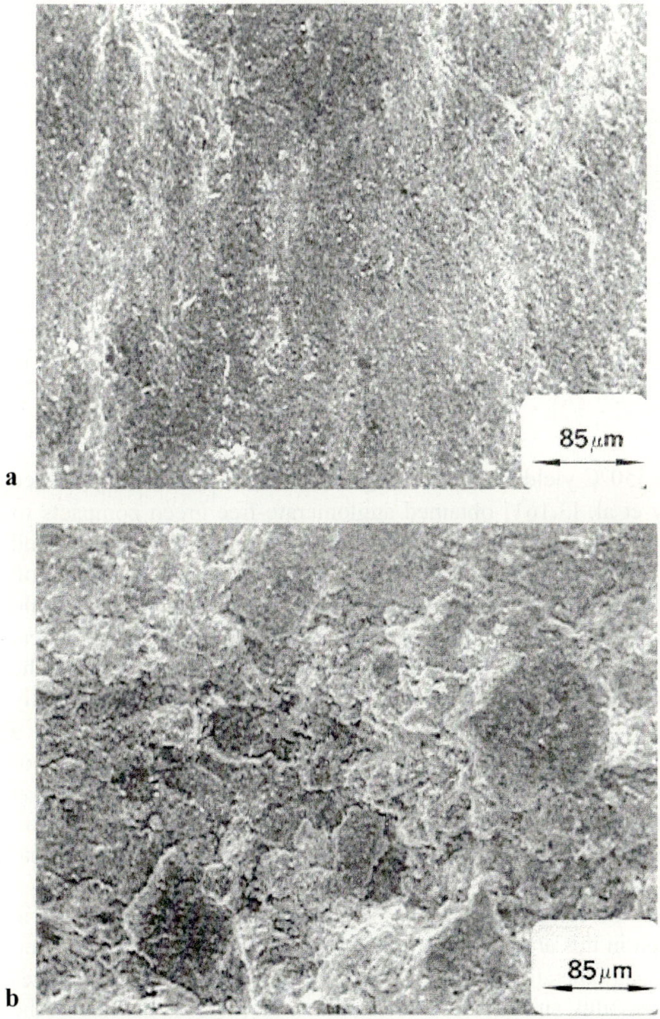

Fig. 3.43. SEM micrographs of unagglomerated **a** and agglomerated **b** Al_2O_3–5vol%ZrO_2 powder mixtures after sintering at 1600°C for 1 h, showing higher porosity and regions of discontinuity in the latter case (From M.Bengisu, Ph.D. Thesis, New Mexico Institute of Mining and Technology Socorro NM, 1992)

Fig. 3.44. SEM micrograph Al_2O_3–5vol%ZrO_2 composite exhibiting local grain growth at the interior of agglomerates (From M.Bengisu, Ph.D. Thesis, New Mexico Institute of Mining and Technology Socorro NM, 1992)

pressure of 1 GPa at 550°C yielded densities higher than 95% TD and suppressed grain growth. Aksay et al. [3.161] obtained agglomerate-free green compacts of submicron Al_2O_3–ZrO_2 powders by colloidal filtration and proper electrostatic and steric stabilization techniques. This resulted in sintered compacts with flexural strengths about 1.6 times that of parts consolidated by isostatic dry pressing and sintered under similar conditions. Zhou et al. [3.162] synthesized CeO_2 powders with an average crystallite size of 10–14 nm by electrochemical synthesis. The sintering behavior of these powders was compared with that of a commercially available powder with a 0.6-μm nominal particle size. During sintering at a constant rate of 10°C/min, the nanocrystalline powder reached a density of 97%TD at 1400°C, while the coarser grained powder reached the same density at 1650°C. Isothermal sintering at 1300°C for 2 h resulted in a grain size of 350 nm and 99.8%TD, whereas microcrystalline powders usually require isothermal sintering temperatures above 1500°C.

Recently, it has been perceived that pore properties can play an important role in sintering. One direction in this area is to relate sintering to pore size distribution and hence the degree of agglomeration [3.163, 3.164]. Another approach emphasizes relations between sinterability and *pore coordination number*, *R*. *R* is defined as the number of touching particles that form a pore in the consolidated state or the number of surrounding grains in the partially sintered state [3.165]. This idea is based on the theory introduced by Kingery and Francois [3.166] that shows that pores will disappear only if their surfaces can develop the right type of curvature. According to this theory, *R* must be smaller than a critical value, i.e., $R<R_c$ to avoid pore or grain boundary separation. R_c increases with the dihedral angle. The ideal green

microstructure would have a periodic arrangement of a monodispersed powder. Local densification is favored in agglomerates where islands of powders with $R<R_c$ are formed. This process increases the size and coordination number of pores between interconnected agglomerates and, thus, renders the remaining pores harder to remove. Increasing the bulk density increases the proportion of pores with $R>R_c$, which may explain why compacts with higher bulk density yield higher final densities for a given sintering schedule.

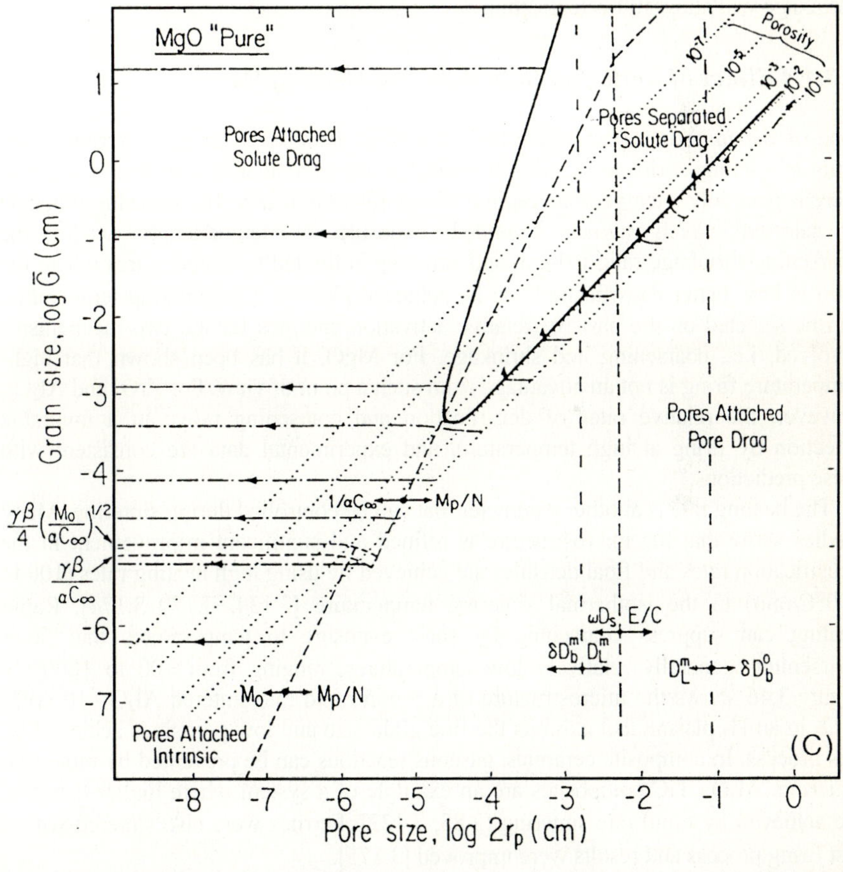

Fig. 3.45. Experimental sintering map for pure MgO (From C.A. Handwerker, R.M. Cannon, and R.L. Coble. *Final-Stage Sintering of MgO.* In: *Advances in Ceramics* Vol.10: *Structure and Properties of MgO and* Al_2O_3 *Ceramics*. W.D.Kingery (ed.) (The American Ceramic Society Columbus OH, 1984), pp 619–643 Copyright 1984 American Ceramic Society Columbus OH, reprinted by permission)

During shrinkage, separation of a boundary from a pore occurs if the boundary velocity exceeds the pore velocity during movement of these features. Once a pore is detached from the boundary and is entrapped in a grain, it can shrink only by the usually much slower mechanism of lattice diffusion compared to grain boundary diffusion [3.167]. The possibility of pore or grain boundary separation can be expressed in a separation diagram in pore and grain size space. Handwerker et al. [3.168] experimentally determined such maps for sintering MgO, as illustrated in Fig. 3.45. It can be determined from this figure that pore or boundary separation becomes likely if the sintering schedule allows coarsening along with grain growth. On the other hand, if pore coarsening can be avoided, even though grain growth occurs, pore shrinkage will allow full densification.

3.3.1.4 Effect of Sintering Schedule and Heating Rate

One of the important parameters that should be monitored during sintering is the grain coarsening/shrinkage ratio. Depending on the particular system that is sintered, there is generally a temperature regime where this ratio is low. If a sintering schedule is selected which avoids extended exposure to temperatures where the coarsening/shrinkage ratio is high, and sintering is limited to temperatures where the ratio is low, better densification can be achieved [3.169]. The sintering temperature can be selected on the basis of relative activation energies for the two mechanisms involved, i.e., coarsening and shrinkage. For MgO, it has been shown that high-temperature firing is not an advantage from such a point of view. For ZrO_2 and Al_2O_3, however, the relative rates of densification and coarsening move in a favorable direction by firing at high temperature, and experimental data are consistent with these predictions.

The heating rate is another parameter that can be controlled during sintering. Many studies show that the microstructure is refined and significant improvement in the densification rates and final densities are achieved by using high heating rates (100 to 500°C/min) to the isothermal sintering temperature [3.134, 3.170–3.174]. Rapid heating can suppress coarsening by short exposure to temperatures that favor coarsening (generally relatively low temperatures, ranging from 700 to 1000°C). Figure 3.46 shows the microstructure of a 5-min rapid rate sintered Al_2O_3–10 vol% ZrO_2 in an H_2 plasma that exhibits the fine grain size and low porosities achieved by this process. In composite ceramics, gaseous reactions can be prevented by rapid-rate sintering. Al_2O_3–TiC composites are an example of a system where higher densities are achieved by rapid rate sintering [3.85, 3.122]. Ferrites were also sintered with a fast firing process and results were improved [3.175].

It was shown that *rapid-rate sintering* and its benefits are uniquely associated with a fine initial particle size [3.176]. Fast firing of Al_2O_3 by soaking only 20 min at 1350°C yielded 99%TD whereas conventional firing using a heating rate of 15 K/min and a soaking time of 20 min resulted in a density of 95.4%TD. It was revealed that a dense outer layer migrates to a porous inner core from the surface to the center [3.177]. It was suggested that this is associated with the formation of transient stresses due to temperature mismatch, which may accelerate

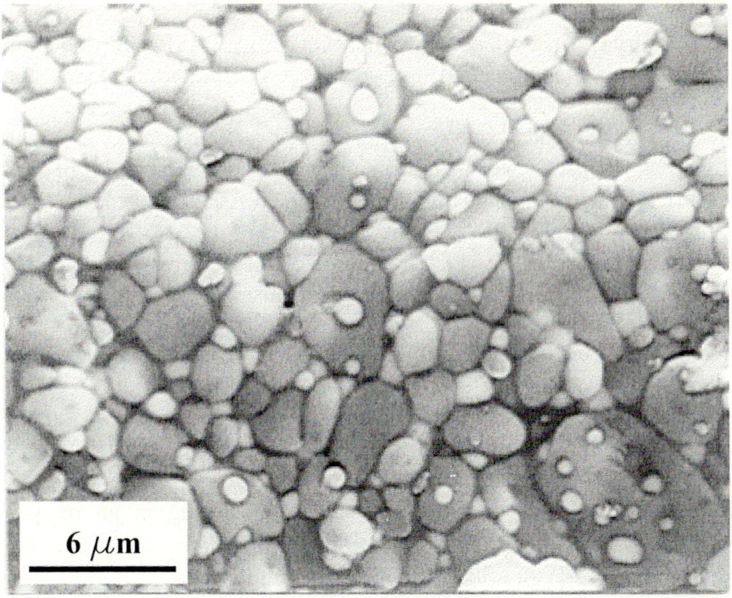

Fig. 3.46. SEM micrograph of plasma-sintered Al_2O_3–10%ZrO_2. (Reprinted with permission from: M. Bengisu and O.T. Inal: In: *Ion Nitriding and Ion Carburizing, Proceedings of ASM's 2nd International Conference (1990)* 44073-0002, Fig. 5, p. 135. ©1990 ASM International® Materials Park OH)

densification. Other important factors that play a role during fast firing are enhanced diffusion due to temperature gradients and material transport by dislocation motion or plastic flow.

3.3.1.5 Effect of Sintering Atmosphere

The atmosphere in which sintering is conducted can greatly affect sintering kinetics and pore size evolution [3.2, 3.178]. Gas trapped inside pores can resist shrinkage unless it is soluble in the crystal. Al_2O_3 doped with MgO can be sintered to full density in an atmosphere of H_2 or O_2, which are soluble in Al_2O_3. The same material cannot be sintered to full density in air, N_2, He, or Ar due to the limited solubilities of N, He, and Ar in Al_2O_3 [3.179]. Impurities that form insoluble gases such as SO_2 or Cl_2 are also sources of reduced shrinkage.

The effect of water pressure on the sintering of MgO has been studied in depth [3.179]. Sintering in dry Ar provided pore shrinkage. A small amount of water vapor in Ar, on the other hand, caused pore growth. The degree of pore growth increased with the partial water vapor pressure. Apparently, water molecules interact with MgO and increase the surface diffusion rate. In contrast, CaO showed little change in pore size with water vapor pressure, and it was found that densification is greatly accelerated by the presence of water vapor. Tin oxide samples do not densify during sintering in dry Ar or dry O_2. Instead, pore growth occurs. Surface diffusion and grain boundary migration or evaporation/condensation are considered mechanisms that cause grain and pore growth.

During sintering of non oxide ceramics such as SiC at temperatures at about 2000°C, a nitrogen atmosphere is necessary to suppress sublimation of SiC or B_4C (added as a dopant). High-pressure sintering (>5 MPa) in an Ar atmosphere does not yield densities higher than 90% since Ar is also insoluble in SiC [3.180]. Instead of high pressures, vacuum can be used, which in fact yields the best densities, but also high sublimation rates. A logical compromise is to use an inert gas at atmospheric pressure.

Reducing atmospheres encountered in graphite heating element furnaces can decrease sinterability due to the reaction of oxide ceramics with C-rich gases. In ceramic systems where sintering is controlled by oxygen ion diffusion, a reduction of oxygen partial pressure in the sintering atmosphere would mean increased amounts of oxygen vacancy, thus promoting densification [3.181]. Oxygen deficiency has reportedly turned ZrO_2 to a gray or black color, and annealing in an oxygen atmosphere restored the white color [3.132,3.182]. These studies show that the sintering atmosphere should be controlled to achieve optimum results.

3.3.1.6 Effect of Defect-Generating Treatments

A high density of defects such as vacancies, dislocations, twins, and stacking faults can be generated in ceramic materials by special processes. Among these processes, ion implantation [3.183,3.184], self-propagating synthesis [3.185,3.186], extended milling [3.187], and shock treatment [3.188–3.189] can be mentioned.

Studies of sintering combustion-synthesized TiC and TiB_2 have not revealed any enhanced sinterability over similar powders that are available commercially, although activated sintering was expected due to high defect concentrations [3.185, 3.186].

Significant effort has been dedicated to studies of the effect of *shock treatment* on sintering kinetics in ceramic powders [3.188–3.193]. The passage of a shock wave through the material introduces point defects, dislocations, higher order defects, and thus residual strain [3.188,3.193]. In addition, particle comminution, exposure of fresh surfaces, and phase changes are potential phenomena associated with shock processing. It is natural to assume that ceramic materials that are subjected to shock waves might be more responsive to sintering than unshocked counterparts, due to one or more of the previously mentioned features. Yet, this is not always true. Some ceramics that reportedly exhibit significant activation due to dynamic loading include Al_2O_3 [3.188], AlN [3.194], Si_3N_4 [3.189], ZrO_2 [3.190], LaB_6 [3.190], and TiC [3.191]. Investigation of different ceramics reveals different levels of activation. For example, it was reported that ZrO_2 powders show a linear increase in sintered density with increasing shock pressure and sintered densities of LaB_6 and AlN densities go through maxima and then decrease, whereas those of Y_2O_3 decrease with increasing shock pressures [3.190].

Bergmann and Barrington [3.188] showed that shocked Al_2O_3 powders densify much more rapidly than unshocked powders, and they attributed this difference to the introduction of a large number of defects into the material. Prummer and Ziegler [3.195] examined the effect of shock loading on the microstrain and substructure of shocked alumina powders. They found that lattice distortion and dislocation densities in shock-treated fine powders are twice as high, compared to 157 hours of milling. In coarse powders, milling for 157 hours produced more dislocations and a larger reduction in particle size than shock treatment.

Yust and Harris [3.192] observed dislocations and deformation twins in explosively loaded alumina by transmission electron microscopy (TEM). The significantly larger appearance of the compacted mass of single crystals, compared to the original crystals, was taken as evidence of recrystallization and grain growth.

Morosin et al. [3.196] studied the sintering rates of alumina powder after high-pressure (22 GPa) shock treatment. They observed a significant increase in densification in the early stages of sintering but a rapid decrease as the process continued. The net result was a lower final density in the shocked powder than in the unshocked powder. It was concluded that strong aggregates produced during shock processing inhibit densification of the whole body. In contrast, local densification in the unshocked powders is avoided since such strong aggregates do not exist. Grinding of both shocked and unshocked powders to break up agglomerates resulted in identical densification profiles, implying that shock treatment affected sintering only by generating dense aggregates. Similar results were found by Hare et al. [3.197] and Bengisu [3.159]. Hare et al. [3.197] found that hard agglomerates responsible for the reduction in sinterability are produced by shock pressures above 5 GPa. Bengisu observed that the sinterability of Al_2O_3 and Al_2O_3–ZrO_2 powders decreases as the shock pressure increases. Figures 3.47 and 3.48 show this effect through dilatometer (sample shrinkage–time) plots of sintering. As seen from these figures, the amount of shrinkage during sintering decreases as the shock pressure is increased. In other words, the attainable density decreases with increased shock pressure because of hard agglomerates produced during shock treatment. Therefore, to optimize shock-activated sintering, it seems essential to control the shock pressure so that powders benefit only from the introduction of lattice strain and defects, and localized bonding or compaction is avoided. It is also important to increase the initial compact density and thus reduce frictional heating and the risk of interparticle bonding.

3.3.2 Liquid-Phase Sintering

Liquid-phase sintering was first analyzed by Kingery [3.198] who proposed a model that involves three stages:

1. particle rearrangement,
2. solution–precipitation, and
3. coalescence (Fig. 3.49).

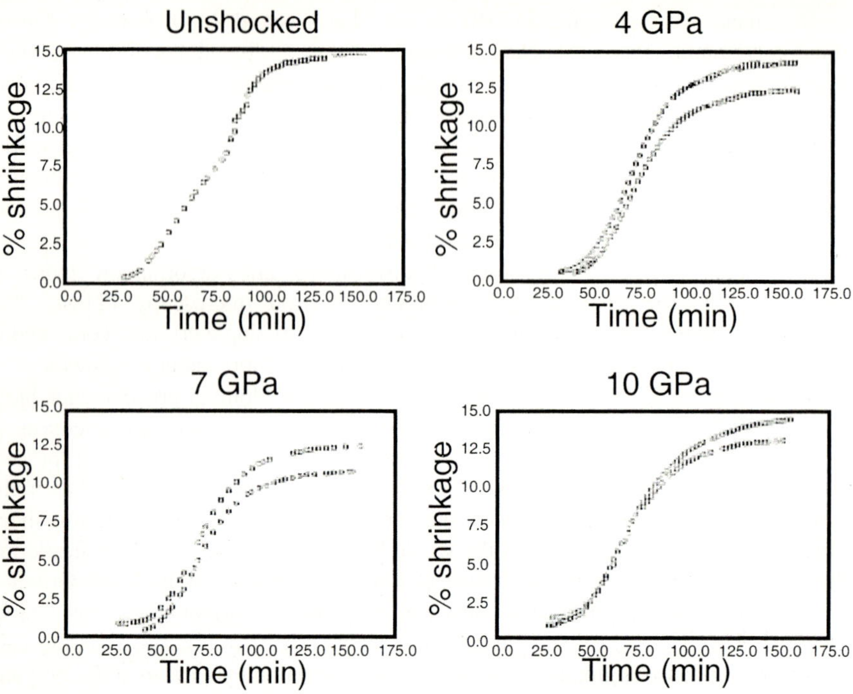

Fig. 3.47. Dilatometer plots for sintering of untreated and shock-treated alumina (From M.Bengisu, Ph.D. Thesis, New Mexico Institute of Mining and Technology Socorro NM, 1992)

According to this theory, solid particles *rearrange* themselves during the initial stages of sintering due to capillary forces. This can take place by particles sliding over one another. The rate of densification corresponds approximately to viscous flow and follows the relation [3.198],

$$\Delta L / L_0 = (1/3)\,(\Delta V / V_0) \approx t^{1+y}\,, \tag{3.51}$$

where $\Delta L/L_0$ and $\Delta V/V_0$ are linear and volumetric shrinkage, respectively, t is time, and the exponent $1+y$ is an experimental constant larger than unity. Processes that can occur during rearrangement are melting, wetting, spreading, and liquid redistribution [3.198]. Rearrangement can be responsible for full densification at a liquid content of 30–35 vol% in some ceramics [3.200].

Once bridges are built up and further rearrangement is not possible, densely packed solid particles separated by thin liquid films carry the major part of the compressive stress at the contact points. The solubility at contact points is larger than

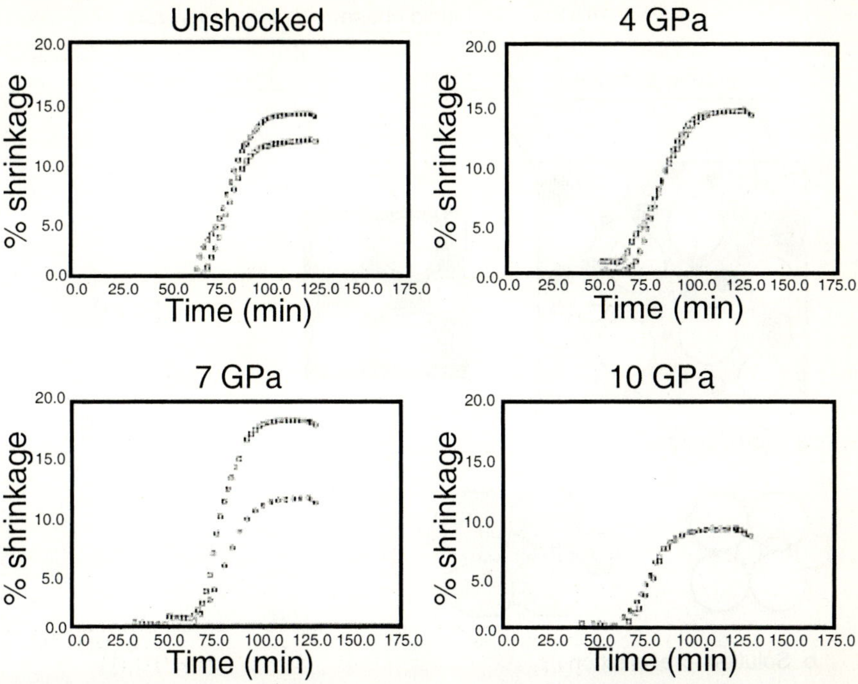

Fig. 3.48. Dilatometer plots for sintering untreated and shock-treated Al_2O_3–10 vol%ZrO_2.(From M.Bengisu, Ph.D. Thesis, New Mexico Institute of Mining and Technology Socorro NM, 1992)

the solubility of other solid surfaces; this results in mass transfer away from contact points and allows further shrinkage. This is called the *solution–precipitation* process.

For spherical particles, Kingery proposed the following shrinkage equation relevant to this stage [3.198]:

$$\frac{\Delta L}{L_0} = \frac{\Delta V}{3V_0} = \left(\frac{6k_2\delta D c_0 \gamma_{LV} V_0}{k_1 RT}\right)^{1/3} r^{-4/3} t^{1/3} , \qquad (3.52)$$

where r is the initial particle radius, k_1 and k_2 are proportionality constants, δ is the liquid film thickness, D is diffusivity, c_o is the initial element concentration (in the diffusion flux equation), γ_{LV} is the liquid–vapor surface energy, V_o is the initial volume of the body, R is the gas constant, and T is the absolute sintering temperature. As an example, for the Fe–Cu system at 1150°C, taking $k_1 = 1/2$, $k_2 = 1$, $\delta = 10^{-7}$ cm, $D = 10^{-5}$ cm^2/s, $c_o = 0.04$ cc/cc, and $\gamma_{LV} = 1280$ ergs/cm^2 as approximations,

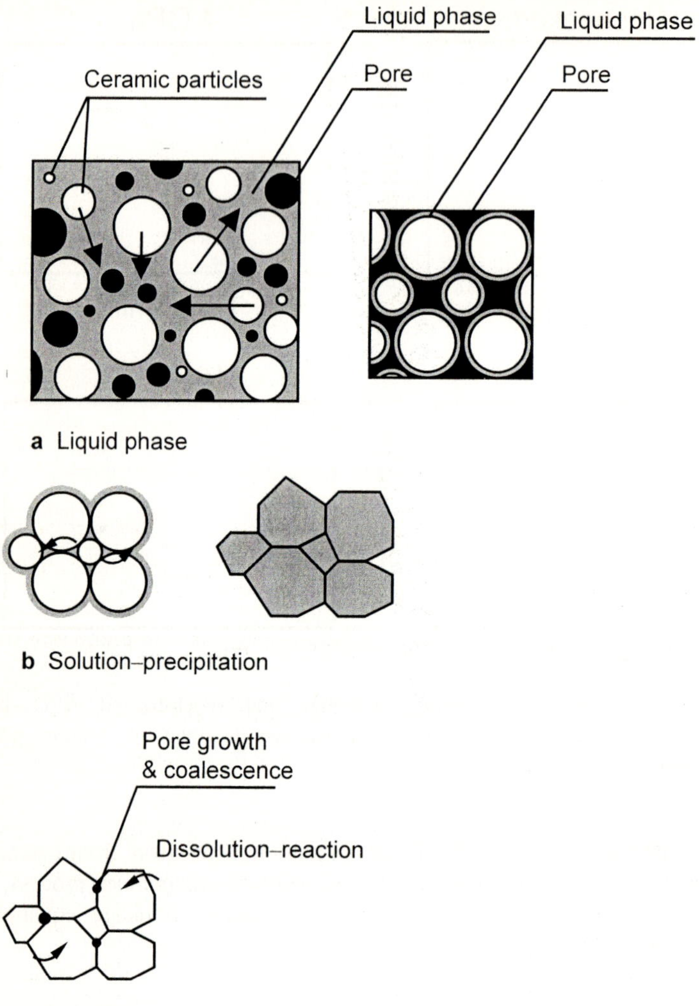

Fig. 3.49 a–c. Stages of liquid-phase sintering

$$\frac{\Delta L}{L_0} = \frac{\Delta V}{3V_0} = (0.45 \times 10^{-6} cm^4 / s)^{1/3} r^{-4/3} t^{1/3} . \tag{3.53}$$

The final-stage of liquid phase sintering is *coalescence*, where permanent solid–solid contacts between particles form and liquid-phase sintering essentially stops. Processes that can take place during this stage are pore growth and coalescence, dissolution of liquid into solid, phase transformations, and reaction between liquid and solid phases [3.199].

Studies indicate that predictions made by Kingery's model are in good agreement with experimental observations for MgO–kaolin, WC–Co, and TiC–Ni systems [3.201]. On the other hand, agreement was not satisfactory for Al_2O_3 in the presence of a liquid phase [3.202]. Huppmann et al. [3.203] proposed that particle rearrangement is the primary mechanism of densification in liquid-phase sintering of alumina. Huppmann and Petzow's study [3.204] indicated that liquid-phase sintering of many systems consists of these stages:

1. particle rearrangement,
2. solution–precipitation by the simultaneous or overlapping occurrence of (a) grain boundary attack by the liquid phase and particle disintegration, (b) directional grain growth, and (c) normal grain growth, and
3. coalescence (Figs. 3.50–3.52).

One important prerequisite of liquid-phase sintering is wetting solid particles by a liquid at the fabrication temperature. A small contact angle favors wetting. The contact angle ψ, formed as a result of the interfacial energy balance at the solid–liquid–vapor intersection [3.205],

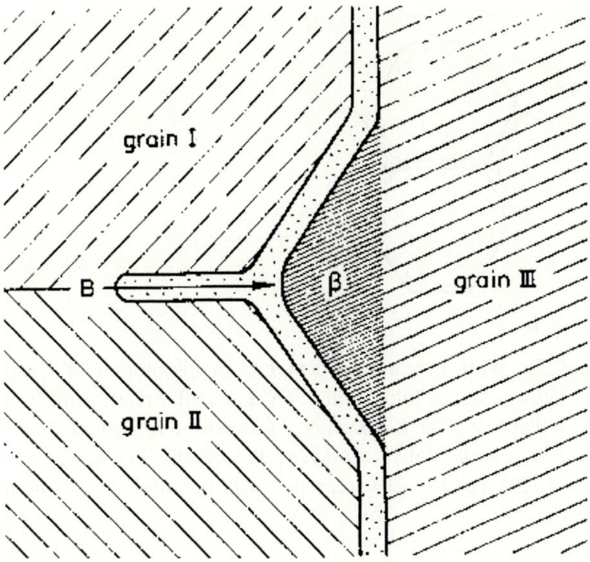

Fig. 3.50. Schematic representation of grain boundary attack by a liquid phase. (From: W.J. Huppmann and G. Petzow: *The Elementary Mechanisms of Liquid Phase Sintering.* In: *Materials Science Research,* Vol.13: *Sintering Processes.* G.C. Kuczynski (ed.) (Plenum Press New York, 1980), pp. 189–201, reprinted with permission of Plenum Publishing Corp. New York)

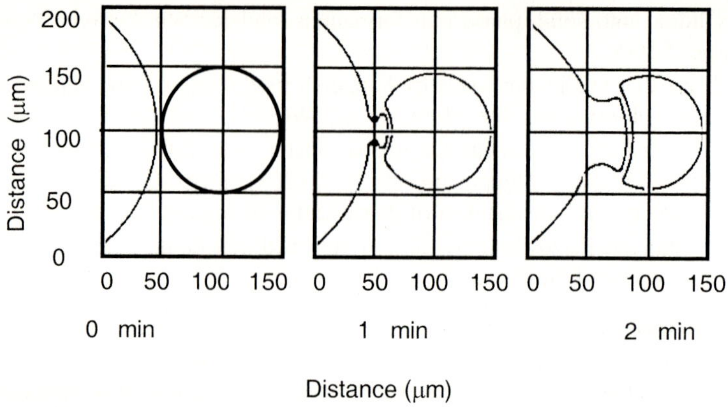

Fig. 3.51. Computer simulation of directional grain growth in sintering W–Ni at 1640°C. $D = 10^{-5}$ cm²/s, shrinking grain 100% W, growth region W–0.15 wt% Ni alloy. The initial distance between the two grains is 5 μm (From: W.J. Huppmann and G. Petzow: *The Elementary Mechanisms of Liquid Phase Sintering.* In: *Materials Science Research,* Vol.13: *Sintering Processes.* G.C. Kuczynski (ed.) (Plenum Press New York, 1980), pp. 189–201, reprinted with permission of Plenum Publishing Corp. New York)

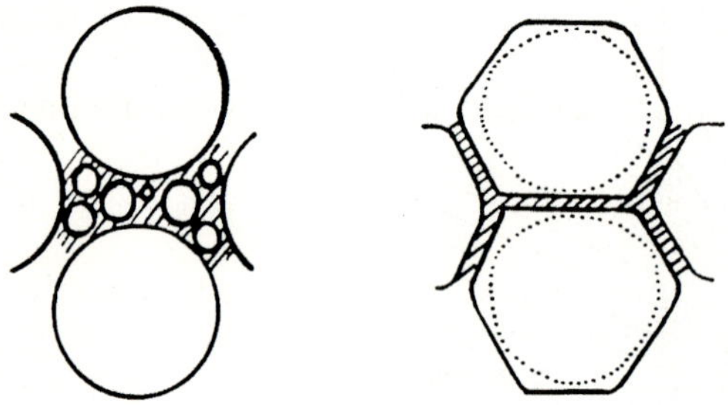

a **b**

Fig. 3.52. Model for shrinkage by particle rearrangement due to dissolution of small particles and reprecipitation on large particles. Initial condition **a** and after rearragement **b** (From: W.J. Huppmann and G. Petzow: *The Elementary Mechanisms of Liquid Phase Sintering.* In: *Materials Science Research,* Vol.13: *Sintering Processes.* G.C. Kuczynski (ed.) (Plenum Press New York, 1980), pp. 189–201, reprinted with permission of Plenum Publishing Corp. New York)

$$\psi = \cos^{-1}[(\gamma_{SV} - \gamma_{SL}) / \gamma_{LV}] \ , \tag{3.54}$$

must be less than 60° for high product densities [3.205]. In Eq. (3.54), γ_{SV}, γ_{SL}, and γ_{LV} represent surface energies between solid–vapor, solid–liquid, and liquid–vapor phases, respectively. The liquid between particles can cause capillary pressures in the range of several MPa's. Smaller particles result in higher capillary pressures and therefore the driving force for densification is higher. This is reflected in Eq. (3.53) which predicts that densification in the solution–precipitation stage is inversely proportional to the four-thirds power of particle size.

As in solid-state sintering, agglomerates may limit densification, especially due to liquid penetration into them. Regions of solidified liquid phase may appear in a sintered microstructure due to inhomogeneous liquid distribution and liquid penetration into pores [3.206].

Intentionally or not, many important commercial sintering operations involve liquid-phase sintering. Typical commercial solid-state sintering practices involve liquid-phase sintering, to some degree, due to the presence of impurities [3.152]. On the other hand, the cheapest and sometimes even the sole way to sinter covalent ceramics is liquid-phase sintering. For example, pressureless sintering of Si_3N_4 is generally performed with liquid-forming sintering aids due to the low solid-state diffusion rate of the highly covalent ceramic. SiO_2 on the surface of Si_3N_4 contributes to the formation of a liquid phase (Fig. 3.42). Y_2O_3, Al_2O_3, and MgO are typically added in the range of 5–10 wt% to increase the viscosity [3.138, 3.207]. It is also possible to crystallize the intergranular glassy phase, allowing stabilization of the mechanical properties up to around 1400°C [3.207].

Additives such as Al_2O_3 and Y_2O_3 are used for liquid-phase sintering of SiC. During sintering in this system, predominantly yttrium–aluminum garnet is formed. In long enough sintering times, the liquid phase can be concentrated at triple grain junctions which leads to clean grain boundaries [3.207]. A finer grain size is usually attained by liquid-phase sintering of SiC compared to that of solid-state sintering. The strength of liquid-phase sintered SiC is about twice that of SiC sintered with B and C (Fig. 3.53). This is related to the higher fracture toughness and a change of the fracture mode from transgranular to intergranular fracture [3.208]. Liquid-phase sintering of SiC is typically carried out at 1900°C in Ar and 2100°C in N_2. Thermodynamic and experimental studies revealed that in liquid-phase sintering of SiC with yttria and alumina additives under Ar at 1900°C, silica on the surface of SiC can be removed by adding small amounts of C. This results in second phases of Y–Al–O and Y–Al–O–C, free of detectable silicon. C additions also result in lower weight loss since the latter will be favored among the following two reactions [3.208]:

$$2SiO_2 + SiC \rightarrow 3SiO + CO \tag{3.55}$$

$$SiO_2 + 3C \rightarrow SiC + 2CO \tag{3.56}$$

thus fewer gaseous reactions will occur. Sintering in N_2 at 2100°C results in forming of AlN [3.208],

$$3SiC + 2Al_2O_3 + N_2 \rightarrow 4AlN + 3SiO + 3CO \tag{3.57}$$

Higher weight loss occurs in this case, consequently sintering in Ar is favored over sintering in N_2. AlN dissolves in the SiC grains during sintering and precipitates at the grain boundaries upon cooling.

AlN is another important covalent ceramic, which has been used in recent years for its thermal and electronic properties. AlN can be sintered at 1750–1900°C by using liquid-phase forming additives, including CaO, EuO_2, or Y_2O_3. These oxides react with Al_2O_3 or aluminum hydroxide present at the surface of AlN and form a glassy phase. The microstructure of AlN that plays a key role in electronic and thermal properties can be controlled by carefully selecting sintering parameters. The second phase can be left at grain boundaries, triple points, or eliminated through special treatments. An important advantage of using sintering aids in AlN is the removal of oxygen from grain interiors that leads to improved thermal conductivity [3.207].

3.3.3 Hot-pressing

Hot-pressing involves simultaneous application of pressure and heat to a powder system. Application of pressure at the consolidation temperature accelerates densification kinetics by increasing the contact stress between particles, by

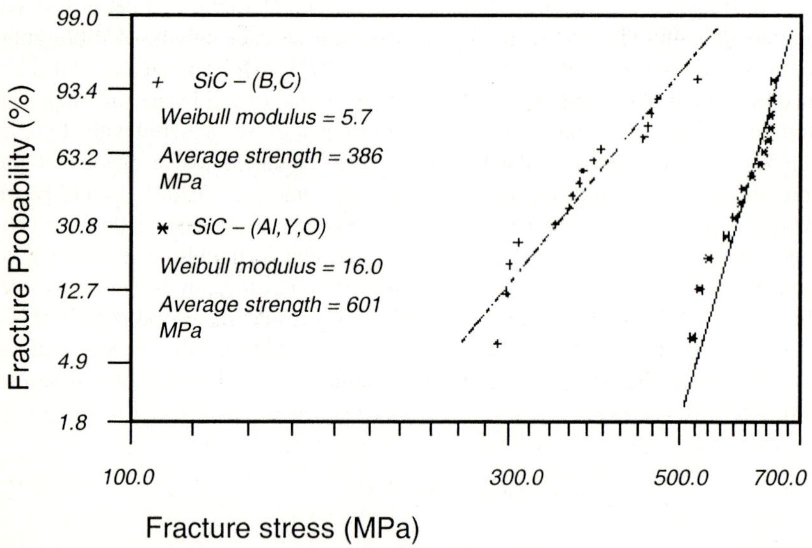

Fig. 3.53. Strength and Weibull modulus of solid state (+) and liquid-phase sintered (x) SiC materials (Adapted from W.D.G. Boecker, R. Hamminger, J. Heinrich, J. Huber, and A. Roosen: *Covalent High-Performance Ceramics.* Adv. Mater. **4 [3]**, 169–178 (1992) used with permission of Wiley-VCH)

rearranging particles to improve packing, and by stress-aided diffusion. Sintering temperatures and times are usually decreased significantly by hot-pressing. Porosity can be totally eliminated in most ceramics. Mechanical properties achieved by hot-pressing are superior in many cases to pressureless-sintered bodies due to reduced porosity and finer grain size. Good densification and a fine microstructure depicted in Fig. 3.54 can be achieved by hot-pressing even in composite ceramics.

The principal mechanisms that control hot-pressing may be summarized as follows:

1. An increase in compact density occurs with applied pressure. Depending on that pressure, densities up to 70% TD can be reached at this stage.
2. Sliding and rearrangement of particles occur with possible particle fragmentation.
3. Plastic flow at particle contacts takes place. It is suggested that this step occurs within minutes. The possible contribution of plastic flow to densification is limited to less than 84% TD in ceramics possibly due to strain hardening, which occurs, for example, in alumina [3.209].

Various other mechanisms have been proposed for densification during the intermediate and final stages of hot-pressing. The most significant are the following:

1. *Nabarro-Herring creep* [3.210], where deformation occurs by vacancy motion through the lattice along a stress-induced concentration gradient from grain boundaries in tension to boundaries in compression. The deformation rate is given by [3.211,3.212]

$$\dot{\varepsilon} = \frac{\dot{\rho}}{\rho} = \frac{\lambda \sigma_{\mathrm{a}} \Omega D_{\mathrm{L}}}{kTd^2}. \tag{3.58}$$

In this equation, $\dot{\rho}$ and ρ represent densification rate and density, respectively, λ is a constant dependent on the shape of the grains (for practical purposes $\lambda = 13.3$), σ_a is the applied stress (erg/cm^3), Ω is the volume of a vacancy (cm^3), k is Boltzmann's constant, D_L is the lattice diffusion coefficient of the slowest moving species (cm^2/s), T is the absolute temperature, and d is the grain size.

2. *Coble creep*, where stress-aided diffusion occurs primarily along grain boundaries. The strain rate may be expressed by [3.212,3.213]

$$\dot{\varepsilon} = \frac{\dot{\rho}}{\rho} = \frac{\lambda \delta D_{\mathrm{GB}} \Omega \sigma_{\mathrm{a}}}{kTd^3}, \tag{3.59}$$

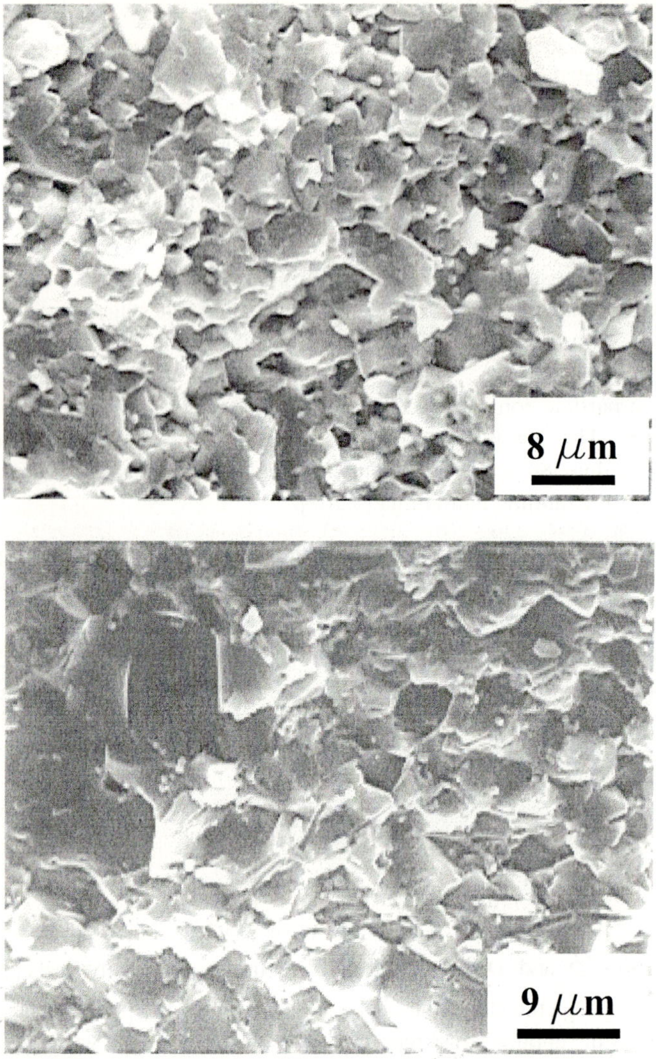

a b

Fig. 3.54. Examples of microstructure achieved by hot-pressing: Al_2O_3-10 vol%ZrO_2 hot pressed at 1500°C, 30 min, 30 MPa **a** and Al_2O_3-10 vol%SiC_w hot pressed at 1800°C, 30 min, 30 MPa **b** (From M.Bengisu, Ph.D. Thesis, New Mexico Institute of Mining and Technology Socorro NM, 1992)

where δ is the grain boundary width, D_{GB} is the grain boundary diffusion coefficient, and $\lambda = 47.5$ for practical purposes.

3. *Dislocation creep*, where deformation occurs by climb and glide of dislocations. A commonly accepted semiempirical formulation for creep rate in this case is given by [3.214]

$$\dot{\varepsilon} = \frac{\dot{\rho}}{\rho} = \frac{KD_c \mu \vec{b}}{kT} \times \left(\frac{\sigma_a}{\mu} \right)^n , \tag{3.60}$$

where K is a materials constant called the Dorn parameter, n is the stress exponent, μ is the shear modulus, $\vec{b}$ is Burger's vector, and D_c is the diffusion coefficient for dislocation creep.

Densification during hot-pressing generally occurs by more than a single mechanism, as in pressureless sintering. Various densification mechanisms can be combined graphically by using deformation or hot-pressing maps developed by Ashby and co-workers [3.215, 3.216]. Such maps can be used to predict the hot- pressing behavior of a given material under any combination of static pressure and temperature. An example is shown in Fig. 3.55. At a given temperature and pressure, one mechanism for densification dominates. The various mechanisms differ in their dependence on internal parameters such as particle size, green density, and purity. Thus, different hot-pressing maps may be developed even for one type of material. The construction of hot-pressing maps is based on the densification rate equations (3.58–3.60). The field boundaries are constructed by equating the rate equations in pairs and solving for ρ as a function of T. In other words, a field boundary is a curve along which two mechanisms contribute equally to densification (Fig. 3.55). Hot-pressing maps are constructed in the form of relative density versus T/T_m (where T_m is the melting point) at constant pressure, relative density versus pressure at constant temperature, or three dimensionally where relative density is plotted against T and P. Apart from their predictive capabilities, such maps can also be helpful in explaining hot-pressing data by showing the limits of extrapolation for a certain mechanism. This may help to interpret experimental results [3.214]. The maps can also be helpful in designing new experiments.

Figure 3.56 schematically shows a *uniaxial hot press*, which is widely used for industrial applications. Inductive heating is advantageous over resistive heating because insulation of the hot-pressing zone is much easier with the former.

Pressure is usually applied by a hydraulic press. Hot-pressing is mostly performed in the pressure range of 5 to 50 MPa. Commonly used press and die materials are graphite, refractory metals, superalloys, Al_2O_3, and SiC. Graphite is frequently used for hot-pressing alumina-based compositions, nitrides, and carbides. Graphite dies are usually coated with BN or Al_2O_3 to prevent reaction with the powders of interest. Vacuum or an inert atmosphere is necessary when graphite is used for hot-pressing; otherwise graphite will oxidize.

Alternatively, Al_2O_3 can be used for hot-pressing in oxidizing atmospheres up to 1200°C.

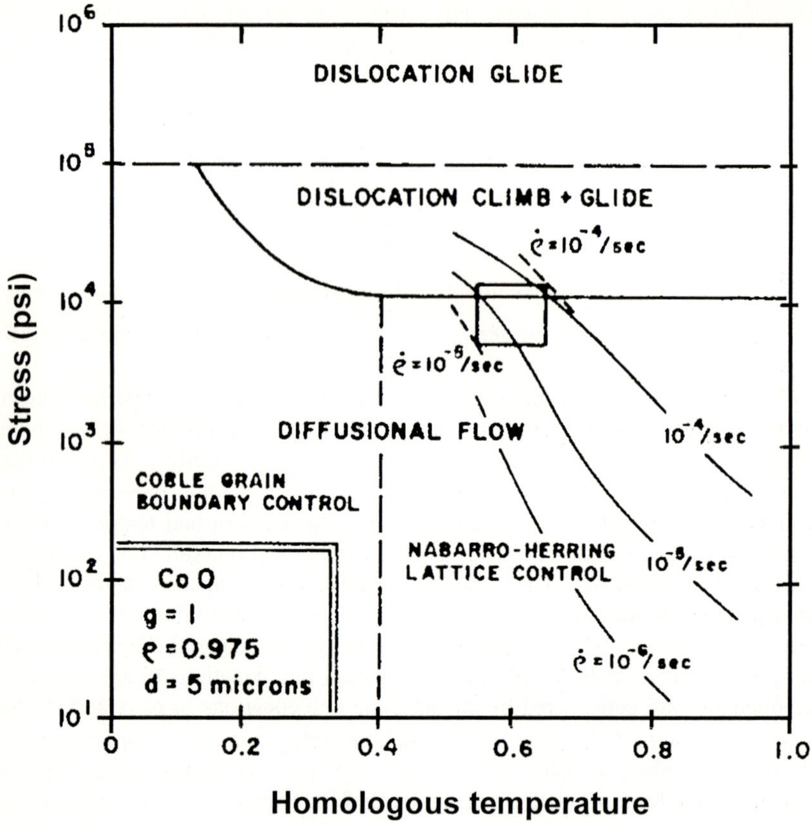

Fig. 3.55. Hot-pressing map for CoO, ρ = 0.975, showing contours of constant densification rate (From: M.R. Notis, R.H. Smoak, and V. Krishnamachari: *Interpretation of Hot-pressing Kinetics by Densification Mapping Techniques*. In: *Materials Science Research* Vol.10, *Sintering and Catalysis*. G.C. Kuczynski (ed.) (Plenum Press New York , 1975), pp 493–507, reprinted with permission of Plenum Publishing Corp. New York)

Although hot-pressing has important advantages over pressureless sintering, it also has some drawbacks. One of these is the shape and size limitation of pressed parts. Only simple shapes such as disks, rods, or tubes can be produced by hot-pressing due to the uniaxial nature of pressure. The aspect ratio is also limited by the die size, problems associated with part removal, and pressure gradients that develop along the length of the part that can result in density and property variations inside the product.

Density gradients may also occur if there are temperature gradients in the hot-pressing zone. Since a loose powder has a lower thermal conductivity than that of a precompacted part, the regions in the vicinity of the dies may heat and consolidate faster, compared to the interior. Precompaction of powders may resolve such a problem. Some of the shape and size limitations can be overcome

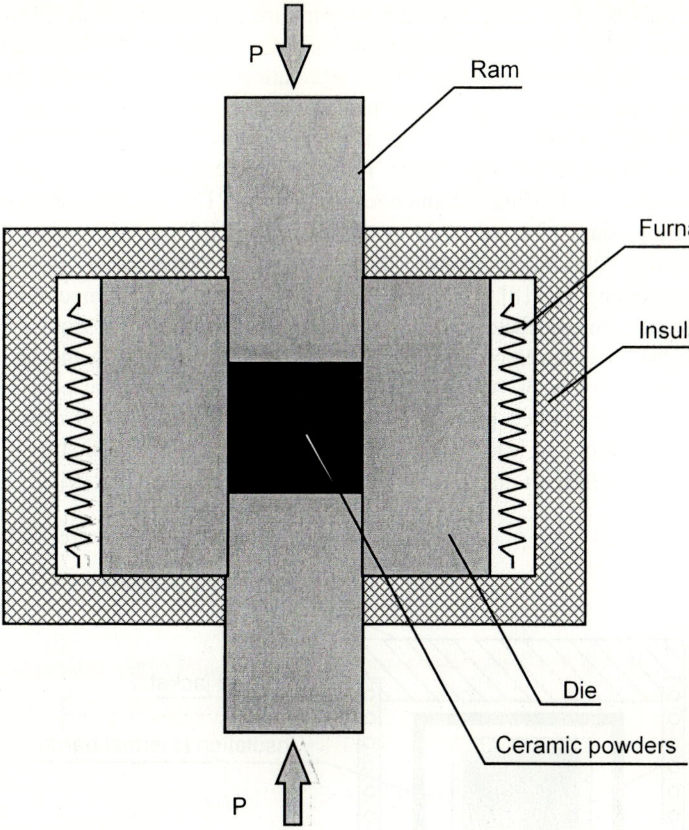

Fig. 3.56. Schematic view of uniaxial hot-pressing

by using hydrostatic instead of uniaxial pressure. This type of pressure is achieved in hot isostatic pressing, described in the next section. Other important drawbacks of hot-pressing are high initial costs, high production costs, and the difficulty to adapt it to continuous production.

3.4 Modern Production Methods

3.4.1 Hot Isostatic Pressing

Some of the limitations of hot-pressing such as restricted shape and aspect ratio can be eliminated by using hot isostatic pressing (HIP). In this method, pressure is applied isostatically by an autoclave via inert gases. Oxygen is used in some cases for oxide ceramics and nitrogen to prevent decomposition of nitrides at high temperatures. A high-temperature furnace enclosed in the autoclave provides a uniform hot zone temperature up to 3000°C (Fig. 3.57). Elevated temperatures are usually obtained by graphite or refractory metal resistive heating elements. Pressures up to 300 MPa can be reached in HIP. To achieve densification of a ceramic preform, typically the preform is evacuated and sealed in a gas-impermeable envelope made of Ta or glass. Ta reportedly reacts with carbide ceramics such as SiC and form TaC during HIP'ing. In such cases, glass encapsulation is preferred. Partially sintered parts with no open porosity are also suitable for HIP'ing without encapsulation [3.217]. Complex shaped green bodies can be partially sintered or encapsulated before HIP'ing. HIP makes it possible to lower the consolidation temperature and obtain better grain growth control, as in hot-pressing [3.218]. On the other hand, pressure and temperature gradients may become limiting factors for large parts. Densification starts from the exterior, and a dense shell develops. Pressure can be transmitted only to the less dense interior if the shell creeps inward. The pressure transmitted to the core is usually different from the applied pressure. This can result in gradients of elastic strain after pressure release, that lead to residual stresses. Similarly, residual stresses develop during cooling of the part due to differential thermal contraction [3.219].

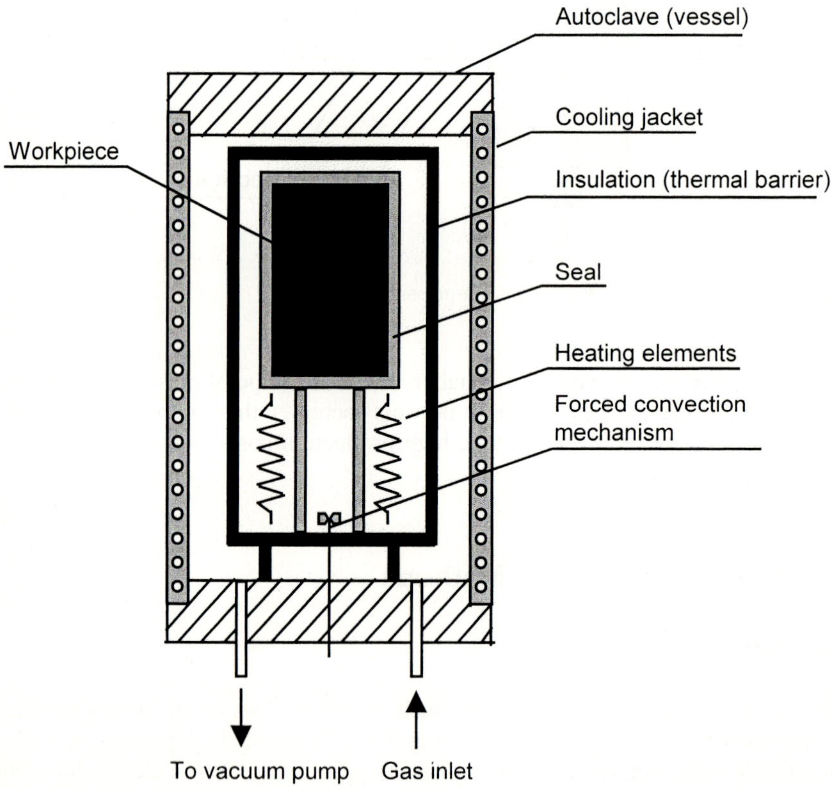

Fig. 3.57. Schematic view of a hot isostatic press

The high external pressure in HIP is an efficient means of preventing decomposition of some technically important materials such as SiC and Si_3N_4 at the densification temperature. Another advantage of this process is the elimination of internal cracks that develop during green processing, particularly dry pressing and binder burnout. Elimination of preferred orientation makes HIP a suitable method for producing whisker-toughened ceramics (WTCs). Unlike hot pressed WTCs, where whiskers align perpendicular to the hot-pressing axis, resulting in anisotropic properties, whiskers are randomly aligned in HIP'ed WTCs and providie uniform microstructures and isotropic properties [3.220].

HIP has become a widely accepted industrial process in the last 30 years, especially for producing complex shapes with high accuracy and near-net shape capability. The yield of the HIP process is also economically feasible. For example, it has been reported that Si_3N_4 parts with a maximum length of 60 mm have been produced by HIP at a rate of about 700 pcs/wk [3.221]. Adlerborn et al. [3.222] reported that more than 26,000 components were processed in only one charge. Uematsu et al. [3.223] showed that very fine grained Al_2O_3 (d < 0.86 μm) can be fully densified at 1300°C in 1 h, but the same is not possible for coarser grained Al_2O_3, even at higher pressures. Helle et al. [3.224] studied the HIP behavior of Al_2O_3 using HIP diagrams, which are similar to hot-pressing maps. According to their study, boundary diffusion is the dominant densification mechanism at pressures of 100 to 200 MPa. The diagrams show that temperatures of 1500°C or higher are required to achieve full density.

3.4.2 Dynamic Consolidation

Dynamic consolidation processes use energetic materials, i.e. explosives or other energy sources, to consolidate powders at high velocities. Figure 3.58 is a photograph of a high-pressure compaction setup using only plastic explosives (Detasheet®). The container, usually made from mild steel or copper, is filled with powders and surrounded by an explosive (Fig. 3.59). Figure 3.60a shows a steel container filled with ceramic powders before and after shock compaction.

Figure 3.60b shows examples of recovered and heat-treated compacts of Al_2O_3–SiC_w. Detonation of the explosive generates a shock wave that travels downward. This shock wave rearranges particles to denser packing and brings them into contact at high pressures to bond at contact surfaces, provided that the energy is sufficient. Some of the possible mechanisms that give rise to bonding and localized deformation at surfaces are jet trapping (mostly in the case of metal powders), frictional sliding, and point deformation [3.225].

In general, ceramic powders can be densified to full density by dynamic consolidation methods. However, unlike metal powder compaction, cracking is almost always associated with dynamic compaction of ceramic powders [3.226–3.229]. For small, cylindrical samples, densities close to theoretical have been achieved with several ceramic materials, including Al_2O_3, ZrO_2, Si_3N_4, and B_4C [3.230]. Problems remain in extrapolating to larger sizes and eliminating micro and

Fig. 3.58. High-pressure explosive compaction setup (From M. Bengisu, Ph.D. Thesis, New Mexico Institute of Mining and Technology Socorro NM, 1992)

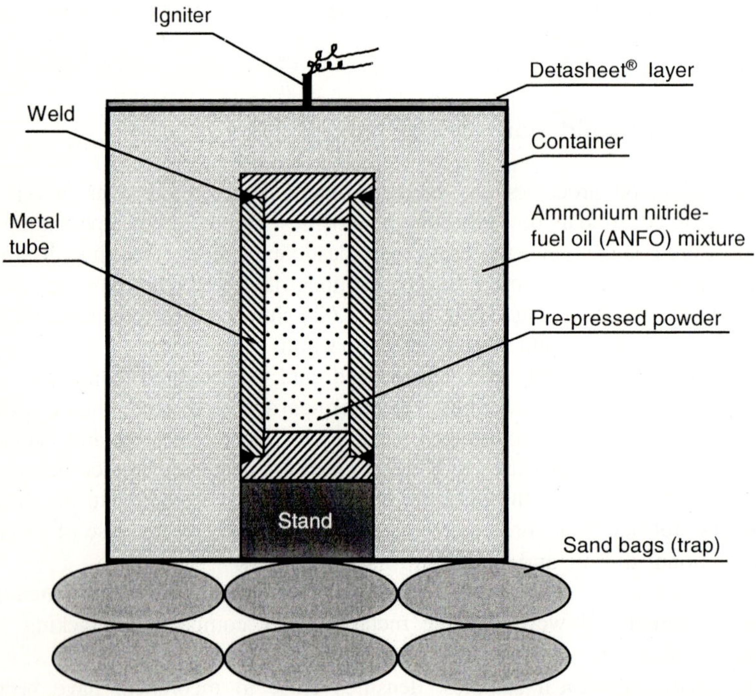

Fig. 3.59. Explosive compaction setup (From M. Bengisu, Ph.D. Thesis, New Mexico Institute of Mining and Technology Socorro NM, 1992)

a

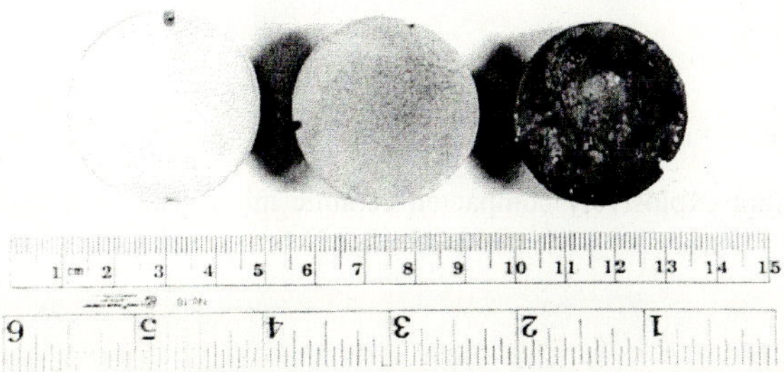

b

Fig. 3.60. Steel can with ceramic powder inside, before and after shock compaction **a** and example of explosively compacted composite preform **b** (From M. Bengisu, Ph.D. Thesis, New Mexico Institute of Mining and Technology Socorro NM, 1992)

macrocracks in dynamically compacted ceramics. Stress development during compression and decompression stages can be very disruptive, and the source and development of such stresses are poorly understood. Until the processes that cause cracking are well understood and eliminated, dynamic compaction cannot be a reliable fabricating method for advanced ceramic materials. There are, however, some exceptions where crack-free ceramic compacts can be obtained by shock compaction. For example, at relatively low shock pressures, crack-free compacts have been fabricated, although a compromise was made in the final density [3.231]. Ceramics like AlN, which exhibit a certain degree of plasticity at high pressures [3.226], can be shock compacted without crack formation. Al_2O_3–ZrO_2 composites were compacted with a two-stage gas gun. Pressures below 12.6 GPa produced compacted pellets with no visible cracks but low relative density (up to 88% TD). Higher pressures increased the density but fragmented the compacts [3.232]. Heating of the ceramics before shock compaction can also prevent shock-induced cracks in some cases [3.233, 3.234]. Such a process is called hot explosive compaction. Figure 3.61 shows the microstructure of a hot explosively compacted Al_2O_3–ZrO_2 composite that reveals a crack-free zone.

Fig. 3.61.. Example of hot explosively compacted ceramic microstructure (Al_2O_3–ZrO_2). (From M. Bengisu, M.Sc. Thesis, New Mexico Institute of Mining and Technology Socorro NM, 1987)

3.4.3 Reaction-Bonding

This consolidation method, also called reaction sintering, involves reacting a solid with one or more species to form a solid mass. Three types of reaction-bonding methods can be distinguished according to the type of reactants. One method involves the reaction of a porous preform and a gas. A second type of reaction bonding occurs when the solid preform is reacted with a liquid phase. In the third type of reaction bonding, which can be called an exchange reaction, solid particle mixtures chemically react with each other at elevated temperatures (1300–1600°C) to obtain a thermodynamically preferred phase mixture. Processes such as chemical vapor deposition (CVD), polymer transformation, and self-propagating, high-temperature synthesis (SHS) can also be considered reaction-bonding processes, but they will be discussed separately.

A good example of the *porous preform method*, depicted in Fig. 3.62, is the fabrication of reaction-bonded silicon nitride (RBSN). Preforms made from Si powder are exposed to nitrogen gas at 1300–1500°C for up to 150 h. Silicon particles expand 22 vol% upon reaction with N_2, decreasing the pore size by two or three orders of magnitude [3.235]. Once shrinkage of the pores closes off pore channels, the diffusion of nitrogen stops. Therefore, fully dense and fully reacted bodies are not possible. Only up to 85 % TD has been achieved under highly controlled conditions

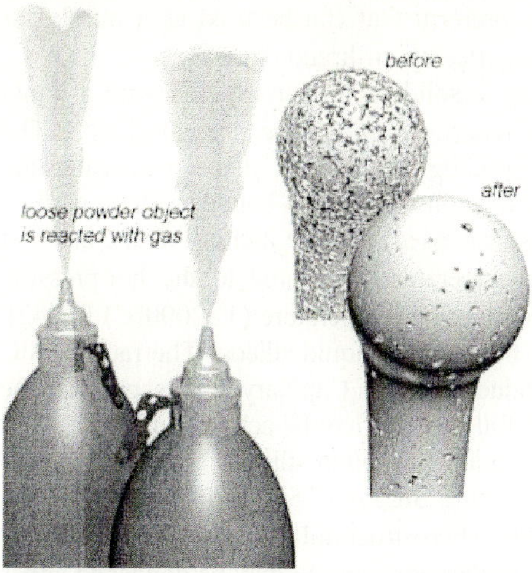

Fig. 3.62. Schematic illustration of the porous preform reaction forming method. (Illustration reprinted with permission from www.designinsite.dk.)

[3.236]. α, β, or $\alpha+\beta$-Si_3N_4 may form during reaction bonding, depending on the temperature, the nature of the Si and gaseous phases, and the impurities. Lower temperatures (<1350°C) favor the formation of α-Si_3N_4 with a fine grain size, and vice versa. α-Si_3N_4 forms from the reaction of solid Si particles with nitrogen gas. Densification proceeds as the pores are filled with Si_3N_4 needles, which grow together to form a mat. Near and slightly above the melting point of Si, the mat becomes denser, and crystallites of β-Si_3N_4 grow into molten silicon. β-Si_3N_4 forms above the melting point of silicon. The major portion of the growth of this phase occurs in a liquid phase, and a minor portion occurs by the reaction of solid silicon and nitrogen gas.

Reaction-formed materials, in which one of the reactant is gaseous, generally do not shrink. The reacted material slowly fills the pores and increases the relative density of the body. The principal disadvantages of RBSN are low strength and limited resistance to oxidation. Some improvement was achieved, however, by using ultrahigh-purity powders [3.237, 3.238] and SiC whisker additions [3.239, 3.240]. The advantages are near-net shape processing capability and high performance without liquid-phase sintering aids [3.241]. High-purity silicon preforms can be reacted completely in much shorter times, e.g., 4 h at 1150° or 10 min at 1250°C [3.242].

SiC and Al_2O_3 can also be produced by reaction bonding, using metal preforms and a vapor phase. Other nitrides, carbides, oxides, borides, or even composite ceramics can be formed using similar techniques. For example, TiN parts were produced by the reaction bonding of porous, high surface area Ti powders and nitrogen gas. Although a TiN coating prevents full conversion of Ti particles under normal conditions, specially tailored porous Ti powders were fully converted to TiN at 1000°C. The end product is a porous preform that can be used as a matrix for further applications, or it can be further densified by infiltration [3.243].

In the second type of reaction bonding, a solid preform is reacted with a liquid phase. This method is also referred to as *reactive liquid-phase sintering* [3.49]. The advantages of this variation of reaction bonding are near-net shape processing and low process temperature capabilities. Fabrication of reaction-bonded silicon carbide (RBSC) falls into this category. RBSC is processed near the melting temperature of Si (~1410°C), which is a much lower temperature compared to the hot-pressing temperature of SiC (usually ~1800°C) or sintering temperature ($T \geq 2000°C$) [3.242]. RBSC is fabricated by infiltrating SiC-C bodies with liquid silicon. The ratio of SiC to C varies, depending on the final product desired. Capillary forces provide the driving force for infiltration. Bodies with full density may be achieved by controlled reaction of Si and C combined with infiltration of molten silicon. The final product typically consists of three components: starting SiC, new SiC formed by the Si–C reaction, and free Si (10–15 vol%) [3.244]. Microstructural studies indicate that SiC formed by the reaction bonding of α-SiC grit/graphite and Si may be deposited either as an epitaxial layer of α on the original grit or as a fine dispersion of β-SiC nucleated on graphite [3.245]. Polymers that convert to graphite by pyrolysis may be used instead of graphite. This provides better extrusion and injection molding capabilities [3.246].

Meissner and Chiang [3.247] studied a new approach in RBSC processing to eliminate free Si from the structure. In this process, initial liquid-phase formers were selected such that refractory silicides compatible with SiC at high temperatures were formed in the product. As a model system, Si–Mo alloys were used to follow the reaction [3.241],

$$(x+2y)\mathrm{Si} + y\mathrm{Mo} + x\mathrm{C} \rightarrow x\mathrm{SiC} + y\mathrm{MoSi_2} \tag{3.61}$$

Upon formation of SiC, the solutes were rejected into the remaining melt until the silicide was produced. Using this approach, RBSC free of Si could be formed [3.247].

Another example in this category is silicon oxynitride. It is believed that orthorhombic crystals of Si_2ON_2 grow from a liquid phase that is saturated with dissolved nitrogen. A glass former such as CaO or MgO forms a liquid phase with SiO_2. Si_2ON_2 precipitates from this glassy phase. The following reactions occur during the formation of Si_2ON_2 in the presence of CaO [3.235]:

$$3\mathrm{Si} + (1+x)\mathrm{SiO_2} + x\mathrm{CaO} \rightarrow 2\mathrm{Si_2O}\cdot x\mathrm{CaO} + x\mathrm{SiO_2} \tag{3.62}$$

$$2\mathrm{Si_2O}\cdot x\mathrm{CaO} + x\mathrm{SiO_2} + 2\mathrm{N_2} \rightarrow 2\mathrm{Si_2ON_2} + x\mathrm{CaO}\cdot\mathrm{SiO_2} \tag{3.63}$$

The final structure is a network of Si_2ON_2 crystals in which pores are partially filled with nitrogen-rich glass.

In the third category of reaction bonding, *exchange reactions*, reactions are usually less exothermic than SHS or thermite processes, and in some cases they are endothermic [3.248]. These processes have been used to produce mullite bodies [3.249, 3.250] and recently for various ceramic matrix composites, including zirconia-toughened mullite [3.251–3.253], BN-toughened composites [3.254], AlN–TiB_2 composites [3.255], and Al_2O_3–TiB_2 composites [3.256]. A transitory liquid phase is generally formed in these systems, although reaction in the solid state followed by sintering or hot-pressing is another route. During mullite production by reaction sintering of alumina and silica at T>1650°C, the nucleation and growth of mullite occurs by an interfacial reaction between alumina and silica (Fig. 3.63) [3.257]. In this case, the growth rate is controlled by Al and Si diffusion through the mullite layer. Mullite formation slows down with time since this is a parabolic rate process. Consequently, eliminating a siliceous liquid phase becomes difficult. At lower temperatures (T<1300°C), a different mechanism controls the nucleation and growth of mullite. In diphasic systems, although alumina is present before the mullite forms, the nucleation site is within the siliceous amorphous matrix. The apparent mechanism is nucleation of mullite within the matrix as the matrix reaches a saturation concentration necessary to support mullite nucleation. The subsequent growth of mullite is controlled by the dissolution of alumina in the matrix but not by interdiffusion through mullite.

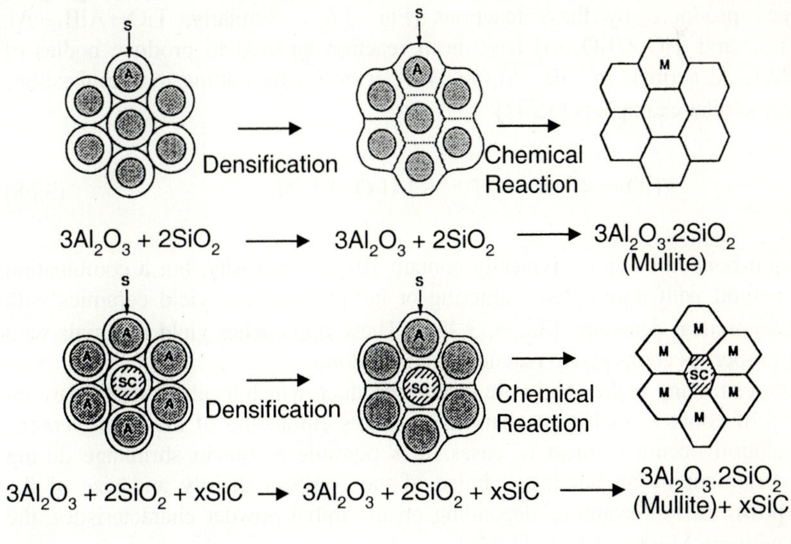

Fig. 3.63. Schematic illustration of transitory viscous sintering to form single-phase mullite and a mullite–SiC composite (From M.D. Sacks, N. Bozkurt, and G. Scheiffele: *Fabrication of Mullite and Mullite-Matrix Composites by Transient Viscous Sintering of Composite Powders.* J. Am. Ceram. Soc. **74 [10],** 2428–2437 (1991). Copyright 1991 American Ceramic Society Columbus OH, reprinted by permission)

The reaction sintering of zircon and alumina powders can yield ZrO_2-toughened mullite according to the following reaction [3.251]:

$$2ZrSiO_4 + 3Al_2O_3 \rightleftarrows 2ZrO_2 + 2SiO_2{\cdot}3Al_2O_3 \quad (3.64)$$

The addition of CaO can speed up the reaction-sintering process and lower the sintering temperature by forming a transitory liquid phase [3.252].

BN-toughened ceramics have been prepared by reacting B_2O_3 with Si_3N_4 and AlN, or both [3.254]:

$$Si_3N_4 + 2B_2O_3 \rightarrow 3SiO_2 + 4BN \quad (3.65)$$

$$B_2O_3 + 2AlN \rightarrow Al_2O_3 + 2BN \quad (3.66)$$

$$13B_2O_3 + 2Si_3N_4 + 18AlN \rightarrow 3(3Al_2O_3{\cdot}2SiO_2)\ \text{(mullite)} + 26BN \quad (3.67)$$

The low melting point of B_2O_3 (450°C) and the low viscosities of liquified boric oxide and borates provide the opportunity for low temperature consolidation of these composites. The reactions that produce the BN particulate phase take place at about 1600°C. Homogeneous composites where BN platelets act as a toughening agent, have been produced by these reactions (Fig. 3.64). Similarly, TiO_2+AlB_{12}+Al, AlB_2+TiO_2, and TiO_2+B_2O_3+Al have been reaction sintered to produce bodies of TiB_2–Al_2O_3–Al cermets or TiB_2–$Al_2O_3{\cdot}2B_2O_3$ ceramics for cutting tool applications [3.256]. A similar example is [3.241]

$$3TiO_2 + 4AlN \rightarrow 3TiN + 2Al_2O_3 + \tfrac{1}{2}N_2 \quad (3.68)$$

Reaction-bonded ceramics typically contain 10–20% porosity, but a combination of this method with liquid-phase sintering or hot-pressing can yield ceramics with close to theoretical densities [3.254, 3.256]. These approaches yield materials with superior properties compared to reaction bonding alone.

Reaction sintering in the solid state forms a product, which localizes itself between the grains of contact. Such a mechanism provokes elongation of the grain centers, thus, dilatation occurs in most of cases. It is possible to obtain shrinkage during thermal treatment by an adequate choice of the compact density and size of the starting powders. For example, depending on the initial powder characteristics, the reaction between MgO and TiO_2 [3.258],

$$MgO + TiO_2 \rightarrow MgTiO_3 \quad (3.69)$$

may result in dilatation or densification during sintering. Using $Mg(OH)_2$ and TiO_2 (anatase) as the starting powders results in densification and shrinkage. On the contrary, starting with a mixture of MgO (periclase) and TiO_2 (rutile) results in dilatation. The difference is explained in terms of the contact angle between the grains of the starting powders.

The reaction-bonding process was also exploited to produce TiB_2-toughened SiC ceramics. The following in situ chemical reactions developed during hot-pressing of the reactants at 2000°C in Ar [3.259]:

$$TiH_2 \rightarrow Ti + H_2 \tag{3.70}$$

$$x\mathrm{SiC} + \mathrm{Si} + 2\mathrm{Ti} + \mathrm{B_4C} \rightarrow (1+x)\,\mathrm{SiC} + 2\mathrm{TiB_2} \tag{3.71}$$

The density was improved by increasing the TiB_2 content. The presence of about 30 vol%TiB_2 obtained by this process improved the relative density from 87 to 99.3%. The mechanical properties were also improved, as discussed in Sect. 4.1.3.3.

3.4.4 Plasma Sintering

Plasma sintering is a process that uses a high-energy plasma produced by microwaves [3.171], RF induction [3.171], or hollow cathode discharge (HCD) [3.260]. Dugdale [3.261] first described the application of glow discharge to heating

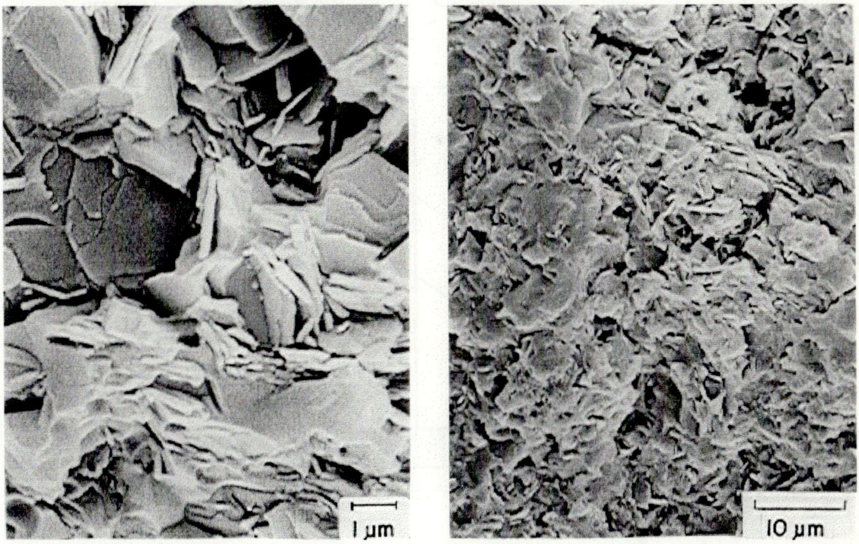

Fig. 3.64. Microstructure of mullite–30vol% BN composite, reaction hot pressed from Si_3N_4, AlN, Al_2O_3, and H_3BO_3 (90 min at 1600°C) at two magnifications (courtesy of Dr. William S. Coblenz, Naval Research Laboratory Washington DC)

refractory materials. Bennett et al. [3.170] used a microwave-generated gas discharge to sinter ceramic and metallic materials. The sintering rate of alumina was significantly higher than in conventional sintering. Cordone and Martinsen [3.260] used a hollow cathode discharge apparatus to sinter alumina. They achieved densities of 96% TD in less than 5 minutes. Johnson and Rizzo [3.171] sintered β"-alumina in an induction-coupled plasma apparatus. Nearly full densities were obtained in very short sintering times. Al_2O_3–10 vol% ZrO_2 composites were sintered to 97% TD at 1300°C by the HCD plasma sintering method in 5 minutes, and similar densities were obtained by conventional sintering in air at 1700°C in 20 min [3.134]. A schematic representation of the HCD process is given in Fig. 3.65, and the typical fine microstructure achieved by the process is shown in Fig. 3.66. Rapid sintering of various ceramics, including Al_2O_3 [3.170–3.172, 3.260–3.264], Y–PSZ [3.172], BeO_2 [3.170], HfO_2 [3.170], UO_2 [3.265, 3.266], and ZrO_2 [3.267] has been demonstrated by various researchers using plasma sintering. Uniform, fine microstructures and better mechanical properties are commonly observed in plasma-sintered ceramics, and fractional densities range from 96 to 100%.

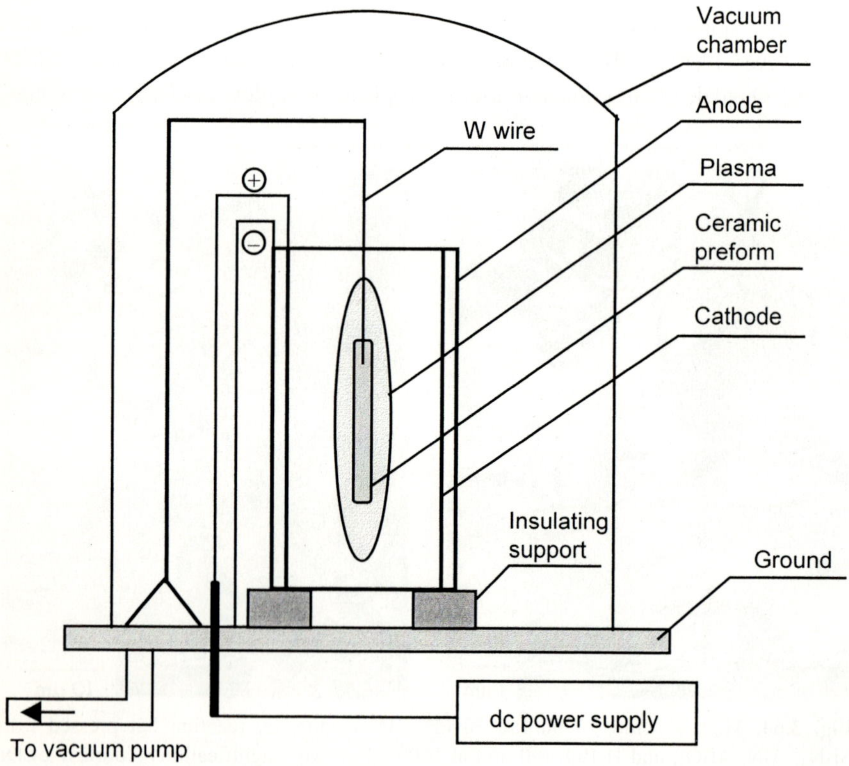

Fig. 3.65. Schematic illustration of the HCD process.

Fig. 3.66. SEM of plasma-sintered (1600°C, 2 min) Al_2O_3–10vol%ZrO_2 microstructure **a** (From M. Bengisu, Ph.D. Thesis, New Mexico Institute of Mining and Technology Socorro NM, 1992) and SEM of plasma-sintered MgO microstructure (1300°C, 4 min) **b** (From M. Bengisu and O.T. Inal: *Sintering of MgO and MgO–TiC Ceramics by Plasma, Microwave, and Conventional Heating.* J. Mater. Sci. **29 [20]**, 5475–5480 (1994) reprinted with permission of Kluwer Academic Norwell MA)

Young and McPherson [3.268] noted the unusually high shrinkage rates of more than 4%/s during rapid pass-through plasma sintering. It was proposed that the high shrinkage rates occur due to temperature gradients that develop during plasma sintering. These gradients, it is thought, drive a vacancy flux that accelerates diffusion and therefore shrinkage. Johnson [3.113] argued that these shrinkage rates can be explained by existing sintering models and that, although temperature gradients can drive mass transport, the resulting transport would have a minimal effect on overall densification. Apparently, more detailed studies will be useful in this field. Although plasma sintering is an attractive technique for research purposes, further development is required for adaptation to industrial purposes, fabrication of reasonably large pieces, and mass production.

3.4.5 Microwave Sintering

Thermal processing by microwave heating is not a new idea: it has been known for two decades. There has been a significant effort in the development and industrialization of microwave processing of materials in recent years [3.269–3.271]. One of the interesting applications of microwave processing is sintering of ceramic materials. A large number of ceramic materials have been investigated in microwave sintering experiments, including Al_2O_3 [3.173, 3.272], Al_2O_3–ZrO_2 [3.273], Al_2O_3–SiC_w [3.173], PSZ [3.174], $BaTiO_3$ [3.274], Al_2O_3–TiC [3.274], Y_2O_3 [3.276], ZrO_2–SiC_w [3.277], ZrO_2–Si_3N_4 [3.277], B_4C [3.278], MgO and MgO–TiC composites [3.39], and B or C doped SiC [3.275]. For a ceramic to be heated effectively, it must exhibit significant loss mechanisms at microwave frequency (300 MHz–300 GHz). Near room temperature, energy losses in dielectric ceramics, it is thought, result from three primary mechanisms [3.173]:

1. ion migration losses.
2. ionic vibration losses.
3. electronic polarization losses.

Microwave processing is generally characterized by uniform heating on a macroscopic scale and rapid heating rates, as opposed to conventional processing. Microwave heating involves the internal heating of a ceramic material volume. Therefore, a potential exists for sintering large ceramic components with uniform microstructure and properties. The risk exists however, of local thermal runaway, melting, or cracking during rapid heating, because the microwave absorption coefficient in many ceramics increases exponentially with temperature.

Microwave processing can selectively heat microstructural constituents due to the different microwave absorption characteristics of these phases. Thus, the possibility exists, for example, that a particle, whisker, or fiber in a matrix might heat more rapidly than the surrounding matrix or vice versa, depending on the phases involved. This aspect of microwave sintering can provide some advantages such as tailoring of interfacial properties in sintering CMCs. Other advantages of microwave sintering include the ability to work with large and complex-shaped ceramic components, nominally uniform volume heating rates of up to 200°C/min, uniform sintering, and fine microstructural features. The microwave sintering furnace is easily adapted to work under different atmospheric conditions, and pressures and cooling times are negligible compared to those in conventional furnaces [3.279].

Vodegel [3.272] compared the sintering of alumina powders with different impurity levels, by conventional and microwave heating. At the initial stages of densification, higher densities were reached at the same temperature and sintering time with microwave heating. Equal final densities of 98%TD were obtained at the end of the sintering schedule with unagglomerated powders. To reach this value, 60 min of isothermal heating at 1650°C was necessary in conventional sintering, whereas 1600°C was sufficient with microwave sintering.

Tian et al. [3.280] studied microwave sintering of Al_2O_3–TiC composites for potential cutting tool applications. Although high densities were achieved upon rapid heating with microwaves, gas breakdown and arcing at high power levels limited the sintering temperature to 1750°C and less. Al_2O_3–10%TiC composites were more difficult to sinter, compared to Al_2O_3–30%TiC and 50%TiC, due to thermal runaway in the former.

Bengisu and Inal [3.39] studied microwave sintering of MgO and MgO–10 vol%TiC composites. The equipment used for this purpose is shown schematically in Fig. 3.67. MgO indirectly heated by microwaves was sintered to higher densities, compared to conventional sintering under similar conditions (Fig. 3.68). However, MgO–TiC composites reacted and formed large pores due to the short time for diffusion of gaseous reaction products, unlike conventional sintering (Fig. 3.69). Thus, microwave sintering is disadvantageous compared to conventional sintering if the possibility exists of chemical reaction between the constituents, such as in MgO–TiC composites.

Park and Meek [3.281] sintered ZrO_2 (+3 mol%Y_2O_3) –20wt%Al_2O_3 composites in a microwave oven operating at 2.45 GHz. Comparison with conventionally

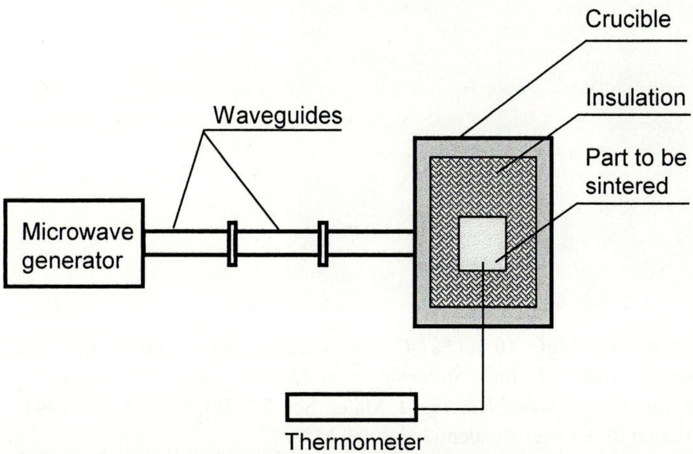

Fig. 3.67. Schematic illustration of a microwave furnace

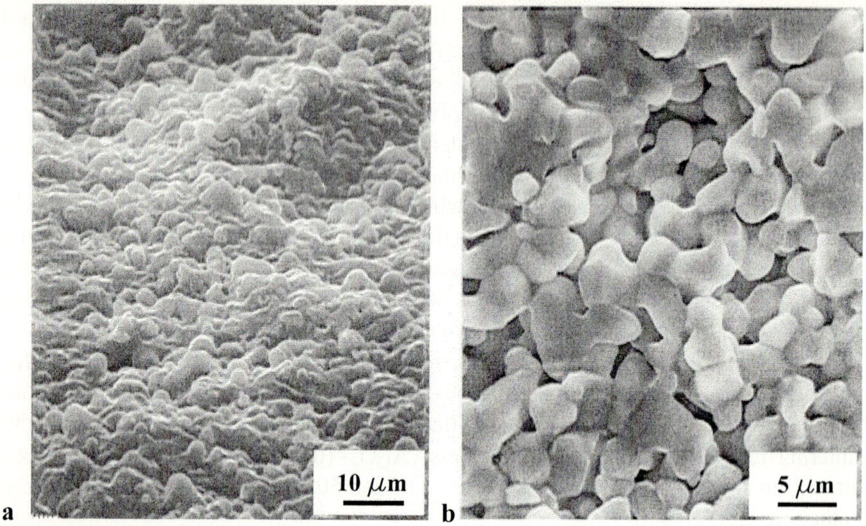

Fig. 3.68. Microstructure of MgO sintered by microwaves at 1500°C for 5 min in air **a** and by conventional heating at 1600°C for 4h in air **b** (From M. Bengisu and O.T. Inal: *Sintering of MgO and MgO–TiC Ceramics by Plasma, Microwave, and Conventional Heating.* J. Mater. Sci. **29 [20]**, 5475–5480 (1994), reprinted with permission of Kluwer Academic Norwell MA)

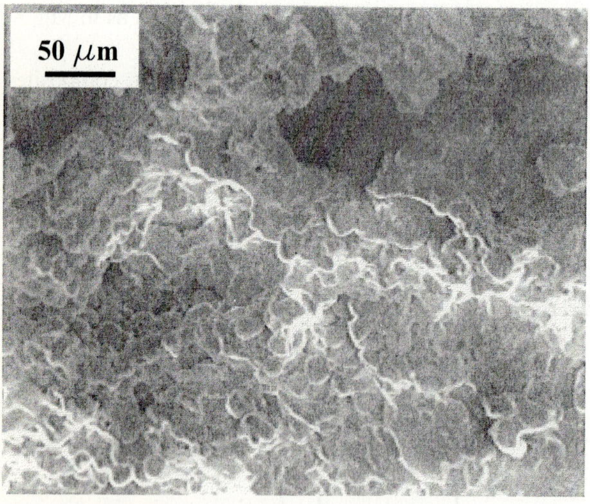

Fig. 3.69. SEM micrograph of MgO–10 vol %TiC microwave sintered at 1600°C for 5 min in air (From M. Bengisu and O.T. Inal: *Sintering of MgO and MgO–TiC Ceramics by Plasma, Microwave, and Conventional Heating.* J. Mater. Sci. **29 [20]**, 5475–5480 (1994), reprinted with permission of Kluwer Academic Norwell MA)

sintered composites with identical composition revealed that densification rates during microwave sintering are much higher than those of conventional sintering. The difference in the shrinkage rates was attributed to enhanced diffusion in an electromagnetic field and rapid heating. It was also observed that the amount of tetragonal zirconia retained upon sintering was higher compared to conventionally sintered samples. The microstructures of microwave-sintered samples were finer than those of their counterparts because of the penetrating action of microwaves and shorter sintering times.

Less success was achieved in Al_2O_3–10vol%SiC_w composites [3.282]. Both 2.45- and 60-GHz microwave sintering yielded a maximum of 70%TD at various sintering temperatures. The low sinterability was explained by the formation of a constrained whisker network, which caused crack-like voids and large pores during consolidation.

These studies show that, although microwave sintering is a suitable method for some monolithic ceramics and particulate CMCs, it is not suitable for consolidating whisker-CMCs. Many potential applications exist for microwave sintering of ceramics in the industry.

3.4.6 Self-Propagating Synthesis

This process takes advantage of the self-propagating exothermic reactions that involve two or more elements or compounds. Many ceramic compounds and particulate composites can be produced by reactions in solid–solid or solid–gas systems. Typical examples are given in Table 3.9 with their corresponding adiabatic temperatures [3.234]. Most of the early work in this area is that of formerly Soviet researchers. The process is used for direct production of ceramic components [3.234, 3.283], fine powders [3.233], or powder mixtures [3.234].

Typically, a solid–solid exothermic mixture such as Ti+C or Zr+2B can be ignited from one end by a suitable heat source, e.g., by a laser beam or a resistive wire (see Fig. 2.6). Once part of the mixture reacts, a combustion wave forms and travels through the whole mixture. The combustion wave may be planar, oscillatory, or spiral, and velocities range from 0.1 to 15 cm/s, depending on the chemical system. The most important advantages of this process are energy savings associated with self-sustaining reactions, simplicity, purity provided by volatilization of most impurities at very high temperature combustion fronts, and the possibility of simultaneous synthesis and densification of the products.

If no pressure is applied to a body during self-sustaining reactions, the product will be a porous and fragile mass. The porosity is mostly generated intrinsically due to higher density products relative to initial powders [3.284]. Sources of extrinsic porosity are pores in the starting powder, outgassing or evaporation from high-temperature reactions, and incomplete transfer of a gas-producing anion from one metal to another during reaction.

There are several techniques, that form dense products via self-propagating synthesis. Dense compacts of TiB_2 and TiC have been produced by hot-pressing thermally ignited powder mixtures [3.234]. Problems associated with this method are the presence of a liquid phase and violent ejection of gases that can break the

die. Rice et al. [3.285] carried out experiments with mixtures of Ti and B to form TiB_2 by a rolling mill. The powder mixture was sealed in steel cans and ignited at one end. At the same time, passing the cans through a rolling mill resulted in dense compacts. Hot isostatic pressing attempts were not successful due to large pressure buildup and destruction of Ta containers [3.234].
Another technique utilizes dynamic pressures to densify bodies just after a self-propagating exothermic reaction. Keckes et al. [3.233] fabricated fully dense TiB_2 and TiC bodies by combustion synthesis followed by dynamic compaction of the still hot, porous ceramic body. A similar process was used by Holt [3.234] to densify TiC, TiB_2, and TiC + Al_2O_3. Some cracks developed in the dense compacts due to dynamic pressures in the case of TiC and TiB_2, whereas TiC + Al_2O_3 did not contain any cracks.

These studies suggest that, if a suitable pressurizing method can be developed, self propagating synthesis can be a powerful method for producing ceramic bodies of various compositions.

Table 3.9. Examples of Self-Sustaining Reactions and Their Calculated Adiabatic Temperatures [a]

Reaction	Product	Adiabatic Temp. (°C)	Melting Temp. (°C)
Ti + 2B	TiB_2	2930	2930
Zr + 2B	ZrB_2	3040	3040
Ti + C	TiC	2940	2940
Hf + C	HfC	3630	3630
Si + C	SiC	1500	2700
Al + $1/2N_2$	AlN	2230	2600 (sublimes)
3Si + $2N_2$	Si_3N_4	1900	4000 (sublimes)
B + $1/2N_2$	BN	3000	3400 (sublimes)
Ti + $1/2N_2$	TiN	4600	2950
5Ti + 3Si	Ti_5Si_3	2200	2120
SiO_2 + 2Mg + C	SiC + 2MgO	2300	
TiO_2 + 4Al + C	TiC + $2Al_2O_3$	2050	

[a] Reference:

J.B. Holt: *The Use of Exothermic Reactions in the Synthesis and Densification of Ceramic Materials. MRS Bulletin* **12 [7],** 60–64 (1987). (Adapted with permission of the Materials Research Society Pittsburgh PA)

3.4.7 Chemical Vapor Deposition (CVD) Processes

CVD processes use chemical reactions of two or more gases, where the product deposits and coats the substrate inside a reaction chamber. CVD and *chemical vapor infiltration* (CVI) are two CVD processes suitable for fabricating ceramic components. In CVD, the deposition rates are low, typically less than 250 μm/hr. Another problem encountered in CVD is residual stress, which seems to arise from variations in stoichiometry and deposition conditions [3.286]. Although mostly used to produce thin coatings, e.g., AlN or Al_2O_3 on WC cutting tools, freestanding shapes can also be fabricated by CVD techniques [3.287]. Some examples of reactants and deposition temperatures for CVD/CVI are [3.286]

1. $SiCl_4 + NH_3$ at 1100–1300°C for Si_3N_4.
2. $SiCl_4 + CO_2 + H_2O$ at 800–1100°C for SiO_2.
3. $AlCl_3 + CO_2 + H_2O$ at 800–1200°C for Al_2O_3.
4. $AlCl_3 + NH_3$ at 1000–1200°C for AlN.
5. $B_4Cl_3 + NH_3$ at 800–1100°C for BN.

Deposition may occur by one of the following processes: chemical reduction, thermal decomposition, or displacement reactions [3.288]. The most commonly practiced CVD process is hydrogen reduction of a metal chloride or fluoride, for example [3.288],

$$TiCl_4(g) + 2BCl_3(g) + 5H_2(g) \rightarrow TiB_2(s) + 10HCl(g) \tag{3.72}$$

Examples of thermal decomposition and displacement reactions are [3.286,3.287],

$$Ni(CO)_4(g) \rightarrow Ni(s) + 4CO(g) \tag{3.73}$$

$$B_2H_6(g) + 2NH_3(g) \rightarrow 2BN(s) + 6H_2(g) \tag{3.74}$$

Deposits grow by nucleation and growth. Adsorption and diffusion of atoms form stable nuclei at preferred sites on the substrate surface. Addition of new atoms provides growth of crystals. Depending on the process parameters such as the substrate type, reaction temperature, and reactant supersaturation, various deposit morphologies can be obtained. Here, supersaturation means the ratio of the partial pressure of the reactant gas adjacent to the substrate to the partial pressure in equilibrium with the substrate. These include amorphous, columnar, epitaxial, equiaxed fine, or coarse-grained structures. Powders, platelets, and whiskers may also form by CVD [3.290].

Components are produced by depositing a thick coating onto a mandrel and subsequently removing the mandrel. An application of this technique is the production of transparent windows for missile guidance systems. Chemically vapor deposited polycrystalline ZnS and CdS in the 8–12 μm range are suitable for this purpose. Freestanding BN components, such as crucibles and cylinders, are also fabricated by CVD. Boron nitride is applied to graphite mandrels and, if properly designed, the graphite mandrel can easily slip off [3.291]. Toughened ceramic composites made by simultaneous CVD of two phases in the SiC–$TiSi_2$ system were produced by Stinton et al. [3.292]. Figure 3.70 shows the microstructure achieved by this process. This composite, which has a microstructure similar to that of a whisker-reinforced ceramic, exhibited high fracture toughness (7.1 MPa $m^{1/2}$). A similar effort was the codeposition of Si_3N_4–TiN by Hirai and Hayashi [3.293].

CVI involves depositing species inside the pores of a preform. CVI processes were developed for fabricating fiber-reinforced composites. A fiber preform is infiltrated by the desired matrix-material precursors as suitable reactant gases. CVI is one of only a few processes capable of incorporating continuous ceramic fibers in a ceramic matrix without chemically, thermally, or mechanically damaging fragile fibers. Extensive research has been conducted on CVI of carbon–carbon composites; other composite systems, including BN–BN, SiC–SiC, SiC–C, Al_2O_3–Al_2O_3, and Al_2O_3–SiC were also studied. A number of important applications of the CVI process exist. The main ones are production of large and/or complex-shaped parts, including carbon–carbon disk brakes for the Concorde, Al_2O_3–SiC heat exchanger tubes, Al_2O_3–SiC thermocouple wells, and C–SiC rocket nozzle and exit cones [3.286].

SiC matrix–SiC or aluminosilicate fiber systems are produced by infiltrating SiC from methyl trichlorosilane (MTS). Densities up to 90% can be achieved by this process. One problem with the CVI process is density gradients that occur due to higher deposition rates near preform surfaces than those on core fibers. To reduce these gradients, the process is repeatedly interrupted for intermediate diamond

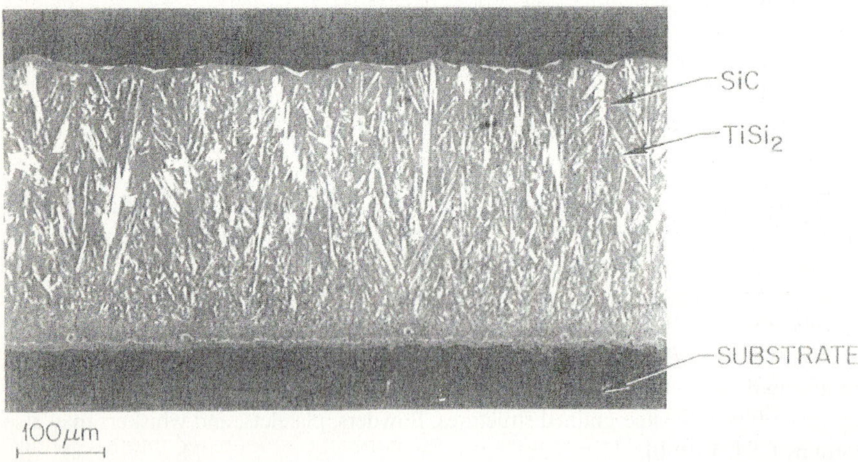

Fig. 3.70. Microstructure of SiC–$TiSi_2$ composite formed by CVI (Courtesy of Dr. David P. Stinton, Metals and Ceramics Division, Oak Ridge National Laboratory Oak Ridge Tennessee)

machining operations to remove material from the surface and open channels to the interior of the preform. The cyclic process requires several weeks, and severe density gradients can still occur in composites thicker than 3 mm [3.287].

An improved CVI process uses thermal and pressure gradients to reduce fabricating time from several weeks to < 24h. In this process, the deposition zone is moved progressively from hotter to cooler regions, eliminating density gradients [3.294,3.295]. Si_3N_4–SiC_f (subscript denotes fibers) and SiC–SiC_f tubes were produced with diameters up to 50 mm and ~10-mm wall thicknesses. Another recent development in this area is a process hybridizing CVD and CVI methods. Fibers are fed into the reactant gas stream from the exterior, building up the composite at vastly increased rates, without the problems of closed channels [3.286].

In summary, CVD processes are generally slow, unless specialized and thus expensive methods are incorporated. The possibility of producting virtually limitless types of composites and microstructures and the relatively low fabricating temperatures are attractive features of CVD processes.

3.4.8 Eutectic Solidification

As previously treated in Sect. 1.6, eutectic solidification of a liquid results in the formation of two solid phases that have no mutual solid solubility,

$$\text{liquid} \rightleftarrows \text{solid 1} + \text{solid 2}. \tag{3.75}$$

This principle can be used to produce eutectic composites. There are limited studies on ceramic eutectics due to the limited availability of phase equilibrium data and the very high eutectic points of many ceramic systems.

Several methods can be used to produce directionally solidified ceramic eutectics. These include the use of gradient furnaces with forced gas flow [3.296], Bridgman type crystal growing furnaces [3.297], the floating molten zone technique [3.298], and edge-defined film-fed growth [3.299] (see Sect. 2.3 for details of some of these techniques). Some oxide ceramic composites that have been prepared by eutectic solidification include Al_2O_3–ZrO_2 [3.297], Al_2O_3–$Y_3Al_5O_{12}$ [3.300], Al_2O_3–TiO_2 [3.301], NiO–CaO [3.302], ZrO_2–$CaZrO_3$, ZrO_2–$SrZrO_3$, MgO–CaO, $MgTi_2O_5$–TiO_2 [3.303], MgO–$MgAl_2O_4$, and ZrO_2–MgO [3.304, 3.305]. According to Minford et al. [3.302], the basic microstructures of eutectic ceramics can be generalized on the basis of the relative interfacial surface area per unit volume (A_v). For fibers, $A_v = 2/\lambda\,(\pi V/0.866)^{-1/2}$ and for lamellae, $A_v = 2/\lambda$, where λ is the fiber/lamellae spacing and V is the volume fraction of the minor phase. Figure 3.71 shows A_v as a function of V. The basis for this representation is the concept of minimized internal surface area in the development of a lamellar or fibrous eutectic microstructure [3.305]. Geometrically, assuming that the surface energy is isotropic, a hexagonal array of fibers is favored when the volume fraction of one phase is less than 0.28. Otherwise, a lamellar microstructure is preferred. Hunt and Jackson [3.306] refined this theory by considering the minimum undercooling for each morphology and found that fibrous structures can be produced up to 32 vol% of the minor phase. If substantial anisotropy in surface energies exist, the lamellar structure can be stable at all volume fractions. The solidification rate is another parameter that determines the type of eutectic solidification microstructure. Hulse and Batt [3.298] investigated a number of oxide eutectics. In the Al_2O_3–Y_2O_3–stabilized ZrO_2 system, cellular growth occurred at all solidification rates above 0.8 cm/h, where the colony size increased with the solidification rate. Below this threshold, plane front growth occurred.

Kennard et al. [3.303, 3.304] obtained well-oriented microstructures when the solidification rate was below 2 cm/h in MgO–$MgAl_2O_4$ and ZrO_2–MgO systems. At 1 cm/h, the microstructure of the ZrO_2–MgO eutectic is composed of fully aligned MgO rods in a cubic ZrO_2 matrix (Fig. 3.72). Sorrell et al. [3.307] prepared ZrC–ZrB_2, ZrC–TiB_2, and TiC–TiB_2 eutectics by directional solidification. At growth rates from 0.4–0.5 cm/h, morphologies were obtained that consisted of columnar grains of parallel lamellae with interlamellar spacings obeying the $\lambda^2 R = C$ law,

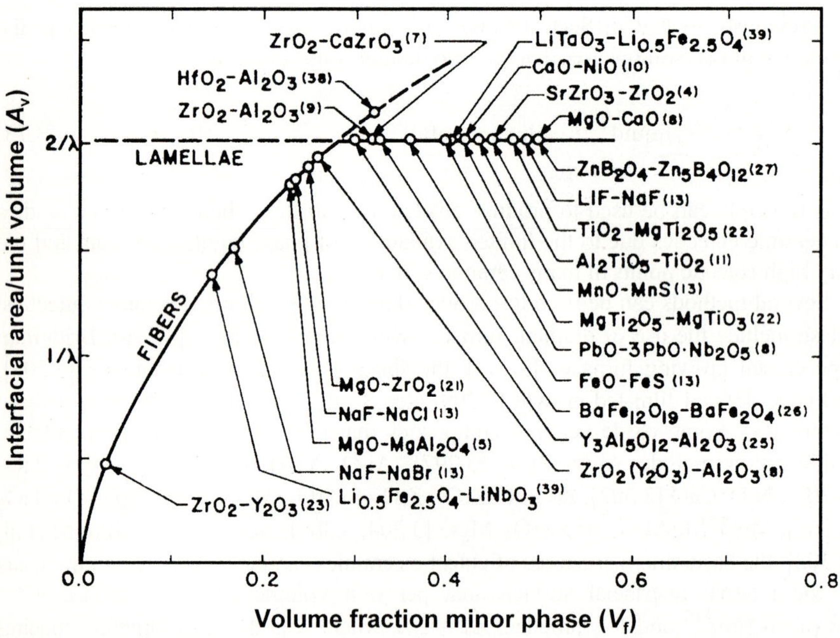

Fig. 3.71. Correlation between the volume fraction of observed nonmetallic eutectic microstructures and surface area in eutectic composites (From W.J. Minford, R.C. Bradt, and V.S. Stubican: *Crystallography and Microstructure of Directionally Solidified Oxide Eutectics.* J. Am. Ceram. Soc. **62 [3–4],** 154–157 (1979). Copyright 1979 American Ceramic Society Columbus OH, reprinted with permission)

where R is the growth rate and C a constant. This relationship is usually obeyed in both the fibrous and lamellar microstructures with a substantial variation in C. SiC–B_4C eutectics were studied by DerHong et al. [3.308]. A lamellar structure was obtained of solidification rates below 2 cm/h.

There are limited studies on the mechanical properties of eutectic composites. The hardnesses of nearly all directionally solidified eutectics are greater than their constituents in carbide–boride systems. Indentation fracture toughness values also show the same trend. Hall–Petch type microhardness behavior was observed in oxide systems where microhardness values are proportional to the interlamellar spacing [3.309].

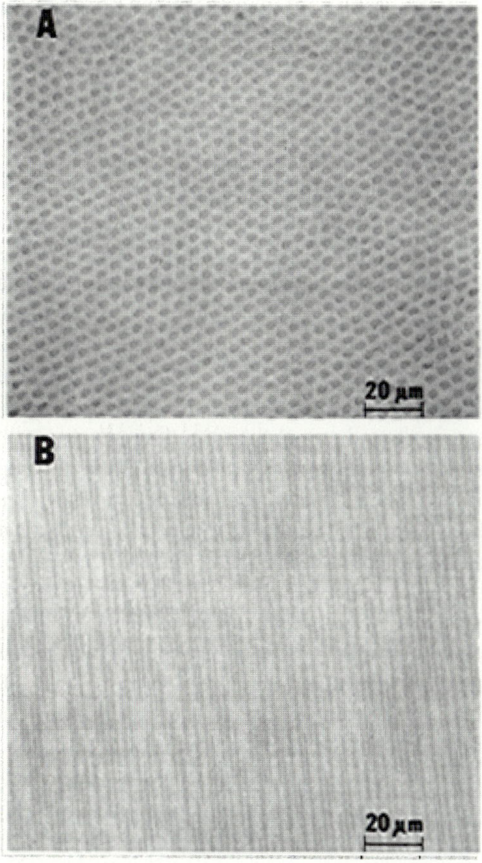

Fig. 3.72. Microstructure of $MgO–ZrO_2$ eutectic solidified at 1.0 cm/h: Transverse **a** and longitudinal section **b** (From F.L. Kennard, R.C. Bradt, and V.S. Stubican: *Directional Solidification of the ZrO_2–MgO Eutectic.* J. Am. Ceram. Soc. **57 [10],** 428–431 (1974), copyright 1974, American Ceramic Society Columbus OH, reprinted by permission)

The eutectic solidification process has found limited application due to high process temperatures and high production costs. The high degree of mechanical attachment of constituent phases and the high relative densities are advantages that should be considered for special applications.

3.4.9 Melt Infiltration

As seen in the previous section, handling ceramics at their melting temperatures poses various difficulties and disadvantages. Nevertheless, some systems have been studied to analyze the potential advantages of composite processing via the melt infiltration process. In this process, porous fibers or powder preforms are brought into contact with a molten matrix phase [3.286, 3.310]. If the matrix phase wets the preform, surface tension can be sufficient as a driving force to fill the voids. Particulate additions can be used in a fiber preform to increase this driving force. Pressure can be used to improve the infiltration process, such as glass transfer molding, where a porous preform is filled by molten glass in a die cavity.

In one study, B_4C+C preforms were infiltrated with molten Si to obtain B_4C+SiC composites [3.286]. Hillig et al. [3.310, 3.311] studied the production of Si/SiC, CaF_2/SiC, $SrSiO_3$/SiC, and $SrAl_2Si_2O_8$/SiC composites by the melt infiltration process. Their studies indicated that the fibers have to sustain temperatures in excess of 1400°C in these systems and they must resist degradation during processing. The infiltration time and behavior depend on factors such as the pore size of the preform, the viscosity of the melt, and the wetting characteristics of the preform by the melt. For example, it was observed that CaF_2 readily infiltrated into both SiC powder and whisker preforms, but not into SiC (Nicalon) fiber preforms.

SiC/carbon fiber preforms have been infiltrated and at the same time reacted with molten Si to produce SiC/Si composites, or the Silcomp™ material [3.310]. Upon reaction with molten Si, C fibers retain their shapes, but convert into SiC fibers with a coarse structure. This composite exhibits good resistance to thermal shock, creep rupture, and oxidation. Thus, it has been employed as gas turbine combustor components, equipment for high-temperature processes, and other components for applications that require exposure to gaseous environments up to 1370°C. The bend strength of the Silcomp composite in the longitudinal direction is about 410 MPa and remains constant until 1100°C, in the transverse direction, it varies from 200–340 MPa. The longitudinal elastic modulus, on the other hand, exhibits a constant but gradual decrease from 410 GPa at ambient temperature to 350 GPa at 1300°C.

3.4.10 Polymer Pyrolysis

Polymer pyrolysis is a process used for various end products, including ceramic powders, fibers, and single-phase and composite bulk forms. The production of polymer-based fibers was discussed in Sects. 2.4.3 and 2.4.4. Just like the fiber process, bulk ceramics can be produced by converting a suitable polymer to a ceramic phase or a composite [3.286]. The usual steps involved in ceramic fabrication by polymer pyrolysis are [3.312]

1. chemical or thermal cross-linking of precursors of low molecular weight compounds to achieve a polymer network structure,
2. pyrolysis of the polymer to generate the desired ceramic in the amorphous form, and
3. transformation of the amorphous intermediate polymer into a polycrystalline ceramic by an additional heat treatment.

Polysilazanes $[–R^1R^2Si–NH–]_n$ are among the better studied polymers for ceramic processing. R_2SiCl_2 groups are converted to oligosilazane by the using ammonia. The oligomer is cross-linked with a base catalyst to yield polysilazanes. R can be a methyl, alkyl, aryl, or vinyl group. Following this step, the polysilazane is pyrolyzed in a temperature range of 400–500°C, and decomposition is completed at above 800°C. The main gaseous by-products generated during pyrolysis were hydrogen and methane [3.312]. The final step is to transform the amorphous $Si_3N_{3.4}$, Si–C–N, or Si–B–C–N phases into α-Si_3N_4, ternary Si–C–N, and quaternary Si–B–C–N crystalline ceramics by suitable heat treatment at 1000, 1450, and 1700°C, respectively. Figure 3.73 shows the microstructure of a poly(vinyl silane)/Si_3N_4 powder blend derived Si_3N_4-SiC composite, gas pressure sintered at 1850°C at 10 MPa N_2 pressure, as an example of the microstructures achieved by the process.

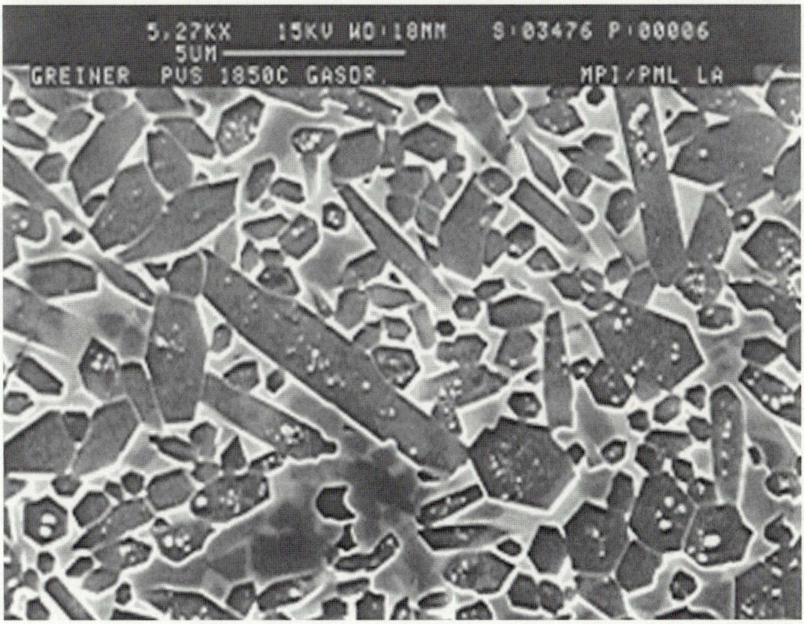

Fig. 3.73. Microstructure of poly(vinyl silane)/Si_3N_4 powder blend derived Si_3N_4–SiC composite (Reprinted from R. Riedel and W. Dressler: *Chemical Formation of Ceramics.* Ceram. Intern. **22 [3]**, 233–239 (1996) with permission of Elsevier Science Oxford UK)

The advantages of polymer pyrolysis include low processing temperatures (800–1500°C) and vast compositional variability. Low processing temperatures are very desirable since they are more economical and practical, and they also provide compatibility of whiskers or fibers with a polymeric precursor during composite processing. Polycarbosilanes and polysilazanes can be converted directly to nearly dense monolithic ceramic bodies at low temperatures (1000°C) without the use of additives [3.313, 3.314]. For example, additive-free amorphous silicon carbonitride bodies were produced with up to 94%TD by pressureless pyrolysis of compacted infusible polysilazane powders at 1000°C in Ar. The highest ceramic yields (up to 72 wt%) are obtained using infusible polymers [3.313]. Further advantages of this process are high purity and generation of novel, metastable, and amorphous materials that cannot be produced by conventional means [3.286, 3.312].

Better mechanical properties can be obtained by liquid-phase sintering of amorphous, polymer-derived powders, instead of direct, pressureless pyrolysis. For example, densification of polysilazane-derived amorphous silicon carbonitride powder by liquid-phase sintering with Al_2O_3 and Y_2O_3 additives in nitrogen atmospheres at 1900°C yielded 99% dense, polycrystalline β-Si_3N_4/β-SiC composite ceramics [3.313]. The average strength and fracture toughness obtained were 650 MPa and 10.2 MPa $m^{1/2}$, respectively.

Apart from polysilazanes, poly(silyl carbodiimide) $[–R_2Si–NCN–]_n$ polymers were also studied as precursors to ceramic materials in the Si–C–N ternary or Si–E–C–N quaternary systems (E=B, Al, Ti, or P). The ceramic yields of both families of polymeric precursors were usually in the range of 60 to 75 wt%. Various methods have been developed for producing multicomponent particulate ceramics by pyrolysis, as reviewed by Riedel and Dressler [3.312].

Chopped fibers can be accommodated in polymer-derived ceramic matrices before reaction, by injection molding or other preconsolidation processes, or copyrolysis can be used for the simultaneous pyrolysis of the matrix and a polymeric fiber [3.286]. The second alternative, however, has been limited so far, due to extensive chemical reaction of the fiber and the matrix during the conversion process. SiC fiber-reinforced SiC composites were fabricated by infiltrating Nicalon fiber preforms with suitable polymers and successive conversion to SiC. Carbon, oxide, or Nicalon fiber-reinforced BN composites were prepared by converting borazine oligomers with up to 90% yields. The low viscosity oligomers of borazine facilitated impregnation of fiber layups. Excellent mechanical properties were achieved with resultant C/BN composites up to 850°C. Less success was achieved with oxide and Nicalon fibers due to reaction at the fiber/matrix interface.

One of the main limitations of polymer pyrolysis is the large density changes involved, for example, from 1 g/cc for the polymeric precursor to about 3 g/cc for SiC as the product. Hence, high shrinkage ratios, porosity, and microcracking are associated features. One successful attempt to reduce shrinkage and related microcracking involved dispersing of about 15 vol% fine graphite particles in a SiC-producing polymer matrix. Another limiting factor is gas evolution during the crystallization phase. In spite of these disadvantages, much interest exists in polymer pyrolysis due to the vast product possibilities and potentially feasible production methods.

3.4.11 Biomimetic Processes

Apart from the many processes developed in laboratories, nature has had its unique way of forming high precision, complicated ceramic structures for millions of years. Shells of mollusks such as the nautilus, sea urchin, and abalone are created by a low temperature process that results in hard, relatively tough, and sometimes extremely intricate shapes. The key to these processes lies in the biological polymers produced by the organisms [3.316]. The polymers, made from proteins and polysaccharides, provide nucleation sites where inorganic ions within the watery cells can settle, agglomerate, and grow into ceramic crystals with preregistered shape and orientation. Currey and Kohn [3.317] and Laraia and Heuer [3.318] studied the microstructure and mechanical properties of various seashells. It was noted that mollusc shells exhibit remarkably high flexural and compressive strength, sometimes in excess of 300 MPa. They are also reportedly very resistant to fracture. Fracture toughness values between 5 and 11 MPa $m^{1/2}$ were measured [3.317]. Fractography of such specimens revealed that a number of familiar toughening mechanisms operate together to produce some flaw tolerance. Even more unusual is the fact that mollusk shells are harder than the native minerals from which they are built. Laraia and Heuer's observations on damage zones and hardness anisotropy suggested that the anomalous hardness is based on microstructural design, rather than intrinsic hardening of the mineral phase.

Figures 3.74 and 3.75 show three examples of mollusk shells. All three possess a certain type of layered structure. The *Murex* microstructure is typical of the crossed-lamellar structure [3.319] as seen in Fig. 3.74. The *Abalone*, instead, has a foliated texture (Fig. 3.75a). The shells of most mollusks are made of calcium carbonate with a small amount of organic material, mainly protein and occasionally chitin. As the microstructures in Figs. 3.74 and 3.75 indicate, nature produced some of the most complex and efficient composite structures long before today's scientists developed similar structures for better mechanical properties.

Researchers are attempting to simulate the natural ceramic process by modifying the surfaces of synthetic polymers to create active ends like the natural material [3.316]. The polymer template is treated with a chemical solution to create a charge density distribution, which, it is thought, is the key to orienting growing crystals. Thin ceramic films for microelectronic applications have been grown by this process. For ceramic parts, a loosely woven template of polymer fibers is created and placed in a supersaturated oxide solution, where ceramic crystals fill in the shape and make it rigid. Although this process is attractive because heating is avoided, there is much to be learned about it before it can be significant as an artificial ceramic manufacturing method.

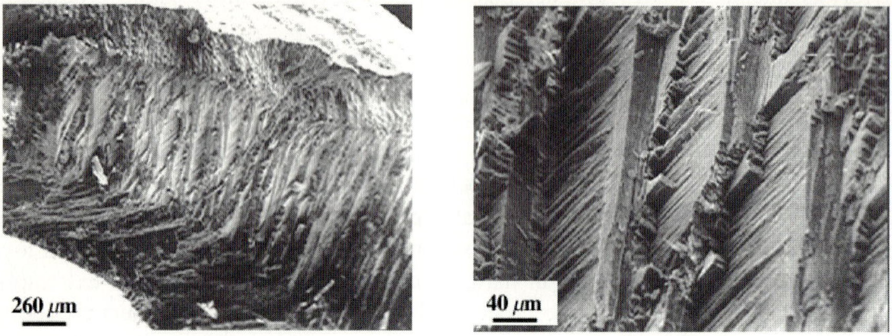

a b

Fig. 3.74 a-b. Structure of *Murex* at two magnifications (From M.Sc. Thesis of R. Baker, New Mexico Institute of Mining and Technology (1993), courtesy R.Baker)

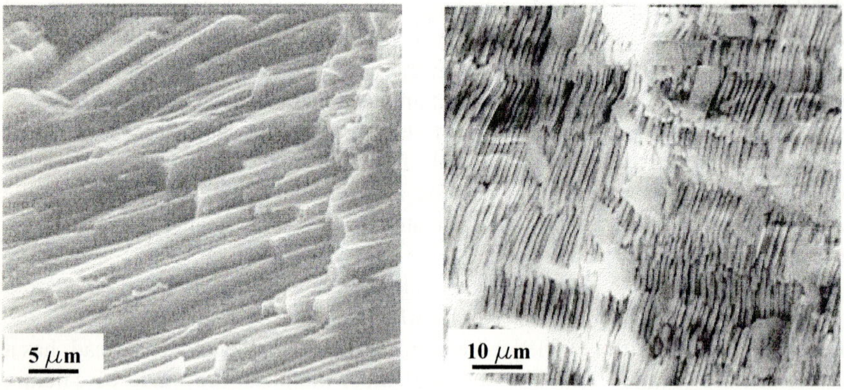

a b

Fig. 3.75. Structure of *Mussell* **a** and *Abalone* **b** (From M.Sc. Thesis of R. Baker, New Mexico Institute of Mining and Technology (1993), courtesy of R.Baker)

The microstructure of coral was reproduced in metals, ceramics, and polymers by the replamineform process [3.320]. In this process, pieces of coral are vacuum impregnated with wax. After hardening the wax, the calcium carbonate skeleton is leached away with hydrochloric acid. The remaining wax has the negative structure of the coral. The negative is impregnated with a lead zirconate titanate (PZT) slip by vacuum impregnation. Upon burning the wax at 300°C, a PZT preform is obtained whose pore characteristics are similar to those of coral, i.e., a narrow pore size distribution of about 50% porosity and complete pore interconnectivity. The preform is sintered, leading to minimal shrinkage (~13%) and a robust PZT skeleton. The skeleton can be further processed by impregnating highly flexible elastomers such as silicone rubber and poling of the PZT. The composite can be crushed to yield a highly flexible transducer with low permittivity, or it can be used in bulk as a low-density, high-coupling resonator.

Another biomimetic process (process imitating nature or life) is the conversion of charcoal into ceramic wood. Wood has a structure made up of vessels and cells filled with water. The porous structure of wood provides an attractive template for special products such as catalyst supports, machinable ceramics, and light structural ceramics [3.321]. The continuous pore structure is another feature that is almost impossible to achieve by artificial processes. The natural ceramic model material is petrified wood. Although the fossilization process is not fully understood, some observations can be made. At the beginning of the process, trees became covered with mud, sand, or volcanic ash due to climatic conditions. With time, water seeped through the mud and sand into the buried trees and filled the empty cells. After millions of years, the wood degenerated and left a silica-rich fossil with the external and internal structure of the original wood. Ota et al. [3.321] attempted to produce a small SiC piece by using an approach similar to fossilization. A piece of charcoal was vacuum infiltrated with tetraethyl orthosilicate (TEOS) and hydrolyzed with an ammoniacal solution to form SiO_2 in situ. The charcoal filled with SiO_2 gel was fired at 1400°C for 2 h in Ar to form SiC. This treatment was repeated up to six times since the yield of SiC was low. A composite of C–SiC with one-dimensionally oriented pores was obtained.

4 Properties of Ceramic Materials and Their Evaluation

4.1 Mechanical Properties

The mechanical properties of a ceramic material must be thoroughly studied before it is considered for any application that imparts certain stresses. Mechanical properties such as hardness, strength, elastic modulus, and fracture toughness are key properties in a ceramic material's performance. Fatigue behavior is important when cyclic stresses are present. At elevated temperatures, the creep behavior of a ceramic is of utmost importance. Erosion and wear phenomena are system-specific. A ceramic component's erosion and wear behavior have to be known especially if the component comes into contact with other solids, liquids, and high pressure gases. Any combination of fatigue, creep, erosion, wear, and corrosion phenomena is possible, which may complicate the analysis of material response under those circumstances. This chapter describes the mechanical properties of ceramic materials and their evaluation, in general, with some specific examples.

4.1.1 Hardness and Elastic Modulus

Hardness is a material's resistance to permanent deformation [4.1]. It is a function of the crystal structure, crystal defects, bond type, fractional density, grain size, and purity of the ceramic. External variables that may affect hardness are temperature and the presence of reactive species, e.g. acids, alkalis, or even water in some cases. Hardness is directly related to the elastic or bulk modulus of a material. In ionic crystals, it can be shown, from first principles that the elastic modulus is a function of the electron charge, *e*, and interatomic distance, r_0:

$$E = K\,e^2 / r_0^{\,4}\,, \tag{4.1}$$

where *K* is a constant [4.2]. In isotropic materials, the *bulk modulus* is related to the elastic modulus by

$$B = E / 3(1-2\nu)\,, \tag{4.2}$$

where v is Poisson's ratio. From (4.1) and (4.2)

$$B = K' e^2 / r_o^4, \tag{4.3}$$

where K' is a proportionality constant different from K.

In ceramic materials with ionic and covalent bonding, relationships similar to Eq. (4.3) can be found. For example, an empirical model shows that for tetrahedral compounds that share eight valence electrons, the scaling relation is [4.3]

$$B = (1971 - 220\lambda) / r_o^{3.5}, \tag{4.4}$$

where B is in GPa, r_o is in angstroms, and λ is a constant, representing the ionicity of the compound. For homopolar semiconductors (group IV), $\lambda = 0$, and for heteropolar III–V and II–VI zinc blende materials, $\lambda = 1$ and 2, respectively. This relationship, it was shown, is accurate within a few percent for group IV and group III–V compounds [4.4].

The largest bulk moduli belong to covalently bonded materials. Diamond, a covalently bonded material, has the largest bulk modulus (1035 GPa) [4.5] known so far. Based on Eq. (4.4) and the related model, Cohen [4.3] suggested that a covalent solid formed from C and N could have a larger bulk modulus and a hardness higher than diamond. Liu and Cohen [4.6] predicted that such a compound might be produced by substituting C atoms for N in Si_3N_4.

As the temperature increases, the hardness generally decreases due to a slight decrease in elastic modulus [4.2] and due to grain growth. Porosity significantly reduces the hardness (Fig. 4.1), as well as the elastic modulus. Various equations have been proposed to relate elastic modulus to porosity, but none of them is universally applicable. For randomly distributed, small, closed spherical pores up to 50 vol% in ceramics that have Poisson ratios of about 0.3, the related equation is given as [4.7]

$$E = E_0(1 - 1.9V + 0.9V^2), \tag{4.5}$$

where E_0 is the elastic modulus of the fully dense ceramic and V is the volume fraction of porosity. Appendix 1 lists typical values of hardness and elastic modulus reported for various ceramic materials.

The hardness of an isotropic composite usually obeys the rule of mixture given by

$$H_{comp} = H_a V_a + H_b V_b + ... + H_i V_i, \tag{4.6}$$

where H_i represents the hardness and V_i the volume fraction of phase i, respectively. The hardness of Al_2O_3–TiC and Si_3N_4–TiC composites, for example, exhibit a roughly linear "rule of mixtures" dependence [4.8].

4.1.1.1 Hardness Testing

Among the most popular hardness testing methods are static indentation tests in which an indenter is forced to penetrate a flat surface. The size of the indentation is a measure of the hardness. There are two groups of indentation tests; macrohardness and microhardness.

Macrohardness tests use relatively large loads and produce large impressions. The indentations usually cover many grains of the materials surface at a time. Therefore, macrohardness tests provide a useful indication of the hardness of the bulk that contains pores, grain boundaries, and other discontinuities. The most practical macrohardness tests for ceramics are the Vickers (diamond pyramid) test and the Rockwell test with a Brale (diamond cone) indenter. Since most ceramics are harder than hard steel, the typical material of choice for metal hardness evaluation is diamond.

Two types of microhardness tests are the *Vickers* and *Knoop* tests; both use diamond indenters. Both tests employ loads ranging from several grams to one kilogram. The Vickers indenter leaves a square impression on flat surfaces, whereas the Knoop indenter leaves a very shallow, diamond-shaped impression. Under conditions where cracks form along with an indentation, hardness readings may be not very useful, since part of the applied load is used up by crack formation. The shallow Knoop indentation may prevent crack formation and therefore is better suited for microhardness testing of brittle materials [4.2].

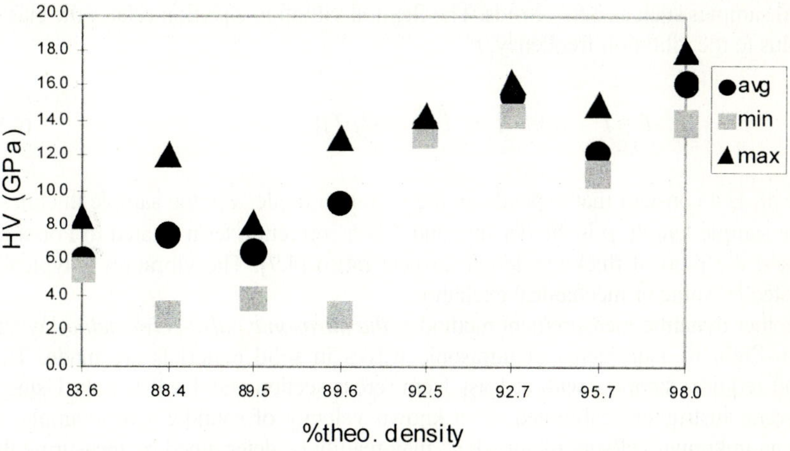

Fig. 4.1 Variation of microhardness with density in Al_2O_3–15vol%ZrO_2 (From M. Bengisu, Ph.D. Thesis, New Mexico Institute of Mining and Technology Socorro NM, 1992)

The Vickers hardness can be calculated from the indentation load, F_i, and the average length of the diagonals, d, of the impression by

$$HV = 1.8544\, F_i / d^2 . \tag{4.7}$$

The Knoop hardness is calculated from

$$KHN = 14.228\, F_i / h^2 , \tag{4.8}$$

where h is the length of the longer diagonal. Microhardness testing can provide information from single grains or from a few grains at a time if the grain size is considerably small. To find the average hardness that is representative of the bulk material, a series of indentation measurements is necessary. Since microhardness testing is done with an optical microscope and a measuring eyepiece, the surface of the material needs to have a very good polish. Surface preparation techniques used to obtain smooth surfaces are discussed along with optical microscopy in Sect. 5.2.1.

4.1.1.2 Testing of Elastic Modulus

Three basic methods for determining E are the strain measurement technique, the resonance technique, and ultrasonic techniques. *The strain measurement method* requires using strain gauges, extensometers, or optical methods during tensile or flexural strength testing to measure strain versus applied stress. The elastic modulus is calculated from the elastic portion of the stress–strain (σ–ε curve). The σ–ε curve will be discussed in detail in Sect. 4.1.2.

Resonance techniques involve determining natural resonant frequencies of simply shaped samples such as bars or rods. The flexural vibration equation relates the elastic modulus to the vibration frequency, f:

$$E = f^2 / \{[m_1^2\, d / (\pi\, l^2)]^2 / (48 \rho T)\} , \tag{4.9}$$

where m_1 is a constant that depends on the vibration mode, d is the sample thickness, l is the sample length, ρ is the density, and T is a correction term related to Poisson's ratio and the ratio of thickness/length (aspect ratio) [4.9]. The vibration is typically generated by sonic or mechanical excitation.

Another dynamic measurement method is *the ultrasonic pulse-echo method* where time-of-flight measurements of ultrasonic waves in solid materials are made. This method requires samples with at least 1 cm^2 cross section and 1 cm length. Using a pulse-echo instrument calibrated for a known velocity of sound c (for example, in steel), an unknown velocity of sound, c_x, may readily be determined by measuring the related (e.g., steel) equivalent thickness, d_s, of the specimen [4.10]. Then, the unknown transverse or longitudinal sound velocity, c_x, is determined from,

$$c_x = cd / d_s , \tag{4.10}$$

where d is the actual thickness measured mechanically or otherwise. The accuracy of the sound velocity measurement is determined by the calibration accuracy of the instrument, which usually does not exceed 1%. Using interferometers in conjunction with pulse-echo equipment, the accuracy can be improved to 0.1%. Once either of transverse or longitudinal sound velocities, c_t and c_l, is determined, the other can be calculated from

$$c_t = c_l \{ (1-2\nu) / 2(1-\nu) \}^{1/2} , \tag{4.11}$$

where ν is Poisson's ratio of the material. From these, Young's modulus can be calculated using

$$E = 4\rho c_t^2 \{[(3/4) - (c_t / c_l)^2] / [1 - (c_t / c_l)^2]\} , \tag{4.12}$$

where ρ is the density of the material. The shear modulus can also be determined from sound velocities:

$$G = \rho c_t^2 . \tag{4.13}$$

4.1.2 Strength

Strength may be loosely defined as the resistance of a material to deformation. Different strength measurement methods result in different strength values. Therefore, a strength value cited for a material makes sense only when the type of test is known. Among the most important strength tests used for ceramics are tensile, compression, three-point bending, and four-point bending tests. Figure 4.2 illustrates schematically these tests and the types of loads encountered.

4.1.2.1 Strength Testing

Tensile Strength

Tensile testing applies a uniform uniaxial load on a sample. Many brittle ceramics behave like ideal elastic materials, i.e., their stress–strain curves have the characteristic given in Fig. 4.3. In such materials, strain is reversible. In other words, upon loading, some elongation (generally negligible) occurs, and when the stress is removed, the material regains its original dimensions. At a certain stress level, the material can no longer resist the load, and fracture occurs. The strength corresponding to the fracture load is called the fracture strength or tensile strength. This type of

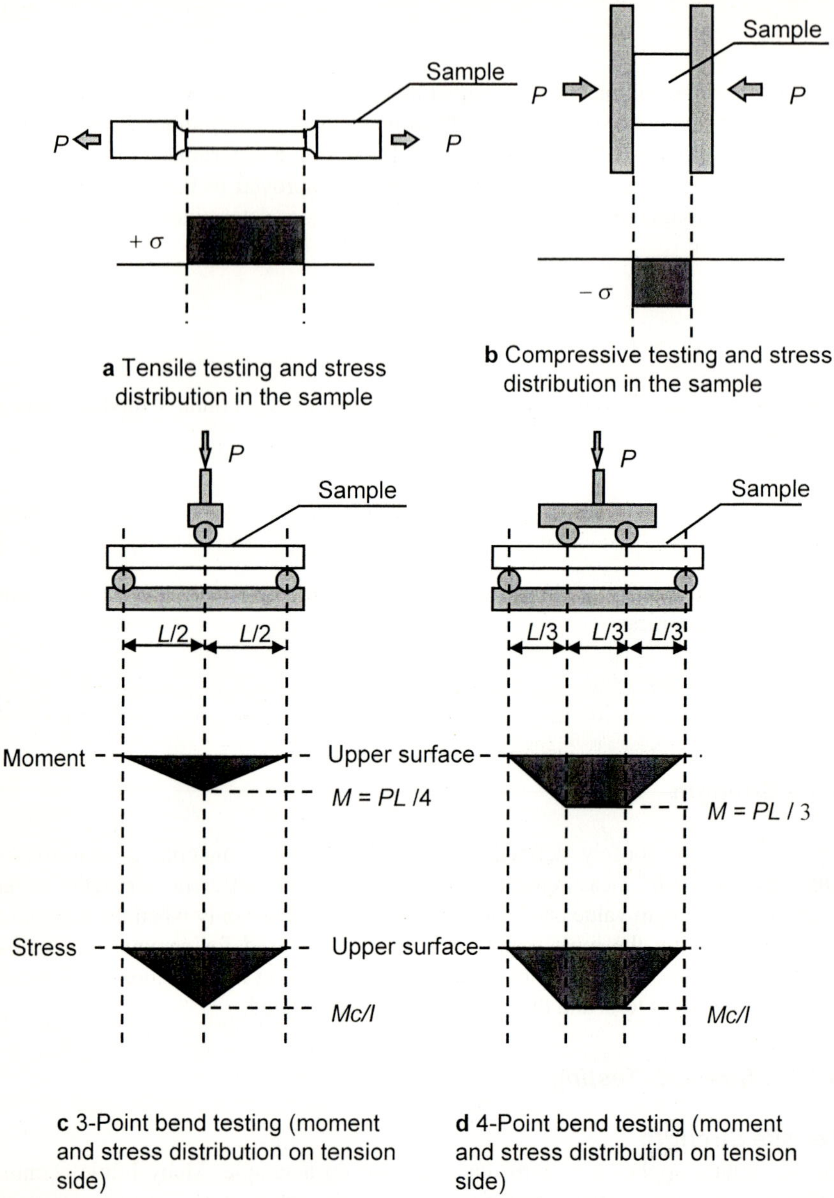

Fig. 4.2 a–d. Strength testing of ceramics

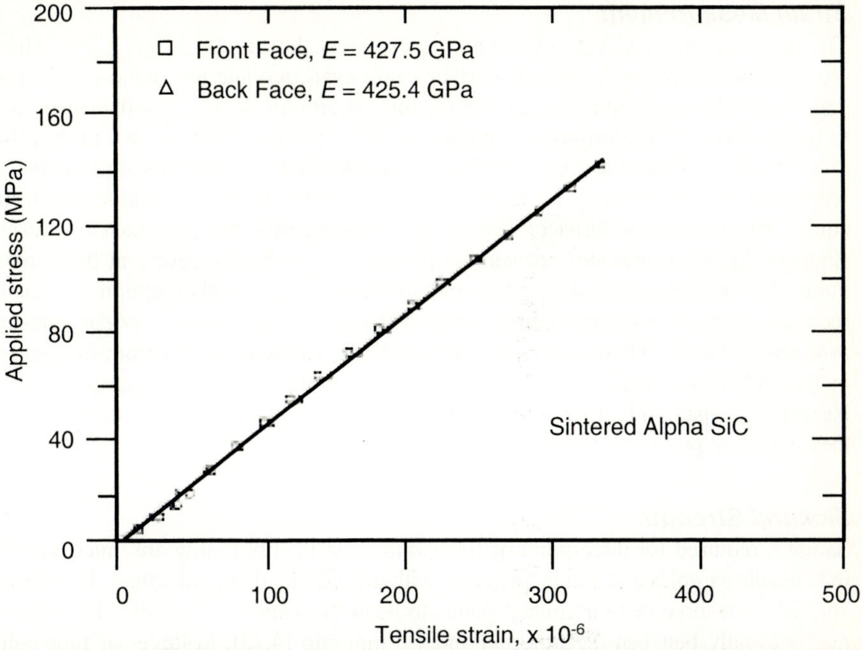

Fig. 4.3. Characteristic stress–strain diagram of an ideal elastic material (From S.G. Seshadri and K.Y. Chia: *Tensile Testing of Ceramics*. J. Am. Ceram. Soc. **70 [10],** C-242–244 (1987) copyright 1987 American Ceramic Society reprinted with permission)

material behavior is also called linear elastic behavior since stress (σ) and strain (ε) are linearly related:

$$\sigma = E\varepsilon, \tag{4.14}$$

where the proportionality constant E is termed the elastic or Young's modulus. The tensile strength can be found from the load, F, and the cross-sectional area, A, using

$$\sigma = F / A. \tag{4.15}$$

Sample preparation for tensile testing is difficult and costly for ceramic materials. Another problem is achieving a uniaxial stress distribution throughout the sample. Even a small misalignment may result in erroneous results due to undesired bending stresses [4.11].

Strain Measurement

The most accurate and versatile strain measurement devices are *strain gages*. These are normally glued on the sample surface using cyanoacrylate type adhesives. Strain gages exist that can operate up to 1000°C for a short term [4.12]. Another method is to use *mechanical extensometers*. These consist of a clip-on unit or two probes that touch on the test piece at a certain pressure. At elevated temperatures, extensometers with extended probes may be used. This option helps to keep sensitive electronic equipment outside the furnace. The contact pressure must be optimized to prevent slippage due to insufficient pressure or fracture due to high pressure at the contact points between the ceramic probes and the test piece. Another option especially recommended for elevated temperatures is the use of *noncontact optical tracking instruments* [4.13]. These instruments normally determine the strain from the change in the diffraction behavior of an optical beam as the specimen is strained. They basically consist of a high-energy laser beam and a detector. Thus, they are costly, but they eliminate problems of contact and high temperature.

Flexural Strength

Samples required for three-point or four-point bend flexure testing are much simpler than tensile samples and can be prepared with the aid of a diamond wheel. The load is applied from three or four contact points to bend the sample. The preferable loading rate is usually between 0.5 mm/min and 1.0 mm/min [4.12]. In three- or four-point bending, the flexural strength (also called *modulus of rupture*, *MOR*) is given by [4.5]

$$\sigma_{fl} = M(d/2)/I, \tag{4.16}$$

where M is the moment and I is the moment of inertia. For three-point bending,

$$M = (L/2)(F/2), \tag{4.17}$$

$$I = (bd^3)/12, \tag{4.18}$$

where L is the length between the two contact points, b is the width, and d is the thickness of the sample (Fig. 4.3). In four-point bending,

$$M = (F/2)\,a, \tag{4.19}$$

and I is the same as given by Eq. (4.18). The flexural strength test has been standardized for electronic-grade ceramics by ASTM under standard No. F417-78 (1984).

Although bend tests can also be used in composite ceramics, some caution is necessary. The stress state in CMCs, especially in fiber CMCs is nonuniform. Flexural strength results have little quantitative meaning if multiple matrix cracking occurs. Monolithic ceramics usually fail by tensile stresses at their surface, but CMCs may fail by tensile or compressive stresses, typically if the aspect ratio of the specimen is greater than 20. Shear failure will play an important role if the ratio is less than 5 [4.13]. Mixed mode fracture occurs in between. Once inelastic effects such as matrix cracking become active, the neutral axis in a flexural specimen shifts toward the undamaged side, and the stress distribution becomes indeterminant. Another problem is that in long fiber CMCs, lengthwise mode II shear failure or compressive failure by delamination and buckling may occur before tensile failure and thus may prematurely terminate the test [4.14]. Thus, flexure testing for fiber CMCs is comparative at best and totally inappropriate for design.

Surface preparation of ceramics is important in strength testing that involves tensile stresses. Imperfections such as cracks, pores, large grains, or even scratches may lead to stress concentration and crack growth from these origins. Therefore, well-polished surfaces are required for less scattered data. In compressive stresses however, such imperfections do not play important roles since cracks or crack-like imperfections tend to be closed under these stresses.

Frictional forces should be minimized in bend tests using rollers at contact points. Otherwise, these forces may give rise to bending moments that oppose those due to the applied load and lead to an artificial increase in fracture loads. Wedging or crushing at loading points may be minimized by using span to depth ratios of 8 or larger. Care should also be taken in arranging load points; asymmetrical load points lead to unequal stress distribution [4.15]. Other precautions that should be considered in bend testing are as follows:

1. Crosshead speeds of 1 mm/min or less should be avoided since at this range, corrosion effects at ambient and second phase effects at high temperatures are the highest.
2. Very high loading rates, e.g., 10^4 MPa/s may be used to avoid corrosive or high-temperature, low-viscosity effects.

Note that three-point bend testing ideally results in a maximum load at the center of the sample, whereas four-point bend testing creates a zone of maximum stresses. Generally, due to the higher probability of stress concentration sites in a maximum load zone, four-point bend strengths are substantially lower than three-point bend strengths [4.15].

Compressive Strength

Ceramic materials typically exhibit high compressive strength since cracks tend to close up and preexisting flaws have a decreased effect, compared to that under tensile stresses. Compression testing is not commonly used for ductile materials but is widely used for bricks, concrete, and to a lesser extent, for engineering ceramics. It is very difficult to measure the compressive strength without interference from tensile or flexural stresses since the applied stress is usually very high. Thus, precise alignment is necessary for realistic results. Specimens used for compression testing are usually short cylinders whose diameter to height ratio is 0.8–1.3 [4.16]. Contact surfaces of loading plates must be strong and hard. Thin ductile (Al or polyethylene) sheets can be inserted between contact surfaces to prevent microscopic misalignment.

In uniaxial compression testing, the compressive strength can be calculated by

$$\sigma_{\mathrm{comp}} = F_{\mathrm{f}} / A \,, \tag{4.20}$$

where F_{f} is the load at fracture and A is the area of the disc surface. Failure during compression testing occurs by growth and coalescence of defects followed by disintegration. Therefore, the compression strength does not depend on the largest, sharpest flaw, but on the total defect population [4.17]. Due to this fact, Weibull statistics (described in Sect. 4.1.2.3) are not useful for compression testing. A rough dependence of compressive strength on grain size can be found since the defect size is proportional to it. Compression testing of fiber-reinforced CMCs is not recommended (except for 3-D reinforced materials) because various problems tend to provoke premature failure due to noncompressive modes, including elastic spreading of the loading platens, nonaxial loading, and buckling instability. Some indirect compression testing methods exist that can be used to overcome the problems found in direct compression testing of fiber CMCs [4.14].

4.1.2.2 Theoretical Strength

The theoretical strength of a material can be estimated from the forces necessary to break atomic bonds. During the separation of two surfaces, attractive forces increase with the distance of separation to a maximum, and then they decrease. The maximum gives the theoretical cohesive strength. The stress–separation function can be derived from a potential energy–separation approximation described by Morse or Born potential functions. The *Morse function* describes the potential energy per unit area of fracture surface separated by a distance, x,

$$U = U_0 \left[e^{-2a(x-x_0)} - 2e^{-a(x-x_0)} \right], \tag{4.21}$$

where U_0 is the maximum potential at distance x_0 and a is a constant [4.18]. Derivation of U yields the stress distribution. The maximum stress is

$$\sigma_{\max} = (\gamma E / 4x_0)^{1/2} \,, \tag{4.22}$$

where γ is the surface energy, E is the elastic modulus, and x_0 is the equilibrium distance or interatomic spacing. From Eq. (4.22) and using appropriate properties, the calculated theoretical strengths of fused silica and sapphire were 2.2×10^{10} N/m^2 and 3.1×10^{10} N/m^2, respectively.

Strengths approaching theoretical values are achieved only in single-crystal fibers, i.e., whiskers. Table 2.3 lists strength levels for various ceramic whiskers. The strengths of polycrystalline fibers are generally higher than those of bulk ceramics but lower than those of whiskers because of the existence of some grain boundaries and surface imperfections. The sources of reduced strength levels in bulk ceramics are flaws such as pores, large grains, agglomerated grains, and various types of cracks. Such flaws act as sites of stress concentration when an external load is applied. Griffith [4.19] calculated the stress to fracture a brittle material assuming that two new surfaces are created during the generation or propagation of a crack. For a flat, isotropic plate that contains a straight crack of length 2 c, the calculated fracture stress was [4.2]

$$\sigma_f = [2E\gamma / (\pi c)]^{1/2} , \tag{4.23}$$

under plane stress conditions. Here, E represents the elastic modulus, and γ is the surface free energy. Comparison of fracture stresses calculated from this equation with measured values indicates that the Griffith equation yields much lower strengths. The discrepancy is usually attributed to the presence of a barrier initiating a crack or to processes that consume additional energy during crack propagation.

4.1.2.3 The Weibull Distribution

The Weibull distribution is one type of statistical distribution. Like the Gaussian distribution (the bell curve), it is used to represent the frequency of a phenomenon for the whole range of values that it exhibits. Metals generally have uniform structures and their plastic deformability enables minimizing the adverse effect of any defect. Thus, their strengths exhibit a Gaussian distribution. In designing with metals, a *deterministic approach* can be used. In this approach, the designer takes into account the maximum stress in the metal component (calculated by finite element analysis or other analytical or numerical methods) and applies a safety factor to account for situations such as a dynamic load [4.5]. Then, a suitable metal, for example, a class of steel, is selected, whose average strength (corresponding to the peak in the relevant Gaussian curve of strength values) is higher than the calculated maximum stress. In ceramic materials, strength values exhibit marked dependence on specimen size and geometry and wide scatter due to randomly distributed flaw populations. In such cases, a *statistical design approach* is most suited [4.5,4.20]. The Weibull distribution and related analysis is used mostly for designing with ceramics since it adequately fits a wide range of data.

The Weibull distribution, unlike the Gaussian distribution, is asymmetric; one tail extends more than the other (Fig. 4.4). The peak does not correspond to the arithmetic average. The Weibull distribution is derived from a model developed by Weibull [4.21] to describe the strength of filaments statistically. The cumulative probability of fiber failure due to applied stress, σ, has been derived as [4.22],

$$P = 1 - \exp\,[-L(\sigma/\sigma_0)^m] , \tag{4.24}$$

where L is the filament length, σ_0 is the scaling parameter, usually taken as the characteristic stress at which P = 0.632 (i.e., the stress corresponding to 63.2% cumulative probability of failure), and m is the shape parameter, whose value determines the shape of the Weibull curve [4.23]. The parameter "m" is better known as the *Weibull modulus*. Similarly, the probability of failure of a tensile or flexural specimen can be expressed as [4.17]

$$P = 1 - \exp\left[-\int_V (\sigma - \sigma_u)/\sigma_0\right]^m \mathrm{d}V\,, \tag{4.25}$$

where σ_u is the threshold stress below which failure will not occur and V is the specimen volume. In three-point bending, the effective volume becomes $V/[2(m+1)^2]$, and in four-point bending, it becomes $V(m+2)/[4(m+1)^2]$ [4.17]. A more simplified equation is derived for uniform uniaxial stress:

$$P = 1 - \exp\left[-V(\sigma - \sigma_u)/\sigma_0\right]^m\,. \tag{4.26}$$

For ceramic materials, σ_u is considered to be zero, resulting in the common form of the two-parameter Weibull distribution,

$$P = 1 - \exp\left[-V(\sigma/\sigma_0)\right]^m\,. \tag{4.27}$$

In practical terms, P represents the fraction of samples that fails at or below σ. Then, the reliability or probability of survival becomes [4.23]

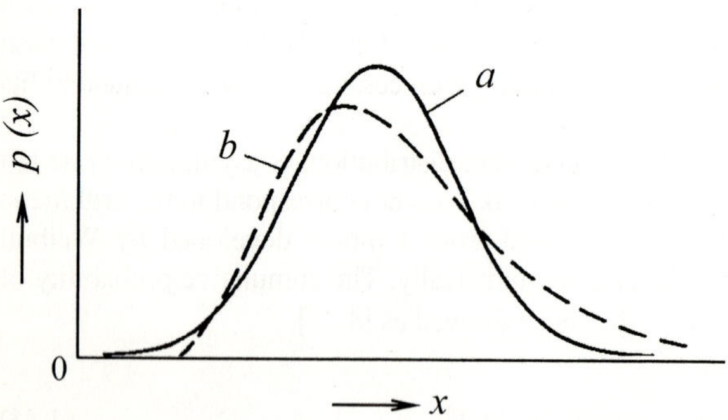

Fig. 4.4. The Gaussian and Weibull distribution (Reprinted from J. Mencik: *Glass Science and Technology 12: Strength and Fracture of Glass and Ceramics.* Elsevier Amsterdam Holland, 1992 with permission from Elsevier Science)

$$R = 1 - P = \exp\left[-V(\sigma/\sigma_0)\right]^m \quad . \tag{4.28}$$

Figures 4.5 and 4.6 show the variation of P versus applied stress for various values of m and σ_o, respectively. Figure 4.5 shows that a high value of m corresponds to a sharp switch from high reliability (zero probability of failure) to low reliability

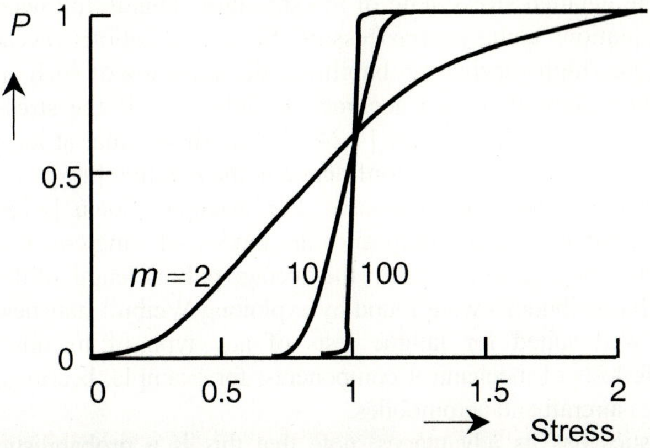

Fig. 4.5. Variation of cumulative failure probability, P, vs. applied stress, σ, for various values of m ($\sigma_o = 1$, L = 1) (From S.Van der Zwaag: *The Concept of Filament Strength and the Weibull Modulus.* J. Test. Eval. **17 [5]**, 292–98 (1989), copyright ASTM, reprinted with permission)

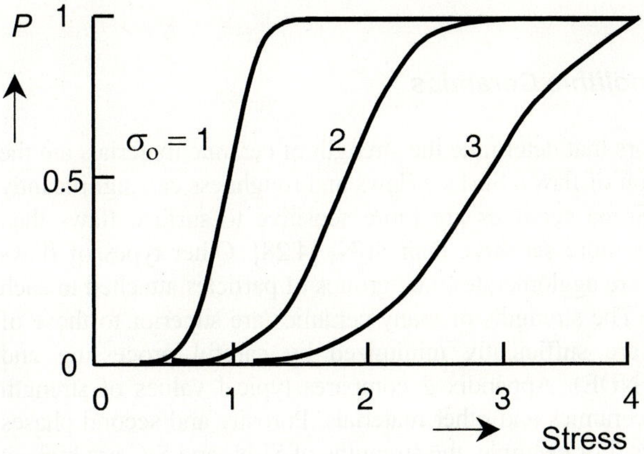

Fig. 4.6. Variation of cumulative failure probability, P, vs. applied stress, σ, for various values of σ_o ($m = 5$, $L = 1$). (From S.Van der Zwaag: *The Concept of Filament Strength and the Weibull Modulus.* J. Test. Eval. **17 [5]**, 292–298 (1989), copyright ASTM, reprinted with permission)

(definite failure), whereas a low value of m corresponds to a wide fracture stress distribution function. High values of m indicate a low scatter of strength values and vice versa. From figures similar to 4.5 and 4.6, it is possible to predict the survival probability of a ceramic sample for a given stress level. A more practical representation of failure probabilities is to plot $\ln[-\ln(1-P)] - \ln(L)$ versus $\ln(\sigma)$ since this yields a straight line with slope m and can be directly used to predict the survival probability [4.22] (Fig. 4.7). To determine the failure probability of a structure in which the stress is not uniform, finite element analysis may be performed to determine the detailed stress state of the structure. This is followed by solving the Weibull equations using postprocessors. Such calculations reveal that for high values of m, i.e., high survival probabilities, the accuracy of such an approach is high. For low values of m, the accuracy is only high if the stress volume integral is evaluated with high accuracy [4.24]. It was shown that at least 60 measurements are necessary to obtain 90% confidence in the m value [4.5].

Weibull statistics can be used for many purposes. For example, Sutcu [4.25] analyzed the fracture of ceramic matrix composites in modes of single-matrix cracking and multiple-matrix cracking. In this study, the average pullout length of the fibers and the pullout length distributions were found by exploiting Weibull statistics. Weibull analysis is also well suited for fatigue tests of any type of material. Therefore, it is used for life tests of mechanical components, for example, bearings, and structural components in aircraft and automobiles.

Although Weibull statistics has its advantages, note that this is a probabilistic approach not related to the physical nature of the fracture process and connected parameters such as flaw size, and it may not hold for all situations. Extrapolations to low stress values should be treated with caution since their accuracy is not well known [4.26]. Predictions of polyaxial failure from uniaxial data may lead to errors. For example, in predicting the biaxial strength of alumina disks based on four-point bending data, there was a significant discrepancy between predictions and experimental results [4.27].

4.1.2.4 Strength of Monolithic Ceramics

Some of the important factors that determine the strength of ceramic materials are the amount, size, and distribution of flaws. Surface flaws and roughness can significantly decrease strength values. Some ceramics are more sensitive to surface flaws than others. For example, SiC is more sensitive than Si_3N_4 [4.28]. Other types of flaws typically found in ceramics are agglomerates, i.e., groups of particles attached to each other, and pores (Fig. 4.8). The strengths of many ceramics are superior to those of metals when such flaws are sufficiently minimized by careful processing and nondestructive evaluation (NDE). Appendix 2 compares typical values of strength and fracture toughness for ceramics and other materials. Porosity and second phases greatly affect strength values. For example, the strengths of Si_3N_4 and SiC are highest in hot-pressed or hot isostatically pressed parts where little or no porosity is present. Lower strength values are found in liquid-phase sintered samples with larger amounts of second-phase materials such as SiO_2 and Y_2O_3. Generally, the lowest strength

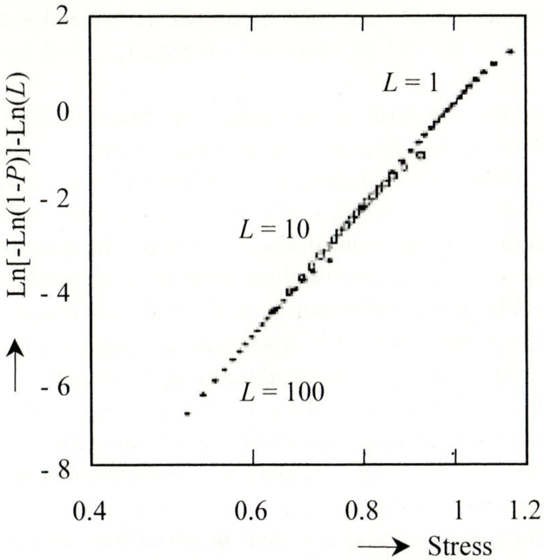

Fig. 4.7. P–σ_o data replotted in a Weibull plot (From S.Van der Zwaag: *The Concept of Filament Strength and the Weibull Modulus.* J. Test. Eval. **17 [5]**, 292–298 (1989), copyright ASTM, reprinted with permission)

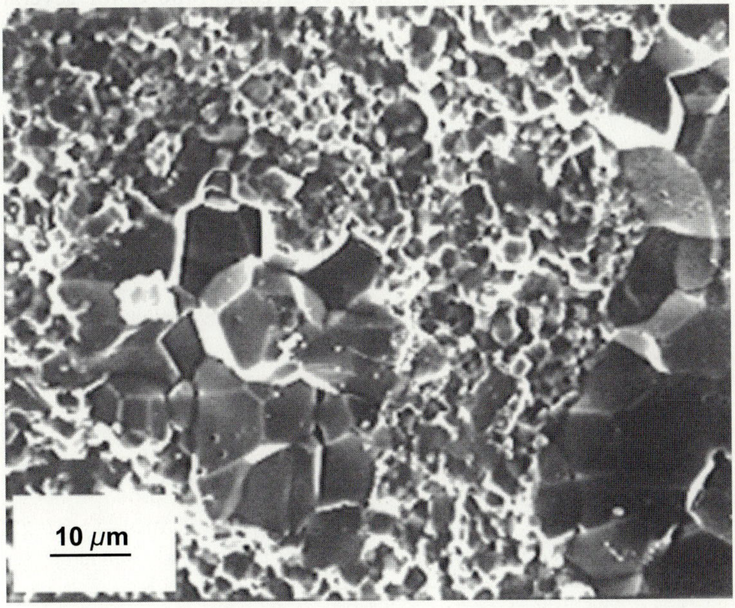

Fig. 4.8. Agglomerated grains and pores in Al_2O_3–YSZ (From M. Bengisu, Ph.D. Thesis, New Mexico Institute of Mining and Technology Socorro NM, 1992)

values are exhibited by reaction-bonded Si_3N_4 and SiC, as shown in Appendix 2. Both of these materials usually contain a significant amount of unreacted Si, and they are quite porous.

The strengths of ceramic materials decrease at elevated temperatures. Three regimes can be ideally distinguished, as represented schematically in Fig. 4.9. At lower temperatures (regime A), linear elastic behavior is exhibited. At higher temperatures, limited plastic deformation is observed. Plastic deformation may occur due to phenomena such as dislocation motion, grain boundary sliding, and viscous flow. The last regime is observed in only a few polycrystalline ceramics such as MgO and UO_2, which undergo considerable plastic deformation at elevated temperatures [4.29]. The critical temperature T_{crit} is 1200°C for UO_2. Above this temperature, slip occurs before fracture. Crack blunting is also possible, leading to slight increases in the fracture stress with increasing temperature [4.30]. Glassy phases may improve the strength up to a certain level above their softening temperatures due to healing of microcracks or cavities in the ceramic. Upon further heating, the glassy phase loses its strength, and a drastic strength decrease will occur [4.31].

Many ceramic materials are subject to *slow crack growth* in various media, including seemingly harmless water. Failure can occur at very low stress levels in materials such as SiO_2, Al_2O_3, MgF_2, and glasses, if exposure to reactive species such as water or ammonia is sustained under prolonged loading conditions [4.32]. Stress corrosion can be more severe at elevated temperatures and leads to strength degradation. The ambient temperature strengths of Si_3N_4 and SiC generally decrease from exposure to oxidizing environments at 1000–1400°C for several hundred hours [4.33].

In some cases, it has been observed that oxidation improves high-temperature strength through modification of glassy grain boundary phases and surface defect

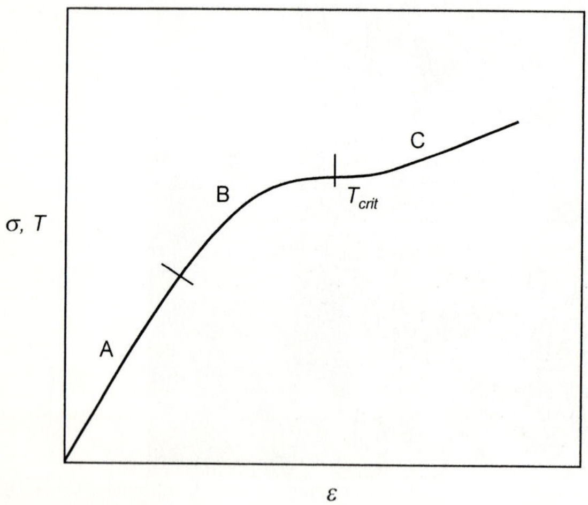

Fig. 4.9. Typical stress-strain behavior of ceramics at different temperature regimes: elastic regime A, plastic deformation B, and strain hardening C

healing [4.34]. Other types of corrosive media generally lead to accelerated pit formation at elevated temperatures that results in severe strength degradation. For example, strength degradation of SiC corroded by Na_2SO_4 and Na_2CO_3 at 1000°C correlates with the corrosion pit depth [4.35].

4.1.2.5 Strength of Ceramic Matrix Composites (CMCs)

4.1.2.5.1 Particulate CMCs

ZrO_2-toughened ceramics, including ZrO_2-toughened Al_2O_3 (ZTA) and ZrO_2-toughened mullite (ZTM) are an important group of particulate CMCs. Figure 4.10 shows the microstructure of an Al_2O_3–10 vol% ZrO_2 composite under the TEM. The strength of ZTA strongly depends on the type of ZrO_2 dispersed in the alumina matrix. If no tetragonal zirconia (t-ZrO_2) phase is present, the strength decrease will be roughly proportional to the amount of ZrO_2. According to the results of Claussen's study [4.36] toughening can still take place in the absence of t-ZrO_2. This has been explained in terms of microcracking, which is discussed in detail under toughening mechanisms. Strength deterioration occurs due to increased average flaw size and in some cases due to the presence of ZrO_2 agglomerates [4.36–4.38].

In contrast to unstabilized ZrO_2, stabilized ZrO_2 improves the and the fracture toughness of alumina. Typically, as the amount of t-ZrO_2 increases, both toughness and strength increase. This can be explained by the stress-fracture toughness dependence in ceramics. At a critical applied stress,

Fig. 4.10. Bright field TEM micrograph of Al_2O_3–10 vol%ZrO_2 composite revealing intergranular zirconia grains (see arrows) in the alumina matrix (From M. Bengisu, Ph.D. Thesis, New Mexico Institute of Mining and Technology Socorro NM, 1992)

catastrophical crack growth occurs in brittle materials. The strength for such a condition is given by [4.39]

$$\sigma_f = AK_c c^{-1/2}, \tag{4.29}$$

where A is a dimensionless constant depending on the crack and stress field geometries, c is the flaw size, and K_c is the fracture toughness. Thus, as the toughness increases, an increase in strength can normally be expected. However, although the highest strength is achieved in ZTA composites that contain the largest content of t-ZrO_2, the highest toughness is usually achieved at an optimal mixture of monoclinic zirconia (m-ZrO_2) and t-ZrO_2 phases. Therefore, a compromise is sometimes required between the two properties for certain applications that require both high strength and toughness.

Very high toughness and high strength have been achieved in tetragonal zirconia polycrystal (TZP) by special means. The simultaneous addition of Al_2O_3 and SrO to CeO_2-TZP (Ce–TZP) results in the formation of strontium hexaluminate ($SrAl_{12}O_{19}$) platelets during air sintering. The strength and hardness of Ce–TZP/Al_2O_3 with $SrZrO_3$ additions were superior to those of Ce–TZP with or without $SrZrO_3$ additions. These platelet-containing Ce–TZP composites have the four-point bend strengths (500-700 MPa) of Ce–TZP/Al_2O_3 particulate composites, and unlike the latter, the high toughness (14-15 $MPam^{1/2}$) can be maintained [4.40].

Certain ceramics can also be strengthened by the dispersion of strong particles such as carbides, nitrides, or borides, which are mainly covalently bonded. In Si_3N_4 matrices, fine particle addition usually does not change the strength significantly [4.41], as seen in Appendix 3. In oxide matrices, however, some strengthening is usually achievable. The matrix material most extensively studied is alumina. Generally, carbide or boride reinforcements decrease the sinterability of alumina [4.42, 4.43], and the resultant sintered composite product suffers from remnant porosity. Therefore, our discussion of strength values in composites will be limited to hot-pressed or HIP'ed parts with low or zero porosity.

The strength of a particulate CMC can be estimated by the rule of mixture. However, mechanisms that lead to a change in toughness such as microcracking or phase transformation may significantly alter the strength. For example, Niihara et al.'s study [4.43] of the Al_2O_3–SiC system indicated that strength at ambient temperatures may be improved by adding small amounts of SiC to Al_2O_3, but strength decreases above a certain volume fraction of SiC. The strength dependence with volume fraction correlated well with the fracture toughness. Deterioration of mechanical properties with increased amounts of SiC was attributed to microcrack linkage. Similar results were obtained in the Al_2O_3–ZrC system. The flexural strength increased from about 300 to 550 MPa upon addition of 5–10 mol% ZrC, above which it decreased to the initial value [4.44].

Al_2O_3–TiC composites are established cutting tool materials with improved strength, compared to Al_2O_3 alone [4.45]. High strength values can be retained at relatively high TiC content. The persistence of improved strength with large TiC content may be explained by the smaller difference in coefficients of thermal expansion (CTEs) between TiC and Al_2O_3, compared to SiC and Al_2O_3. The mechanical properties of Al_2O_3–TiC_p composites can be seen from Appendix 3. Large CTE differences between the matrix and second-phase particles usually increase the possibility of microcracking since the particles will be under high local stresses in such circumstances. Thus, particles with CTEs close to the matrix are preferred for strengthening.

Recent studies using platelets instead of particulates usually resulted in reducing strength. For example, Baril and Jain [4.46] observed that a sharp decrease in strength occurred upon adding SiC platelets to Al_2O_3, regardless of the platelet size. This was attributed to the role of platelets as fracture initiating defects and grain growth agents.

Dispersion or precipitation strengthening of glasses is a method similar to particulate reinforcement. Partial crystallization of separated phases may result in strong dispersions, and the overall strength can be increased thereby. Partial or full amorphous-to-crystalline transformation yields what is known as a glass ceramic. Under suitable conditions, certain glasses undergo phase separation or precipitation. For example, highly efficient nucleation is achieved in Li aluminosilicate glasses when equal amounts of TiO_2 and ZrO_2 are added at a total amount of 4 mol% [4.47]. Heating of glass of this composition to about 50°C above the glass transition temperature results in the precipitation of tiny $ZrTiO_4$ crystals <10nm in diameter. A solid solution of metastable β-quartz nucleates and grows on these precipitates. The resulting material is strong and has a very low CTE (7×10^{-7}/ °C; 0–500°C); thus it has outstanding thermal shock resistance (see Sect. 4.1.4). Glasses can also be strengthened by direct additions of hard particles. Again, the CTE difference between the glass and the reinforcement should be as small as possible. Glass–ZrO_2 [4.48, 4.49] and glass–Al_2O_3 [4.49] composites have been produced with improved strength, compared to the glass matrix.

4.1.2.5.2 Whisker/Short-Fiber-Reinforced CMCs

Mechanical properties of selected whisker-reinforced CMCs are listed in Appendix 4. Three strengthening mechanisms occur in whisker- or short-fiber-reinforced CMCs: *load transfer*, *matrix prestressing*, and *strengthening through toughening* [4.50, 4.51]. For simplicity, we will consider only whiskers here, but the same mechanisms are valid for short fibers, as well.

A certain length at each end of the whisker, l_t, is required for load transfer to the whisker by interfacial shear. Stresses and strains can be very high in the matrix around whisker terminations. Upon matrix cracking, whiskers whose ends are at a distance less than l_t from the major crack, will be pulled out if the whisker–matrix interfacial bond is not very strong. Whiskers with lengths less than the critical length $l_c = 2l_t$ will always be pulled out. The transfer length is given by [4.52]

$$l_t = \sigma_f r / (2\tau) , \tag{4.30}$$

where σ_f is the stress acting on the whiskers, r is the whisker radius, and τ is the interfacial shear strength. According to Homeny and Vaughn [4.50], significant load transfer strengthening by whiskers requires that $E_w / E_m > 2$, where E is the elastic modulus and w and m represent the whisker and matrix, respectively. For example, the ratio E_w/E_m is ~1.6 for Al_2O_3–SiC_w composites. This does not permit any significant strengthening through load transfer. Strong whisker–matrix bonding is necessary for strengthening by this mechanism. Note, however, that strong interfacial bonding is not desired for most whisker-toughening mechanisms to occur.

If $CTE_w > CTE_m$, compressive stresses will develop in the matrix. This can increase the overall fracture strength since initially the compressive stress has to be exceeded before any tensile stress can be applied to the composite. However, compressive stresses should be controlled carefully because, otherwise, they may lead to microcracking and debinding, and degrade strength.

From Eq. (4.29), it is expected that an increase in K_{Ic} may bring about a higher value of σ_f. Such a correlation is usually found in whisker-reinforced ceramics that are studied, as can be seen in Appendix 4.

Whisker-like morphologies can also be achieved in glass ceramics. Interlocking rod or blade-like crystals that serve as strengthening and toughening agents were obtained in certain systems such as potassium fluorrichterite and fluorcanasite glass ceramics. A random acicular microstructure accounts for the high strength and toughness of K–F–richterite glass ceramics. The microstructure consists of a stable assemblage of K–F–richterite plus cristobalite. The flexural strength increases from 150 MPa of the originating glass to 200 MPa upon crystallization. The fracture toughness is 3.2 MPa $m^{1/2}$. Internal nucleation in fluorcanasite is achieved through precipitation of CaF_2 crystallites and acicular growth of canasite upon these nuclei [4.47]. The resulting structure is essentially crystalline and monophase. Strength values near 300 MPa, Young's modulus near 80 GPa, and fracture toughness values of 5.0 MPa $m^{1/2}$ are typical. The high toughness values are contributed to crack branching and deflection by rod-like crystals mainly (see Sect. 4.1.3.2).

4.1.2.5.3 Continuous Unidirectional Fiber-Reinforced CMC's

A composite must have a minimum volume fraction of fibers to be actually reinforced. The critical (or transition) fiber volume fraction that must be exceeded to achieve a strength higher than that of the matrix alone is given by [4.53]

$$V_{crit} = (\sigma_{mu} - \sigma_m{}') / (\sigma_{fu} - \sigma_m{}') , \tag{4.31}$$

where σ_{mu} and σ_{fu} are the ultimate tensile strengths of the matrix and fiber, respectively, and $\sigma_m{}'$ is the matrix stress at the strain corresponding to σ_{fu}. Figure 4.11 illustrates the critical fiber volume fraction in terms of σ_{fu} and σ_{mu}. According to a simple *rule of mixture (ROM) model*, matrix cracking occurs at [4.52]

$$\sigma_{mc} = \sigma_{mu} V_m + \sigma_f^* V_f = E_m \varepsilon_u V_m + E_f \varepsilon_u V_f , \tag{4.32}$$

where $V_m = 1 - V_f$ is the volume fraction of the matrix phase, V_f is the volume

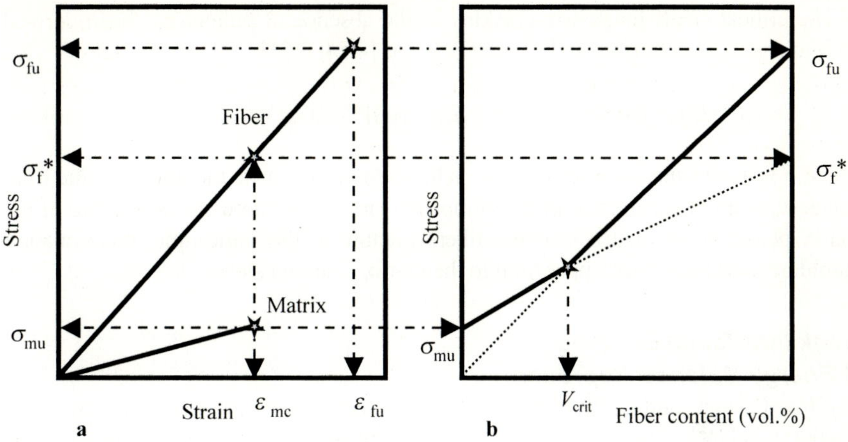

Fig. 4.11. Stress–strain curves **a** and fiber volume fraction–composite strength relationship **b** according to the ROM model

fraction of the fibers, E_m and E_f are elastic moduli of the matrix and the fiber, respectively, and σ_f^* is the fiber strength at $\varepsilon_{mc} = \varepsilon_u$ (ε_{mc} is the matrix cracking strain). After matrix cracking occurs, the stress will be supported by the fiber bundle; thus, the ultimate composite strength can be estimated by,

$$\sigma_{cu} = \sigma_{fu} V_f . \tag{4.33}$$

The ROM model assumes that the presence of fibers does not affect the matrix fracture strength, that matrix cracking does not change the fiber fracture strength, and that the fiber strength is uniform throughout. These, however, are incorrect assumptions in many practical situations. Let us now discuss models that take these factors into consideration.

Strong Interfacial Bonding

Zero or limited pullout will occur in strong fiber–matrix bonding. Figure 4.12 shows typical stress–strain curves for both strong and weak interfacial bonding. In a Saphikon (Al_2O_3) fiber-reinforced glass composite with no fiber coating, strong fiber–matrix bonding leads to high fracture strength with only elastic deformation. Note that in this case, fracture occurs once the fracture stress of the composite is reached. This type of failure is called catastrophic failure. In the second case, in which fibers are coated with SnO_2 to achieve a slippery fiber–matrix interface, considerable fiber pullout takes place. Matrix cracking occurs first, followed by fiber pullout, and finally fiber fracture. This behavior is generally desired in a composite material since catastrophic failure can have unpredictable and undesired results, as the term implies.

The critical stress for matrix cracking in the absence of pullout has been derived by Budiansky, Hutchinson, and Evans (BHE) as [4.54, 4.55]

$$\sigma_{mc} = B\, E_c \,\{(6V_f^2 E_f)\,/\,[(1-V_f)^2\, E_c\,(1+\nu_c)]\}^{1/4}\, \{G_m\,/\,rE_m\}^{1/2}\,, \tag{4.34}$$

where B is a constant close to unity, E_c is the composite elastic modulus, r is the fiber radius, ν_c is the Poisson ratio of the composite, and G_m is the work of fracture of the matrix phase. If the matrix and the fibers fracture at the same time, this equation should be used only as an upper limit to the matrix cracking stress.

Weak Interfacial Bonding

a) Energy Balance Approach

Aveston, Cooper, and Kelly [4.56] (ACK) estimated matrix cracking stress by using an energy balance. Their model takes into account such factors as the frictional work done by fiber pullout, the decrease of strain energy in the matrix, and the increase of strain energy in the fibers. The ACK model gives the matrix failure strain as follows [4.55]:

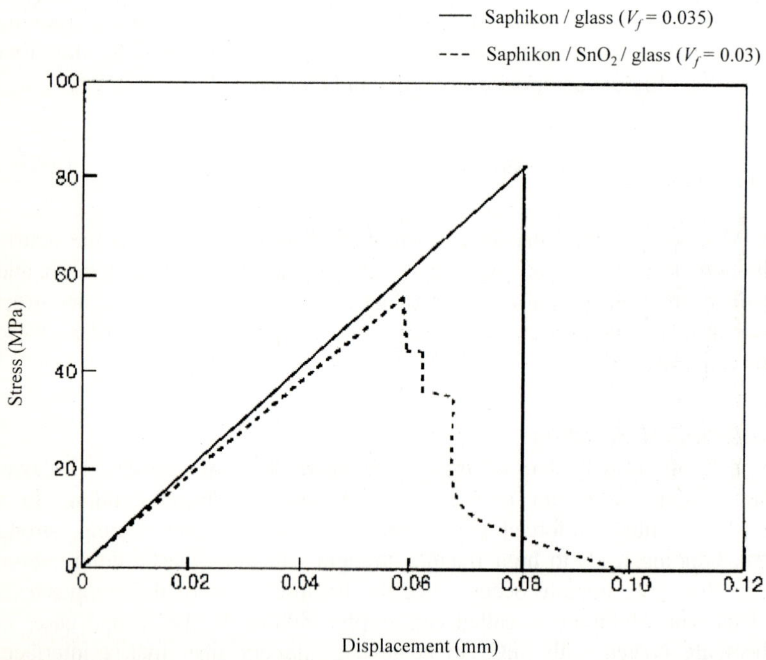

Fig. 4.12. Typical stress–strain curves for strong and weak interfacial bonding (From R. Venkatesh: *Interface Engineering of Alumina Fiber/Glass Composites*. Ph.D. Thesis, New Mexico Insitute of Mining and Technology Socorro NM, 1993, courtesy of R. Venkatesh)

$$\varepsilon_m = [(6\tau_f G_m E_f V_f^2) / (E_c E_m^2 r (1-V_f))]^{1/3}, \tag{4.35}$$

where τ_f is the frictional shear stress between the fibers and the matrix. The matrix cracking stress and the ultimate composite strength can be calculated from the following equations:

$$\sigma_{mc} = E_m \varepsilon_m, \tag{4.36}$$

$$\sigma_{cu} = E_c \varepsilon_m. \tag{4.37}$$

The BHE and ACK models predict that the matrix failure stress and strain above a certain fiber content are improved, compared to the unreinforced matrix, and the improvement is proportional to the fiber volume fraction.

b) Fracture Mechanics Approach

Analysis of a steady-state matrix cracking condition by fracture mechanics has also been performed [4.54,4.57]. In this analysis, three regions are distinguished in a unidirectional composite (Fig. 4.13). In region 1, the matrix crack is fully developed.

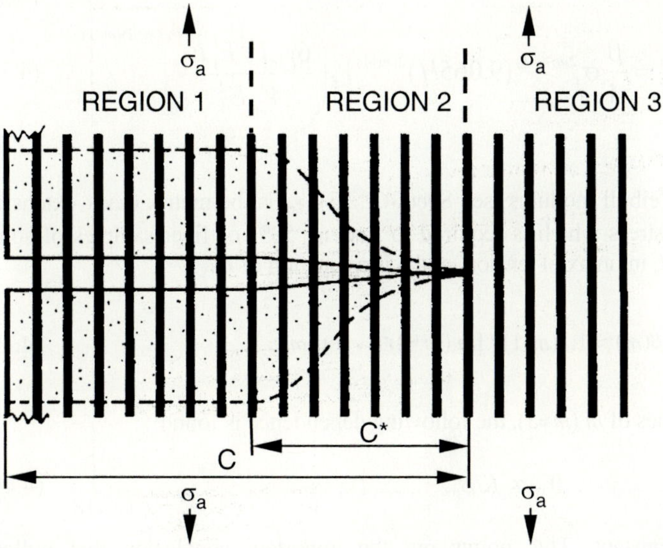

Fig. 4.13. Three regions in a unidirectional composite during steady-state crack propagation (reprinted from: R.W. Davidge and J.J.R. Davies: *Ceramic Matrix Fiber Composites: Mechanical Testing and Performance.* Int. J. High Tech. Ceramics. **4 [2–4]**, 341–358 (1988), with permission from Elsevier Science)

Region 2 is a transition zone where crack opening is still in progress. In region 3, no matrix crack has started yet. The stress required to propagate the crack to region 3 has been calculated and ends up the same as given by Eq. (4.36). Experimental results from various fiber–CMCs were also consistent with Eq. (4.36) [4.58]. The importance of the fracture mechanics approach lies in the fact that it enables defining matrix cracking stress as an intrinsic property of the Region 2 is a transition zone where crack opening is still in progress. In region 3, no matrix crack has started yet. The stress required to propagate the crack to region 3 has been calculated and ends up the same as given by Eq. (4.36). Experimental results from various fiber–CMCs were also consistent with composite since the crack tip stress is independent of the crack length, once the crack is longer than the equilibrium transition length C^*. As the analysis implies, C has to be larger than C^* (Fig. 4.13). The analysis also enables calculating the matrix crack size for the transition between crack-size-dependent behavior and crack-size-independent behavior.

Effect of Variable Fiber Strength Distribution

Fibers typically exhibit variable strength and pullout lengths. This is due to the random distribution and the size of flaws inside the fibers. The statistical nature of fiber failure has been analyzed by Sutcu [4.25, 4.59] and Thouless and Evans [4.60]. The analysis by Sutcu [4.59] yielded a relationship between the energy dissipation due to fiber pullout, W_p, and the critical matrix cracking stress:

$$W_p \sigma_{mc}^{(3m-3)(m+1)} = \frac{\beta}{4} \sigma_f^{2m(m+1)} (9.065H)^{2(m+1)} V_f \left(\frac{6\gamma_m V_f^2 E_f E_c}{V_m E_m^2} \right)^{(m+1)(m-1)} , \quad (4.38)$$

where m is the Weibull modulus (see Sect. 4.1.2.3), γ_m is the matrix crack extension energy, σ_f is the stress which is required to fracture 50% of fiber samples of some standard length, H, in uniaxial tension, and β is a function of m,

$$\beta(m) \approx 1.1(m-1) / [m\,(m+2)^2] \quad \text{for } m \geq 2\,. \quad (4.39)$$

For typical values of m ($m \approx 5$), the following dependence is found:

$$W_p \propto K/\sigma_{mc}^{2} \quad (4.40)$$

where K is a constant. This points out the important conclusion that pullout toughening and fiber strengthening are counteractive phenomena. Thus, an increase of σ_{mc} is possible only at the expense of W_p.

An equation was proposed to predict the ultimate strength of unidirectional composites. It takes into account the variable fiber strength [4.61, 4.62] but ignores interactive effects between failed fibers as well as stress supported by fractured fibers by stress transfer from the matrix through interfacial friction:

$$\sigma_{cu} = V_f \breve{S} \exp\left[-\frac{1-(1-\tau D / r\breve{S})^{m+1}}{(m+1)[1-(1-\tau D / r\breve{S})^{m}} \right], \tag{4.41}$$

where

$$(r\breve{S} / \tau D)^{m+1} = (A_0 / 2\pi r l)\,(r\sigma_0 / \tau D)^m \;[1-(1-(\tau D / r\breve{S})^m)]^{-1}\,. \tag{4.42}$$

In these equations, D is the spacing between adjacent matrix cracks, and A_0 and σ_0 are constants related through [4.63]

$$\overline{\sigma} = \sigma_0 (A_0 / 2\pi r l)^{1/m}\, \Gamma(1+1/m)\ , \tag{4.43}$$

where $\overline{\sigma}$ is the fiber strength of length l, Γ is the gamma function, and $\breve{S}$ is given approximately by

$$\breve{S} = m^{-1/m}\, \sigma / \Gamma(1+1/m)\,, \tag{4.44}$$

where D is defined as

$$D = 1.337\, V_m \sigma_{mu}\, r / (2V_f\, \tau)\,. \tag{4.45}$$

Experimental Studies

Continuous fibers made from various ceramic materials that have a wide range of elastic moduli, densities, and strengths, are commercially available today. The production and properties of such fibers have been discussed earlier in Sect. 2.4. A number of important fiber-reinforced CMCs have been developed for a variety of applications. Among these are C–C, glass–fiber, SiC–SiC_f, Al_2O_3–SiC_f, and Si_3N_4–SiC_f, where the subscript f represents the fiber. Some of these composite systems will be discussed in this section.

C–C composites have found a wide range of applications such as brake shoes in jet planes, leading edges in the space shuttle [4.64], and bioceramic implants [4.65, 4.66]. Three types of C matrices are generally used: thermoplastic (TP), resin-derived thermosetting (TS), and chemically vapor deposited (CVD) carbon. Figure 4.14 shows two examples of C–C composites. In all of the C–C composites, a high degree of bonding exists between the fibers and the matrix, resulting in high strength but low toughness. Several directions of fiber reinforcement may occur, including bi-directional laminates, three directional orthogonal weaves, and multidirectional weaves and braids. Mechanical properties of C–C composites are shown in Appendix 5. The strength of C–C composites can be altered significantly by surface treatment of carbon fibers and heat treatment of the composites [4.67]. The flexural strength of C–C composites with surface-treated C fibers, carbonized at 1000°C and heat treated at 2000°C was higher than those, which were only carbonized at 1000°C. Heat treatment at 2000°C facilitated fiber pullout during failure, probably due to differential shrinkage during cooling and resulting in weak fiber–matrix interfaces. The flexural strength of carbonized composites of untreated fibers was about 100 MPa higher than those carbonized and heat treated at 2000°C. In the latter, extensive microcracking, which may have caused strength degradation, was observed by TEM. The average flexural strength of composites with surface-treated fibers was significantly higher than those with untreated fibers; maximum values were 1325 MPa versus 342 MPa, respectively. Corresponding flexural modulus values were 195 MPa and 131 MPa, respectively. Strength values of 3-D reinforced C–C composites range between 150 and 300 MPa. Strength values of felt and randomly oriented short-fiber reinforcements range between 50 and 100 MPa. A characteristic ratio of 1/1.4 for flexural/tensile strength is usually observed in C–C composites [4.68].

C is readily oxidized at high temperatures, and therefore C–C composites are not suitable for elevated temperature applications in oxidizing atmospheres. To extend the application of C–C composites to such conditions, a mixture of matrix materials has been developed. The oxidation resistance of composites, including C/BN–C_f, C/B_4C–C_f, and C/SiC–C_f is better than C–C alone. A coating is usually applied to their surfaces for improved oxidation resistance [4.69].

Composites with glass or glass–ceramic matrices incorporating various fibers have been extensively studied due to their superior mechanical properties, compared to the corresponding matrix properties. Processing them is generally easier and cheaper than crystalline matrix CMCs due to the viscous deformation behavior of glass, usually at 600–1000°C; this allows using conventional hot-pressing and filament winding techniques.. Studies in glass or glass–ceramic matrix composites indicate that significant fiber pullout can usually be achieved in them, leading to tensile strain values of 0.5% and high fracture toughness. Fiber orientation is critical in these composites since only a 5° misalignment from the tensile axis can lead to 50% strength reduction and a 10° misalignment leads to an 80% strength reduction [4.55]. Typically, mechanical properties are poor in transverse directions in tough glass or glass–ceramic matrix composites due to weak interfaces. Therefore, cross-plied or two-dimensional layups are desired for most applications. The properties of some of these composites are listed in Appendix 6.

Studies of *alumina and mullite matrix-ceramic fiber composites* revealed that uncoated fibers do not improve strength or toughness, but coated fibers provide significant improvements. Four-point bend strengths of composites with uncoated Nicalon or uncoated AVCO SiC fibers were about 150 MPa, whereas 40 vol% coated Nicalon fibers imparted flexural strengths approaching 1000 MPa and those with AVCO SiC fibers had strengths up to 785 MPa [4.70]. A similar trend

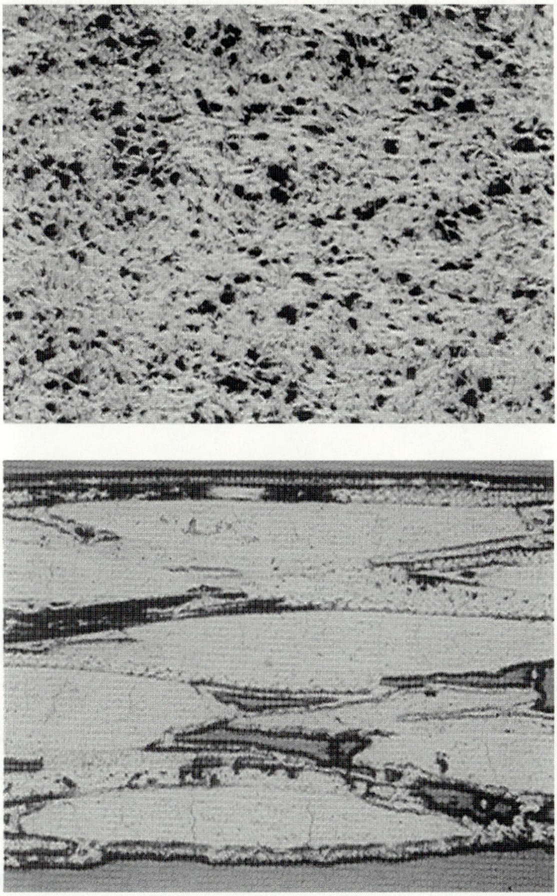

a

b

Fig. 4.14. Microstructure of a pitch-derived C–C composite **a** and continuous PAN-derived fibers in a CVD-derived matrix **b** (Courtesy of Jim E. Sheehan, MSNW, Inc. San Marcos CA) (Magnification: 50X)

is observed in mullite–SiC_f composites. Load–deflection curves of mullite reinforced with as supplied SiC fibers that have a carbon-rich surface and SiC fibers with a BN coating illustrate that the low interfacial shear stress allows more fiber pullout and consequently, a tougher composite [4.71]. Failed cross-sections of the mullite–SiC composite show fiber pullout and crack deflection. A combination of fiber pullout, crack deflection, and microcracking improves overall toughness.

SiC–SiC_f composites were developed especially for high-temperature applications because of the good oxidation resistance of SiC. No strengthening is expected in these composites since they are tough composites with weak interfaces and also since the fiber and matrix strength are similar. Typically, such composites are fabricated via chemical vapor infiltration (CVI) of Nicalon SiC fiber preforms. Unidirectional fiber composites with 40–45 vol% fibers and 25–32% open porosity exhibited MOR values of 210–410 MPa [4.72]. Much higher modulus of rupture (MOR) values (about 900 MPa) were also reported for similar composites [4.73].

4.1.3 Fracture Toughness

Toughness is a term used to define the resistance of a material to crack formation and propagation. It is also defined as the ability of a material to absorb energy before failure:

$$U = \int_0^{\varepsilon_u} \sigma(\varepsilon)\, d\varepsilon \ , \tag{4.46}$$

where ε_u is the failure strain, σ is the applied stress, and ε is the strain [4.74]. The *work of fracture* (*WOF*) is defined as the total energy absorbed by the specimen per unit area, calculated from the area under the load-displacement curve:

$$\mathrm{WOF} = \int_0^{\Delta_{max}} \frac{F(\Delta)}{A}\, d\Delta \ , \tag{4.47}$$

where A is the cross-sectional area, F is the load, and Δ is the elongation. U is a measure of the fracture initiation energy, and WOF is a measure of the total energy absorbed until the crack fully travels through the material. The plane strain fracture toughness, K_{Ic}, can be considered a material parameter. K_{Ic} is the critical stress intensity above which a crack will propagate and result in failure of the material. The subscript I in the K_{Ic} term represents mode I displacement, i.e., the load is perpendicular to the crack. Other displacement modes are mode II, sliding mode, and mode III, tearing mode. K alone is called the stress intensity factor, which is a quantity that depends on the geometry of the crack and the type of loading. K is a measure of the amplitude of the stress field around the crack tip. It will vary from one system to another, but when it reaches the critical stress intensity, K_{Ic}, it will remain constant for that material, regardless of the crack or load type [4.2].

For brittle materials, the Griffith criterion yields the following relationship for plane strain conditions under mode I displacement [4.16]:

$$K_I^{\,2} = G_I\, E \,/\, (1-\nu^2) \ , \tag{4.48}$$

and for plane stress:

$$K_I^{\,2} = G_I\, E\,, \tag{4.49}$$

where G is the energy release rate, i.e., the amount of energy released by a unit increment of fracture surface. For fracture to take place in mode I displacement,

$$K = \sigma Y a^{1/2} \geq K_{Ic}\,, \tag{4.50}$$

where Y is a constant depending on the crack configuration and loading geometry and a is half the crack length. This criterion assumes the following general relationship for oblique cracks:

$$K_{Ic} = \sigma\, Y\, (a_{ef})^{1/2} = \sigma\, Y\, (a \cos \beta)^{1/2}\;, \tag{4.51}$$

where β is the angle between the actual crack plane and the plane perpendicular to the stress direction, whereas a_{ef} is the effective half crack size (the component of the actual crack size on the said plane).

From Eqs. (4.48–4.51) it can be deduced that fracture toughness is proportional to the elastic modulus of the material. This relationship can be clearly observed from Fig. 4.15, where the fracture toughness increases as the elastic modulus within the same glass family (Al_2O_3–P_2O_5) increases.

In an ideally elastic (brittle) material, crack propagation occurs catastrophically if the crack size is above the critical size, a_c. At a crack size $a < a_c$, subcritical crack growth may occur due to environmental effects, i.e., stress corrosion resulting in crack growth to the critical size and failure. The environmental effect on crack growth is best represented by a crack velocity (V)–toughness (K_I) curve [4.75, 4.76] (Fig. 4.16). Segment I corresponds to chemically enhanced crack growth. Below the threshold value, K_{th}, cracks do not propagate. Segment II (plateau region) corresponds to continuum behavior, where minor species in a liquid or gas control the crack tip reaction. The velocity at which this plateau appears is determined by the diffusion of the active constituent to the crack tip and does not depend on the tested material. Segment III is independent of environment. Microstructure, material thickness, and stress have a major influence on the crack growth rate within this segment. Only the last segment of Fig. 4.16 is observed if the environmental effect is negligible.

If the ceramic material is not ideally elastic, in other words, if it is toughened, it generally exhibits R-curve behavior. This behavior manifests itself by a rising toughness (T) vs. crack size (a) curve, as shown in Fig. 4.17. Above a critical crack size, a_c, steady state toughness is reached. R-curve behavior is typical of ceramics with large grain size and ceramics toughened by ZrO_2, whiskers, or fibers. The slope of the R-curve is an indication of the toughening rate during crack opening [4.77].

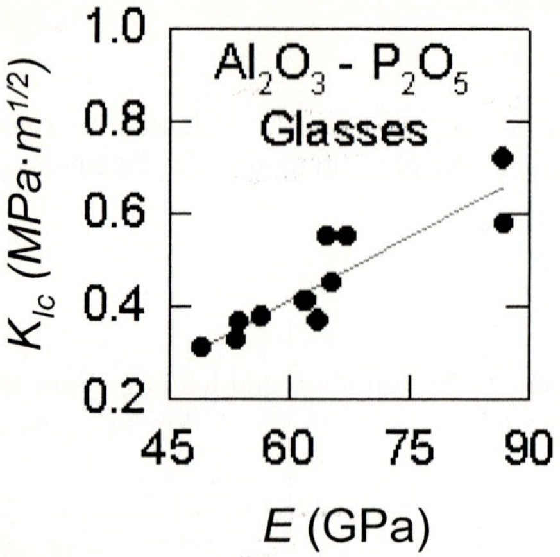

Fig. 4.15 Relationship between elastic modulus and fracture toughness of Al_2O_3–P_2O_5 glasses (From NIST Ceramic Division, Property Data Summaries for Advanced Materials http://www.ceramics.nist.gov/srd/summary/advmatdb.htm - S.W. Freiman, T.L. Baker, and J.B. Wachtman, Jr.: *A Computerized Fracture Mechanics Database for Oxide Glasses*, NBS Technical Note 1212 (National Bureau of Standards, 1985)

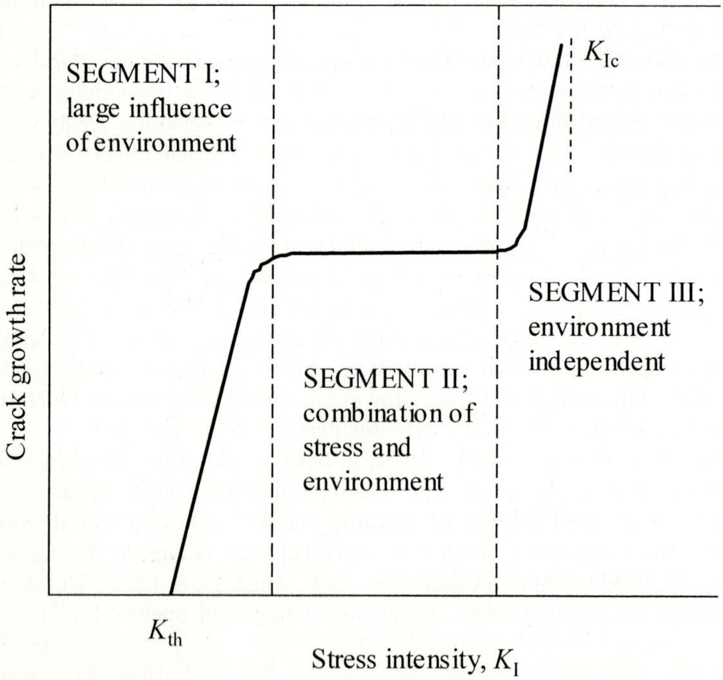

Fig. 4.16. Generalized crack growth rate–stress intensity behavior

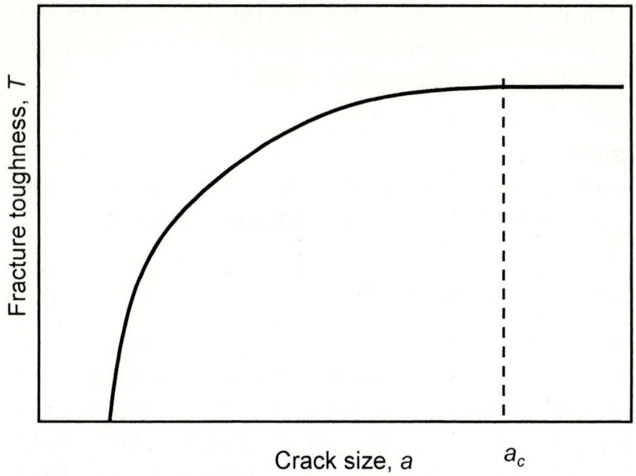

Fig. 4.17. Rising toughness with increasing crack size (R-curve behavior)

4.1.3.1 Fracture Toughness Measurement

A large number of fracture toughness measurement methods for ceramics have been devised. The only method accepted as a standard among these, until now, is the single-edge precracked beam (SEPB) method. This method was adopted in 1990 as a Japanese industrial standard for fine ceramics (JIS R 1607) [4.78]. The most important testing methods together with their pros and cons are explained following.

4.1.3.1.1 Double-Cantilever Beam (DCB)

In this method, the specimen geometry has many variations. A side groove is placed in the specimen to initiate a sharp crack. Upon loading, crack propagation occurs in a constant plane. Two modes of the test are constant load, suitable for the constitution of V–K_I curves in transparent bodies, and constant displacement, suitable for calculating V–K_I curves in opaque bodies. The applied stress intensity is given by [4.79]

$$K_\mathrm{I} = 2F\,(M / ww')\,, \tag{4.52}$$

$$M = 1/w + 3a^2 / h^3\,, \tag{4.53}$$

and the fracture toughness can be calculated from

$$K_\mathrm{Ic} = 3.45Fa\,[1 + 0.7(h/a)] / (bh^3)\,, \tag{4.54}$$

where M is the moment, F is the load, w is the specimen width, w' is the width of the grooved section (if tapered), a is the instantaneous crack length, and h is the distance from the crack plane to the edge of the specimen (Fig. 4.18c).

4.1.3.1.2 Double Torsion

This method requires a considerable amount of material for testing. Specimen preparation is relatively easy. The fracture toughness is determined by [4.5]

$$K_{Ic} = FW_m\,(3(1+v)\,/\,(Wd^3d_n))^{1/2}\,, \tag{4.55}$$

where W_m is the moment arm, v is Poisson's ratio, W is the specimen width, d is the specimen thickness, and d_n is the thickness of the groove (Fig. 4.18d). Precracking by controlled loading is a requirement in this testing method. Unlike the DCB method, no special loading fixtures are required. A disadvantage of this method is that the crack front is curved so that K_I may not be uniform. Nevertheless, the results of this technique closely agree with those of other techniques.

4.1.3.1.3 Short Rod–Short Bar

These methods do not require a precrack due to the unique shape of the notch. A chevron notch shown in Fig. 4.18e–f is machined into the specimen, providing a starting point for the crack under load. Relatively small samples can be used. However, loading of the sample can again be complicated. The following expression was derived for the short rod configuration [4.80]:

$$K_{Ic} = (AF_c)\,/\,[D^{3/2}\,(1-v^2)^{1/2}]\,, \tag{4.56}$$

where $A = 20.8$, a dimensionless constant, F_c is the peak load to fracture, and D is the rod diameter. The fracture toughness of the short bar specimen is calculated from [4.81]

$$K_{Ic} = [\,F_c\,/\,(B\sqrt{W}\,)]\,Y_m^*\,, \tag{4.57}$$

where Y_m^* is given in the following polynomial form:

$$\begin{aligned} Y_m^* = &-0.36 + 5.48(W/H) + 0.08(W/H)^2 \\ &+ [30.65 - 27.49(W/H) + 7.46(W/H)^2]\,(a_o/W) \\ &+ [65.90 + 18.44(W/H) - 9.76(W/H)^2]\,(a_o/W)^2\,. \end{aligned} \tag{4.58}$$

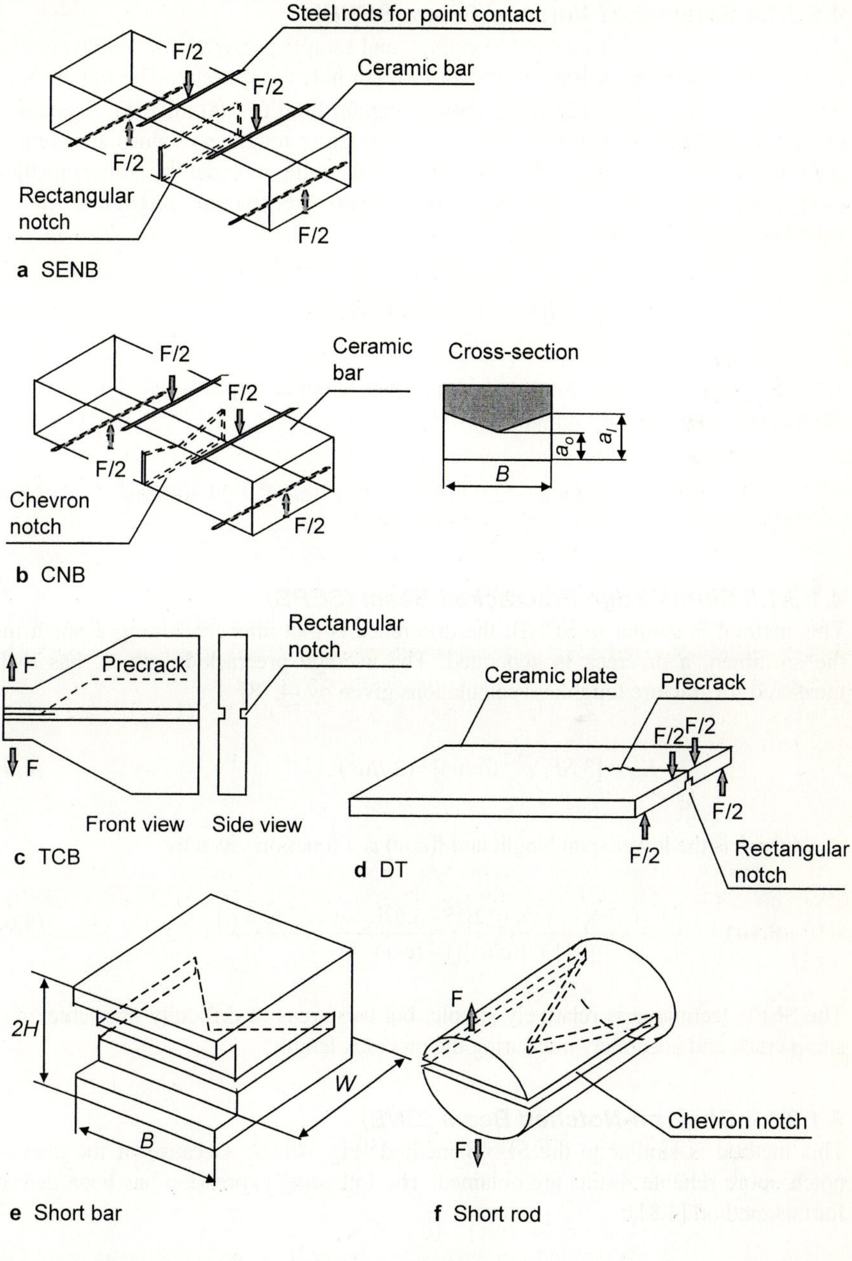

Fig. 4.18. K_{Ic} measurement by single-end notched beam (SENB) **a**, chevron-notched beam (CNB) **b**, tapered double-cantilever beam (TCB) **c**, double torsion (DT) **d**, short bar **e**, and short rod **f** methods.

4.1.3.1.4 Single-End Notched Beam (SENB)

Although this is a simple method to employ and sample preparation is relatively easy, producing a sharp crack from a straight-through notch is difficult. The notch should be as thin as possible [4.82] since above a certain limit (~ 100 μm), K_{Ic} depends on the notch thickness [4.83]. Generally, higher fracture toughness values are obtained with the SENB technique than found in the same material by other methods [4.82,4.84]. In four-point loading (Fig. 4.18a), the fracture toughness can be calculated from [4.84]

$$K_{Ic} = 3\,F_c\,[(L_o - L_i)\,Y a^{1/2}]\,/\,(2\,Bw^2)\,, \tag{4.59}$$

where L_o and L_i are the outer and inner span lengths, respectively, a is the notch depth, and Y is a calibration factor given by

$$Y = 1.99 - 2.47(a/w) + 12.97(a/w)^2 - 23.17(a/w)^3 + 24.80(a/w)^4\,. \tag{4.60}$$

4.1.3.1.5 Single-Edge Precracked Beam (SEPB)

This method is similar to SENB; the difference is that after machining a notch into the specimen, a precrack is generated. The average precrack length, c, has to be measured for fracture toughness calculations given by [4.78]

$$K_{Ic} = [3\,SF_c\,c^{1/2}\,\mathrm{f}(c/w)]\,/\,(2\,Bw^2)\,, \tag{4.61}$$

where S is the lower span length and $\mathrm{f}(c/w)$ is a function given by

$$\mathrm{f}(c/w) = \frac{1.99 - (c/w)(1-(c/w))\,[2.15 - 3.93(c/w) + 2.7(c/w)^2]}{[1+2(c/w)]\,[1-(c/w)^{3/2}]}\,. \tag{4.62}$$

The SEPB technique is relatively simple, but two areas of difficulty are achieving a sharp crack and accurately measuring the precrack length.

4.1.3.1.6 Chevron-Notched Beam (CNB)

This method is similar to the SENB method (Fig. 4.18b). Because of the chevron notch, more reliable results are obtained. The following expression has been derived for this method [4.81]:

$$K_{\mathrm{Ic}} = \frac{F_{\mathrm{c}}}{Bw^{1/2}}(3.08 + 5.00\alpha_0 + 8.33\alpha_0^2)\frac{(L_{\mathrm{o}} - L_{\mathrm{i}})}{w} \times \left[1 + 0.007\sqrt{\frac{(L_0 L_{\mathrm{i}})}{w^2}}\right]\left[\frac{(\alpha_1 - \alpha_0)}{(1-\alpha_0)}\right], \quad (4.63)$$

where $\alpha_0 = a_0/w$ and $\alpha_1 = a_1/w$.

4.1.3.1.7 Direct Crack Measurement (DCM)

Although considerable effort has been spent in developing a theoretical basis for evaluating K_{Ic} by DCM methods, residual stresses in samples, environmental factors, and many other parameters complicate the results of this test. Therefore, generally DCM should not be used to evaluate absolute values of K_{Ic}. However, due to its simplicity, it is advantageous to use it for comparative purposes, for example, to compare the toughness values of similar materials produced by different methods. For well developed cracks formed upon indentation ($c>2a$) (Fig. 4.19), Anstis et al. [4.85] derived

$$K_{\mathrm{c}} = 0.016(E/H)^{1/2}(F_{\mathrm{i}}/c^{3/2}), \quad (4.64)$$

where E is the elastic modulus, H is the microhardness, F_{i} is the indentation load, and c is the average crack length. An accuracy of 30–40% is expected from this

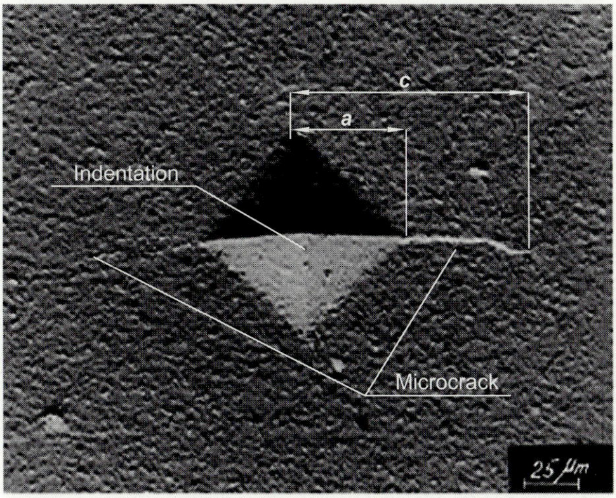

Fig. 4.19. Optical light micrograph of an alumina sample showing a diamond indentation for K_{Ic} measurement by direct crack measurement (DCM)

expression. Many other expressions have been derived. A detailed review of these expressions and their applicabilities was prepared by Ponton and Rawlings [4.86].

4.1.3.1.8 Indentation Strength in Bending (ISB)

This is among the most convenient techniques and has the added advantage of reasonable reliability. A polished surface is indented by a diamond pyramid resulting in sharp corner cracks. The indented surface is then exposed to tensile stresses by three- or four-point bending (Fig. 4.20). The load needs to be selected carefully to ensure that the crack size is above the critical size that yields useful results. This method has also been frequently used to derive R-curves of various ceramic materials [4.87–4.89]. In such studies, the indentation load used on different samples is increased to increase the initial crack size. an R-curve can be constructed from a series of a–K_c measurements including a_c–K_{Ic}. K_{Ic} can be calculated from [4.91,4.92]

$$K_{Ic} = 0.59(E/H)^{1/8} (\sigma_f F_i^{1/3})^{3/4} , \tag{4.65}$$

where σ_f is the three- or four-point fracture strength (see Sect. 4.1.2).

4.1.3.2 Toughening Methods and Mechanisms

Ceramic materials have been known as brittle materials until recently due their intrinsic brittle nature. Appendix 2 lists typical fracture toughness values of monolithic ceramics. The brittleness or lack of plastic deformability in ceramic materials may be explained by the following:

1. Due to ionic/covalent bonding, mostly nonsymmetrical crystal lattice, and the limited number of slip systems that do not fulfill the requirements described by the Von Mises rule [4.92], dislocation motion in ceramics is severely restricted. Only a few ceramics such as MgO, TiC, and UO_2 have cubic structures that allow slip, especially at elevated temperatures and in single crystals. In the polycrystalline form, even such ceramics are brittle at ambient temperatures, due to the insufficient number of slip systems. According to the Von Mises criterion, five independent slip systems are required for general homogeneous plastic deformation, i.e., for unimpeded dislocation motion from one grain orientation to another. However, this number is only two for MgO at 0–1700°C and three for UO_2 at 700–1200°C. It is noteworthy that TiC, which has a cubic NaCl type structure similar to MgO, has five independent slip systems above 900°C, allowing appreciable plastic deformation [4.29].
2. Inherent flaws incorporated during fabrication of ceramic materials act as stress concentration sites and aid brittle failure.

Thanks to the detailed understanding of these problems and development of solutions by continuous research effort, ceramic materials are being considered for applications not imagined before. Apart from low toughness, a main disadvantage

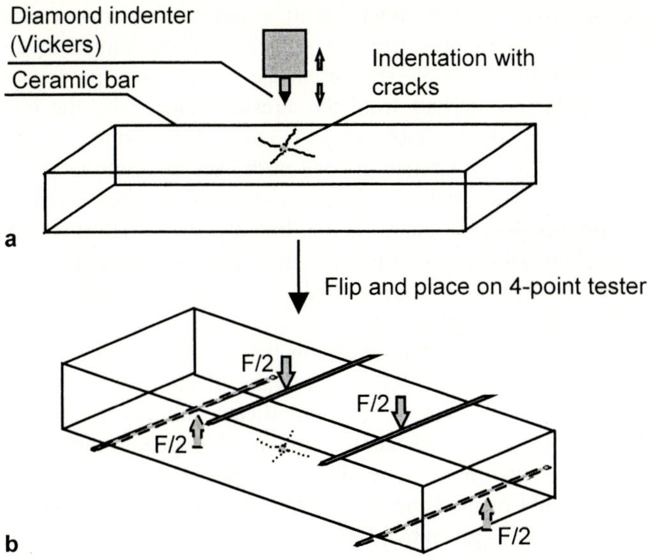

Fig. 4.20. K_{Ic} measurement by the indentation strength in the bending (ISB) method; indentation **a** and four-Point bend testing **b**

of ceramic materials is their high cost. The reason for high costs lies basically in the high reject levels during the production of components [4.93]. Structural imperfections, especially internal and surface flaws, are the major cause of high reject levels. Increased fracture toughness levels can significantly reduce the number of parts rejected during production and increase reliability during service. For this reason, there has been a considerable amount of research conducted on toughening ceramics. The most important contributions to this area are zirconia toughening, discovered by Garvie et al. [4.94] in 1975; whisker toughening, first observed by DeBoskey and Hahn [4.95] in 1967 and implemented by Tamari et al. [4.96] in 1982; and short-fiber [4.97] and continuous fiber toughening [4.69,4.98,4.99], which are extensions of the whisker-toughening concept. Other methods include dispersion toughening [4.100–4.102] and ductile-particle toughening [4.103,4.104]. Various mechanisms are responsible for observed toughness increments. These are discussed in detail in the following sections.

4.1.3.2.3 Transformation Toughening

The martensitic transformation of ZrO_2 from the high temperature tetragonal form to the ambient temperature monoclinic form is accompanied by about a 4% volume increase and 7% shear distortion [4.38]. The spontaneous athermal transformation in

ZrO_2 particles can be prevented by the constraint of a surrounding matrix or by alloying. Small dispersions of tetragonal zirconia particles will transform to their stable monoclinic polymorph under tensile stresses in the vicinity of a propagating crack. Transformation toughening derives from the stresses induced by the transforming particles, due to the volume increase associated. The residual stresses induced in the matrix around the transformation zone push back and act to close the propagating crack (Fig. 4.21). Thus, a higher stress is required to open the crack, rendering the material tougher than the matrix alone. Initially, the transformed zone does not produce any shielding of the crack tip and there is no toughening. However, as the crack grows, the transformation zone extends over the crack surfaces, and toughening increases, as defined by an R-curve.

It is found experimentally that, once the constraint that causes transformation is removed, which is the case for particles left in the wake of an advancing crack, reverse transformation is possible. However, a significant number of particles remains in the monoclinic form. Let us consider the amount of toughening for a crack completely surrounded by a transformed zone. The assumptions are that the mean stress determines the half height of the zone (h) and that the transformation strain is purely dilatational. The value of h is given by [4.38]

$$h = [3^{1/2}\,(1+\nu)^2 / (12\pi)\,]\,[K_{Ic}{}^{m} / \sigma_{mc}]^2\,, \tag{4.66}$$

where ν is the Poisson ratio, $K_{Ic}{}^{m}$ is the matrix toughness, and σ_{mc} is the critical mean stress at which transformation can occur. Toughening is calculated as [4.105]

$$\Delta K_{Ic} = \eta E^{c} V e^{T}\,[h / (1-\nu)]^{1/2}\,, \tag{4.67}$$

where η is a factor that depends on the zone shape at the crack tip and has a typical value between 0.22 and 0.38, E^{c} is the composite modulus, V is the volume fraction of transforming tetragonal zirconia particles, and e^{T} is the transformation (dilatational) strain. As suggested by Eq. (4.67), toughening is proportional to the

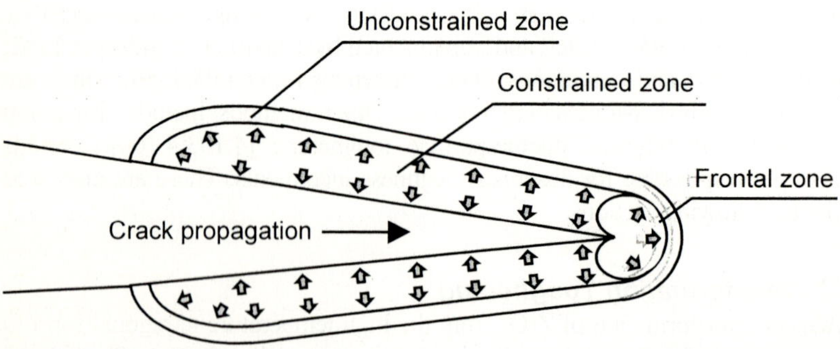

Fig. 4.21. Stresses arising in a dilatant transformation zone within a ZrO_2-toughened matrix

volume fraction of tetragonal zirconia particles that transform to the monoclinic polymorph under induced stress. This trend can be observed in Fig. 4.22, which illustrates measurements made in *zirconia-toughened alumina (ZTA).*

Several conditions are required for transformation toughening to occur [4.106]. First, the matrix and zirconia should not react to form a new phase. Second, the dispersed zirconia phase should not be soluble in the matrix and vice versa. In addition, the following parameters must be optimized to maximize toughening:

1. the particle size of the zirconia,
2. the stabilizing phase concentration,
3. the particle size distribution, and
4. the particle–matrix thermal expansion mismatch.

Unstabilized particles that are larger than a critical size will transform spontaneously during cooling, whereas particles that are too small will not transform, even under stress. The critical particle size for ZTA is 0.6 μm [4.107].

Many ceramic materials can be toughened by incorporating tetragonal zirconia. These include Al_2O_3 [4.108], ZrO_2 [4.94], MgO [4.106], mullite [4.109], TiB_2 [4.110], and whisker-reinforced mullite [4.111] and alumina [4.112]. Very high strength (>2 GPa) ceramics with toughness values of 5–8 MPa $m^{1/2}$ have been fabricated in ZTA and tetragonal zirconia polycrystal (TZP) systems [4.38]. In the MgO–partially stabilized zirconia (MgO–PSZ) system, toughness values in the range of 15 MPa $m^{1/2}$ have been achieved, but the strength is limited to 600 MPa [4.39]. Therefore, a trade-off between high strength and high toughness exists.

The fracture toughness of zirconia-toughened ceramics (ZTCs) decreases as the service temperature is increased, due to a decrease in the amount of t-ZrO_2. This behavior is the major limitation of the transformation-toughening mechanism [4.113].

4.1.3.2.4 Microcrack Toughening

The concept of microcrack toughening is derived from studies of ZTCs. Although a number of theoretical treatments relate the toughness increment in brittle solids to thermally or stress-induced microcracks [4.114-4.117], there is only one fully validated example of this mechanism, which is the Al_2O_3–ZrO_2 system [4.118]. Ceramic structures that contain localized residual stresses are susceptible to microcracking. Such stresses can arise as a result of phase transformations, thermal expansion anisotropy in monolithic ceramics, and thermal expansion or elastic mismatch in composite ceramics. It has been shown that there is generally a critical particle size for microcracking [4.119]. Spontaneous microcracking will not occur below the critical size. However, microcracking can be induced at subcritical particle size by applied tension. For microcrack toughening to occur, stress-induced microcracking must be generated within a limited zone of stress concentration near the crack tip.

Toughening arises from two types of crack shielding processes. One is due to the reduced elastic modulus of the microcracked material and the other to dilatation caused by microcracking [4.117]. Evans and Faber showed that total toughening due

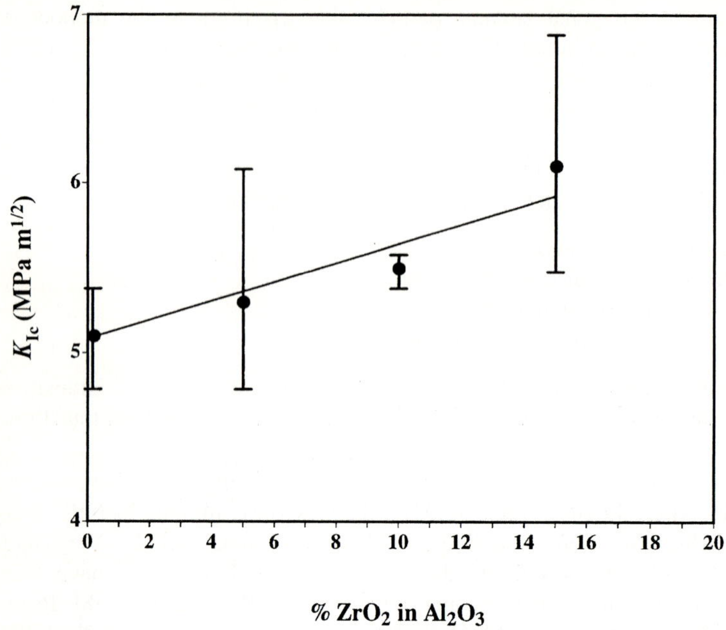

Fig. 4.22. Increase in fracture toughness with increasing ZrO_2 content in ZTA composites (From M. Bengisu, Ph.D. Thesis, New Mexico Institute of Mining and Technology Socorro NM, 1992)

to these phenomena can be expressed as [4.117]

$$K_c^{comp} / K_c^M = (1 + a f_s) + (bEf_s h^{1/2}\, \theta \, / K_c^M)\,, \tag{4.68}$$

where K_c^{comp} and K_c^M are toughnesses of the composite and the matrix, respectively, a and b are constants, f_s is the saturation microcrack density, E is the modulus of the material without microcracks, h is the microcracking process zone width, and θ is dilatation. For noninteracting spheres that influence microcracking under a residual stress, σ_R, toughening is given by [4.38]

$$\Delta K_{Ic} = 1.12 V \sigma_R\, h^{1/2} (1 + \nu)\,, \tag{4.69}$$

where V is the volume fraction of spheres. Microcracks associated with a phase transformation in which there is a volume increase, such as the martensitic transformation of zirconia, are expected to form radially from second-phase particles into the matrix. For this situation, toughening is derived as

$$\Delta K_{Ic} = 0.21 E e^{T} h^{1/2} (V + 0.6 V^{1/3}) / (1-\nu) , \qquad (4.70)$$

where e^{T} is the volumetric strain of the transformation. For similar process zones, simultaneous microcracking and transformation are a more effective shielding process than transformation alone. However, overall, transformation shielding is expected to be more effective according to these equations.

Rühle et al. [4.107,4.118] showed by TEM analysis that microcracking can occur in ZTA. It was reported that when the amount of t-ZrO_2 increased relative to m-ZrO_2 in a 15vol% ZrO_2-Al_2O_3 composite, the fracture toughness decreased slightly, indicating the complementary nature of microcracking and transformation toughening in ZTA [4.107]. At the other extreme, Karihaloo [4.120] showed that, when the ZTA composition contains mostly t-ZrO_2 precipitates, the primary source of toughening is stress-induced transformation and only a very minor contribution comes from thermally formed m-ZrO_2. It was also shown that no microcracking could occur in this material during stress-induced transformation because of the reduction in the hydrostatic stress generated by the transformation. However, if the critical hydrostatic stress needed to nucleate microcracks is far less than that initiating the martensitic transformation, microcracking can still occur.

4.1.3.2.5 Ferroelastic Domain Switching

Recent studies of tetragonal ZrO_2 polycrystals resulted in the development of a new theory to explain toughening in these materials. According to this theory [4.121], stress-induced alignment of the c axis of t-ZrO_2 along the maximum principal stress axis results in a pure shear type shape change. Virkar and Matsumoto [4.121] suggested that the unusual toughness levels in nontransforming TZP could be due to this phenomenon.

Li et al. [4.122] reported that domain switching occurs only in biaxial compression and that the critical stress for this mechanism under bending can exceed the bend strength of the material. They concluded that thermal activation may lower the critical stress sharply, but at lower temperatures, domain switching is not likely to be an important toughening mechanism.

Contrary to Li et al.'s opinion, Srinivasan et al. [4.123] and Virkar and Matsumoto [4.124] showed that domain reorientation occurred on fracture surfaces of bend bars made of Y_2O_3-stabilized TZP (Y–TZP) as well as on ground surfaces of Y–TZP and CeO_2-stabilized TZP (Ce–TZP). These observations were based on X-ray diffraction. They also reported that only trace amounts of m-ZrO_2 were observed on these surfaces, eliminating the possibility of transformation toughening. Considering these concepts, a Ce–TZP ceramic with toughness in excess of 16 MPa $m^{1/2}$ has been fabricated. It has also been suggested that a contribution to transformation toughening might occur due to ferroelastic domain switching in transformable zirconia.

As the contradictory results from different studies suggest, the actual contribution of the ferroelastic domain switching mechanism in TZP is unclear, and further study of this subject will probably resume. For the time being, ferroelastic domain switching is considered a possible high-temperature mechanism since it is expected to be less temperature sensitive than transformation toughening [4.105].

4.1.3.2.6 Crack Bowing

Crack bowing originates from resistant second-phase components in the path of a propagating crack. Lange [4.125] suggested that the interaction of a crack front with two or more inhomogeneities in a brittle matrix can increase its length and therefore the fracture energy along with the strength. Green [4.126] related the change in fracture surface energy to the fracture toughness change when crack bowing occurs, by the following equation:

$$\sigma_c/\sigma_m = K_c / K_m = (E_c\Gamma_c / E_m\Gamma_m)^{1/2} , \tag{4.71}$$

where σ_c is the stress needed to propagate a crack through a series of obstacles, σ_m is the matrix fracture strength, K_c and K_m are critical stress intensity factors , E_c and E_m are elastic moduli, and Γ_c and Γ_m are fracture surface energies, whose subscripts c and m denote the composite and matrix, respectively. Experimental data were fitted to an analytical expression by a polynomial,

$$\sigma_c/\sigma_m = K_c/K_m = \sum_{n=0}^{n=7} = A_n\,(r'/c)^n , \tag{4.72}$$

where A_n are numerically derived coefficients for the polynomial fit, r' is the distance from the primary crack front to the origin of the secondary crack at the breakaway position, and $2c$ is the average interparticle distance. r' is a function of obstacle dimension and can be modified to account for obstacle penetrability. Faber and Evans [4.127] further developed the crack bowing theory. Their predictions for crack bowing between spheres, discs, and rods reveal that the highest toughening is achieved in high aspect ratio disks within a brittle matrix.

The crack bowing theory is not fully developed and not very predictive. Experimental studies related to this theory are rare, and a detailed theoretical model does not exist at this stage. Instead, empirical solutions are available.

4.1.3.2.7 Crack Deflection

Faber and Evans [4.127] analyzed crack deflection phenomena in brittle materials and calculated possible toughness increases for composites containing dispersions of various shapes. In their theory, crack deflection is differentiated from crack bowing in that deflection produces a nonplanar crack, whereas bowing produces a nonlinear crack front. Two conditions can produce a nonplanar crack: residual strain in the composite and weak matrix/second-phase interfaces. Residual strain can be generated due to elastic modulus and/or thermal expansion mismatch between the matrix and second phase and anisotropy of thermal expansion in a monolithic or composite material. The sign of residual strain determines the direction of deflection, although its magnitude has no significant influence on the magnitude of deflection toughening.

The strain energy release rate for a deflected crack is given by [4.127]

$$G = (1/E)\,[k_1^{\,2}\,(1-\nu^2) + k_2^{\,2}\,(1-\nu^2) + k_3^{\,2}\,(1+\nu)]\,, \tag{4.73}$$

where k_i are local stress intensity factors. Comparison of G with the corresponding strain energy release rate for an undeflected (planar) crack, G^m, yields relative toughening as follows;

$$G_c^{\,T} / G_c^{\,m} = G^m / G\,, \tag{4.74}$$

where $G_c^{\,m}$ is the strain energy release rate of the matrix and $G_c^{\,T}$ is the strain energy release rate of the composite. If it is assumed that twist and subsequent tilt occur along one-half of the crack front, the total strain energy release rate of whisker-reinforced composites is given by

$$G_{rod} = (\alpha/2)G^T + (\beta/2)G^t\,, \tag{4.75}$$

where α and β are ratios of undeflected to deflected crack front lengths and G^T and G^t are strain energy release rates due to twist and tilt of the crack front, respectively.

Crack deflection is commonly observed in many ceramic matrix composites and is mostly accompanied by other mechanisms such as crack bridging and whisker pullout that can lead to toughening. Figure 4.23 shows various phenomena, including crack deflection, that can typically occur in whisker-reinforced CMCs. Figure 4.24a shows a SiC whisker-reinforced Al_2O_3 composite where the deflection of an indentation-produced crack was observed by SEM. The typical crack path in a brittle ceramic with no deflection is shown for comparison in Fig. 4.24b. Liu et al. [4.128] modeled toughening by crack deflection for whisker-reinforced CMCs with whiskers randomly oriented in two dimensions to account for whisker orientation in hot-pressed composites. Carter and Hurley [4.129] measured deflection angles in $MoSi_2$–SiC_w composites. Their measurements show that a 20vol% whisker addition to the matrix increases median deflection angles from 7.2° to only 13°, but toughness is increased from 5.3 to 8.3 MPa $m^{1/2}$ (≈50%). Experiments by Faber and Evans [4.130] on samples of Al_2O_3- and ZrO_2-modified SiC showed that 50 and 100% increases in fracture toughness correlated with increases in the average deflection angle from 7° to 22° and from 7° to 35°, respectively. It can be concluded from the relatively small differences in deflection angles that crack deflection alone cannot account for the observed toughness increases.

4.1.3.2.8 Whisker and Fiber Pullout

4.1.3.2.8.1 Whisker Pullout

Composites that exhibit fiber debonding and frictional sliding have pullout contributions to toughness. The same is true for whisker-reinforced ceramics. Two recent analyses that describe pullout toughening in brittle matrices will be considered here, those of Becher et al. [4.131] and Wells and Beaumont [4.133].

Becher et al. considered the energy change or the work done by sliding whiskers in their calculations and derived the following equation for pullout toughening [4.131, 4.132]:

$$\Delta K_p = [(G_o + \Delta G_p)E_c / (1-\nu)]^{1/2} - K_o \ , \tag{4.76}$$

where E_c and ν are Young's modulus and Poisson's ratio of the composite, respectively, K_o is the matrix toughness, G_o is the crack extension force of the matrix , and ΔG_p is the toughening term due to pullout which has the form,

$$\Delta G_p = (4V_f \tau_i^2 l_p^3) / (3E_w r^2) \ . \tag{4.77}$$

In this equation, V_f is the fiber (whisker) volume fraction, τ_i is the interfacial shear strength, l_p is the average whisker pullout length, E_w is Young's modulus of whiskers, and r is the whisker radius. According to this equation, toughening by the whisker pullout mechanism can be maximized by increasing the whisker content, the interfacial shear strength between whiskers and the matrix, the pullout length, and by decreasing the whisker radius. Note that trying to increase the interfacial shear strength would always decrease the pullout length since some interfacial movement has to be possible for pullout to occur. Otherwise, premature whisker failure will occur during matrix microcracking.

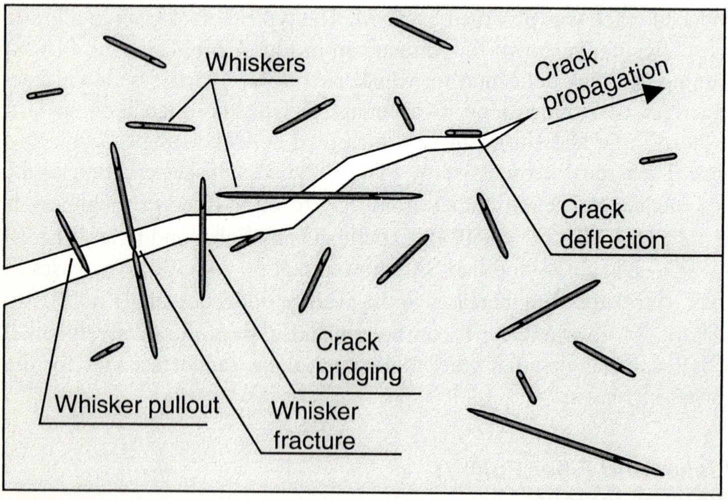

Fig. 4.23. Schematic illustration of various toughening mechanisms that occur in whisker CMCs

The analysis of Wells and Beaumont [4.133] considers composite systems that contain short brittle fibers. Their calculations are based on a brittle polymer–whisker composite system, but the same analysis can conveniently be applied to whisker CMCs as well, since the matrix type has no direct effect on toughening in their model. Three contributions are addressed in this analysis: toughening by interfacial debonding, fiber fracture, and fiber pullout. Toughening by debonding is given by

$$\Delta G_d = 8\, l_d\, \Gamma_i\, V_f / d \qquad \text{for } b > l_d/2\,, \tag{4.78}$$

$$\Delta G_d = 8\, b\, \Gamma_i\, V_f / d \qquad \text{for } b < l_d/2\,, \tag{4.79}$$

where d is the whisker diameter, Γ_i is the interfacial fracture energy, b is the depth of the fiber end buried beneath the surface of the crack plane, and l_d is the maximum length of debonded crack. Toughening by fiber fracture is calculated from

$$\Delta G_f = (V_f/E_f)\frac{\{\sigma_p^2\, l_d/2 - [(\sigma_p - \sigma_d)^2\, (e^{-\beta l_d} - 1)]}{2\beta + 2\sigma_p\,(\sigma_p - \sigma_d)(e^{-\beta l_d} - 1)/\beta}\,, \quad \text{for } b > l_d/2 \tag{4.80}$$

where β is an elastic coefficient, σ_p is the maximum stress built up in the debonded fiber by interfacial friction, and σ_d is the fiber stress at the tip of the debonded crack. Equation (4.80) has to be modified when $b < l_d/2$, by replacing l_d with b. Finally, pullout toughening is given by

$$\Delta G_p = V_f \sigma_p \left[l_p + \frac{e^{-\beta l_p} - 1}{\beta} \right]. \tag{4.81}$$

In parameters that affect whisker toughening, a similar trend can be found in Eqs. (4.78) to (4.81) to that said for Eq. (4.77). Total calculated toughening contributions for various parameters are listed in Table 4.1. A comparison of the two analyses indicates that, for relatively large pullout lengths, they both give strikingly similar values. However, for small pullout lengths, which are more typical in whisker–CMCs, the equations given by Wells and Beaumont overestimate pullout toughening, and Becher et al.'s approach gives reasonable results.

Although some whisker pullout has been observed in Al_2O_3–SiC_w [4.131,4.134] (Fig. 4.24), mullite–SiC_w [4.131,4.134,4.135], and glass–SiC_w composites [4.131], Si_3N_4–SiC_w composites generally do not exhibit pullout [4.102,4.136,4.137]. Average pullout lengths measured were 1.2 to 2.0 μm in Al_2O_3–SiC_w, 0.4 to 0.8 μm in mullite–SiC_w, and 0.2 to 0.4 μm in glass–SiC_w composites by Becher et al. [4.131]. Shaw and Faber [4.134] observed an increase in pullout lengths in

Table 4.1. Calculated Pullout Toughening Contributions for Various Combinations of Whisker Content and Pullout Lengths [a]

	$V_f = 0.1$ $l_{po} = 1.5\ \mu m$	$V_f = 0.5$ $l_{po} = 1.5\ \mu m$	$V_f = 0.1$ $l_{po} = 10\ \mu m$	$V_f = 0.5$ $l_{po} = 10\ \mu m$
ΔK_p (1)	0,35	1,65	16.50	44.40
ΔK_p (2)	2,03	7,21	16,86	45.38

[a] References:

1. P.F. Becher, C.H. Hsueh, P. Angelini, and T.N. Tiegs: J. Am. Ceram. Soc. **71 [12],** 1050–1061 (1988)
2. J.K. Wells and P.W.R. Beaumont: J. Mater. Sci. **23 [4],** 1274–1278 (1988)

mullite–SiC_w composites with increasing test temperatures, although no difference was observed for whisker-toughened alumina (WTA). Tiegs [4.138] demonstrated that the fracture toughness of an Al_2O_3–20 vol% SiC_w composite can be increased from 4.7 to 6.5 MPa $m^{1/2}$ by using a carbon (C) coating on the whiskers. The presence of C results in a weak interfacial bond that enhances pullout and possibly other toughening mechanisms. This shows that toughening by whisker pullout can be an important mechanism if pullout lengths can be increased.

4.1.3.2.8.2 Fiber Pullout

Toughening by the pullout mechanism can be very effective in fiber-reinforced CMC's. In a uniaxial fiber-reinforced composite wherein the crack propagates perpendicular to the fiber orientation (Fig. 4.25), WOF is given by [4.55],

$$\mathrm{WOF} = (V_f/6)\,(L_1^2/L_2)\,\sigma_f\,, \tag{4.82}$$

where L_1=L and L_2=L_c if L<L_c, or else L_1=L_c and L_2=L . In Eq. (4.82), σ_f is the fiber tensile strength, V_f is the fiber volume fraction, L is given by

$$L/r = [E_m\,\varepsilon^*(1-V_f)\,/\,V_f]\,/\,(2\tau_f)\,, \tag{4.83}$$

and the maximum pullout length, L_c, assuming uniform fiber strength is given by

$$L_c = r\sigma_f\,/\,2\tau_f\,. \tag{4.84}$$

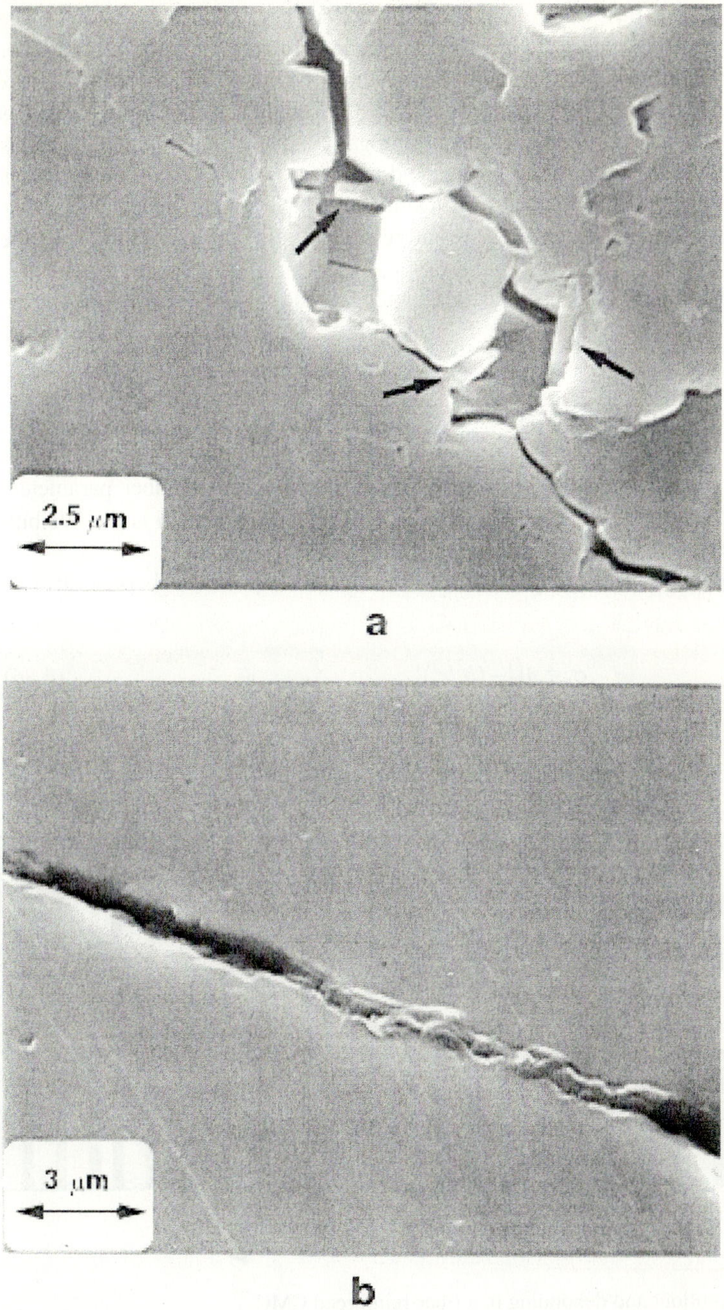

Fig. 4.24. Crack deflection in Al_2O_3–$10SiC_w$ **a** compared to no deflection in unreinforced ceramic **b** (From M. Bengisu, Ph.D. Thesis, New Mexico Institute of Mining and Technology Socorro NM, 1992)

Fiber lengths above this value do not permit pullout since fracture occurs before pullout. In Eqs. (4.83) and (4.84), E_m is Young's modulus of the matrix, r is the fiber radius, ε^* is the strain at the point of matrix cracking, and τ_f is the fiber–matrix interfacial shear strength. According to Eq. (4.82), toughening by fiber pullout is proportional to the fiber content, fiber strength, and r/τ_f. Comparison of this dependence and the matrix cracking threshold stress (see Eqs. (4.35,4.36)) indicates that pullout toughening and the matrix cracking stress have inverse dependences on r and τ_f. Therefore, an increase in WOF would result in easier matrix cracking and vice versa [4.55].

Thouless and Evans' analysis indicates that toughening due to fiber pullout has the form [4.60],

$$\Delta G_p \propto \langle h \rangle^2 (\tau_f/r)\,, \tag{4.85}$$

where $\langle h \rangle$ is the mean pullout length. Incorporation of statistical fiber parameters indicates that toughening is proportional to $[r^{m-3}/\tau^{m-1}]^{1/(m+1)}$, where m is the Weibull modulus of the fibers.

Continuous fibers in a matrix of different CTE will experience thermal stresses [4.53],

$$\sigma_r = A[1-(b^2/r^2)]\,, \tag{4.86}$$

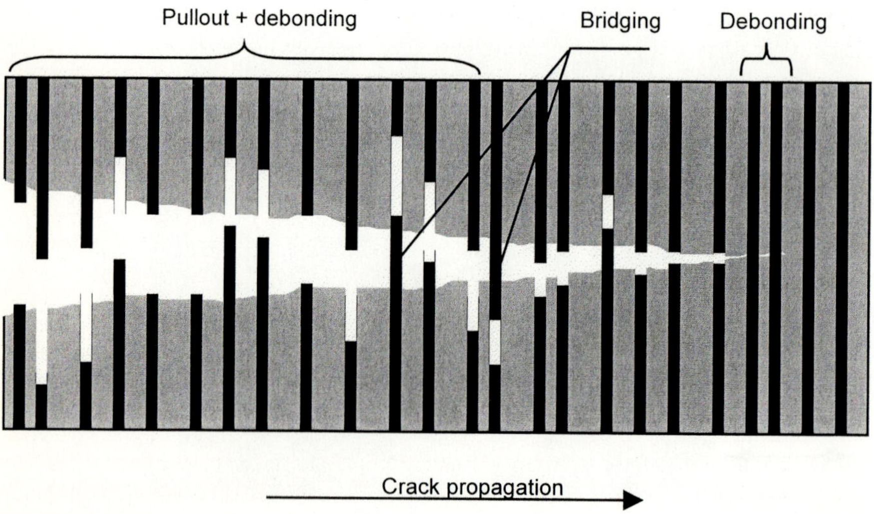

Fig. 4.25. Fiber pullout and debonding in a fiber-reinforced CMC

$$\sigma_t = A[1+(b^2/r^2)]\,, \tag{4.87}$$

and

$$\sigma_z = [A/(a/b)^2]\frac{[2\nu(a/b)^2]+\{1+(a/b)^2(1-2\nu)+(a/b)^2(1-2\nu)[(b/a)^2-1](E_m/E_f)\}}{1+[(b/a)^2-1](E_m/E_f)}\,, \tag{4.88}$$

where

$$A = -[E_m(\alpha_m-\alpha_f)\Delta T\,(a/b)^2]\,/\,\{1+(a/b)^2(1-2\nu)[(b/a)^2-1](E_m/E_f)\}, \tag{4.89}$$

and

$$\nu_m = \nu_f = \nu \tag{4.90}$$

In these equations, r is the distance in the radial direction, a is the fiber radius, and b is the outer radius of the matrix shell, as depicted in Fig. 4.26.

For a composite to exhibit increased fracture toughness by the fiber pullout mechanism, the radial stresses should be compressive. The compressive thermal mismatch stress should be high enough to provide resistance to fiber pullout so that the pullout action can consume some of the energy during crack growth and increase the fracture toughness. However, the stress should not be too high since otherwise interfacial shear stresses would oppose pullout and premature fiber failure would occur.

4.1.3.2.9 Crack Bridging

Crack bridging phenomena have been observed in various composite systems and well documented in rubber-toughened polymers [4.139], cermets [4.140,4.141], and recently in alumina with a large grain size [4.142,4.143]. The perception of toughening in coarse-grained ceramics by crack bridging is very new. In single-phase, coarse-grained ceramics, it is hypothesized that toughening occurs by two mechanisms. The first mechanism is *frictional interlocking* of opposing fracture surfaces, as the crack tends to propagate and the crack opening displacement increases [4.144] (Fig. 4.27). A crack wake region of up to 100 μm behind the crack tip can contribute to the toughness increase. Coarse-grained ceramics provide rough surfaces and thereby promote such a mechanism. Studies of Al_2O_3 [4.142,4.145] and ZrO_2 [4.145] have confirmed this thesis. The amount of toughening depends on the number of bridging sites, and this results in R-curve behavior in toughness. Thus, a direct toughening equation is not given [4.143].

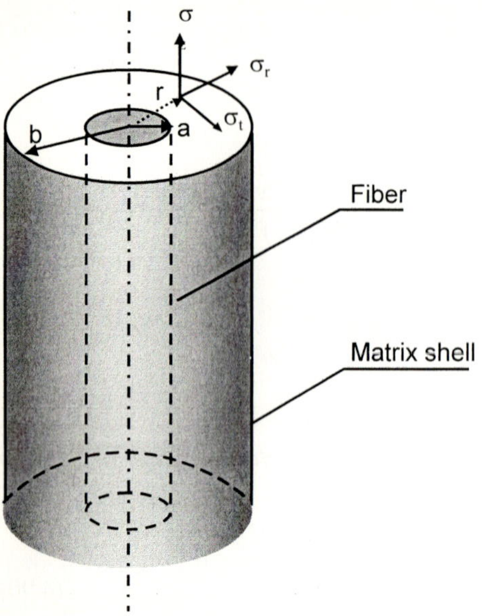

Fig. 4.26. Thermal stresses in a fiber CMC

The second mechanism involves the *formation of ligamentary bridges* [4.144] (Fig. 4.28). The fracture resistance is increased by this phenomenon since propagation of the crack will have to overcome the barrier of fracturing these ligaments. If the weakest links are not interconnected linearly, but deflect from a linear path, additional toughening may occur due to crack deflection. Both grain interlocking and bridging mechanisms are more effective in noncubic materials than in cubic materials due to thermal expansion anisotropic stresses in the former [4.146,4.147].

A similar mechanism can operate in fiber- or whisker-reinforced ceramics. In this case, debonded fibers or whiskers can act as ligaments. Becher et al. [4.131] derived toughening by crack bridging in whisker-reinforced ceramics as

$$\mathrm{d}K = \sigma_\mathrm{f}^{\mathrm{w}}\,[(V_\mathrm{f}\, r E^\mathrm{c} G^\mathrm{m}) / [6(1-\nu_\mathrm{c}^2) E^\mathrm{w} G^\mathrm{i}]^{1/2}, \tag{4.91}$$

where $\sigma_\mathrm{f}^{\mathrm{w}}$ is the fracture strength of whiskers, r is the whisker radius, E^c and E^w are elastic moduli of the composite and the whiskers, respectively, and G^m and G^i are crack extension forces of the matrix and the whisker–matrix interface, respectively.

Evans and McMeeking derived the following relationship describing toughening due to crack bridging by fibers [4.148]:

$$\Delta G_\mathrm{b} = (\sigma_\mathrm{f}^3 r A_\mathrm{f}) / (6\tau_\mathrm{f} E_\mathrm{f}), \tag{4.92}$$

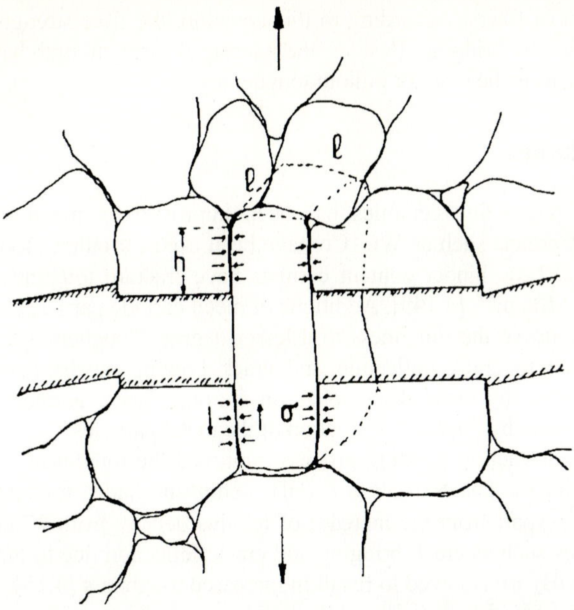

Fig. 4.27. Frictional interlocking in coarse-grained or composite ceramics (reprinted from G. Vekinis, M.F. Ashby, and P.W.R. Beaumont: *R-Curve Behavior of Al_2O_3 Ceramics.* Acta Met. **38 [6]**, 1151–1162 (1990), with permission from Elsevier Science)

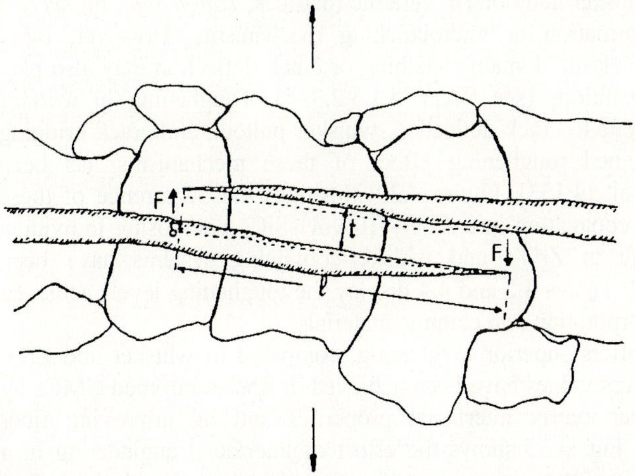

Fig. 4.28. Formation of ligamentary bridges in coarse-grained or composite ceramics (reprinted from G. Vekinis, M.F. Ashby, and P.W.R. Beaumont: *R-Curve Behavior of Al_2O_3 Ceramics.* Acta Met. **38 [6]**, 1151–1162 (1990), with permission from Elsevier Science)

where A_f is the area fraction of fibers. According to this equation, the fiber strength strongly influences toughening by bridging. Besides, the interfacial shear strength has to be low for high toughness, as is the case for pullout toughening.

4.1.3.3 Experimental Studies

The fracture toughness of monolithic ceramics has been improved by metal or ceramic *particle additions*. Cermets such as WC–Co have been used for cutting tool applications for a long time. Low binder content cermets have fracture toughness values ranging from 5 to 15 MPa $m^{1/2}$ [4.149]. Additions of metal carbide particles to oxide ceramics or glasses improve the toughness to a lesser degree. Toughening in such composites is attributed to crack deflection and crack bowing mechanisms [4.101,4.126,4.150]. Toughening by metal dispersoids, on the other hand, is due to crack bridging and crack blunting by ductile deformation of the metal particles.

TiB_2 additions to SiC by a reaction bonding process improved the toughness of SiC to twice its initial toughness. The SiC–30 vol%TiB_2 composite had a fracture toughness of ~6.2 MPa $m^{1/2}$. Apart from the increase of relative density from 87 to 99%, toughening mechanisms such as crack bridging and crack deflection due to the residual stress field around TiB_2 are believed to result in increased toughness [4.151]. Additions of metal carbides, nitrides, borides, and silicides increase the fracture toughness of Si_3N_4 (Appendix 3). Generally, nitrides are more compatible with Si_3N_4 than carbides since the latter can react with Si_3N_4 and form SiC. K_{Ic} values from 4 to 10 MPa $m^{1/2}$ were reported for TiN particle-reinforced ceramics [4.152].

High levels of toughening suitable for intermediate temperatures (<1000°C) can be achieved by ZrO_2 and whisker additions to ceramic matrices. *Toughening by ZrO_2* is typically due to transformation or microcracking mechanisms. However, other mechanisms such as ferroelastic domain switching or crack deflection may also play a role under certain conditions (see Sects. 4.1.3.2.3–5). Toughening in *whisker reinforcement* is largely due to crack deflection, whisker pullout, and crack bridging mechanisms. The combined toughening effect of these mechanisms has been modeled by Bengisu et al. [4.153]. Figure 4.29 illustrates the occurrence of these phenomena upon crack propagation in a Al_2O_3–10 vol%SiC_w composite. Individual mechanisms that operate in ZrO_2- and whisker-containing systems have been described in Sect. 4.1.3.2. Tables 4.2 and 4.4 display the toughening levels achieved by ZrO_2 and whisker incorporation into ceramic materials.

Fiber reinforcement offers superior toughening, compared to whisker and ZrO_2 toughening. High toughness values have been achieved in fiber-reinforced CMCs by carefully controlling fiber–matrix interfacial properties and by improving fiber properties. For example, Fig. 4.30 shows the effect of interfacial engineering in a ZTA (PRD, DuPont) fiber–glass composite. Coating of fibers by SnO_2 before composite fabrication produced a low friction coefficient, hence fiber pullout and higher toughness compared to the composite with uncoated fibers were realized. Although K_{Ic} values are commonly reported for fiber-reinforced CMCs, one should keep in mind that this practice is fundamentally wrong. K_{Ic} is a property derived from linear elastic fracture mechanics (LEFM) assuming that crack growth and fracture occur under essentially elastic conditions, as the name implies. Another assumption is that a single crack exists. However, fiber-reinforced ceramics exhibit multiple-matrix cracking and considerable deformation before fracture. Therefore, K_{Ic} or K_c values for such composites should be used with caution. Properties such as the total energy absorption until failure, the stress at the onset of multiple-matrix cracking, and the failure stress are better suited to define the toughness of a fiber-reinforced ceramic matrix composite [4.74]. Some examples of toughness values of glass and ceramic CMCs are given in Appendix 6.

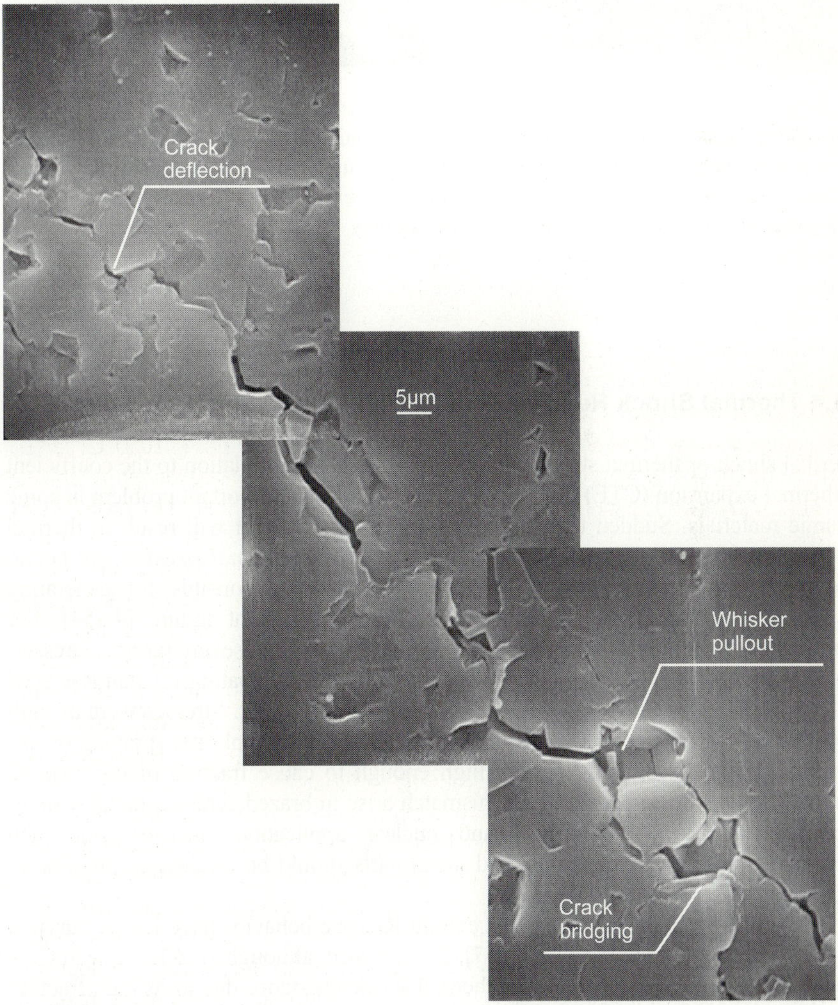

Fig. 4.29. SEM micrograph depicting various toughening mechanisms in whisker-toughened alumina (From M. Bengisu, Ph.D. Thesis, New Mexico Institute of Mining and Technology Socorro NM, 1992)

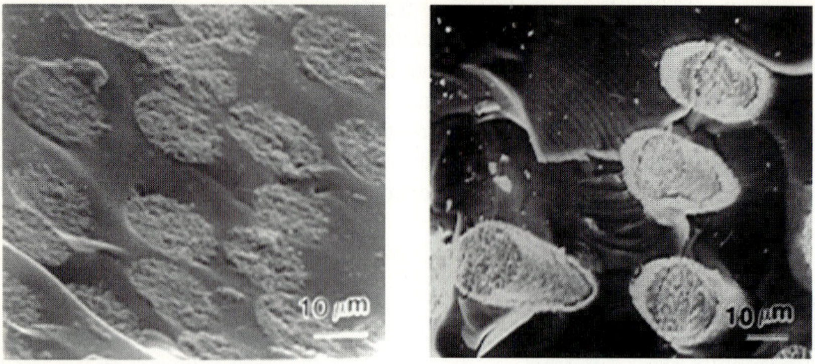

Fig. 4.30. Effect of interfacial engineering on fiber pullout and debonding in PRD fiber-reinforced glass: before coating **a** and after SnO_2 coating applied on fiber surface **b** (From R.Venkatesh: *Interface Engineering of Alumina Fiber/Glass Composites*, Ph.D. Thesis, New Mexico Insitute of Mining and Technology Socorro NM, 1993 courtesy of R. Venkatesh)

4.1.4 Thermal Shock Resistance and Thermal Fatigue Resistance

Thermal shock or thermal stress failure, which has a strong relation to the coefficient of thermal expansion (CTE) discussed in Sect. 4.2.4, is an important problem in some ceramic materials. Sudden temperature changes in a material will result in thermal strains. If these strains cannot be accommodated freely, *thermal stresses* will occur. Thermal expansion behavior is the primary property responsible for generating thermal stresses and if not controlled carefully, subsequent failure [4.154]. For example, plasma-sprayed coatings on metals usually undergo severe thermal stresses. Residual thermal stresses in a plasma-sprayed ZrO_2–Y_2O_3 coating on stainless steel and a NiCrAlY bond coat in between were calculated [4.155]. Stresses were as high as 60 MPa. The surface of the coating is under tension after plasma spraying on the substrate. Usually, such stresses are high enough to cause fracture of the coating. Similarly, large stresses due to CTE mismatch arise in brazed ceramic–metal couples used for electronic, automotive, and nuclear applications [4.156]. For such applications, CTEs of the ceramic and metal parts should be as close as possible to each other.

Ceramics with high toughness that exhibit R-curve behavior have higher survival possibility during thermal shock [4.157]. For instance, although PSZ has a high CTE (12×10^{-6} 1/°C), it exhibits very good thermal shock resistance due to its high fracture toughness [4.5].

Thermal shock is a type of catastrophic failure, which occurs if the thermal stresses exceed the failure stress. As a typical result of thermal shock, spalling, i.e., surface failure, occurs. Resistance to thermal shock can be formulated by different thermal shock resistance parameters for various conditions [4.16,4.31]:

$$R = \sigma_t (1-\nu) / E\alpha = \Delta T_c\,, \tag{4.93}$$

$$R' = \sigma_t (1-\nu)\lambda / E\alpha = R\lambda\,, \tag{4.94}$$

$$R'' = \sigma_t (1-\nu)a / E\alpha = Ra\,, \tag{4.95}$$

where σ_t is tensile strength, ν is Poisson's ratio, E is Young's modulus, α is the coefficient of thermal expansion, λ is the thermal conductivity, and a is the thermal diffusivity of the material. R should be used when the surface heat transfer rate is high but the heat transfer rate to the interior (λ) is low. For slower surface heat transfer rates, R' or R'' are the valid thermal shock resistance parameters. These two parameters can also be used simultaneously to determine the critical cooling rate. Note that R is only an indicator, which can be used for comparative purposes. High values of R indicate relatively high resistance to thermal shock. Table 4.2 lists R values for some ceramic materials. Apparently, LAS (lithium aluminum silicate) glass has a much better chance of surviving sudden thermal changes than, for example, Al_2O_3. Although Si_3N_4 has much higher strength compared to LAS, the CTE of LAS is so small that it practically does not expand or contract to cause any thermal shock. Materials such as Pyroceram® and Pyrex® glass, well known for their thermal shock resistance, also have very low CTEs and high thermal shock resistance.

Table 4.2. Calculated Values of *R* for Various Ceramics

Material	Tensile Strength σ (MPa)	Coefficient of Thermal Expansion α (1/°C)	Elastic Modulus E (MPa)	Resistance to Crack Initiation [a] R (°C)
AlN	125	0,0000042	310,000	72
Al_2O_3	290	0,0000074	380,000	77
B_4C	350	0,0000056	420,000	112
SiC	280	0,0000038	400,000	138
RBSN	200	0,0000025	240,000	250
LAS Glass	80	0,0000003	100,000	2000

[a] For data used in the calculation of R values, see tables in this chapter related to each parameter (Poisson's ratio was taken as 0.25 for all materials.)

Thermal fatigue occurs when a material is cycled through a temperature range several times. The cycle number needed to cause failure is called the critical cycle number [4.158]. During thermal cycling, subcritical crack growth occurs up to a point at which the stress intensity at the crack tip, K_I, exceeds the fracture toughness, K_{Ic}, corresponding to that crack length. The number of cycles that will lead to failure is governed by the parameters that describe the size and geometry of the crack as well as those that describe the kinetics of crack growth. Important parameters are ΔT for the cycle, the peak temperature, thermal conductivity, and thermal diffusivity. Materials with low thermal diffusivity undergo a greater extent of crack growth per cycle, which results in a shorter cyclic thermal fatigue life than materials of the same size but with higher thermal diffusivity. Although some empirical methods exist for calculating thermal fatigue performance, they are subject to considerable error. Thus, it is preferable to use field tests if service conditions can be anticipated. A theoretical understanding of the effect of various parameters on the phenomenon will assist in interpreting collected experimental data [4.154].

Measurement

The conventional method of measuring thermal shock resistance is *quench testing*. This is a test in which a ceramic rod is heated to a high temperature and then quenched in a liquid, usually water. The minimum difference between the initial sample temperature and the temperature of the immersion medium that leads to fracture or significant strength deterioration is used as an indicator of the thermal shock resistance of a ceramic. Another indicator is the decrease in strength upon quenching. A different procedure for thermal shock testing involves a batch of specimens quenched under exactly defined conditions. The degree of damage is determined in terms of crack length or density, the number of samples destroyed, or the strength of the samples after testing. The same procedure is then repeated at a higher temperature difference, until all samples are damaged or destroyed. A relationship between the temperature difference and the number of damaged products or other thermal shock resistance indicators can be drawn from these tests [4.16].

For a plate with infinite area, the maximum permissible temperature change ΔT_{max}, which omits failure, was given by Eq. (4.93). For a solid cylinder subjected to quenching in a linearly convective (Newtonian) heat transfer environment, ΔT is given by [4.154,4.159]

$$\Delta T_{max} \approx 1.45\sigma_f (1-\nu)\,\{1 + [3.41k/(bh)]\}\,/\,(\alpha E)\,, \qquad (4.96)$$

where σ_f is the tensile strength, b is the cylinder radius, k is the thermal conductivity, and h is the convective heat transfer coefficient. Similar equations are given for different sample geometries.

The main problem in quench testing is the lack of an accurate model of heat transfer conditions, especially due to boiling and associated vapor blankets which depend on surface conditions of the sample (roughness, etc.). Alternative methods for quench testing have been developed. Air-jet testing is one of them. In this method, a controlled jet of air is used to cool the surface of a ceramic disk, which makes heat transfer modeling plausible. One disadvantage is that the cooling rate is slow, necessitating high initial specimen temperatures to cause failure. Such high initial temperatures introduce complexities due to heat transfer by radiation.

The resistance to thermal shock can be evaluated in terms of *elastic modulus variations* and *retained strength* [4.160] (Fig. 4.31). At a critical temperature change, both elastic modulus and strength decrease. Acoustic emissions (AE), which provide in-situ information on crack initiation, are detected by a piezoelectric sensor (Fig. 4.32). The complexity of the AE response of a material makes quantitative analysis difficult, but a clear correlation exists between AE events and degradation of mechanical properties, as seen in Fig. 4.31.

Another alternative to traditional water quench tests is to change the quenching medium. Oils are more dependable than water, especially because the superficial heat transfer coefficient increases linearly with the temperature of the medium. For water, there is a parabolic dependence [4.161]. That complicates the interpretation of collected data. Liquid metal media are also used instead of water. In this case, the dominant heat transfer mode is conduction. The main drawbacks are the same as those of other quench tests: stress concentrations occur at specimen corners, the onset of fracture is difficult to define, and transient specimen temperature and stress are assumed rather than measured.

A newer technique is based on heat transfer from a hot specimen to a cooling rod upon contact. The rapid cooling and contraction of the center of the specimen result in uniform, biaxial, tensile thermal stress. Significant acoustic emissions are detectable during the first fracture events. The advantages of the contact thermal shock test are well-defined thermal boundary conditions, sufficiently high cooling rates, and the capability of measuring the transient specimen temperature as well as the time-to-failure via acoustic emission [4.162].

Finally, the *ribbon thermal shock test* should also be mentioned here. In this test, prismatic samples are heated for 15 min to about 1000°C with an oxyacetylene flame and then cooled for another 15 min in air. The thermal shock resistance is determined with an ultrasonic sensor that measures variations in Young's modulus [4.161]. The advantages are similar to those described in contact thermal shock testing.

4.1.5 Fatigue Resistance

Two types of fatigue phenomena seen in ceramic materials are static and cyclic fatigue. The term *static fatigue* is applied to failure under constant load that is generally used interchangeably with stress corrosion and subcritical crack growth.

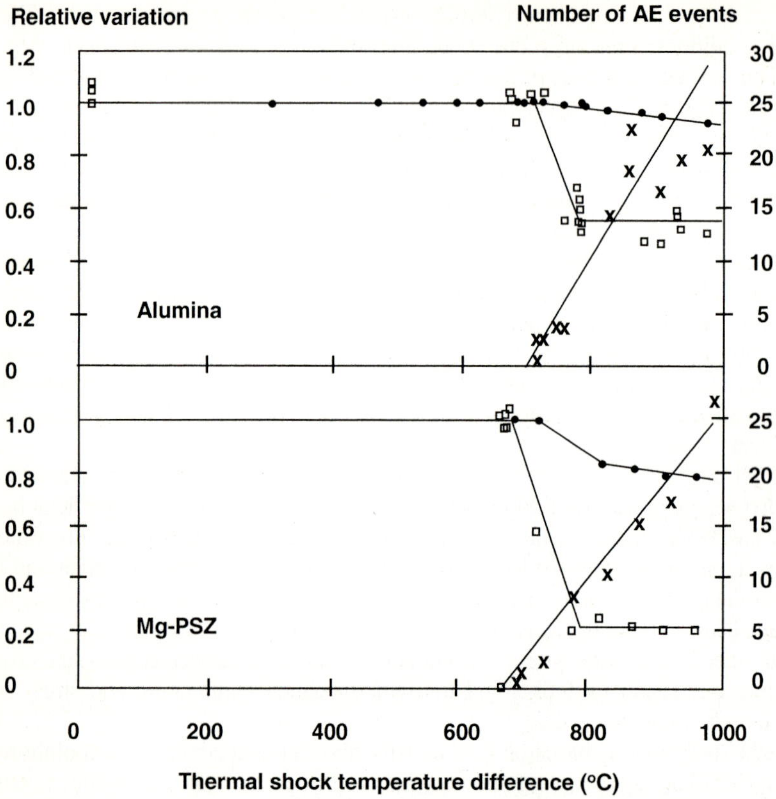

Fig. 4.31. Relative variation of retained strength ( ), elastic modulus (λ), and number of events (x) as a function of thermal shock temperature difference for alumina and Mg–PSZ (reprinted from F. Mignard, C. Olagnon, and G. Fantozzi: *Acoustic Emission Monitoring of Damage Evaluation in Ceramics Submitted to Thermal Shock.* J. Eur. Ceram. Soc. **15 [7]**, 651–653 (1995), with permission from Elsevier Science)

This phenomenon is explained in Sect. 4.6. *Cyclic or mechanical fatigue* is the phenomenon of damage accumulation and eventual failure of a material upon cyclic loading-unloading. Although ceramic materials exhibit very limited plasticity at ambient temperature, evidence of mechanical fatigue effects in them has been noticed [4.163–4.165]. Fatigue may occur in structural ceramic components exposed to a large number of oscillatory loading and/or thermal cycles. Such conditions arise in applications such as medical prosthetic implant devices, high-temperature engine components [4.166], e.g., the adiabatic diesel engine [4.167], and catalytic converters.

Various models have been proposed to account for the fatigue of ceramics in the absence of environmental effects. Evans [4.163] suggested that two mechanisms can lead to fatigue. The first of these mechanisms is the formation of a plastic zone around asperities located on crack surfaces during compression and subsequent lateral

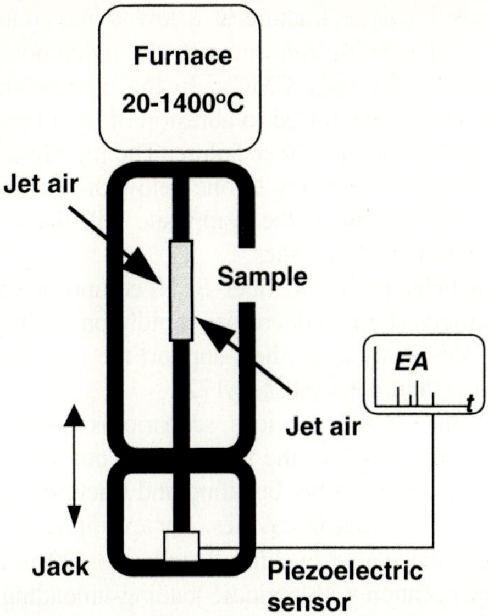

Fig. 4.32. Schematic illustration of thermal shock measuring device (reprinted from F. Mignard, C. Olagnon, and G. Fantozzi: *Acoustic Emission Monitoring of Damage Evaluation in Ceramics Submitted to Thermal Shock.* J. Eur. Ceram. Soc. **15 [7]**, 651–653 (1995), with permission from Elsevier Science)

crack formation during unloading. It has been suggested that this mechanism is relevant in fine-grained ceramics. The second model pertinent to coarse-grained anisotropic ceramics is based on the phenomenon of microcrack formation and coalescence.

Lewis and Rice [4.167] proposed various qualitative models that incorporate microcrack growth and coalescence upon cyclic loading. Ewart and Suresh [4.168] and Suresh and Brockenbrough [4.169] further developed the microcracking model for cyclic compression loading. According to these studies, an important contribution to fatigue arises from residual tensile stresses induced locally at notch tips due to strain irreversibility upon stress reversal. It has been suggested [4.122,4.124,4.125] that during cyclic tensile loads, other sources of strain irreversibility may exist that lead to fatigue crack growth. For example, the relaxation of internal stresses upon microcracking and the wedging and frictional sliding of debris on crack surfaces have been proposed as alternative sources. All of these would produce compressive closure loads across macrocrack faces and shear loads at the crack tips. An important observation by Guiu et al. [4.165] is that compressive loads, either residual or directly applied across crack faces, tend to increase fatigue crack growth rates.

An interesting implication of the model by Suresh and Brockenbrough [4.169] is that ceramic matrix composites, which exhibit nonlinear deformation, would have a greater tendency to generate notch-tip residual stresses under far-field cyclic compression. Experimental studies in SiC whisker-reinforced Si_3N_4 indicated that an increase in whisker content leads to mode I fatigue fracture at a lower maximum compressive stress than required for monolithic Si_3N_4; this supports the predictions. Lewis [4.170] proposed a fatigue mechanism for fiber CMC's. In the mechanism proposed, matrix cracking and minimal fiber extraction lead to abrasion of the fibers. This results in local weakening of fibers and eventually, fiber failure. The recurrence of these events will lead to shorter fiber lengths, eventually to one below the critical length for load transfer to fibers. Thus, the strength of the composite will decline progressively as the number of load-carrying fibers diminishes.

Holmes et al. [4.171] studied the fatigue behavior of SiC fiber–Si_3N_4 composites at high temperatures and observed that the fatigue strength decreased rapidly on cycling into the region of matrix microcracking. This study and others support the idea that matrix microcracking can cause early fatigue failure in CMCs [4.172].

The buckling of fibers in a case similar to the previous scenario is another mechanism that can lead to fatigue. Here, if the length of the fibers pulled out during loading is larger than the critical length for buckling, fiber buckling and fracture will occur instead of their being forced back into the matrix cavities. For example, at a stress level of 500 MPa, SiC fibers have a calculated buckling length of ~ 90 μm. This process will again lead to gradual deterioration with periodic loading-unloading; hence fatigue will occur [4.172].

Experimental results indicated that cyclic loading produces failure after a much shorter time than seen in static loading. The ratios of time to failure for each case, t_s/t_c, increase with increasing peak stress, revealing the existence of true mechanical cyclic fatigue effects. The cyclic fatigue behavior of Al_2O_3 and Si_3N_4 was studied in air and under vacuum to distinguish the static fatigue effect from mechanical fatigue. A clear distinction between the fatigue behavior in air and under vacuum was observed [4.173]. It was shown conveniently that the observed cyclic fatigue effect in air was the superposition of the pure cyclic loading effect and static fatigue effects.

The toughening capacities of bridging grains, whiskers, or fibers under cyclic loading are diminished due to frictional wear degradation of the bridging zone behind the crack tip. Recent models based on this mechanism provide good predictions of cyclic crack growth rates [4.174]. Studies have shown that in ceramics toughened by grain bridging mechanisms, such as coarse-grained alumina or silicon nitride with elongated grains, repetitive sliding wear of the frictional grain bridges can cause premature debonding of grains and reduced frictional pullout stresses. These factors significantly reduce the toughening capacity under cyclic fatigue [4.166]. Measured crack growth rates and fractography studies of silicon nitride supported the notion of cyclic crack growth promoted by diminished shielding. A larger crack opening displacement is developed in whisker-reinforced Si_3N_4 than in the unreinforced matrix under cyclic compression [4.175]. The fatigue resistance is seen to depend strongly on the maximum applied stress intensity, K_{max}, and rather weakly on the applied stress intensity range ΔK [4.166,4.176]. The cyclic fatigue behavior of ceramics is apparently different from that of metals and polymers, for which crack growth is determined mainly by ΔK. It is postulated that in ceramics where crack bridging is important, a mechanical equilibrium develops during crack growth between degradation in shielding due to cyclic damage in the crack wake and increased shielding with growing crack length (i.e., *R*-curve behavior) [4.176].

Similarly, with increasing *R*, the sliding distance between the grain bridges is expected to decrease, leading to less frictional wear, hence less degradation in shielding per loading cycle [4.166]. Experimental studies partly support these theories [4.176].

4.1.5.1 Test Methods

Among various test methods used to analyze the cyclic fatigue resistance of ceramics are direct pull-push, repeated indentation, rotary bending, and plane bending tests. The *axial pull-push test* is the most appropriate technique for finding fundamental fatigue properties. However, alignment of samples is critical for satisfactory results. In this method, a cylindrical sample is subjected to cyclic tensile-compressive loading. Tension–tension or compression–compression tests are also possible. The number of cycles to failure is then recorded. From a series of experiments with different load ranges, a curve of log σ (stress) versus log N (number of cycles) can be constructed. A method for ensured alignment is one where spherical bearings are used in the gripping sections; these help the self-alignment of tensile samples in the axial direction.

In the *rotary bending method*, a cylindrical sample is loaded in bending while being rotated. After both ends of the sample are secured by chucking rods, the applied stress is increased gradually. This method is particularly useful for simulating cyclic loading encountered in rotary motion of gas turbines and turbocharger rotors. For example, Si_3N_4 is used for such applications, and its fatigue properties have been evaluated by the rotary bending method [4.178].

Fatigue testing can also be applied in the flexural mode. Soma et al. [4.177] used an electrodynamic vibrator at a frequency of 3 KHz; the cyclic stress amplitude in the specimen was measured by strain gauges affixed to specimens. At this frequency, accelerated fatigue testing of Si_3N_4 was possible to up to 10^{10} cycles at ambient temperature.

Recently, a simple method was proposed that is based on *cyclic microindentation* [4.179]. A standard Vickers microindenter is used to produce indentation cracks. The cavity produced by the preload, P, is then repeatedly indented with a smaller load, P', until chipping or spalling of the material around the indentation occurs. The occurrence of chipping is detected by an acoustic signal monitor. A plot of the effective compressive force P'–$0.4P$ versus the number of indentations represents the cyclic fatigue behavior. Although the method is simple and requires little sample preparation and small volumes, it is restricted to materials on which fairly straight cracks form upon indentation. Another difficulty arises due to the measurement of load rather than stress. Therefore, the data cannot be compared with results from other cyclic fatigue tests.

4.1.6 Creep Resistance and Superplasticity

Creep and superplasticity are interrelated phenomena. For ceramic materials, both require high temperatures, typically equal to or higher than 0.5 T_m, where T_m is the melting temperature [4.180,4.181]. Superplasticity is defined as one mode of deformation found in fine-grained polycrystalline solids at elevated temperatures [4.182]. The creep mode in superplastic deformation is mostly diffusion-assisted grain boundary sliding with little or no grain elongation and no cavity formation [4.180]. In superplastic deformation, plastic deformation occurs due to both applied stress and creep. The following sections describe the main features of both phenomena.

4.1.6.1 Creep Resistance

Creep is the progressive deformation of materials at constant load over a period of time [4.183]. Two broad categories of creep are those that involve lattice mechanisms and those that involve boundary mechanisms [4.180]. Among lattice mechanisms that can lead to creep are dislocation glide, dislocation climb, and dissolution of dislocation loops. Boundary mechanisms are those associated with the presence of grain boundaries. These are Nabarro–Herring creep, Coble creep, grain boundary sliding with or without a liquid phase, and nucleation and growth of flaws along grain or phase boundaries. The lack of grain boundaries in single crystals imparts high creep resistance to them. Single crystals such as NaCl, CaO, and MgO have highly symmetrical cubic crystals, which permit easy dislocation glide at elevated temperatures and causes low creep resistance. Other single crystals with less symmetrical crystal structures have fewer slip systems available; thus, their creep resistance is higher. Some oxides such as alumina, mullite, spinel, $Be_2Al_2O_4$, and yttrium alumina garnet (YAG) have very high creep resistance in certain crystal directions but exhibit a high degree of anisotropy [4.184]. Binary oxides with complex structures exhibit higher creep strength compared to single-phase oxides. Table 4.3 gives the creep strength of some single-crystal ceramics. There is no obvious relationship between creep resistance and melting point.

Creep may occur by more than a single mechanism, as for densification during hot pressing (Sect. 3.3.3). Ashby et al. demonstrated this by deformation mechanism maps developed for various single-phase ceramics [4.185–4.187]. These are stress versus temperature diagrams divided into fields within each of which a particular deformation mechanism is dominant. For a certain combination of stress and temperature applied to a single-phase ceramic, it is possible to predict the creep mechanism using these maps. Most mechanisms of high-temperature creep predict a *steady-state creep* rate. This can be represented by the *Norton creep equation* [4.5,4.188],

$$\dot{\varepsilon} = A\, d^{-p}\, \sigma^{n} \exp\left(-Q/RT\right), \qquad (4.97)$$

where A is a constant, d is the grain size, p is the inverse grain size exponent, Q is the apparent activation energy for creep, R is the gas constant, σ is the applied stress, n is the stress exponent, and T is the absolute temperature. The activation

Table 4.3. Relationship of Creep Strength to Crystal Structure and Melting Point [a]

Oxides	Crystal System and (Structure)	Melting Point (°C)	Highest Creep Strength (MPa)	Lowest Creep Strength (MPa)
Mullite	Orthorhombic	1850	480 [b]	–
$BeAl_2O_4$	Orthorhombic (olivine)	1870	>280	2
YAG	Cubic (Complex BCC)	1950	135	94
Sapphire (Al_2O_3)	Hexagonal (HCP)	2001–2050	63	4
Spinel ($MgO–0.9Al_2O_3$)	Cubic (complex cubic)	2135	33	25
Y_2O_3	Cubic (fluorite)	2401–2450	36	–
BeO	Hexagonal (HCP)	2551–2600	41	<1
CaO	Cubic (NaCl)	2601–2650	1.2	0.6
YSZ	Cubic (α)	2715	14	9
MgO	Cubic (NaCl)	2751–2800	11	0.2
UO_2	Cubic (fluorite)	2801–2850	32	10
ThO_2	Cubic (fluorite)	3050	46	39

[a] Reference:
Reprinted from S. Deng and R. Warren: *Creep Properties of Single Crystal Oxides Evaluated by a Larson–Miller Procedure.* J. Eur. Ceram. Soc. **15 [6],** 513–520 (1995), with permission from Elsevier Science

[b] 1400°C, 200 h, no plastic deformation

energy for creep can be determined by measuring the slope of $\dot{\varepsilon}$ versus $1/T$. The dominant creep mechanism can be determined from that value and n.

Nabarro–Herring creep occurs by vacancy diffusion through individual grains and the creep rate is proportional to $1/d^2$ [4.189,4.190],

$$\dot{\varepsilon} = \dot{\rho}/\rho = \lambda\sigma_a\Omega D_L/kTd^2 \ . \tag{3.58}$$

Coble creep occurs mainly by grain boundary diffusion. The creep rate in this case is proportional to $1/d^3$ and to σ:

$$\dot{\varepsilon} = \dot{\rho}/\rho = \lambda\delta D_{GB}\,\Omega\sigma_a/kTd^3 \ . \tag{3.59}$$

Dislocation creep usually governs creep at high temperatures and high stress levels where deformation occurs by climb and glide of dislocations,

$$\dot{\varepsilon} = \dot{\rho} / \rho = [KD_c \mu \vec{b} / kT](\sigma_a / \mu)^n . \tag{3.60}$$

For explanations of symbols in Eqs. (3.58) – (3.60), refer to Sect. 3.3.3.

At high temperatures and high stresses, creep usually occurs by *grain boundary sliding* where adjacent grains become displaced with respect to each other. This mechanism can be accommodated by vacancy movement or transgranular plastic flow, leading to grain elongation in the tensile direction, which is termed *Lifshitz sliding* [4.180]. Grain boundary sliding without grain elongation, termed *Rachinger sliding*, may occur with the aid of a thin glassy intergranular phase or by cavity formation and growth.

The creep rate increases with porosity [4.31]. The increase of porosity in Al_2O_3 from 5 to 50% increases the creep rate by a factor of 50. Similarly, the creep rate of MgO increases by a factor of 6 with an increase of porosity from 2 to 12% [4.5]. This can be explained by the increase of the effective stress in the remaining solid structure.

At intermediate strain rates and high temperatures, the growth of intergranular flaws can be responsible for the creep of ceramics. Deformation in fine-grained, single-phase ceramics of high purity is believed to take place by grain boundary cavitation and microcracking in addition to diffusional and dislocation flow mechanisms. In ceramics with glassy grain boundary phases, such as reaction-bonded SiC and Si_3N_4, the flow of the intergranular phase may lead to the growth of microdefects. The growth and coalescence of such defects will change the elastic modulus of the material with the consequence of nonlinear deformation [4.189]. In ceramics with high glass content and in glasses, creep is controlled by the viscous flow of the glass. Viscosity and in turn, the creep rate, depends on the composition of the glassy phase and on temperature [4.5].

The creep resistance of creep-susceptible ceramics can be improved by creep-resistant reinforcements. For example, the creep rate of alumina has been markedly increased by incorporating 15wt% SiC whiskers [4.191], as seen in Fig. 4.33. The improvement can be explained by the superior creep resistance of the second phase and inhibition of grain boundary sliding, among other possibilities. It has been noted that these mechanisms aid the creep resistance of similar composites [4.192]. Unlike whisker reinforcement, second-phase particle additions to certain ceramic matrices reportedly decrease creep strength, for example, in Si_3N_4–TiN_p and Si_3N_4–SiC_p composites [4.8, 4.152]. Cavitation at the particle–matrix interface was observed, which may explain the weakening.

The creep behavior of duplex microstructures was analyzed by French et al. [4.188]. Duplex microstructures were found that possess creep rates lower than their single-phase constituents, for example, in Al_2O_3/c-ZrO_2 and Al_2O_3/$Y_3Al_5O_{12}$ (YAG). This was explained by the segregation of Y^{3+} to alumina grain boundaries and the hindrance of boundary mechanisms such as creating and annihilating point defects or

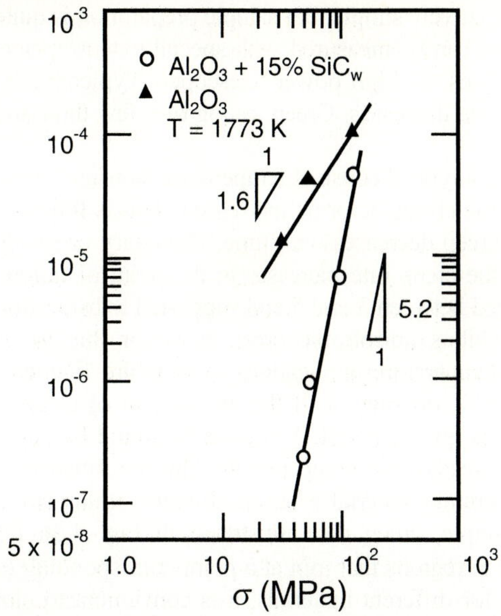

Fig. 4.33. Strain rate vs. stress relations for creep deformation of polycrystalline alumina and SiC-reinforced composite at 1773 K (From A.H. Chokshi and J.R. Porter: *Creep Deformation of an Alumina Matrix Composite Reinforced with Silicon Carbide Whiskers*. J. Am. Ceram. Soc. **68 [6],** C-144–145 (1985). Copyright 1985, American Ceramic Society, reprinted with permission)

transporting them along grain boundaries. Similar segregation effects were observed upon additions of 500 ppm Y^{3+} to MgO-doped Al_2O_3, where the creep rate decreased by a factor of 5–10 and 1000 ppm Zr^{4+} to Al_2O_3, where the creep rate decreased by a factor of 50. Doping by Ti^{4+}, on the other hand, increased the creep rate. The different reaction to doping is explained by the ionic radius of the dopants. The ionic radius of Ti^{4+} is only slightly larger than Al^{3+}, whereas those of Y^{3+} and Zr^{4+} are significantly larger. Thus, they hinder Al^{3+} ion diffusion along the grain boundaries, resulting in a lower creep rate.

Creep Testing

Until recently, creep testing has been conducted in the flexural mode due to the simplicity of testing and sample preparation. This type of testing involves three- or four-point bending at elevated temperatures and deflection measurements via transducers. However, although the strain distribution in the initial stages may be symmetrical about the geometric center of the bar during flexural creep testing, it was realized that the strain axis shifts toward the compressive surface in the course of testing. This occurs because ceramics tend to creep faster under tension than in compression [4.193,4.194]. Tension and compression creep tests are becoming more common for these reasons. Figure 4.34 illustrates a tensile loading apparatus for creep testing. Compression creep testing is relatively simple, but sample preparation is quite difficult for tensile creep testing. Strain can be measured with special extensometers or by direct monitoring of reference flags via high power telescopes. Typically, the time to rupture increases as the creep rate decreases. Creep testing of a few thousand hours is not uncommon [4.17].

Four regions can be distinguished in a typical constant temperature–constant stress creep diagram (Fig. 4.35). The first is the elastic deformation region. This is followed by primary or transient creep, where creep decreases with time, secondary or steady state creep, and tertiary creep where the creep rate increases to the point of failure. The stress exponents determined, ranged between 3 and 5 and suggested a dislocation creep mechanism in YSZ. Al_2O_3 exhibits anisotropic creep behavior due to its hexagonal structure. Therefore, high-temperature applications of sapphire filaments require careful control of crystallographic orientation. If the mechanism of creep is known, related parameters for equations given in Sect. 4.1.6. can be found by curve fitting experimental results using the steady-state creep region. This information is important for predicting the life of a ceramic material when used under similar stress and temperature conditions. For example, creep data illustrated in Fig. 4.36 for sintered α-SiC show two different creep regions that join at a point corresponding to 1600°C. When comparing creep data for different materials, it is convenient to plot the data as creep rate versus 1/T (as given in Fig.4.36) or versus logσ [4.5].

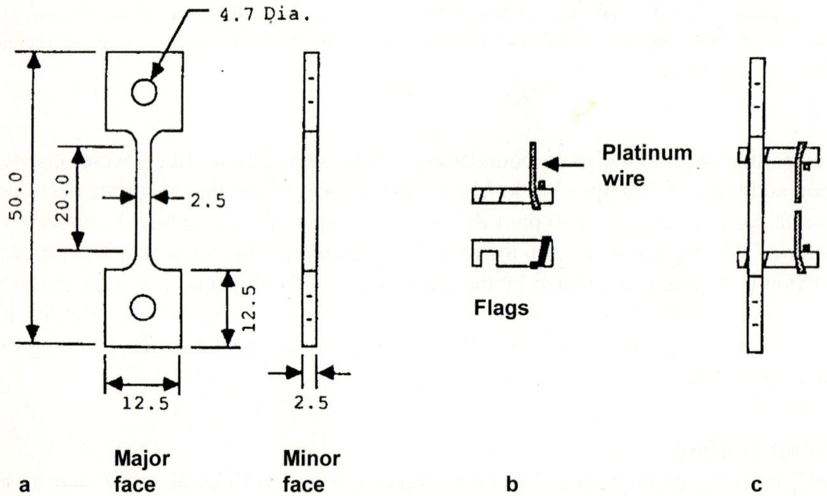

Fig. 4.34. Schematic illustration of major and minor faces of the flat dogbone-shaped tensile specimen **a**. Tensile specimen is attached to loading rods using a pin and clevis arrangement. Flags used in creep measurement **b**. Flags attached to the gauge section of specimen **c**. The flags contact the specimen along two lines to form a well-defined gauge length (From D.F. Carroll and S.M. Wiederhorn: *High Temperature Creep Testing of Ceramics*. Int. J. High Tech Ceramics **4 [2–4]**, 227–241 (1988), with permission from Elsevier Science)

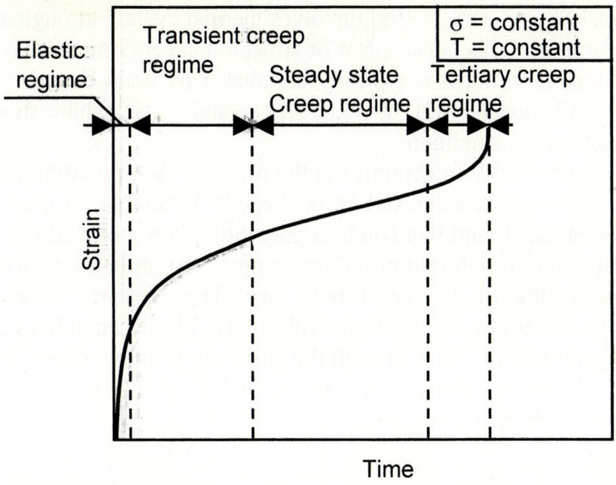

Fig. 4.35. Typical constant temperature–constant stress creep curve

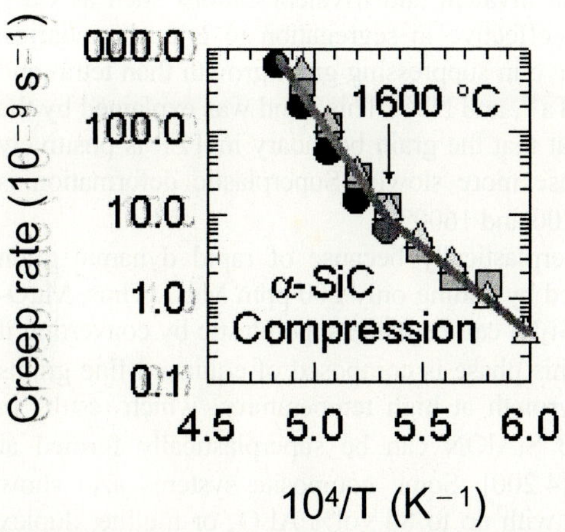

Fig. 4.36. Creep rate under compression in sintered α-SiC as a function of inverse temperature (From NIST Ceramic Division, Property Data Summaries for Advanced Materials http:// www.ceramics.nist.gov/ srd/summary/advmatdb.htm.) Not copyrightable in the United States.

4.1.6.2 Superplasticity

Superplasticity is defined as the ability of materials to undergo substantial (50 to 100 %) tensile elongation [4.195,4.196]. Superplasticity is exhibited by some ceramics only at high temperatures. Two types of superplasticity are known: transformational and structural superplasticity. Transformational superplasticity, which is observed in materials such as Bi_2O_3 [4.197] and ZnS [4.198], involves thermal cycling through a polymorphic phase transformation. The mechanism of structural superplasticity may vary from one ceramic to another, but it is typically dominated by grain boundary sliding accommodated by diffusion [4.182]. Y–TZP, Al_2O_3, and Si_3N_4 exhibit this type of superplasticity under certain conditions.

A common rule for superplasticity in ceramics is the presence and retention of ultrafine grains in the microstructure. Equiaxed grain shape is preferable for grain boundary sliding to occur. It was found that solute segregation plays a crucial role in superplasticity through grain growth inhibition, lowering of the grain boundary energy, and thus, strengthening of the grain boundary. The presence of an intergranular glassy film is not a necessity for superplasticity, but it can enhance the strain rate [4.182]. High diffusion rates through the lattice or grain boundary or through a liquid grain boundary phase have to be achieved [4.196]. The mechanical requirements for superplasticity are

1. plastic stability in tension, and
2. suppression of brittle intergranular fracture or cavitation during deformation [4.188,4.199].

Studies of Y–TZP have shown that divalent and trivalent cations such as Ca^{2+}, Mg^{2+}, Y^{3+}, and Yb^{3+} are much more effective in segregation to grain boundaries. Correspondingly, they are more effective in suppressing grain growth than tetra- and pentavalent cations such as Ce^{4+}, Ti^{4+},Ta^{4+}, and Nb^{5+}. This trend was explained by the space charge model, which pointed out that the grain boundary in TZP is positively charged and that larger cations diffuse more slowly. Superplastic deformation is exhibited in these ceramics between 1200 and 1600°C.

Pure alumina fails to deform superplastically because of rapid dynamic grain growth. However, this can be prevented by adding only 200 ppm MgO. Thus, MgO-doped Al_2O_3 exhibits superplasticity. Si_3N_4 can be made superplastic by converting it to α'- SiAlON by careful alloying. This phase is composed of equiaxed fine grains that resist static and dynamic grain growth at high temperatures, which results in severe strain hardening. β and $\alpha+\beta$ SiAlON can be superplastically formed at 1550°C, and α-SiAlON at 1600°C [4.200]. Some composite systems also show superplasticity. These include Y–TZP with up to 80 vol% Al_2O_3 or mullite, duplex zirconia/mullite, and duplex zirconia/alumina composites [4.196]. Si_3N_4–SiC composites also exhibit superplasticity at 1600°C at a strain rate of 8×10^{-5} 1/s [4.200]. Similarly, a Si_3N_4–SiC composite exhibited superplastic elongation of more than 150% at 1600°C at an initial strain rate of 4×10^{-5} 1/s [4.182]. The superplasticity is probably achieved through the presence of small amounts of intergranular liquid phase.

If there is a continuous network of a glassy phase, it may enhance grain boundary diffusivity up to several orders of magnitude. Also, the glassy phase itself usually shows easy viscous flow under stress, aiding superplasticity. Examples of superplastic ceramics with glassy grain boundary phases are MgO with small additions of LiF and glass ceramics such as lithium aluminosilicate and β-spodumene [4.181]. Examples of the maximum strain achieved in some superplastic ceramics are shown in Table 4.4.

Table 4.4. Structural Superplasticity in Crystalline Ceramics [a]

Material	Initial Grain Size (μm)	Tempe–rature (K)	Testing Conditions	Strain–Rate Sensitivity [b]	Strain Rate (1/s)	Maxi–mum Strain (%)
Al_2O_3 (+0.05%MgO)	~1	1723	Compression	0.5–1.0	$\sim 10^{-4}$	>39
Al_2O_3 (+500 ppm MgO + 500 ppm Y_2O_3)	0,7	1823	Tension	–	$\sim 10^{-4}$	65
Al_2O_3 (+0.05%MgO)	0,75	1723	Compression	~0.75	$\sim 10^{-6}$	>39
Al_2O_3 (+0.25%MgO)	1.0	1693	Compression	–	$\sim 10^{-4}$	>45
$BaTiO_3$	0.45	1423	Compression	~0.5	$\sim 10^{-4}$	>39
Hydroxyapatite {$Ca_{10}(PO_4)_6(OH)_2$}	0,7	1323	Tension	~0.22	$\sim 10^{-4}$	153
MgO	0.1–1.4	1327	Compression	~0.83	$\sim 10^{-5}$	>80
α–SiC (+1.5%Al)	1.5–2.8	2173	Compression	~1	$\sim 10^{-4}$	>40
2Y–TZP	0,55	1723	Tension	0,5	$\sim 10^{-4}$	200
3Y–TZP (3 mol%Y_2O_3 – TZP)	0.3–2.2	1723	Compression	~0.5	$\sim 10^{-4}$	78
3Y–TZP	0,3	1723	Tension	~0.5	$\sim 10^{-4}$	>120
3Y–TZP	0.3–0.4	1723	Tension	0,53	$\sim 10^{-4}$	>160
3Y–TZP/20wt%Al_2O_3	0.9	1772	Compression	0.53–0.83	$\sim 10^{-4}$	>84
3Y–TZP	0.5	1570	Compression	–	$\sim 10^{-5}$	39
3Y–TZP	~0.2	1573	Compression	~0.7	$\sim 10^{-4}$	>41
3Y–TZP/20wt%Al_2O_3	~0.5	1723	Tension	~0.5	$\sim 10^{-4}$	>200
3Y–TZP	0.54	1653	Compression	~0.5	$\sim 10^{-4}$	60
3Y–TZP	0,3	1823	Tension	~0.3	$\sim 10^{-4}$	350
3Y–TZP	0,3	1723	Tension	~0.5	$\sim 10^{-5}$	246
3Y–TZP/20wt%Al_2O_3	0,5	1923	Tension	~0.5	$\sim 10^{-3}$	~500
3Y–TZP	0,53	1600	Compression	~1	$\sim 10^{-4}$	99
3Y–TZP	0,3	1723	Tension	~0.5	$\sim 10^{-4}$	148
3Y–TZP	0,3	1673	Tension	~0.4	$\sim 10^{-4}$	>330
3Y–TZP/80wt%Al_2O_3	<1.0	1823	Tension	~0.5	$\sim 10^{-4}$	>120
3Y–TZP (+3mol%Mn)	0,3	1673	Tension	~0.4	$\sim 10^{-4}$	230
3Y–TZP	0,2	1673	Compression	–	$\sim 10^{-4}$	70

(Continued)

Table 4.4. *(continued)*

Material	Initial Grain Size (μm)	Temperature (K)	Testing Conditions	Strain–Rate Sensitivity [b]	Strain Rate (1/s)	Maximum Strain (%)
3Y–TZP	0,3	1823	Tension	~0.3	$\sim10^{-4}$	800
3Y–TZP	0,5	1723	Tension	~0.3	$\sim10^{-5}$	355
4Y–TZP	0,75	1723	Tension	0,56	$\sim10^{-4}$	160
ZTA (18–22 wt%ZrO_2)	<0.9	1723	Tension	–	$\sim10^{-4}$	>100
UO_2	2	1623	Compression	~1	$\sim10^{-5}$	68,6

[a] Reference:
(Adapted with permission from JOM, 42 [7], a publication of The Minerals, Metals & Materials Society Warrendale, PA 15086)
[b] Strain rate sensitivity is the inverse of the stress exponent, n

Deformation processing of dense ceramics, for example sheet forming, blow-molding, extrusion, stamping, and forging, have the advantages of greater shape flexibility, dimensional accuracy compared to conventional green forming and consolidation methods, smooth surface finish, and improved mechanical properties by forging. Superplasticity can be very advantageous for such near-net shape processes.

Although the number of ceramics that can be deformed superplastically is limited at present, there are prospects for superplasticity in a wide spectrum of fine-grained ceramics. The development of new superplastic ceramics will result in many industrial applications in different ceramic families such as bioceramics and ceramics for structural, electronic, magnetic, and optical applications.

4.1.7 Abrasive and Sliding Wear Resistance

Material degradation and loss are significant problems in applications requiring prolonged contact of abrasive particles or surfaces with the workpiece exterior. Such conditions arise in applications such as high efficiency gas turbines, high-temperature bearings, combustion engines, and high-performance cutting tools. Abrasive wear occurs by sliding contact of hard particles with the surface that leads to damage and material loss. Wear processes may involve chemical reactions, or they may be purely mechanical. Chemical interactions may lead to local welding of surfaces, which break apart during motion of the wear couple. Irregular fracture leads to surface roughness, plowing, gouging, and related processes [4.201]. *Mechanical wear* may occur by the formation and subsequent growth of lateral cracks that intersect the surface and cause material loss. Localized plastic deformation, work hardening, and fracture is another mechanism . This mechanism was proposed for situations in which hard ceramics are worn by softer particles [4.202]. Plastic deformation and work hardening of surfaces of cubic crystals (MgO and TiC) by cone indentation were demonstrated by Brookes and Parry [4.202]. In abrasive wear by brittle fracture, material loss is modeled by the proportionality [4.203],

$$V_i \propto F_i^{7/6} K_c^{-2/3} H^{-1/2} L_i \,, \tag{4.98}$$

where V_i is the volume of material removed per unit time, F_i is the indentation load, K_c is the fracture toughness, H is the hardness, and L_i is the length of the groove produced by particle i per unit time. Note that the wear mechanisms and wear rates are not intrinsic material properties, but they depend on the material's triboenvironment [4.204].

Sliding wear is the wear mode when two ceramic surfaces slide over each other, and no abrasive particles are in the system. A number of mechanisms may cause wear during sliding. At low sliding velocities that lead to no significant temperature rise, brittle fracture is the dominant wear mechanism. Typical Hertzian cracks were observed in glass, worn by a spherical diamond pin at a velocity of 0.03 m/s and a load of 135 N. In polycrystalline alumina, cracks were observed that penetrated grain boundaries or surface cavities formed under less severe conditions and grew by edge-spalling upon successive passes [4.205]. At high sliding velocities, temperature increases can occur due to frictional heating. The combined effects of thermal and mechanical stresses can cause thermal cracking at the surface. Thermal cycling can also eventually lead to fatigue and fracture. Furthermore, under severe frictional conditions or fine grain size, plastic deformation and even local melting may occur. Crack opening by viscous flow has been observed on glass surfaces worn by a diamond pin at a velocity of 0.5 m/s and a load of 80 N. At even higher velocities it was observed that deformation occurs by viscous flow [4.205].

The slurry abrasion resistance of various ceramics was studied, and relevant wear mechanisms were analyzed [4.204]. Two types of wear mechanisms were observed: microploughing and microcracking (Fig. 4.37). Si_3N_4, Mg–PSZ, and SiAlON exhibited microploughing-dominated wear. Microcracking-dominated wear was observed in Al_2O_3 and Al_2O_3–ZrO_2 (Fig. 4.38). Microcracking was not observed when K_{Ic}>6.5 MPa m$^{1/2}$, microploughing was not observed when K_{Ic}<4.1 MPa m$^{1/2}$, and both phenomena took place in between. Thus, the material property that controls the mechanistic response to abrasive slurry wear was K_{Ic}. In any case, the wear rate decreased with increasing hardness.

In general, *wear resistance* improves with higher hardness values. Reinforcement of oxide ceramics with harder carbides, borides, or nitrides may improve their wear resistance. For example, reinforcing ZTA with TiC and TiN particles reduced the wear rate of the matrix by a factor of 3 [4.206]. Similarly, addition of SiC whiskers to Al_2O_3 improved its wear resistance by a factor of 4 [4.207]. The superior wear resistance was attributed to improved hardness and fracture toughness upon whisker reinforcement. The addition of ZrO_2 to Al_2O_3 also increased its wear resistance considerably (from 0.12 mm^3 to 0.03 mm^3 in wear volume for 0 and 20 vol%ZrO_2 additions, respectively), along with a decrease in microhardness from 22 to 19 GPa and an increase in toughness from 5 to 11 MPa $m^{1/2}$. At higher ZrO_2 content (45 vol%), the wear resistance decreased (0.07 mm^3 wear volume), with a further decrease in hardness (to 17 GPa) and a further increase in toughness (to 13 MPa $m^{1/2}$) [4.208]. Wu et al. [4.209] observed substantial improvement in wear resistance in Al_2O_3 reinforced with SiC, TiC, or TiB_2 particles, compared to Al_2O_3 alone. AlN and Si_3N_4 also exhibited wear resistance superior to Al_2O_3, although their hardness was lower; the difference was attributed to microstructural effects such as grain size. Figure 4.39 compares the wear resistance of various ceramics and some metals for valve lining applications and demonstrates the superior wear resistance of SiC over Al_2O_3 and other materials.

The mechanism of wear was studied in detail for Y–TZP ceramics [4.210]. It was found that plastic deformation along with rolled wear debris, microcutting, and abrasion were the primary wear modes at ambient temperature. It was also observed that increased fracture toughness results in reduced wear rates. Evidence of transformation of the zirconia phase was not observed on the worn surfaces, but the wear debris exhibited evidence of this transformation. In $MgO–ZrO_2$ and $Y_2O_3–ZrO_2$,

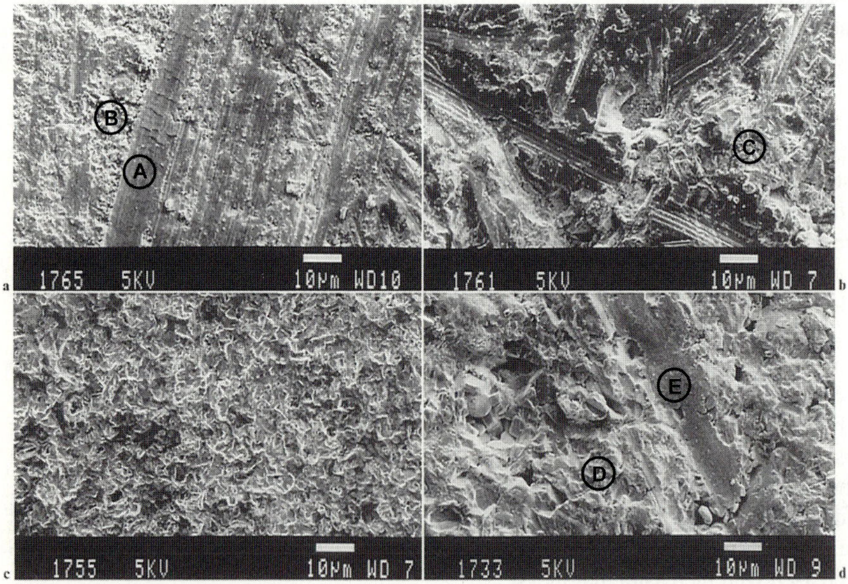

Fig. 4.37. SEM images of worn surfaces. Si_3N_4 **a**, Mg-PSZ **b**, SiC **c**, and Al_2O_3 **d**. Legend: A- delamination cracks, B, C- nondirectional wear features in microploughed ceramic, D- grain facets, and E- microploughing track, (From J.R. Alcock and O.T. Sorensen: *Slurry Abrasion Resistance of Engineering Ceramics*. Br. Ceram. Trans. **95 [1]**, 30–34 (1996), reprinted with permission of The Institute of Materials London UK)

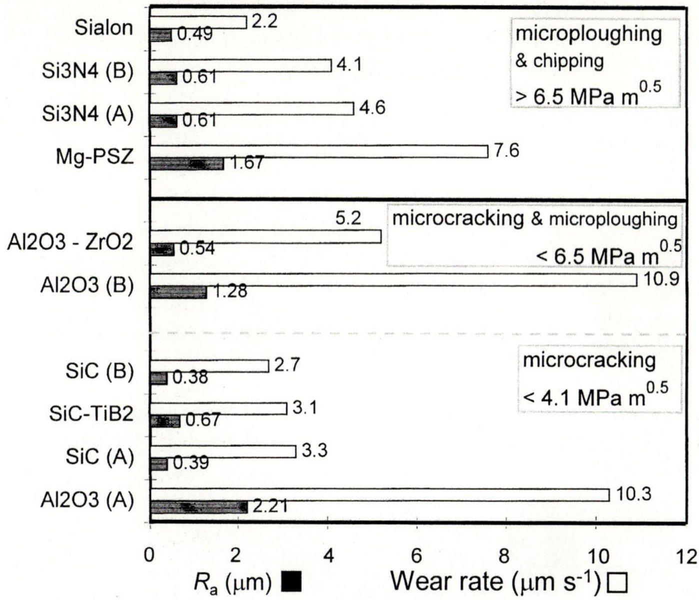

Fig. 4.38. Wear rate and mean roughness of worn ceramics, grouped according to wear mechanism and therefore fracture toughness (From J.R. Alcock and O.T. Sorensen: *Slurry Abrasion Resistance of Engineering Ceramics*. Br. Ceram. Trans. **95 [1]**, 30–34 (1996), reprinted with permission of The Institute of Materials London UK)

it was observed that the wear rate decreased with increasing temperature up to 200°C and then rose at higher temperatures. In the presence of water or high humidity, stress corrosion fracture becomes an important wear mechanism for zirconia materials. Wear maps that were developed for Y–TZP under various lubricating conditions exhibited wear rates at different loads and contact speeds.

Porosity is another material parameter that influences wear. Typically, higher porosity results in higher wear rates. For example, infiltration of reaction-bonded silicon nitride (RBSN) with Al, Al alloys, or Si improved the wear resistance significantly. Similarly, hot-pressed Si_3N_4 exhibited superior wear resistance compared to RBSN [4.149]. Pin-on-disk wear tests of Al_2O_3 and B_4C showed exponential wear rate dependence on porosity, and the slope of the line increased sharply at around 20% porosity for Al_2O_3. An increased amount of debris that leads to a change in the wear mode may cause the observed slope change [4.211]. Wu et al. [4.212] studied the dependence of wear rate on the grain size of various ceramics, including Al_2O_3, MgO, $MgAl_2O_4$, and ZrO_2. Wear data approximated a Hall–Petch type correlation:

$$1/w \propto Kd^{-1/2} \ , \tag{4.99}$$

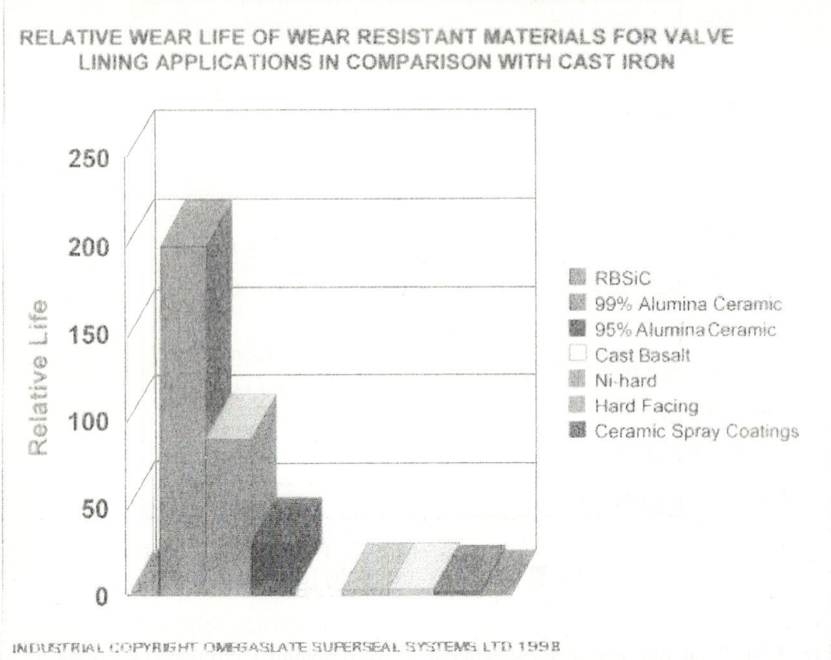

Fig. 4.39. Relative wear life of various ceramic and conventional materials for valve lining applications (Courtesy of Omegaslate SuperSeal Systems Ltd. Worcestershire UK)

where *w* is the wear rate (mm^2), *K* is a constant, and *d* is the grain size (mm). The same study showed that the amount of Y_2O_3 in ZrO_2 does not affect the trend, as long as no detectable amounts of m-ZrO_2 are in the ceramic. If, however, m-ZrO_2 is present, microcracking usually affects wear behavior [4.212].

Test Methods

Various laboratory tests have been devised for ceramics, including pin-on-ring, pin-on-disk, and ball-on-three-balls (ASTM D 2266-86) tests. Field tests may be of higher practical value for ceramic components whose wear performance needs to be studied under conditions that simulate the actual application.

Pin-on-Ring Test

In this method, load is applied onto a specimen with the form of a pin that is pressed against the rim of a rotating abrasive ring (Fig. 4.40). The relative displacement at the pin/ring interface can be measured by a linear variable differential transformer (LVDT) and plotted directly [4.213]. However, since both the pin and the ring are worn during testing, the periodically measured weight of the sample or in-situ measured pin lengths can be used for specimen-specific wear determinations [4.206].

Pin-on-Disk Test

This is a method similar to the former test; the major difference is the use of an abrasive disk rather than a ring (Fig. 4.41). In both configurations, grinding paper of various grit sizes, which are readily available, may be used. The wear rate is determined in the same manner as described. A dry sand–rubber wheel abrasion test standard is described in ASTM G 65-81. ASTM G 99-90 describes conventional *pin-on-disk methods* typically used but not limited to metal wear by ceramics. A diamond pin can be used as the abrasive for ceramics.

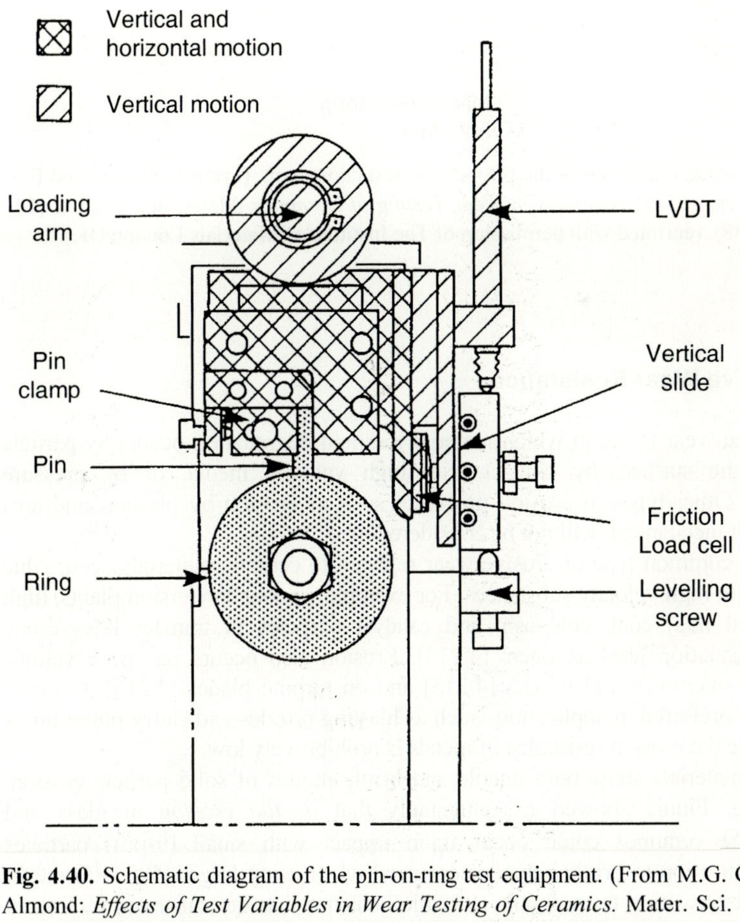

Fig. 4.40. Schematic diagram of the pin-on-ring test equipment. (From M.G. Gee and E.A. Almond: *Effects of Test Variables in Wear Testing of Ceramics.* Mater. Sci. Tech. **4 [10],** 877–884 (1988), reprinted with permission of The Institute of Materials London UK)

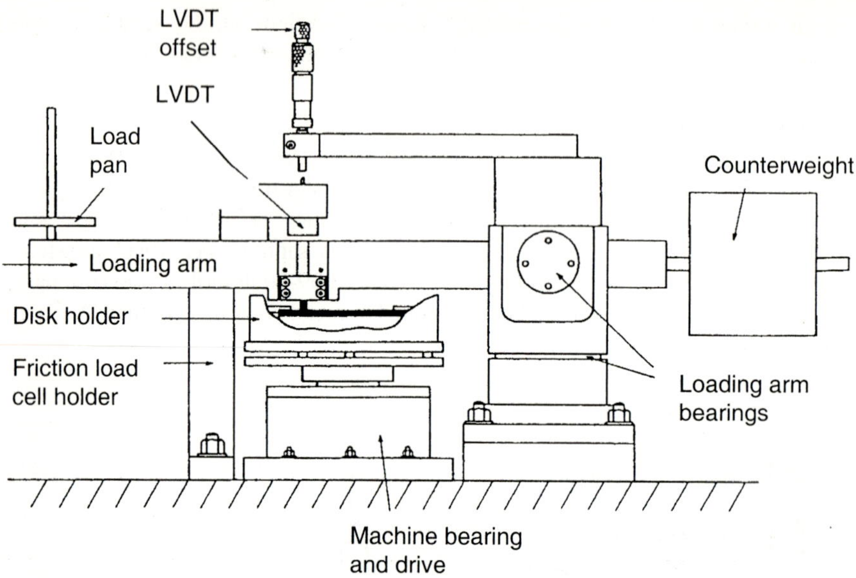

Fig. 4.41. Schematic diagram of the pin-on-disk test equipment. (From M.G. Gee and E.A. Almond: *Effects of Test Variables in Wear Testing of Ceramics.* Mater. Sci. Tech. **4 [10]**, 877–884 (1988), reprinted with permission of The Institute of Materials London UK)

4.1.8 Erosive Wear Resistance

Erosion is the wear mode in which material loss and degradation occurs by particle impact on the surface, by contact with high velocity media, or by pressure fluctuations. Other types of erosion processes such as erosion by plasmas and arcs which are not mechanical, will not be considered here.

The most common type of erosive wear relevant to ceramic materials occurs due to contact with high velocity substances. For example, in coal conversion plants, high velocity (~20 m/s) coal, coal ash, and catalyst particles in transfer lines erode refractory insulation used as liners [4.214]. Erosion also occurs on space vehicle windows by micrometeroid impact [4.215] and on turbine blades [4.216]. Ceramic materials are preferred in applications such as blasting nozzles and slurry pump liners [4.217] where the erosion resistance of metals is prohibitively low.

Ceramic materials show both ductile and brittle modes of solid particle erosion. For example, Finnie showed experimentally that *ductile erosion* in glass and polycrystalline ceramics could occur upon impact with small (9-μm) particles [4.214]. A transition to brittle behavior occurred when using larger (500–100 mesh) particles. The transition from ductile to brittle behavior was demonstrated by impact angle versus erosion curves. Microstructural studies provided evidence of such a transition. Usually, ductile materials, including most metals, undergo maximum erosion rates at impact angles of about 20–25°. Brittle materials, on the other hand, typically show maximum erosion at ~90° (normal impact).

Various models have been proposed for *brittle material erosion*. Sheldon and Finnie [4.218] developed a model assuming that erosion occurs from sharp particles that cause lateral cracks and subsequent chipping. According to the *classical model* for erosion in brittle materials, a sharp erodent particle produces an indented plastic zone upon impact (Fig. 4.42) [4.219]. This results in a lateral crack below the plastic zone. The erosion rate is given by [4.219]

$$V_0 = \alpha\, H_t^{\,q}\, K_t^{\,-4/3}\, R^m\, v^n\,, \tag{4.100}$$

where α is a proportionality constant which includes the effect of impact angle and target density, H_t and K_t are target hardness and toughness, respectively, R is the particle size, v is the particle velocity, and q, m, and n are exponents that vary according to the indentation fracture models used. n varies between 2.0 and 3.2, depending on erodent shape and contact conditions.

Evans et al. [4.220] considered the effect of localized plastic deformation in their model and obtained the following equation:

$$V_e = v^{19/6}\, R^{11/3}\, \rho^{19/12}\, K_t^{-4/3}\, H_t^{-1/4} \tag{4.101}$$

where ρ is the particle density.

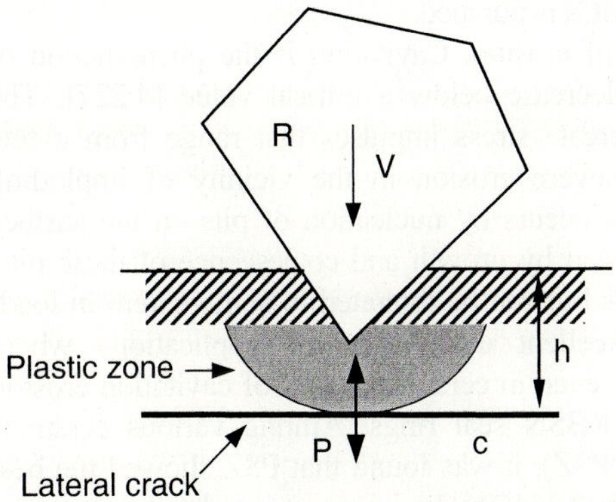

Fig. 4.42. Schematic of an erosion impact on a brittle material that causes lateral cracks and subsequent chipping (From L. Murugesh, S. Srinivasan, and R.O. Scattergood: *Models and Material Properties for Erosion of Ceramics.* J. Mater. Eng. **13 [1]**, 55–61 (1991), copyright 1991 Springer-Verlag Heidelberg Germany)

Recent models take into account such phenomena as grain boundary fracture and ejection of individual grains in polycrystalline ceramics [4.221] and plastic shear controlled micromachining at low impact angles [4.222]. All of these equations are satisfactory for monolithic ceramics but are less successful in describing the erosion rate of composites [4.223]. Experimental studies usually confirm the proportionality of the erosion rate (V_e) and the inverse of hardness [4.224], whereas toughened ceramics such as partially stabilized zirconia (PSZ) and whisker-toughened ceramics (WTCs) do not always exhibit distinct dependences of V_e and material toughness [4.224,4.225], as opposed to model predictions. The inconsistency is generally explained by the R-curve behavior of ceramics. The toughness used for calculations represents values that correspond to fully developed large cracks. Since indentation cracks by individual impacts do not produce such large cracks, toughness values that pertain to erosion should be those at the beginning of the R-curve that correspond to small crack sizes. These toughness values are substantially lower than steady-state toughness values. It has also been observed that whisker–matrix interfaces in WTCs act as easy crack propagation sites and aid crack growth during erosion. In some systems, increased whisker content resulted in higher erosion rates. It can be expected that better erosion resistance would be obtained in ceramics from surface hardening, but the use of toughened materials does not always reduce erosion rates. In a comparative work, the erosion rates of different ceramic materials (Al_2O_3, Al_2O_3/TiC, Al_2O_3/SiC_w, Si_3N_4, Si_3N_4/TiN, ZTA) were studied under similar conditions. Among these, Al_2O_3/TiC and Al_2O_3/SiC_w had the highest erosion resistance, and the composite Si_3N_4/TiN exhibited the lowest. The addition of TiN increased the toughness but decreased the hardness of Si_3N_4. In the former materials, the hardness either increased or remained constant with reinforcement [4.226]. Thus, it might be concluded that both hardness and fracture toughness have to be controlled carefully if increase in the erosion resistance of CMCs is pursued.

Cavitation erosion is another type of erosion. Cavitation is the phenomenon of bubble formation due to a pressure decrease below a critical value [4.227]. The collapse (implosion) of bubbles can create stress impulses that range from a few hundred to 1000 MPa, resulting in severe erosion in the vicinity of imploding bubbles. Damage by cavitation erosion occurs by nucleation of pits on the surface after a certain incubation period, followed by growth and coalescence of these pits. Cavitation is critical in valves and seals but can be tolerated to some extent in load-bearing structures. Ceramics are excellent candidates for applications where cavitation erosion is a threat. However, even in ceramics, cases of cavitation erosion have been reported, for example, in RBSN seal rings. Among various ceramics studied (Al_2O_3, WC–Co, SiAlON, and PSZ), it was found that PSZ showed the best cavitation erosion resistance followed by SiAlON. In a separate study, it was found that SiAlON exhibits the highest erosion resistance among a group of ceramics, including pure SiC, SiAlON, SiC–B_4C, B_4C, RBSC, PSZ, TZP, pure Al_2O_3, and 97.5% Al_2O_3. The second and third most erosion resistant ceramics were TZP and pure Al_2O_3, respectively, and the lowest erosion resistance was observed in 97.5% Al_2O_3. In cavitation erosion, it has been found that four factors control the erosion rate: the kinetic energy of the cavitation cloud, temperature, target hardness, and the fracture toughness of the target. Surface finish and microstructural defects are other important considerations. High resistance to cavitation erosion is achieved in materials with ultrafine grain size, zero porosity, high fracture toughness, high hardness, and a smooth surface finish.

Erosion Testing

Among the various erosion test apparatuses, sand-blaster systems are the most popular due to availability and versatility. These devices use compressed gas such as air, N_2, or Ar to eject erosive particles onto a target at the desired velocity and impact angle (Fig. 4.43). The gas stream and reflecting particles may be sources of uncertainty in experiments using such equipment, especially on normal impact. Other methods include gas guns and rail guns for achieving extreme velocities, particularly to simulate erosion conditions that occur in spacecraft by micrometeroid impact. Particle impact onto rotating specimens and particle ejection that uses centrifugal forces of a rotating device are among other test methods. In all erosion tests, the decrease in the target mass versus the mass of erodent particles spent is measured periodically. Further details of erosion tests using gas jets can be found in ASTM G76-83. Cavitation erosion can be tested using an ultrasonic tip immersed in a liquid medium 1 mm away from a polished sample surface. The erosion rate is measured by periodic weight measurements, for example, every other hour [4.227].

4.2 Thermal Properties

Many ceramics such as alumina, mullite, and magnesia, are known for their stability at elevated temperatures. Not surprisingly, the material with the highest known

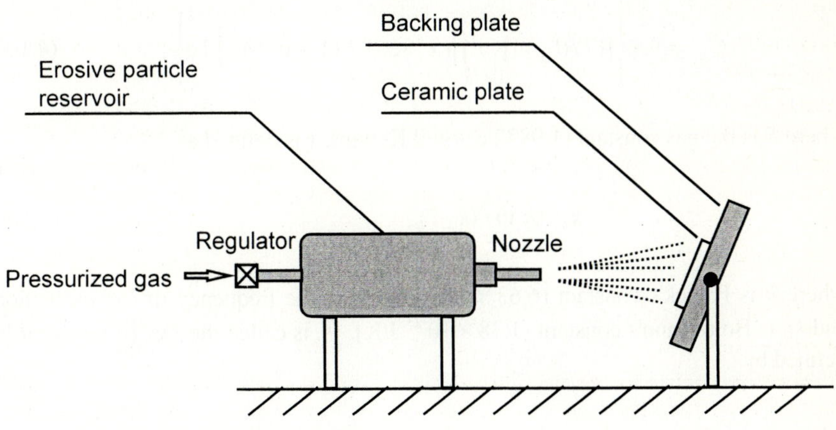

Fig. 4.43. Schematic illustration of sand-blaster erosion testing device

melting temperature, 3890°C, is a ceramic material, HfC [4.5]. Beside refractory applications, there are many cases where thermal properties are of interest. Examples are electronic circuitry, machining of hard materials, and applications that involve frequent heating/cooling cycles such as combustion engines. Among important thermal properties are thermal conductivity, the coefficient of thermal expansion, heat capacity, and melting point. Such physical properties as heat capacity, thermal conductivity, and the coefficient of thermal expansion, may be explained by thermally induced vibrations in the atomic structure of a material, also called phonons [4.228]. Appendix 7 lists the thermal properties of various ceramics.

4.2.1 Heat Capacity

The *heat capacity* of a material is defined as [4.229]

$$C_v = [\partial q/\partial T]_V = [\partial U/\partial T]_V , \tag{4.102}$$

at constant volume, or

$$C_p = [\partial H/\partial T]_P = [\partial U/\partial T]_P , \tag{4.103}$$

at constant pressure, where U is the internal energy, T is the absolute temperature, H is the enthalpy, P is the pressure, and V is the volume. In other words, heat capacity is the energy required to raise the temperature of 1 unit mass of material by 1 K. The heat capacity per unit mass is termed *specific heat*. An approximate general expression for the heat capacity of a solid described by the Debye model is

$$C_V = 9R\left\{[(T/\theta_D)^3] \int_0^{\theta_D/T} \left[(x^4)e^{-x} / (1-e^{-x})^2\right] dx\right\}, \tag{4.104}$$

where R is the gas constant (1.9872 cal/mol Kelvin), x is defined as,

$$x = (h\,v) / (k_B\,T) , \tag{4.105}$$

where h is Planck's constant (6.63×10^{-34} J s), v is the frequency of ion oscillation, and k_B is Boltzmann's constant (1.38×10^{-23} J/K). θ_D is called the *Debye temperature* defined by

$$\theta_D = h v_D / k_B , \tag{4.106}$$

where ν_D (the Debye frequency) is the maximum frequency of ion oscillation. The Debye temperature is the point above which the vibrational energy is large enough to give away ionic displacements such that [4.228],

$$\text{mean potential energy} = k_B T / 2\,, \qquad (4.107)$$

so that classical mechanics can be adopted. Below this temperature, however, quantum mechanics must be employed. At absolute temperatures approaching zero, it may be calculated that C_V also approaches zero. This agrees well with experimental studies at temperatures of $\sim\theta_D/10$ for many materials.

The dependence of specific heat on temperature fits a sigmoidal curve for most materials, including ceramics. This can be seen from Figs. 4.44 and 4.45. These figures demonstrate that specific heat is relatively insensitive to the composition or the crystal structure of ceramics. The continuity of a specific heat–temperature curve can be disturbed by second-order phase changes such as order–disorder transformations, ferromagnetic transitions, ferroelectric transitions (at the Curie temperature), glass transitions, and superconducting transitions (at the critical temperature) [4.230]. Such discontinuities are usually used to pinpoint a second-order transition.

Measurement

The *flash technique*, described by Parker et al. [4.231] is a relatively simple method which can be used to determine thermal diffusivity, heat capacity, and

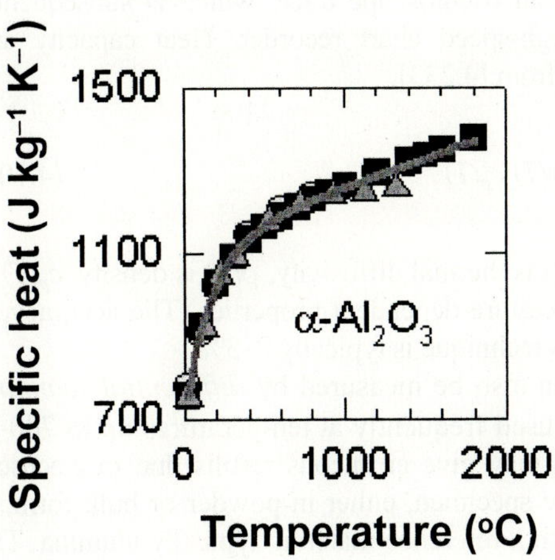

Fig. 4.44. The dependence of specific heat on temperature for α-Al_2O_3 (From NIST Ceramic Division, Property Data Summaries for Advanced Materials http://www.ceramics.nist.gov/srd/summary/advmatdb.htm.) Not copyrightable in the United States.

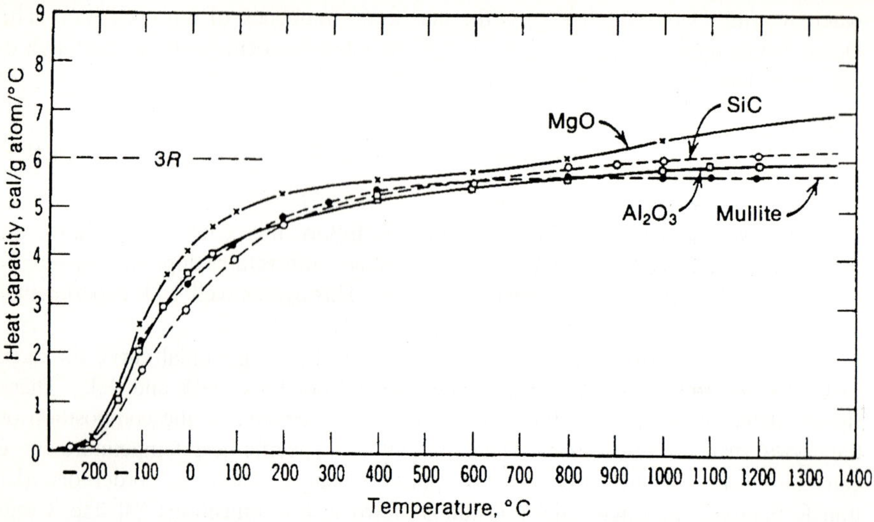

Fig. 4.45. The dependence of heat capacity on temperature for various polycrystalline ceramic materials (From W.D. Kingery, H.K. Bowen, and D.R. Uhlmann: *Introduction to Ceramics*. Copyright 1976 John Wiley & Sons New York, reprinted with permission)

thermal conductivity. The front face of a small disk-shaped sample is irradiated by a short burst of radiant energy, usually xenon flash lamp radiation, a laser beam, or an electron beam [4.232]. Dissipation times are of the order of 10^{-3} s or less. The resultant temperature variations at the rear surface of the sample are recorded (Fig. 4.46), and the thermal diffusivity is calculated from temperature rise versus time data. Generally, data recording is done via an oscilloscope trace, which is subsequently photographed or recorded with a high-speed chart recorder. Heat capacity and thermal conductivity can be calculated from [4.233],

$$k(T) = a(T)\, \rho(T)\, c_p(T)\,, \tag{4.108}$$

where $k(T)$ is thermal conductivity, $a(T)$ is thermal diffusivity, $\rho(T)$ is density, $c_p(T)$ is heat capacity, and (T) represents temperature-dependent properties. The accuracy of thermal diffusivity measurement by this technique is typically 2–3%.

The heat capacity of a material can also be measured by *differential scanning calorimetry (DSC)*. This is a method used frequently at temperatures up to 700°C [4.149]. At higher temperatures, DSC may give erroneous results due to electrical noise and radiation. In this method, the specimen, either in powder or bulk form, is heated in a crucible simultaneously with a reference material, typically alumina. The temperature difference between the sample and reference gives a measure of the

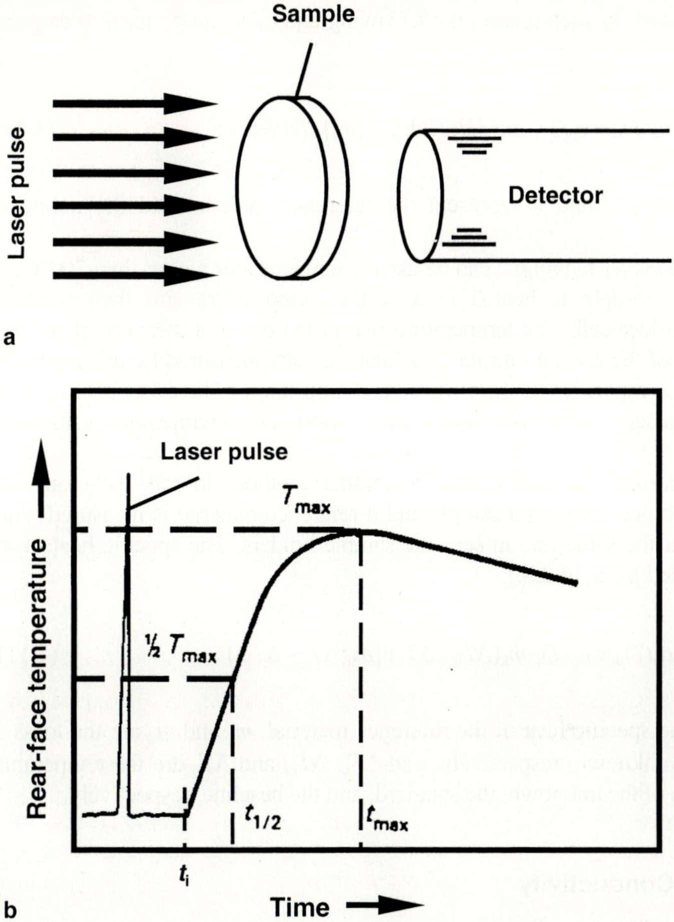

Fig. 4.46 a-b. Flash technique for thermal property measurements. (From: *Engineered Materials Handbook,* Vol.4, *Ceramics and Glasses* (1991), ASM International® Materials Park, OH 44073-0002), p. 614 (Fig.3), reprinted with permission)

difference in heat capacities between the two. The DSC equipment measures and records the rate of heat flow ($\mathrm{d}H/\mathrm{d}t$) to a sample as it is heated at a constant rate. A chart of heat capacity versus temperature is printed by the equipment. The heat capacity is calculated from [4.230]

$$\mathrm{d}H(T)/\mathrm{d}t = m\, c_{\mathrm{p}}(T)\, \mathrm{d}T/\mathrm{d}t \tag{4.109}$$

where m is the mass of the sample. To eliminate errors due, for example to the effect of the thermal resistance between the sample and the heat flow sensor, a reference material can be used. In such a case, the following equation can be used to calculate the heat capacity:

$$c_p(T)_u = c_p(T)_k\, m_k[dH(T)/dt]_u \,/\, \{m_u[dH(t)/dt]_k\}\ , \tag{4.110}$$

where the subscripts k and u represent the reference material and the unknown, respectively.

The *drop calorimetry technique* can be used at temperatures higher than 700°C,. In this technique, a sample is heated to a certain temperature and then suddenly transferred to a colder cell. The temperature rise in the cell is a measure of the heat content variation of the cooled sample. The heat capacity measured by this method is an average of heat capacities in the temperature drop range. The exact heat capacity can be found by a series of measurements with various initial temperatures, followed by curve fitting.

Differential thermal analysis (DTA) is another method. In this technique, the temperature difference between a sample and a reference material is measured while both are heated at the same rate in separate sample holders. The specific heat of the sample is calculated from [4.230]

$$c_p(T)_u = c_p(T)_k\, m_k(\Delta T_u - \Delta T_b)/[m_u(\Delta T_k - \Delta T_b)]\ , \tag{4.111}$$

where $c_p(T)_k$ is the specific heat of the reference material, m_k and m_u are the mass of the standard and unknown, respectively, and ΔT_u, ΔT_k, and ΔT_b are the temperature difference signals of the unknown, the standard, and the baseline, respectively.

4.2.2 Thermal Conductivity

Thermal conductivity, defined as the rate of heat transfer through a substance, is a key property in applications such as thermal insulation and electronic components. The heat transfer through a medium subjected to a temperature gradient dT/dx within a given time, t, and a cross section perpendicular to the heat flow direction, A, is expressed by the *Fourier equation* [4.230,4.234],

$$dQ/dt = kA\ dT/dx\ , \tag{4.112}$$

where k is the thermal conductivity of the medium. A major portion of heat transfer in ceramics at ambient temperatures is carried out by *conduction*, specifically by lattice vibrations (phonons). At elevated temperatures (above 800°C), heat conduction by *radiation* becomes important. The third mechanism, *convection*, can generally be ignored [4.31].

The amplitudes of ionic vibrations increase with temperature. Adjacent ions cause periodic forces that act upon each other. Their energy is transferred by phonons, and a continuous flow of heat occurs until thermal equilibrium is achieved in the solid. The expression for thermal conductivity of nonconductors by conduction through phonons is [4.31,4.228]

$$k_c = C_V \, v \, \lambda/3 \,, \tag{4.113}$$

where C_V is the heat capacity, v is the average velocity of a phonon, and λ is the average wavelength of phonons.

Crystalline materials have higher conductivities than amorphous materials of the same composition because the periodicity of a crystal structure interferes less with phonons than distorted structures [4.235]. Solid solutions decrease crystal regularity and thus diminish thermal conductivity. Structural discontinuities such as pores and glassy phases (common, for example, in refractory ceramics) reduce thermal conductivity for the same reason [4.234]. An approximate relationship for thermal conductivity in porous materials is given by [4.235,4.236]

$$k = k_s(1-P_c) \,, \tag{4.114}$$

where k_s is the conductivity of the fully dense material and P_c is the porosity of the cross-section. This equation is limited to cases where $T \leq 500°C$ and the solid phase is continuous. Other properties that affect thermal conductivity are grain orientation and atomic mass differences in the structure. For example, diamond has one of the highest known thermal conductivities among all materials (900 W/m K), more than twice that of copper. Its high thermal conductivity, among other factors, is due to a single type of atom in the crystal structure. Graphite has a layered structure, so the bulk thermal conductivity is low. However, within the layers, the thermal conductivity is very high (2000W/m K) in strongly aligned, dense pyrolytic graphite. BeO, SiC, and B_4C have relatively high thermal conductivities since their constituent elements have similar atomic mass [4.5].

At temperatures above θ_D, the number of phonons, n, is high. The mean free path of phonons, and thus λ is small. As n becomes larger, phonons start to interact and become scattered. This is called the *umklapp (flip-over) process*. The umklapp process may result in anharmonic vibrations, which further tend to decrease λ. Under such circumstances, thermal transfer is very inefficient. Metals are good thermal conductors because the carriers are electrons, which are not scattered by phonons as much as phonons are. Therefore, their valence electrons are relatively independent. As T approaches zero, the thermal conductivity also approaches zero. In the range ~ $\theta_D/10$ to $\theta_D/2$, impurities and lattice defects play an important role in thermal conduction. The approximate dependence on temperature in this range is given by [4.228]

$$k_c = YT^{1/2} \,, \tag{4.115}$$

where Y is a constant. At higher temperatures ($T>\theta_D$) a constant different from Y is valid for Eq. (4.115). The decrease in thermal conductivity is a result of the umklapp process. A general relationship between thermal conductivity and other thermal properties was given before by Eq. (4.108).

At higher temperatures, radiant heat transfer becomes important. The additional thermal conductivity component due to this mechanism is given by [4.31]

$$k_r = 16\sigma\,\eta^2 T^3 \lambda_m / 3\,, \tag{4.116}$$

where σ is the Stefan-Boltzman constant, η is the refractive index of the medium, and λ_m is the mean free path, in this case, of the photons. The influence of k_r in dense ceramics is usually significant at temperatures above 1200°C. However, due to the high porosity of refractories, the effect of k_r on them can be observed at temperatures as low as 500°C. The effective thermal conductivity can usually be given by adding the components due to conduction and radiation [4.31],

$$k = k_c + k_r\,. \tag{4.117}$$

Measurement

The flash method described in Sect. 4.2.1 and the radial heat flow technique, which will be described in Sect. 4.2.3 determine thermal conductivity along with other thermal properties. In addition to these methods, the hot wire method, the guarded hot plate method, and the calorimeter method can be used for thermal conductivity measurements.

The hot wire method [4.237] is a transient method that involves a thin wire embedded in the material to be tested, which is usually in the form of two rectangular blocks with appropriate grooves. Various types of the hot wire method exist. The general equation for the hot wire method is

$$k = Q \ln (t_2/t_1) / [4\pi (T_2 - T_1)]\,, \tag{4.118}$$

where Q is the power input per unit length of wire (W/m) and t_2 and t_1 are times corresponding to temperatures T_2 and T_1 (in Kelvin). In one type, a thermocouple is used as the hot wire and at the same time measures the temperature. In the version using the electrical resistance technique, direct current is supplied to the hot wire by a power supply. The current is determined by measuring the voltage drop, V_s, across a reference resistance, R_s, in series with the hot wire. Thus, thermal conductivity is calculated from experimentally measured quantities as

$$k = V\,V_s \ln (t_2/t_1) / [4\pi\, l R_s(T_2 - T_1)]\,, \tag{4.119}$$

where V is the voltage drop across the hot wire, and l is the length of the wire. The hot wire method is being considered as a standard ASTM test method. Commercial equipment for this type of test is available [4.237].

The *guarded hot plate method* determines thermal conductivity under steady-state conditions. This is a comparative method where the test cylinder with an aspect ratio of about 1 is sandwiched between two nearly identical reference cylinders with known conductivity and two heaters. A temperature gradient and, after a certain time, a steady-state heat flux are established between the upper and lower heaters. The heat flux is calculated for both the top and the bottom reference samples, using the reference material's thermal conductivity and Eq. (4.112). The average value from the two is used to calculate the thermal conductivity of the sample, again from Eq. (4.112). Among all of the methods mentioned, this is generally regarded as the most accurate. Other experimental arrangements also exist for the guarded hot plate method [4.237].

The *calorimeter method* is also a steady-state method where typically a brick test sample is surrounded by guarding bricks and the heat flows through them into a water-cooled copper calorimeter. Thermal conductivity is calculated from the temperature gradient and the rate of heat flow through the sample. The calorimeter and guarded hot plate methods are well suited for heterogeneous samples, but the required sample size is large and the test times are long in both methods.

4.2.3 Thermal Diffusivity

The rate at which a material cools to the temperature of its surrounding is known as thermal diffusivity. From Eq. (4.108),

$$a(T) = k(T) \, / \, \rho(T) \, c_p(T) \quad . \tag{4.120}$$

The term "thermal diffusivity" arises from the consideration of heat as an entity that can diffuse through a substance subject to different boundary conditions, and results in spatial and transient temperature variations.

Measurement

In addition to the flash method described for heat capacity measurement, thermal diffusivity can be measured by the *dynamic radial heat flow method*. This is a transient method in which the sample is cycled through a temperature range [4.237]. The sample is heated and cooled at constant rate, and the temperature gradient is measured by two thermocouples. The thermal diffusivity of a cylindrical sample with diameter d, length l, and an aspect ratio of $l/d > 2$, is found in m^2/s by

$$a = [R^2 \, (dT/dt)] \, / \, (\Gamma \, \Delta T) \quad , \tag{4.121}$$

where ΔT is the temperature gradient within the sample, R is the distance in meters over which ΔT is measured, dT/dt is the heating and cooling rate (K/s), and Γ is a constant (equal to 4 for an infinite cylinder, 6 for a sphere, and 3 for an infinite plate).

Photothermal methods can be used to determine thermal diffusivity on different scales, from the micrometer to the millimeter range [4.238]. Two such methods are the *mirage cell* and *photoreflectance methods*. In the former, the sample is placed on a mirage cell and illuminated with intermittent radiation (Ar laser beam). Heating of the surface creates a temperature gradient that becomes larger in the sample. The temperature gradient creates a refraction index gradient, which can be measured by the deflection of a He-Ne laser beam. With this setup, it is possible to arrive at the limit of photonic disturbance (10^{-11} rad·$Hz^{-1/2}$ for angular deviation, which corresponds to 10^{-5} K $Hz^{-1/2}$ for thermal variation on the surface in air). With such a deflection, it is possible to reveal thermal waves in the range of a few tens of kHz.

The second method uses a photoreflectance apparatus. The sample is illuminated with a focalized spot for heating, and the measuring beam is reflected onto the sample surface in the heated area. The focalized spot with a diameter of a few microns can be obtained by a commercial microscope by focusing the heating beam. A diode laser produces the measuring beam. After reflection on the area of modulated heating, the measuring beam is received by a photodiode attached to a lock-in amplifier at high frequency. This experiment requires high-frequency modulation to prevent heat diffusion into large areas. The quantity measured is the degree of modulation $\Delta R/R$ of the reflectance. This results from the dependence of the refractive index on the modulated temperature. A sensibility of 10^{-5} in $\Delta R/R$ can be achieved within a frequency interval amplified up to 20 MHz.

With photothermal experiments, one can determine the thermal diffusivity on different scales, from millimeter to micrometer, by varying the frequency modulation. To measure thermal diffusivity with the testing methods described, the heated area is determined whose size depends on diffusivity. Thermal diffusivity can be determined with a precision of 5%. Using the photothermal microscope, it is possible to measure thermal diffusivity on a micrometer scale. By decreasing the diffusion length by increasing the frequency, heated areas of only a few tens of microns can be achieved. Thus, determination of the heated area permits measuring local thermal diffusivity with high precision. Thermal diffusivity within minute areas such as a single grain can also be determined by these techniques, which is important in heterogeneous and porous materials, such as refractory ceramics.

4.2.4 Coefficient of Thermal Expansion

When a material is heated, its volume typically increases proportionally with the temperature rise. A three-dimensional array of atoms which vibrate about their equilibrium positions is a well-established model for crystalline solids. As the temperature is increased, atomic vibrations increase in amplitude about a_o, the average amplitude. However, because repulsion at close proximity is stronger than attraction at extreme separation, most solids possess an asymmetric potential energy and thus exhibit anharmonic vibrations. As a consequence, most solids expand with an increase in T, yielding a positive coefficient of thermal expansion, α.

Although anharmonic vibrations have a major effect on thermal expansion, second-order effects such as magnetic, electronic, order–disorder, and ferroelectric effects could result in zero or negative thermal expansion (NTE) in certain temperature ranges. The true coefficient of thermal expansion (CTE) is structure insensitive, but mechanical properties (σ, K_c) are dependent on microstructure and may also influence thermal expansion measurements. Furthermore, microcracking or grain boundary separation may lead to hysteresis and to an apparent NTE even if the true CTE is positive. However, such phenomena result in strength degradation and limit applications. Isotropic NTE is much more preferable for many applications. Isotropic NTE behavior below room temperature reportedly occurs in amorphous SiO_2, cubic Cu_2O, and Si [4.239]. NTE above room temperature is reported in cubic materials such as ZrV_2O_7 (zirconium vanadate) and solid solutions of the type $ZrV_{2-x}P_xO_7$. These materials have a very low CTE at ambient temperature and NTE at higher temperatures. The ZrW_2O_8 (zirconium tungstate) family is practically more important among all NTE materials because it is unique in showing strong NTE over a broad temperature range including ambient temperature. NTE behavior remains strong from 600°C (873 K) down to –263°C (10 K, i.e., close to absolute zero). ZrW_2O_8 is unstable between 600 and 1105°C, and thus, unsuitable for applications. ZrW_2O_8 has a cubic crystal structure in its α (20–430 K) and β (430–950 K) form. The thermal expansions ($\Delta l/l_o$) of α- and β-ZrW_2O_8 are measured as -8.7×10^{-6} and -4.9×10^{-6} (K^{-1}), respectively. The only mechanism that leads to NTE in isotropic oxides is based on open network structures where oxygen has a twofold coordination with respect to metal cations. If the M–O bonds are strong enough to show negligible expansion during heating, the transverse thermal motion of oxygen in the M–O–M linkages can cause NTE phenomena [4.239]. An important prospective application of NTE ceramics is limiting the thermal expansion of composite materials, for example, those used for teeth restorations and substrates for mirrors. Other applications envisioned are various forms of temperature compensators and thin films for sensitive temperature sensors and actuators.

The linear CTE is given by [4.5]

$$\alpha_L = (\Delta l/l_0)/\Delta T, \tag{4.122}$$

where l_0 is the sample length at 0°C, Δl is the change in length, and ΔT is the change in temperature. Similarly, the volumetric CTE is given by

$$\alpha_V = (\Delta V/V_o)/\Delta T, \tag{4.123}$$

where V represents the volume.

The bulk CTE can be estimated from linear CTEs measured by X-ray methods using [4.230]

$$\alpha_{V,avg} = 3\ \alpha_{L,avg}\ . \tag{4.124}$$

The units 1/°C or 1/K are generally used for CTE data. The linear CTE can be approximated by [4.228]

$$\alpha_L = (3Mk_B)\ /\ (4L^2a_0)\ , \tag{4.125}$$

where M and L are coefficients that describe the amplitude of an oscillating particle in the x coordinate,

$$V(x) = Lx^2 - Mx^3\ . \tag{4.126}$$

M and L are related to the bonding energies of the ions. The approximation given by Eq. (4.125) does not suggest dependence between α_L and temperature. However, remember that coefficients L and M are, in fact, functions of temperature. Above θ_D, α increases approximately linearly with temperature. Below θ_D, α varies exponentially with temperature. Examples of the variation of α_L with T for various ceramics, metals, and polymers are shown in Figs. 4.47 and 4.48.

Anisotropic structures have asymmetrical distributions of atomic bonds. Therefore, ceramics that do not have cubic or amorphous structures may exhibit anisotropic thermal expansion. Yet, if a polycrystalline ceramic has randomly oriented grains, the bulk will exhibit approximately isotropic thermal expansion. As an example, polycrystalline alumina has an average CTE of 8.7×10^{-6} 1/°C, whereas $\alpha_L = 8.3 \times 10^{-6}$ 1/°C in the c axis and $9.0\text{x}10^{-6}$ 1/°C perpendicular to the c-axis of a single crystal. Graphite is another example of anisotropic crystals. The CTE in the direction of weak (van der Waals) bonding is high (27×10^{-6} 1/°C), but within the graphite layers, atomic bonding is strong and the CTE is low (1×10^{-6} 1/°C).

The thermal expansion behavior of ceramics is of great concern to those who work with ceramic materials because of the potentially severe thermal stresses caused by cooling or heating. Thermal stress is a very important factor in designing ceramic components where heating, cooling, or thermal cycling will occur. The subjects of thermal stress and thermal shock were treated in Sect. 4.1.4. Thermal stresses also play a significant role in composite materials. The mechanical properties of ceramic matrix composites (CMCs) are directly related to thermal stresses that occur due to CTE mismatch between constituents, as explained in Sects. 4.1.2. and 4.1.3.

Measurement of CTE

The usual method of measuring thermal expansion involves the use of a *dilatometer*. This device is made of two important components, a furnace (or cooling

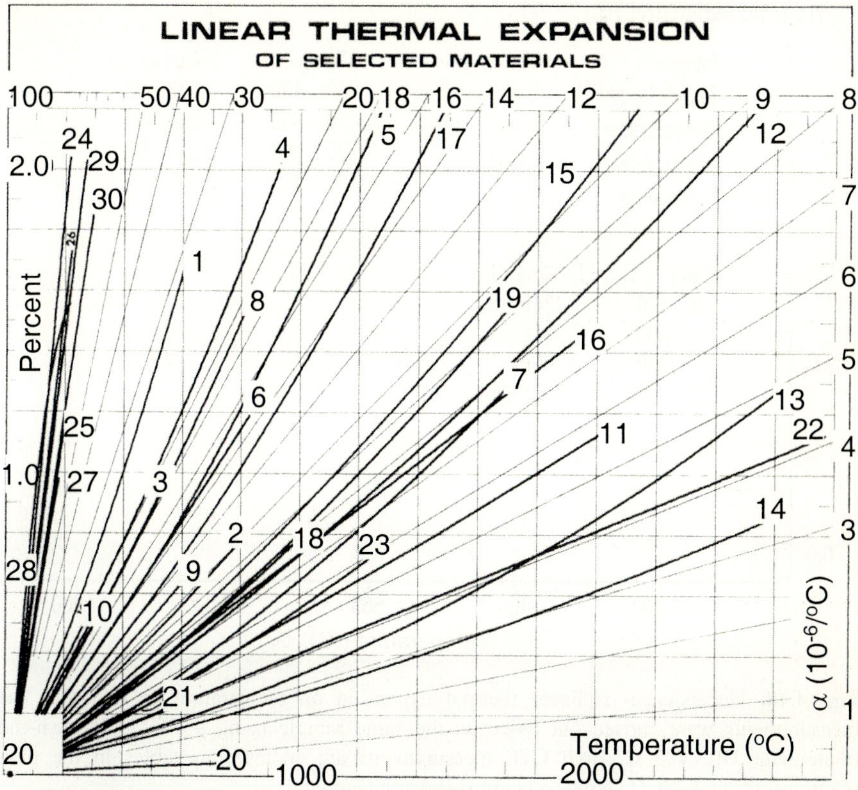

Read % expansion on the left axis, match equi-coefficient lines to scale on right and top

1. Aluminum (99.9% pure)
2. Platinum (99.9% pure)
3. Copper (OFHC)
4. Silver (99.9% pure)
5. Nickel (Grade 600)
6. Iron (99.9% pure)
7. Beryllium
8. AISI 304 Stainless
9. AISI 440C stainless
10. Brass, free machining
11. Tungsten
12. Graphite, AXM-5Q
13. Graphite, ATJ (WG)
14. Graphite, ATJ (AG)
15. Beryllium Oxide
16. Zirconium Oxide
17. Magnesium Oxide
18. Aluminum Oxide
19. Sapphire (59°)
20. Fused Silica (quartz)
21. Borosilicate Glass
22. Boron Nitride
23. Boron Carbide
24. Acetal
25. PVC
26. ABS
27. Acrylic
28. TFE
29. Nylon
30. Polycarbonate

No warranties are expressed or implied for the accuracy of the data.

Fig. 4.47. Dependence of linear thermal expansion on temperature in various materials (courtesy of Anter Corp. Pittsburgh PA)

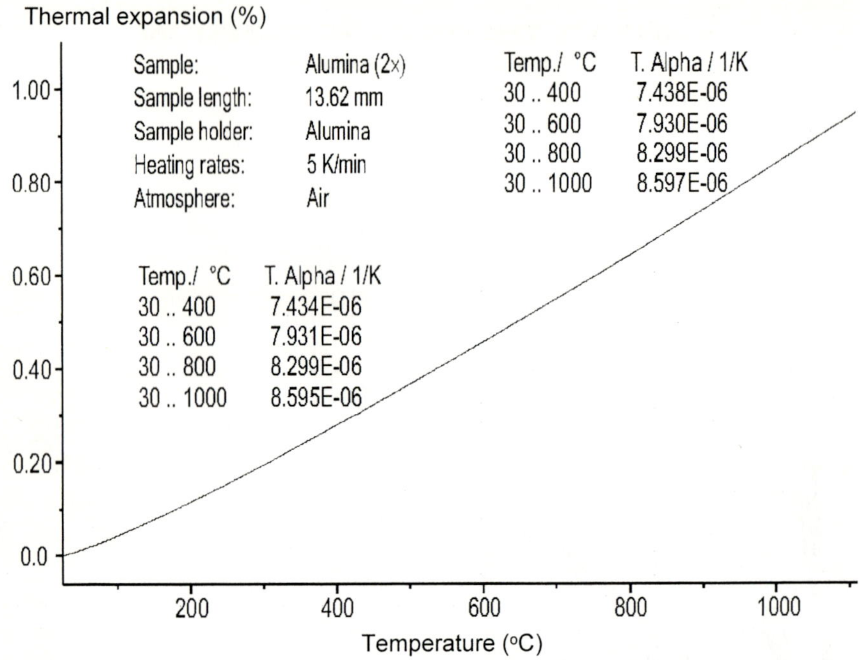

Fig. 4.48. Dependence of linear thermal expansion on temperature in α-Al_2O_3. The measurements were carried out twice on the same sample using a NETZCH push-rod dilatometer DIL 402 C. Both CTE measurements are enclosed as tables in the plot. (Courtesy of NETZSCH Gerätebau GmbH Bavaria Germany)

equipment) and a linear variable differential transformer (LVDT). The dilatometer yields an electrical signal through the LVDT, once shrinkage or swelling of the sample is detected during heating or cooling. This signal may be recorded by a chart recorder or the dilatometer may be connected to a computer for easier data storage and handling. The electrical signal can be converted to dimensional change in percent if necessary calibrations are made. In single push-rod dilatometers, a reference sample with known expansion coefficient is used to calibrate the length versus temperature signal. A second experiment is conducted with the actual sample, and the expansion coefficient is calculated by comparing the measured expansion with that of the standard. In the case of double push-rod dilatometers, the sample and the standard are mounted side by side and are heated simultaneously. The expansion is determined by direct comparison with the standard [4.230].

Interferometry is a more accurate but expensive and slower method. This is an optical technique based on the principle that two parallel reflective surfaces, which are a short distance apart will display interference fringes when illuminated by a monochromatic light source. The fringes will move if the distance between reflecting surfaces changes. If a sample is positioned between two polished, optically flat surfaces and heated, the interference fringes will move according to the equation [4.230],

$$\frac{\Delta L}{L} = \frac{\lambda N}{2L} + \frac{A}{L}, \tag{4.127}$$

where ΔL is the change in sample length, L is the initial length, λ is the wavelength of the light source, and A is a correction factor depending on the surrounding atmosphere ($A = 0$ for vacuum).

X-ray methods for measuring the lattice parameters of crystals at elevated temperatures provide a means for determining a very precise measure of thermal expansion of crystals [4.240]. Powder samples and small single crystals can be used in this technique.

Optical measurement of the sample length during heating is another direct method applicable to bulk materials is. Usually a measuring microscope or telescope through a window in the furnace can be used for this technique. Samples should be as long as possible for satisfactory measurements. They also should be allowed to expand in both directions. Otherwise, the expansion of the fixed end should be taken into account. Window materials are available for high temperature use.

4.3. Optical Properties

4.3.1. Transmittivity and Reflectivity

Most crystalline ceramics are opaque in bulk and generally do not have optical applications. However, glasses are among the most important materials used for optical components. Ceramics with strong covalent bonding (nitrides, borides, carbides) are opaque, although transparent thin films can be prepared from them by using special techniques. Most of the remaining ceramics are composed of O^{2-}, Si^{4+}, Al^{3+} and alkali/alkaline earth ions. These ions do not have any absorption bands in the visible spectrum and thus, the ceramics should be transparent [4.31]. Actually, they are opaque typically. The opacity is due to the presence of grain boundaries, which are sites of reflection. If these boundaries are limited and the material is thin, the material may become transparent. Single-crystal oxide ceramics are transparent in the visible spectrum but show varying transparencies in the infrared and ultraviolet regimes.

According to *Lambert's law*, light intensity decreases exponentially within a material with the distance traveled [4.149,4.234],

$$I = I_0 \exp(-ax), \tag{4.128}$$

where I_0 is the intensity at $x = 0$, x is the distance (or thickness), and a is the absorption coefficient of the medium. When light passes through a ceramic material, the incident light is reflected partially; part of it is absorbed, and the rest will leave the body from the opposite face. The ability of a body to transmit light is described by the *transmittance* or *transmission factor* [4.234,4.241],

$$\tau = I / I_0 = (1-\rho)^2 \exp(-ax), \tag{4.129}$$

where I_0 and I represent incident and transmitted light intensities, respectively, and ρ is the reflected fraction at each interface. The ending *-ance* for optical properties is used to indicate that these are properties of a given sample with a given thickness [4.242]. Transmittivity in a material is related to such factors as porosity, grain size, impurities, and imperfections. Any substance that causes scattering of incoming radiation will decrease transparency and contribute to opacity. However, the size of the feature should also be considered. If the size is smaller than the wavelength of the incident radiation, scattering is negligible.

The reflectivity of a plane wave incident normally to a flat, specular, absorbing surface is given by [4.243]

$$\rho = [(\eta_1 - \eta_2) / (\eta_1 + \eta_2)]^2 , \quad (4.130)$$

where η_1 is the refractive index of material No. 1 and η_2 is the refractive index of material No. 2. If material No. 2 is air, η_2 becomes equal to 1.

Reflectivity determines the fraction of light reflected when a light wave hits an interface. In ordinary glasses where $\eta \approx 1.5$, $\rho = 0.04$, in other words, 4% of the perpendicular incident light becomes reflected at each glass/air interface [4.234]. It is possible to decrease the reflectivity of such a glass by coating it with a glass based on fluoride, which has a lower refractive index. To increase reflectivity, the glass can be coated with oxides of higher refractive index or metallized (which is the case for mirrors, where the reflected fraction of light can reach up to 90%). If the incident light is not perpendicular, the reflected fraction will depend on the angle of incidence.

Measurement

Transmittivity and reflectivity are most commonly measured by a double-beam spectrophotometer [4.243]. Corrected data from single-beam measurements are also used [4.242]. A spectrophotometer consists of four components, as illustrated in Fig. 4.49: a radiation source, a monochromator, the sample, and the detectors [4.244]. Because multiple reflection occurs at various planes of the sample, transmittivity is given by the equation following, instead of Eq. (4.129):

$$\tau = I / I_o = [(1-\rho)^2 \exp(-ax)] / [1-\rho^2 \exp(-2ax)] , \quad (4.131)$$

where x is the sample thickness. This expression is valid only for normal incidence. Note that Eq. (4.131) transforms into Eq. (4.129) when $\rho = 0$.

The reflectivity of the sample can be calculated from

$$\rho \approx \rho_{obs} / [1 + (I/I_o) \exp(-ax)] , \quad (4.132)$$

where ρ_{obs} is the observed (measured) reflectivity, measured by a second detector (see Fig. 4.49).

Entry and exit surfaces, through which the light wave passes, should be perfectly flat and parallel for accurate measurements. The method is applicable from ultraviolet to infrared wavelengths within the spectrum. Transmission ranges and other optical properties of important ceramics are given in Appendix 8.

4.3.2. Refractive Index and Dispersion

When light travels through a transparent material, its velocity decreases compared to its velocity in vacuum. The refractive index is defined as [4.5]

$$\eta = v_{vac} / v_{mat} = \sin(i) / \sin(r)\,, \tag{4.133}$$

where v_{vac} and v_{mat} represent the light velocity in vacuum and in the material, respectively, and i and r represent the incident and refracted light angle with the surface normal (Fig. 4.50). The refractive index varies with the wavelength of incident radiation. This variation is called dispersion. Dispersion usually increases simultaneously with η, and it normally results in a decrease in η as the wavelength increases. η typically varies between 1.3 and 2.7 in the visible spectrum for common oxides. Lower values are found in the infrared spectrum and higher

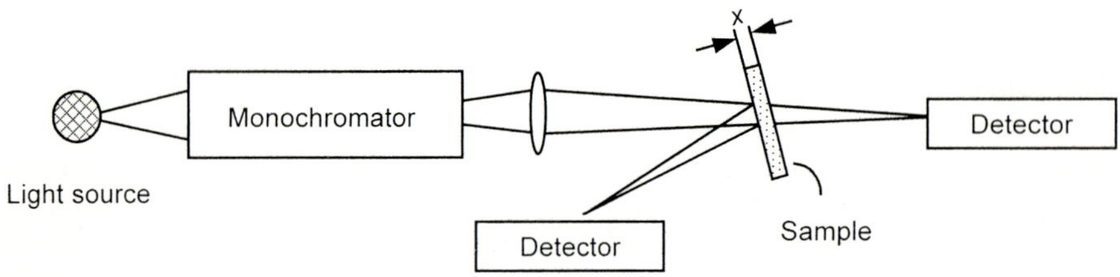

Fig. 4.49. Schematic illustration of spectrophotometer for transmittivity and reflectivity measurements (From M.W. Barsoum: *Fundamentals of Ceramics*. Copyright 1997 McGraw-Hill Companies, Inc. New York, reprinted with permission)

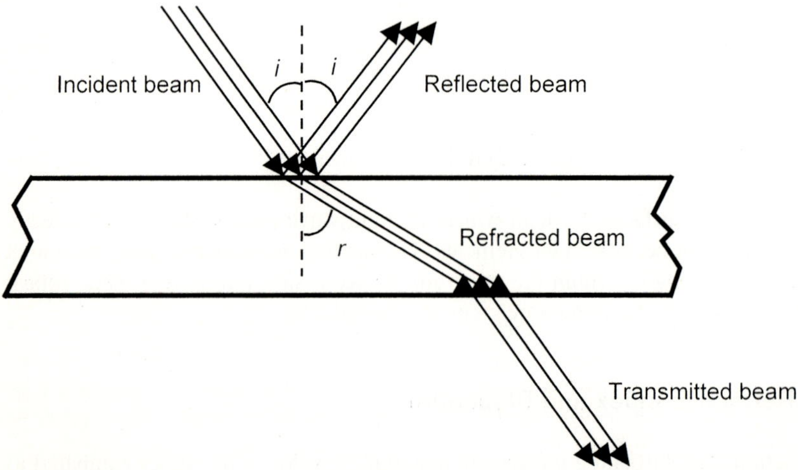

Fig. 4.50. Interaction of an incident light beam with a transparent material

values in the ultraviolet spectrum. In general, η is expressed by a complex quantity [4.244],

$$\eta' = \eta + i\,k, \tag{4.134}$$

where the imaginary component of the equation, k, is called the extinction coefficient or absorption index. k is a measure of the absorption capability of a material. Figures 4.51–4.53 are plots of η and k versus wavelength for a typical semiconductor (SiC) and two insulators (glassy and crystalline SiO_2). η_o and k_o are properties measured for ordinary rays [4.245].

Dispersion is measured mostly by the *Abbe value* given by [4.244]

$$v_D = (\eta_D - 1) / (\eta_F - \eta_C), \tag{4.135}$$

where η_D, η_F, and η_C are indexes of refraction for the 587.56 nm spectral line of sodium, the 486.13 nm line of hydrogen, and the 656.27 nm line of hydrogen , respectively. The Abbe value generally ranges between 20 and 100 in glasses.

Measurement

The *Becke line method* is a simple method for measuring refractive index in transparent materials [4.149]. A few grains of –100 to +200 mesh specimen powder are placed on a petrographic slide with a drop of index oil (used for petrographic specimens). A cover glass is placed on them. By observing the Becke line, one can tell if the oil has a higher or lower refractive index than the powder. The accuracy of this method is about ± 0.004%.

It is also possible to calculate the refractive index and absorption coefficient of a given material from spectrophotometric measurements of ρ and τ, explained in Sect. 4.3.1. For these calculations, one needs to use *Fresnel's formula* [4.244]:

$$\rho = [(\eta - 1)^2 + k^2] / [(\eta + 1)^2 + k^2] \tag{4.136}$$

and Eqs. (4.131) and (4.132).

4.3.3. Emissivity

Emissivity, ε, is defined as the ratio of radiant flux per unit area leaving the surface of a material to that of a blackbody at the same temperature. Emissivity is the emittance from a specimen that is thick enough to be completely opaque and has an optically smooth surface [4.245]. The emissivity of a blackbody is 1. Emissivity is a dimensionless quantity and lies between 0 and 1 for any other type of surface [4.149]. A perfect emitter (a blackbody) emits radiation defined by the *Planck equation*,

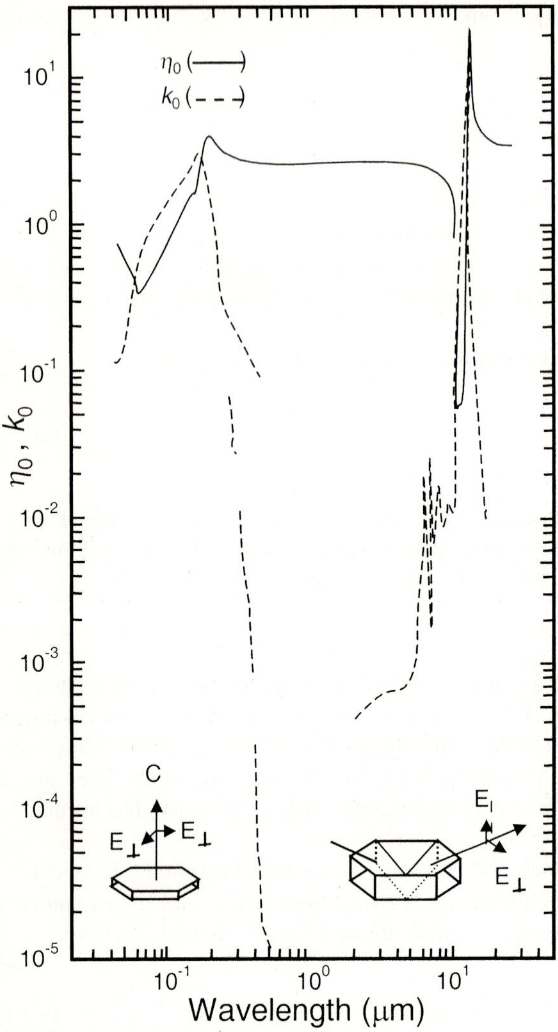

Fig. 4.51. Refractive index and extinction coefficient versus wavelength for a typical semiconductor (SiC). Typical sample orientations are shown in the insets (From *Handbook of Optical Constants of Solids*. E.D. Palik (ed.) (Academic Press Orlando FL, 1985), reprinted with permission)

$$E_b(\lambda) = c_1\, \lambda^{-5}\, [\exp\,(c_2 / \lambda T) - 1]^{-1}\,, \tag{4.137}$$

where $E_b(\lambda)$ is the hemispherical radiant power per unit area at surface temperature T for unit wavelength interval centered on wavelength λ, $c_1 = 3.742 \times 10^{-2}$ Wm^2, and $c_2 = 1.439 \times 10^{-2}$ mK, respectively. The emissivity averaged over all angles, i.e., the *spectral hemispherical emissivity*, is defined as

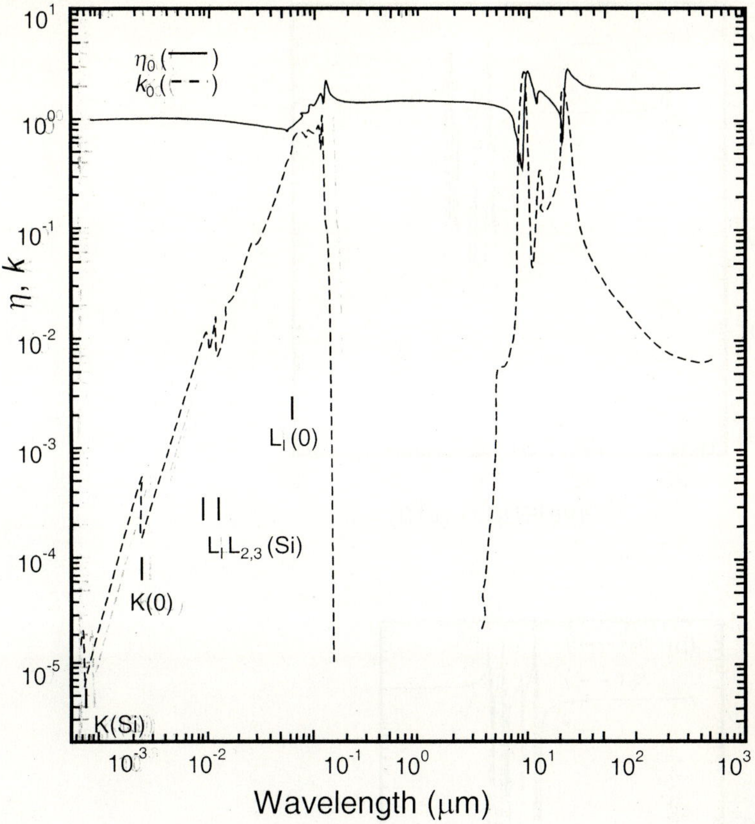

Fig. 4.52. Refractive index and extinction coefficient versus wavelength for silicon dioxide (glass) (From *Handbook of Optical Constants of Solids*. E.D. Palik (ed.) (Academic Press Orlando FL, 1985) reprinted with permission)

$$\varepsilon(\lambda,T) = E_\lambda(\lambda,T) / [E_b(\lambda,T)] . \tag{4.138}$$

On the other hand, emissivity averaged over both wavelengths and angles, called total hemispherical emissivity, is given by

$$\varepsilon_t(T) = E_t(T) / E_{b,t}(T) = E_t(T) / c_3T^4 , \tag{4.139}$$

where $c_3 = 5.67 \times 10^{-8}$ W/m^{2}K^4 and E_t is the total radiant power from a hemisphere.

Emissivity data generally quote total normal emissivity, which may be different from total hemispherical emissivity, and therefore such data should be used with

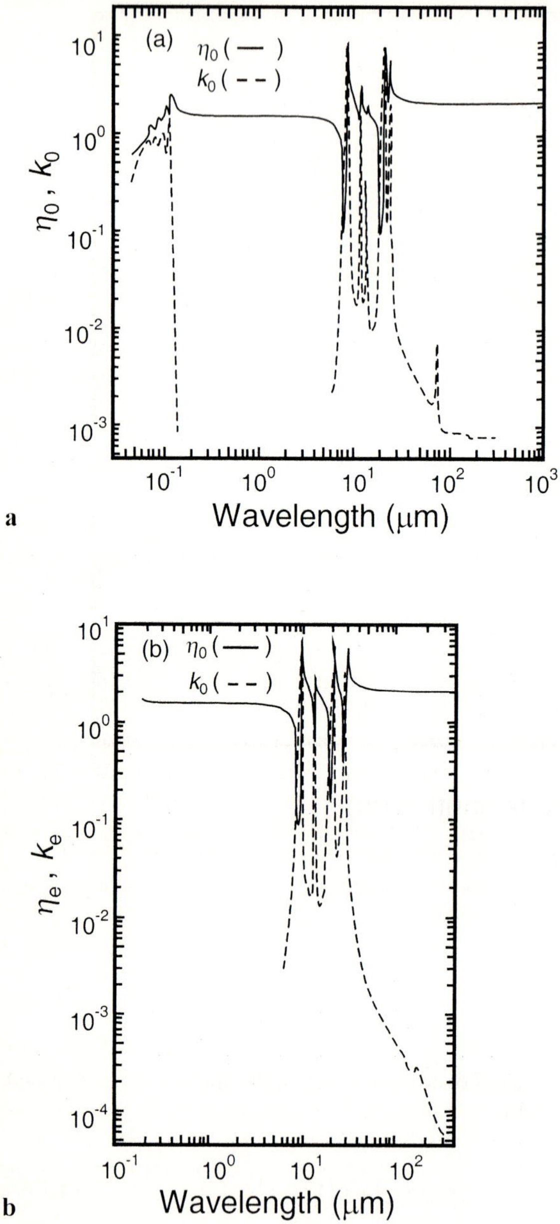

Fig. 4.53. Refractive index and extinction coefficient versus wavelength for silicon dioxide type α (crystalline) **a;** equilibrium refractive index and equilibrium extinction coefficient versus wavelength for the same material **b** (From *Handbook of Optical Constants of Solids*. E.D. Palik (ed.) (Academic Press Orlando FL, 1985), reprinted with permission)

caution. An important use of emissivity is in infrared pyrometry. To make a correct measurement with an infrared pyrometer, calibration is required with respect to the emissivity of the material whose temperature is to be determined.

Emissivity is one of four thermal radiative properties. The other three, reflectivity, ρ, absorptivity, α, and transmittivity, τ, are related by [4.242]

$$\rho + \alpha + \tau = 1, \tag{4.140}$$

from a flux balance.

Measurement

Emissivity can be measured directly by *spectrometers*. Different spectrometers such as ultraviolet, visible, or infrared spectrometers, have varying ranges of sensitivity to the wavelength of illumination. Spectrometric measurements coupled with independent temperature measurements (e.g., with a thermocouple) yield emissivity data.

Calorimetric methods are commonly used for emissivity determinations. In one technique, a sample is electrically heated to a certain temperature. If conduction and convection are practically eliminated, the energy necessary to keep the temperature constant will be that which is lost from the sample by radiation, provided that the heating elements are embedded. Comparison of the emitted energy to the theoretical blackbody radiation at the same temperature yields emissivity.

Another method involves the use of a calorimeter. When a sample in contact with a calorimeter is irradiated with a given heat flux, the change of heat content can be measured. The adsorbed energy measured by this method is compared to the energy of incoming radiation and gives a measure of absorptivity. This is numerically equal to emissivity in opaque materials.

4.3.4. Birefringence

The velocity of light and therefore, the refractive index of a material varies when external stress is applied. This phenomenon is called *birefringence* or *photoelasticity* [4.241,4.243]. The velocity of light that is entering a transparent material such as glass at an angle different from the optical axis or stress will differ in various oscillatory planes. One component of the polarized light travels faster than the other in the stressed material. The resulting phase displacement and path difference Δ are proportional to sample thickness, t, and the difference in velocity, and thus, to the stress,

$$\Delta = B\,t\,(\sigma_1 - \sigma_2)\,, \tag{4.141}$$

where $\sigma_1 - \sigma_2$ is the difference of stresses and B is the stress – optical coefficient (25×10^{-7} MPa^{-1} for alkali-lime-silica glass, 35×10^{-7} MPa^{-1} for fused silica, and 26×10^{-7} MPa^{-1} for aluminosilicate glass [4.241]). (See Sect. 5.4.4 for birefringence as a characterization tool.)

Measurement

Birefringence is measured by *polarimeters* (polariscopes). There are numerous types of polarimeters such as the Savart polariscope, Steel's polariscope, and Tsuruta's polariscope. In polarimeters, a birefringent plate is used as the beam splitter, and the variations in the optical paths are observed through polarization phenomena [4.247]. The details of stress–optical coefficient measurement in glass are given in ASTM standard C 770-77.

4.3.5. Color

Color is an important property in certain applications such as decorative glasses and tableware, glazes, filters for photography, specially colored glass (e.g. sunglasses, tinted glass for automotive/domestic use, etc.), and industrial paints (which usually contain some type of ceramic pigments). Color is directly related to absorption since the human eye sees that part of the optical spectrum, which is not absorbed and which is reflected or transmitted. The color of a material we see is the negative of the colors in the visible spectrum (light with a wavelength of 400 to 700 nm) that is absorbed [4.235]. The difference between the additive effects of pigments and colored light should be noted. For example, mixing red and blue pigments, which are two of the three additive primary colors, yields purple (the third primary color is yellow). Mixing pigments of all colors in equal quantities yields black [4.248]. On the other hand, mixing red and blue lights obtained by a filter yields a muddy black or dull brown color. Mixing all colors in the form of light in equal quantities yields white [4.249].

Absorption that results in color is associated with *electron transitions* [4.5]. Electron transitions can occur as

1. internal transitions within transition-metal, rare-earth, or other ions with incomplete electron shells,
2. charge transfer by electron movement from one ion to another,
3. electron transitions due to crystal imperfections, and
4. band-gap transitions.

Band-gap transitions result in intrinsic color in many semiconductor materials. Materials that contain transition elements with incomplete d shells or rare-earth elements with incomplete f shells thus have multiple valence states and commonly have characteristic colors. Electron transfer of types 1–3 is usually related to small amounts of impurities or crystal defects, whereas type 4 is a bulk property [4.5]. If a transparent material, such as glass, is hit by light radiation, the energy carried by light is normally not sufficient to excite transition electrons to permitted higher energy levels. However, if the glass contains constituents whose electrons are easily excited, such as transition-metal ions (Ti, V, Cr, Mn, Fe, Co, Ni, Cu), the absorption band will extend into the visible spectrum. This is the reason that the glass will appear colored. The color will be complementary to the absorbed radiation. Similarly, the main reason for color in crystalline ceramics is the presence of ions of transition or rare-earth elements in solid solution [4.234].

Another mechanism that gives rise to color is the formation of "color centers" of a crystal [4.5,4.234]. *Color centers* or *F-centers* (from Farbzentrum = Color Center in German) are anion vacancies that contain a trapped electron. These vacancies can be formed in alkali halide crystals such as NaCl, KBr, and KCl when they are heated in the presence of the alkali metal vapors or when they are irradiated with a high energy source such as X rays or neutrons. Trapped electrons absorb certain wavelengths and result in forming of certain colors. By this mechanism, colorless NaCl crystals become yellow, KBr blue, and KCl magenta.

Ceramic pigments are widely used in paints and in high-temperature applications where other pigments are destroyed [4.5]. Applications of pigments in ceramic ware are discussed in Sect. 7.2.5. Two major systems of classifying pigments are the *Color Index* and the *Dry Color Manufacturers Association (DCMA)* classification. The former is the standard reference related to colors, published jointly by the British Society of Dyers and Colorists and the American Association of Textile Chemists and Colorists [4.248]. The latter is a classification of dry colors, according to crystallography, color, application, composition, and Color Index names and numbers that covers most ceramic pigments. Appendix 9 lists some typical pigments used in glazes or paints (organic coatings).

4.4. Electrical Properties

4.4.1. Electrical Conduction

The large difference in the electrical conductivity of different types of materials such as metals and insulating ceramics is conveniently explained by the band theory of solids. According to this theory, electrons in crystals are arranged in energy bands separated by forbidden regions of energy gaps [4.250] (Fig. 4.54). These gaps result from the interaction of conductive electron waves with the ion cores of crystals. If the allowed energy bands are totally full or empty and are separated by a gap, the electrons cannot be easily accelerated into higher energy states in an electric field [4.251]. Such a material is called an *insulator* or a *dielectric material*.

In a metal, no gap exists between occupied and unoccupied electron states. Semiconductors have slightly filled or slightly empty bands. *Electrical conductivity*, defined as the ratio of the electrical current density to the electrical field intensity [4.234,4.251],

$$\sigma = J / E , \tag{4.142}$$

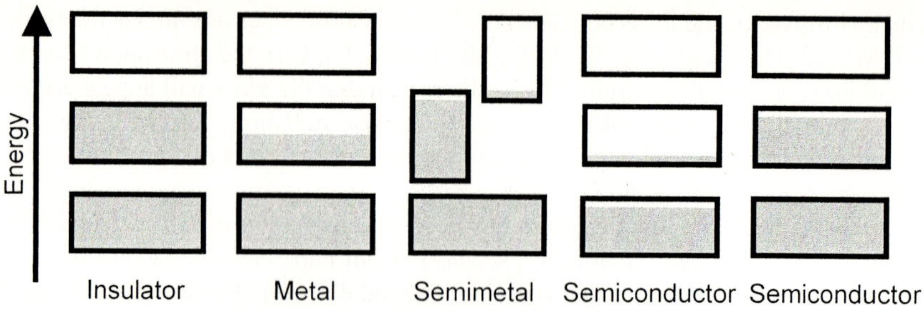

Fig.4.54. Schematic illustration of electron occupancy of allowed energy bands for an insulator, metal, semimetal, and semiconductor. The semiconductor on the left is at a finite temperature, and the carriers are excited thermally. The other semiconductor is electron-deficient because of impurities (From C. Kittel: *Introduction to Solid State Physics*. Copyright 1986 John Wiley & Sons, reprinted with permission)

ranges from $10^{10}\ \Omega^{-1}\text{cm}^{-1}$ for pure metals at low temperatures to $10^{-22}\ \Omega^{-1}\text{cm}^{-1}$ for insulators. *Resistivity* is the reciprocal of conductivity [4.5],

$$\rho = 1/\sigma. \tag{4.143}$$

Appendix 10 lists resistivity and other important electrical properties of ceramic materials. Electrical conduction can be expressed in terms of carrier concentration, n, and carrier mobility, μ,

$$\sigma = nq\mu, \tag{4.144}$$

where q is the charge per unit volume. Charge carriers may be of different types including electrons, holes, anions, and cations. In electrical conduction by more than one mechanism, conductivity is expressed as [4.252]

$$\sigma = \Sigma\sigma_i = \Sigma n_i q_i \mu_i, \tag{4.145}$$

where σ_i represents the partial conductivity of each mechanism. Ceramic materials exhibit the most diverse range of conductivities among all classes of materials. The best insulators are ceramics; under suitable conditions some ceramics are the best conductors (high temperature superconductors). On the other hand, many ceramics are semiconductors, and some exhibit metallic conduction under ambient conditions. A wide range of electrical and electronic applications can thus be found in the market.

Measurement

Conductivity/resistivity and dielectric breakdown strength of a dielectric are determined by current–voltage (*I–V*) measurements made in a measuring cell [4.234, 4.253]. The measurement of electrical properties in insulating materials such as ceramics necessitates some important precautions:

1. adaptation of measuring cells with three terminals, necessary to eliminate errors due to surface currents in dc and parasitical capacitance in ac;
2. adaptation of cells which accurately control the sample and environmental temperature;
3. maximum insulation of the measuring electrode, whose insulating parts should be maintained at ambient temperature;
4. metallization of sample surfaces to obtain efficient contact between the electrodes and the sample; and
5. accurate screening of the measuring system to prevent any possible interference from electromagnetic radiation or similar effects.

4.4.1.1. Insulators (Dielectric Materials)

Insulating ceramics are commonly used in electrical power lines, electronic equipment, particularly very large scale integration (VLSI) circuits, and thermocouples. Beside low electrical conductivity, insulators used in such applications exhibit high corrosion resistance and stability at elevated temperatures [4.234]. Most ceramics have filled conduction bands, and the energy required to raise electrons into the next conduction band is too high. For example, in ionic ceramics such as NaCl, this potential well is of the order of 6 to 10 eV/mole, and in covalent ceramics, of the order of 1.2 to 4 eV/mole. Any conduction that occurs in pure dielectric materials can thus be attributed to lattice mechanisms [4.254]. At elevated temperatures, most ceramics are ionic conductors even if they are insulators at ambient temperature. Common insulators are Al_2O_3, BeO, MgO, AlN, BN, Si_3N_4, SiO_2, cordierite, mullite, porcelains, and most glasses. Pure ZrO_2 is an insulator with a band gap of ~5eV at ambient temperature [4.252]. Pure ThO_2 is an insulator with a band gap of 5.75 eV. α-Al_2O_3 has a large band gap of ~10eV and is one of the most employed insulators. Hexagonal BN is an insulator with a band gap of 5–6 eV. Cubic BN is a better insulator than CaO, α-Al_2O_3, or MgO. Electrical conduction at ambient temperature is dominated by impurities that act as donors or acceptors. Impurity content of as little as 1 ppm can create donors, acceptors, or other point defects of concentration on the order of 10^{16} cm^{-3}. The concentration of intrinsic carriers at 1000°C is on the order of 10^9 cm^{-3} in comparison.

The carrier concentration and mobility can be thermally activated, therefore the temperature dependence of electrical conduction can be described by an Arrhenius type equation [4.252]:

$$\sigma = \sigma_0 \exp(-Q/kT)\,, \tag{4.146}$$

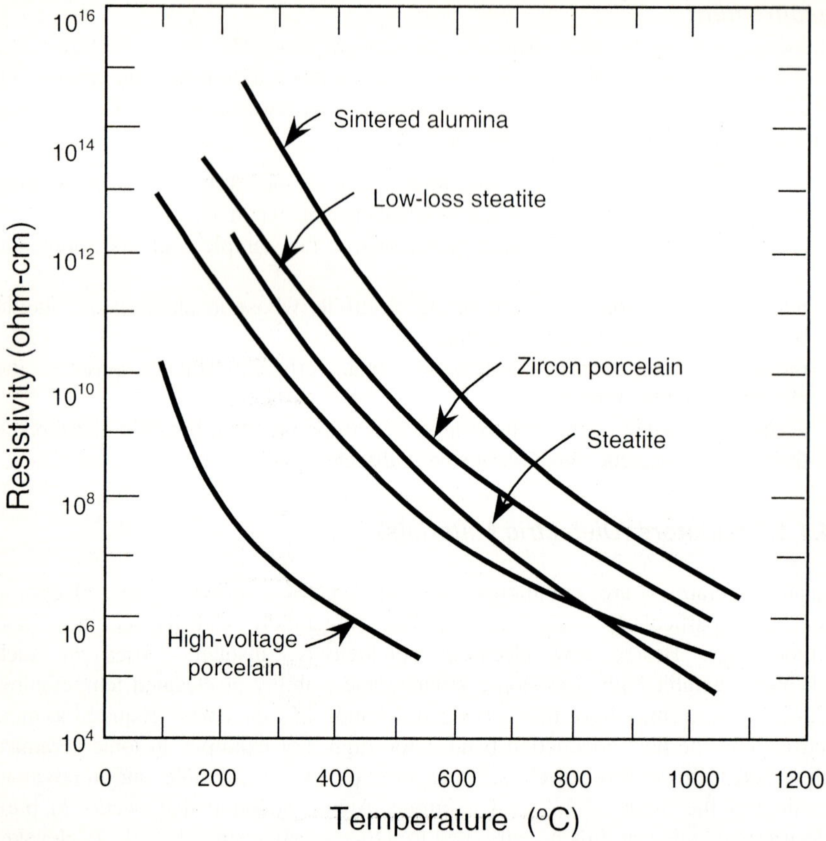

Fig. 4.55. Variation of resistivity with temperature for some ceramics (mostly bonded with a vitreous phase) (From W.D. Kingery, H.K. Bowen, and D.R. Uhlmann: *Introduction to Ceramics*. Copyright 1976 John Wiley & Sons New York, reprinted with permission)

where σ_0 is a factor that may depend on temperature and Q is the activation energy of electrical conduction. Figure 4.55 shows the variation of resistivity with temperature in various insulators. In crystalline ceramics, conductivity is dependent on the concentration of crystal defects and impurities. At high temperatures, the atmosphere, especially the partial pressure of oxygen, can affect the equilibrium number of defects such as ion vacancies. Above ~1200°C, ceramics should not be considered safe insulators since ionic conduction as well as other effects may increase conductivity.

The resistivity of dense porcelains (siliceous clay-based ceramics) generally exceeds 10^{11} Ω·cm at 25°C. Impurities are the limiting factor in achieving higher resistivities [4.255]. Similarly, the resistivity of Al_2O_3 decreases with the amount of impurities. For example, an 85% pure Al_2O_3 has a typical resistivity of 4.6×10^{10} Ω·cm at 300°C, and this increases to 2.0×10^{11} Ω·cm at 99.5% purity and up to

1.0×10^{15} Ω·cm at 99.9% purity [4.256]. Grain boundaries can play an important role in resistivity. For example, a special α-SiC (including a small amount of BeO) ceramic with resistivities over 10^{13} Ω·cm was developed [4.257] for electronic substrate applications, because of high resistivity at grain boundaries. Two possible sources of increased resistivity of SiC were considered. One possibility is that barriers to conduction may be formed by a thin insulating layer at grain boundaries. However, a second phase was not observed at grain boundaries. A second model assumes that depletion layers are formed at grain boundaries by electron carriers trapped there. In these transition regions, mobile carriers are supposed to be completely depleted.

Electrical conduction in glasses is due mainly to mobile ions such as Li^+, Na^+, K^+, H^+, and OH^- in an electric field. The number of conductive ions increases with the temperature and the electric field.

The conductivity of insulating ceramics can be increased by conductive particle additions. For example, composites with very low resistivities can be achieved by adding TiC or TiN dispersoids to Si_3N_4. Above a certain particle content, the resistivity decreases abruptly. Fine grain TiN additions cause a sudden drop in resistivity above 20 vol% reinforcement. This boundary, called the percolation threshold, occurs due to the formation of a conductive network of particles. Above this threshold, conductivity slowly increases due to the improving quality of the conductive network. A certain degree of conductivity may be desirable for specific electrotechnical applications for shaping consolidated ceramics by electrical discharge machining (EDM). Si_3N_4 alone cannot be machined by EDM, but a high material removal rate is obtained in Si_3N_4–TiN composites [4.152].

4.4.1.2. Semiconductors

Semiconductors are essential materials for the electronics industry today. Some semiconductors are temperature sensitive and are used as *thermistors*. Up to a certain temperature, they behave as insulators. At high enough temperatures, electron vacancies are produced, and a significant number of valence electrons gain enough energy to be raised to the conduction band (Fig. 4.56). Some other types of semiconductors are voltage sensitive and are used as *varistors*. These materials also act as insulators below a critical voltage. However, when the voltage is increased above that critical voltage, whose value depends on the material involved, conduction occurs (Fig. 4.57).

The degree of covalent bonding in mixed bond ceramics determines their applicability to semiconductor devices. As the ionic character increases, the material acts more like an insulator. Remember also that the higher the degree of electronegativity of the components, the higher the ionic character, as explained in Sect. 1.2.2.

An important classification of semiconductors is based on the source of conduction. If the concentrations of charge carriers are characteristic of the "pure" ceramic, it is called an intrinsic semiconductor. Examples are NiO, intermetallics such as GaAs, PbTe, PbS, and the elements Si, Ge, Sn. At 0 K, these covalently bonded materials act as insulators since their valence bands are completely filled

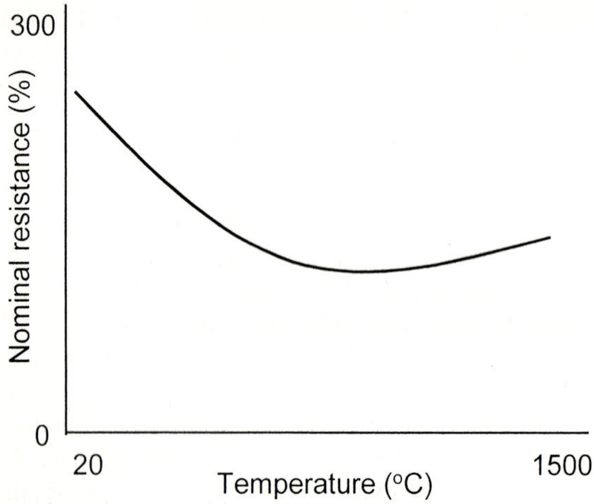

Fig. 4.56. Conduction behavior of a thermistor

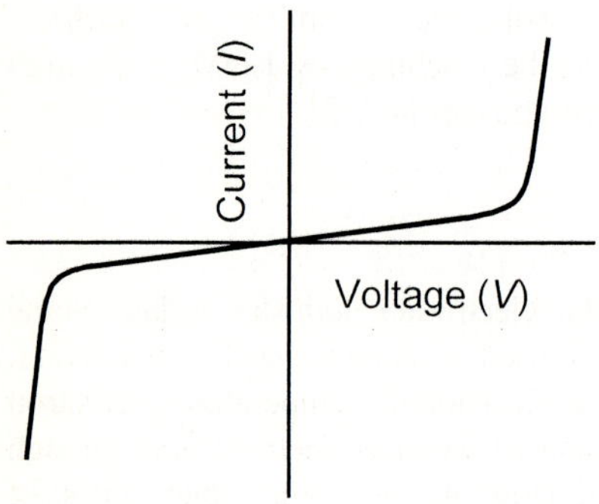

Fig. 4.57. Conduction behavior of a varistor

[4.254]. At higher temperatures, electrons move from the valence to the conduction band and act as charge carriers. At the same time, the number of holes left behind in the valence band, which also serve as carriers and promote conduction, increases. Such materials have resistivities with a negative temperature coefficient.

Semiconductivity can also be achieved by controlled additions of impurities to a material. Such a material is called an extrinsic semiconductor. Because most intrinsic semiconductors do not have low enough resistivity at ambient conditions, they are "doped" or "poisoned" to become extrinsic semiconductors. Two types of impurity atoms (dopants) are generally used: those that have a valence of +3 or +5. Atoms that have a +3 valence cause electron holes in the valence band. Initially, an electron from the neighboring semiconductor ion orbits about the +3 atom as well as the source ion.

This causes a positive charge on the +3 atom, the induced charge acts as a potential well, and eventually leads to the capture of the orbiting electron. Thus, +3 type ions are also called acceptors. The imperfection increases the ambient resistivity, but an increased potential will cause electrons to move via these holes. This is equivalent to the motion of a positive charge carrier and therefore the doped material is called a p-type extrinsic semiconductor. Typical +3 atoms used are B, Al, In, and Ga.

Impurities of +5 (As, Sb, P) add extra electrons to the conduction band. Such ions are called donors because they donate electrons to the conduction process without affecting the bonds of the host lattice. The charge carrier is negative (an electron), and therefore the doped material is called an n-type extrinsic semiconductor. Examples of extrinsic semiconductors are Al (+3) doped Si (p-type semiconductor) and P(+5) doped Si (n-type semiconductor). A third class of dopants is known as amphoteric dopants. These can behave either as acceptors or donors. For example, Cu can act as an acceptor or as a donor in Ge, depending on its degree of ionization [4.254].

In ionic ceramics, nonstoichiometry achieved by any means (e.g., oxygen deficiency in oxides achieved by heat treatment or reduction of Al_2O_3 at high temperatures) can also cause extrinsic semiconduction. Semiconductors usually exhibit conductivities between 10^3 and 10^{-9} $1/\Omega\cdot$cm.

4.4.1.3. Conductive Ceramics

Since metals are easily formed and are generally much cheaper than ceramics, conductivity in ceramics under ambient conditions is not a very useful property. However, conductive ceramics may be preferred over metals in corrosive and/or high temperature environments. Some examples are MHD electrodes used up to 1500°C in plasmas, fuel cell electrodes at 1000°C in oxidizing conditions, graphite electrodes used in electrolysis and pure metal/alloy production, and electrodes used in molten Na–S batteries at 300°C [4.255].

Some transition-metal oxides such as ReO_3, ReO_2, CrO_2, TiO, and VO exhibit metallic (electronic) conduction at ambient temperature [4.5]. Conduction is a result of unfilled d or f electron bands in these oxides. For example ReO_3 has a partially filled 5d conduction band. TiO is another conductive ceramic. When Ti has a valence of +2 in the oxide, the product, TiO, becomes a conductive ceramic due to partially filled 3d levels. Magnetite (Fe_3O_4) is one of the most conductive oxides with $\sigma \approx 10^4$ $(\Omega\cdot\text{cm})^{-1}$, comparable to the conductivity of Mn, one of the least conductive metals, with $\sigma = 72 \times 10^4$ $(\Omega\cdot\text{cm})^{-1}$ [4.256].

It has been observed that if the size ratio of the nonmetal to metal atom is less than 0.59, many transition-metal carbides, nitrides, and borides can form simple structures and exhibit metallic properties [4.252]. Borides of all transition metals exhibit excellent metallic conductivity. Nitrides of transition metals in groups IVA, VA, and VIA are also metallic conductors. Some borides and nitrides have better conductivities, i.e. lower resistivities than their parent metals (Appendix 10). Carbides of transition metals generally show metallic conductivity, but deviations from stoichiometry may dramatically change the conductive behavior. For example, TiC is a metallic conductor in its stoichiometric form, but a large deviation from stoichiometry causes TiC_{1-x} to become a semiconductor [4.252]. Besides their good conductivity, all of the borides, nitrides, and carbides mentioned have very high melting points, high hardness, and good thermal conductivity. However, they are not resistant to oxidation above 1000°C. Their use for high temperature applications thus requires high vacuum or nonoxidizing atmospheres.

4.4.1.4 Ionic Conduction

Some ceramics conduct electricity by ions rather than electrons. Those that exhibit relatively fast conduction are called fast ion conductors or solid electrolytes. Ionic conduction in most ceramics occurs at elevated temperatures, although some are ionic conductors at ambient temperature.

Ionic conduction involves point defects in ionic crystals or special features called tunnels. Two types of point defects that may occur in ionic crystals are *Frenkel* and *Shottky* defects. The former is created by the removal and movement of an ion to an interstitial position to produce a vacancy–interstitial pair. The latter is a group of anion–cation vacancies. In NaCl, an equal number of anion and cation vacancies are necessary for electrical neutrality, whereas in $MgCl_2$, two Cl^- vacancies are required for each Mg^{2+} vacancy. Mechanisms that cause electrical conductivity by ions also produce a net mass transport. Therefore, ionic conductivity can be measured both by conductivity experiments or tracer diffusion analysis. Other methods that are used to study ionic conductivity are nuclear magnetic resonance (NMR), X-ray diffraction, neutron diffraction, and optical/infrared spectroscopy.

One of the often cited criteria for fast ion conductivity in solids is a highly ordered framework, which provides tunnels through which ions can move easily. However, the existence of fast ion conductivity in glasses raises serious doubts about the validity of this criterion [4.258]. Defects that lead to ionic conductivity can be produced extrinsically or intrinsically. Extrinsic fast ion conduction occurs when a compound is doped with a suitable element or compound, similar to extrinsic semiconduction achieved by doping. Doped group IVA oxides give rise to high anionic conductivity. These oxides, ZrO_2, HfO_2, CeO_2, and ThO_2, have high oxygen conductivities when doped with alkaline-earth oxides (Sc_2O_3, Y_2O_3) or rare-earth oxides [4.259]. The presence of divalent or trivalent cations in the crystal structure causes the formation of anion vacancies to preserve electrical neutrality. Oxygen conduction occurs principally due to these vacancies. A special type of extrinsic doping occurs in composite materials. It was observed that a systematic addition of insulating Al_2O_3 powders to LiI can increase the Li^+ ion conductivity up to fifty times [4.258]. In this case, high levels of defect concentration are induced only at interfaces because of the limited solid solubility of the insulating phase in the ion conductor matrix.

Intrinsic defects occur in some compounds with built-in deficiencies, mostly observed at high temperatures. For example, the high-temperature phase of Bi_2O_3 (δ-phase) has a defect fluorite-type structure with an intrinsic 25% deficiency in the oxygen sublattice. This phase is stable between 730°C and its melting point, 825°C. Upon phase transformation during cooling from δ to monoclinic α-Bi_2O_3, the oxygen conductivity decreases by three orders of magnitude. Ta_2O_5 is another ceramic that exhibits intrinsic ionic conductivity at elevated temperatures. At temperatures of about 800°C, the α phase is a fast oxygen conductor. The intrinsic conductivity is attributed to the existence of oxygen vacancies in the structure.

Two types of ionic conductors can be distinguished, anionic and cationic. Most of the anionic conductors studied are oxygen conductors, but some systems also exhibit N, C, and S conduction. For example, CaO exhibits high C diffusion, AlN exhibits N diffusion, and CaS/Y_2S_3 exhibits S diffusion. Ag^+, Cu^+, H^+, and alkali ion conductors are important cationic conductors. The properties of fast ionic conductors are listed in Appendix 11. Ionic conductivity in one important ion conductor from each group is further discussed later.

4.4.1.4.1. Cubic Zirconia

The high-temperature cubic ZrO_2 phase with a fluoride structure can be stabilized at room temperature by a solid solution of lower valence cations such as Y^{3+}, Mg^{2+}, Sc^{3+}, or Ca^{2+}. Substitution with such cations leads to O^{2-} ion vacancies to balance the charges. Typical compositions are ZrO_2–9mol%Y_2O_3 and ZrO_2–15mol%CaO. At 1000°C, ionic conductivity approaches 0.1 1/Ω·cm [4.258]. Cubic zirconia is used in oxygen sensors, which have a variety of industrial applications. Such sensors are used to control the oxygen content of gases and molten metals and to check automobile exhaust gas emissions by detecting of the oxygen content of the exhaust gas mixture. Oxygen sensors for automotive applications are better known as lambda sensors. A typical lambda sensor is shown in Fig. 4.58. Due to restrictions in passenger car gas emissions in the US (especially California), Japan, and recently in Europe, almost every new car produced in these areas uses these sensors. Recent legislation on emissions in many other countries necessitates the use of lambda sensors and catalytic converters since this is the only feasible technology at present.

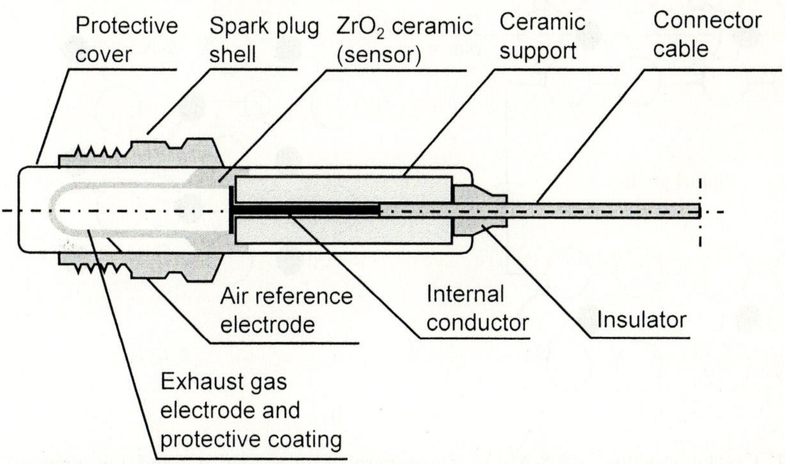

Fig. 4.58. Cross section of a lambda sensor

Other applications of ionically conducting stabilized zirconia are oxygen pumps, solid electrolyte fuel cells, and high-temperature heating elements for oxidizing atmospheres.

4.4.1.4.2. ß-Aluminas

The family of β-aluminas has a layered structure that is basically hexagonal. Three important compositions are β-Al_2O_3 ($Na_2O{\cdot}11Al_2O_3$), β'-Al_2O_3 ($Na_2O{\cdot}8Al_2O_3$), and β''-Al_2O_3 ($Na_2O{\cdot}5Al_2O_3$). The structure of β-Al_2O_3 is shown in Fig. 4.59. The spinel-like layers are separated by mirror planes that contain Na^+ and O^{2-} ions. The spacing is such that it allows the motion of Na^+ ions in two dimensions in the layer, although no motion is possible in the c direction. Due to degradation during service and cost problems with β- and β''-Al_2O_3, searches for new sodium ion conductors were continued. This led to the new compounds $Na_3Zr_2PSi_2O_{12}$ (Nasicon with conductivities similar to β''–$Al_2O_3 \approx 0.4$ $(\Omega{\cdot}cm)^{-1}$), $Na_{1+x}Y_xZr_{2-x}(PO_4)_3$, $Na_{1+x}(Cr,In,Yb)_xZr_{2-x}(PO_4)_3$, and other compounds with Nasicon-type structures [4.260].

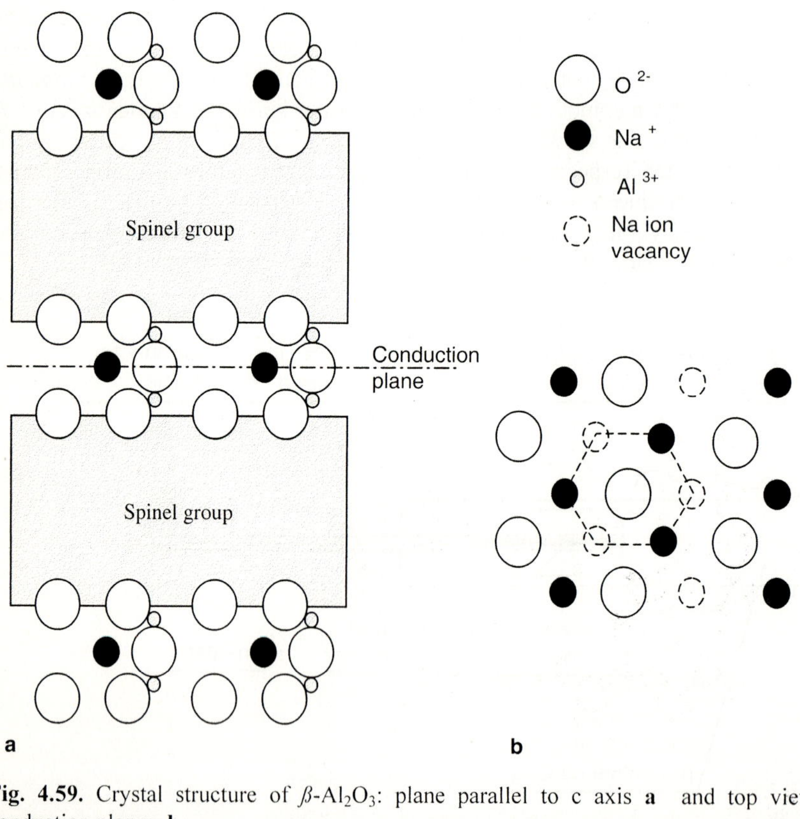

Fig. 4.59. Crystal structure of β-Al_2O_3: plane parallel to c axis **a** and top view of conduction planes **b**

The most important application of β-alumina is as a solid electrolyte in Na–S batteries that used in automotive and satellite applications for high-power-density electric supply. The battery has to be operated around 300°C. A schematic representation of a commercially available Na–S cell is shown in Fig. 4.60. A sodium anode and a sulphur–graphite cathode are used with β or β"-Al_2O_3 as the electrolyte in between. Most types of solid electrolyte batteries operate by the diffusion of an alkali metal ion from the anode across the inert solid electrolyte, where it forms an easily reversible compound with the cathode. In the Na–S battery, sodium ions move to the cathode and form polysulfide compounds during discharge. Typical discharge voltages of 2.08 V are achieved in a cell. During the recharging cycle, which is achieved by applying a reverse voltage, Na^+ ions return to the molten Na electrolyte. An alternative solid electrolyte battery being investigated involves Li anodes, Li_xTiS_2 electrolytes, and V_2O_5 cathodes [4.256, 4.258].

4.4.1.5. Superconductors

Superconductivity is a phenomenon discovered in 1911 by Kammerlingh H. Onnes in mercury [4.261]. In 1986, Bednorz and Müller [4.262] discovered superconductivity in La–Ba–Cu–O Perovskite ceramics at 30 K that subsequently led to a Nobel prize [4.263]. The remarkable discovery of 93 K superconductivity in Y–Ba–Cu–O ceramics by Wu et al. [4.264] immensely increased research activity in this field. The highest critical temperature, so far, has been found in Hg–Ba–Cu–O compounds at 150 K under extremely high pressures [4.265].

Three of the most important properties of a superconductor are the critical transition temperature, T_c, below which the material exhibits zero resistivity

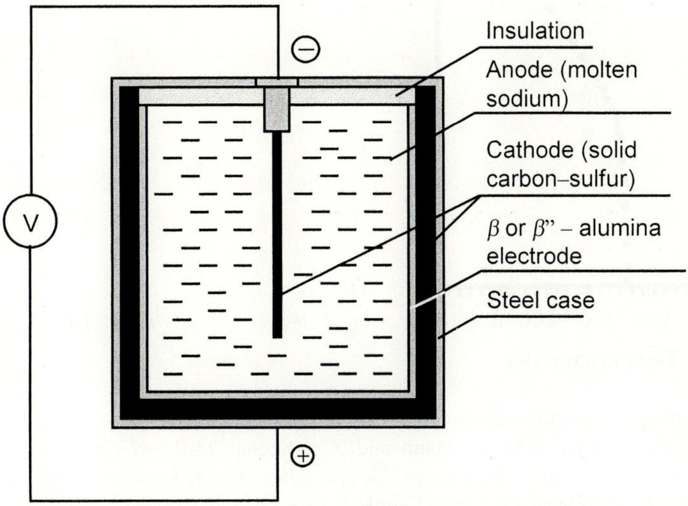

Fig. 4.60. Schematic illustration of a commercial Na–S battery

(Fig.4.61), the critical magnetic field strength, H_c, and the critical current density, J_c. Above the limits of H_c and J_c, superconductivity is destroyed [4.266]. The critical current density, J_c, in magnetic fields, is currently a limiting factor in many superconductor applications. It was established that the important microstructural features that limit current density in polycrystalline oxide superconductors are those that limit current transmission at grain boundaries, weak flux pinning, microcracking, and the absence of flux pinning sites at grain interiors. Current limiting grain boundary features include chemical variations at grain boundaries, for example intergranular phases or impurities and incoherent CuO_2 planes in $YBa_2Cu_3O_{7-\delta}$ compounds [4.267].

A reliable proof of superconductivity is the *Meissner effect*. All superconductors expel magnetic flux, in other words, they are diamagnetic at well below T_c. The internal static electric field, E, is zero at this temperature. The Meissner effect is demonstrated by the levitation of a permanent magnet on top of a superconductor below T_c as depicted in Fig. 4.62.

Superconductors are commonly categorized as type I and type II. All *type I superconductors* are pure metals, such as Hg, Pb, and Sn. In type I materials, the superconductive state is immediately lost above H_c. In *type II superconductors*, however, the magnetic field is accepted by the material above a certain magnetic field strength, H_{c1}, and rejected below that. Above a higher field strength, H_{c2},

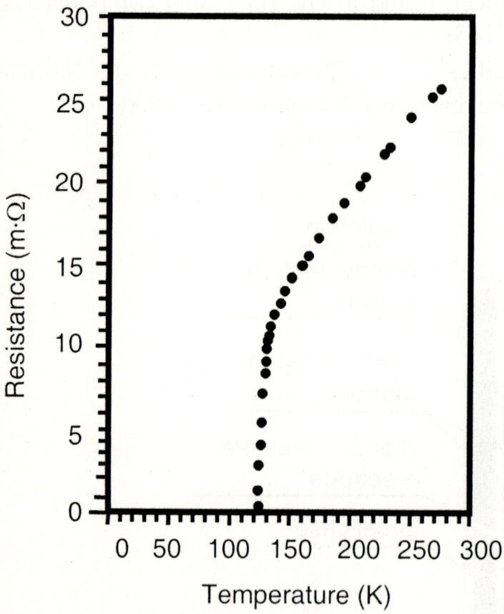

Fig. 4.61. Resistance–temperature dependence of a nominal $Tl_{2.2}Ca_2Ba_2Cu_3O_{10.3+x}$ superconductor (Reprinted from A.M. Hermann and Z.Z. Sheng: *Tl-Based Cu-O High Temperature Superconductors*. In: *Advances in Superconductivity*. K.Kitazawa and T. Ishiguro (eds.) Copyright 1989 Springer Verlag GmbH & Co)

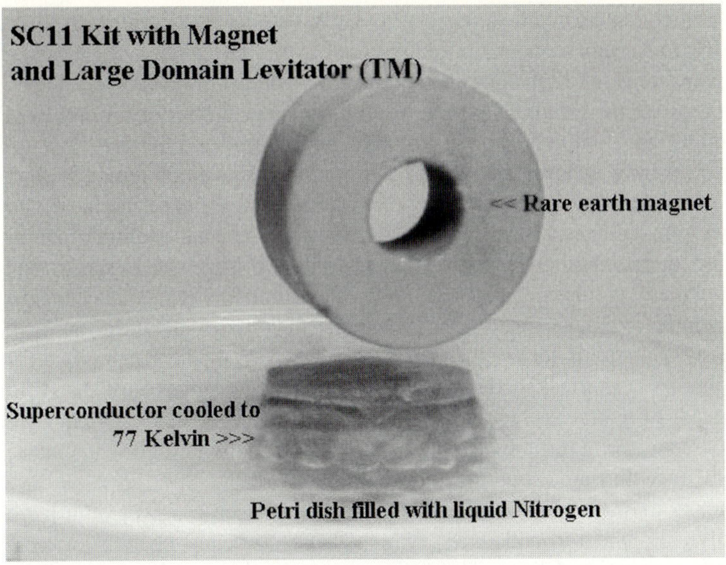

Fig. 4.62. Demonstration of the Meissner effect (photo courtesy of Superconductive Components, Inc. Columbus OH)

superconductivity is again destroyed. However, superconductivity can still be maintained between H_{c1} and H_{c2}. This type of superconductivity is observed commonly in alloys and compounds. Two exceptions are Nb and Va metals. Type II superconductors have a higher chance for practical applications because of their higher H_c.

In 1957, Bardeen, Cooper, and Schrieffer published a theory to explain the superconducting state of materials [4.268]. The main idea of the theory is that below a critical temperature, T_c, electron–electron interactions due to exchange of phonons between two electrons of opposite spins and momenta can form an electron pair. Such electron pairs are known as Cooper pairs, and their net spin is zero. Under an electric field, these pairs move in phase with other pairs without any interference from atoms or electrons. Thus zero resistivity is achieved. Thermal vibrations have to be minimal for Cooper pairs to form. The critical temperature below which superconductivity occurs is given by the *BCS theory* as [4.254]

$$T_c = 1.14\theta_D \exp(-1/AN)\,, \tag{4.147}$$

where θ_D is the Debye temperature, A is a particle-lattice-pair interaction coefficient, and N is the density of states at the Fermi energy level for one spin direction. The BCS theory has been successful in explaining many aspects of superconductivity in conventional superconductors such as NbTi, Nb_3Sn, or V_3Ga, which are used for the majority of superconductive applications today. However, its applicability to high-temperature ($T_c > 95$ K) superconductors is questionable [4.269,4.270].

Practical applications of high T_c superconductors emerge quite slowly due to problems such as brittleness, fabrication into useful shapes, and inferior properties of bulk superconductors, compared to superconducting films and single crystals. Nevertheless, these new materials offer many possibilities never imagined before. Applications of superconductors include medical devices, supermagnets for generators and power lines, supercomputers, electrical energy storage, squids (superconducting quantum interference devices), levitating high speed trains, and magnetic confinement of plasma for thermonuclear fusion [4.271,4.272]. Medical applications include magnets for nuclear magnetic resonance imaging used in medical diagnosis and squids for magnetoencephalography, a medical technique for studying the brain's electrical activity. Superconducting magnets are used in generators instead of conventional electromagnets to increase the efficiency of generators. Superconducting turbine generators can be realized with high temperature superconductors, once they are economically feasible.

Superconductors may one day be used for transmitting electric power on a large scale, especially for underground lines, which have a better chance of cost competition with conventional lines. The main reason for their application will be their zero resistance, and hence, unlike any other conductor (e.g. copper or aluminum), no heat losses. One important problem that has to be tackled before such large-scale industrial applications can be realized, however, is the fabricability of ceramic superconductors into wires and other useful shapes.

4.4.1.5.1. Y–Ba–Cu–O Compounds

The well-known superconductor referred to as 1-2-3 has the composition $YBa_2Cu_3O_{7-\delta}$ where δ ranges from 0 to 0.6. The crystal structure of this compound is shown in Fig. 4.63. This is an orthorhombic oxygen-deficient crystal with distorted Perovskite ($CaTiO_3$) structure. Higher T_c values are achieved with lower δ (higher x), as demonstrated in Fig.4.64. When $\delta > 0.6$ (equivalent to $x<6.4$ in Fig.4.64), a transition to a tetragonal structure occurs. This phase is not superconducting. Therefore, oxygen deficiency is one of the keys to high T_c 1-2-3 compounds. Ordering of vacancies in the basal planes results in the formation of Cu-O chains along the b axis. At one time, the existence of these chains was considered essential for high T_c in 1-2-3 compounds. However, recent studies indicate that the planes rather than chains are the essential structural elements [4.272]. The double CuO_2 layers are common to all Cu-containing high-T_c superconductors and are believed to be the key to high-temperature superconductivity in these oxides [4.274]. An important feature of high-T_c oxide superconductors is their structural anisotropy that leads to anisotropical electrical and superconductive properties. The high-conductivity direction is parallel to CuO planes. These planes have to be doped to be conducting [4.275]. For example, in La_2CuO_4, an antiferromagnetic insulator, doping with Sr, Ba, or Ca results in removing electrons from Cu–O planes and vacancy formation. The resulting compound becomes superconductive, and the T_c depends on the carrier concentration.

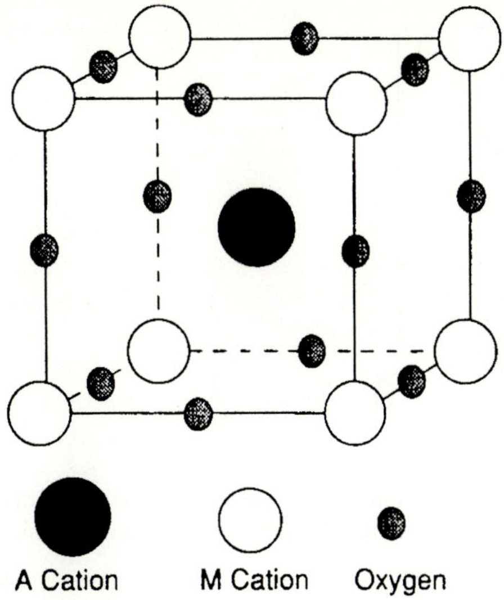

a

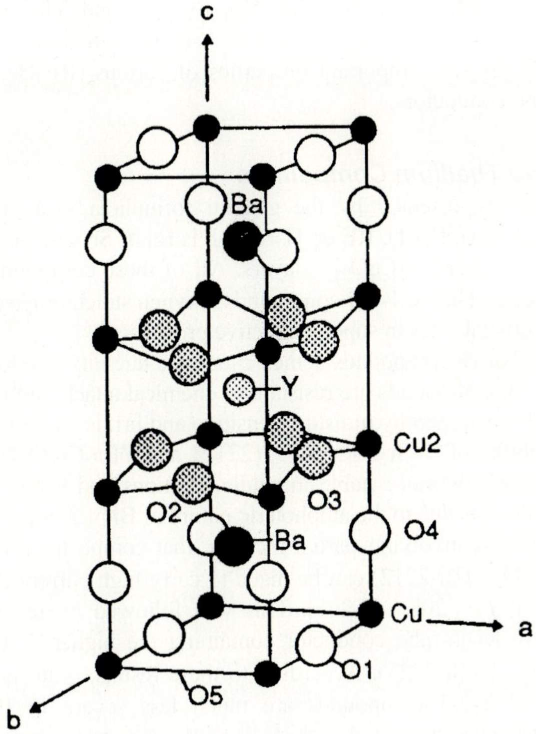

b

Fig. 4.63. The ideal Perovskite structure **a** and the structure of $YBa_2Cu_3O_{7-\delta}$ superconductor **b**. (Reprinted with permission from *JOM*, 41, (1), a publication of The Minerals, Metals & Materials Society Warrendale, PA 15086)

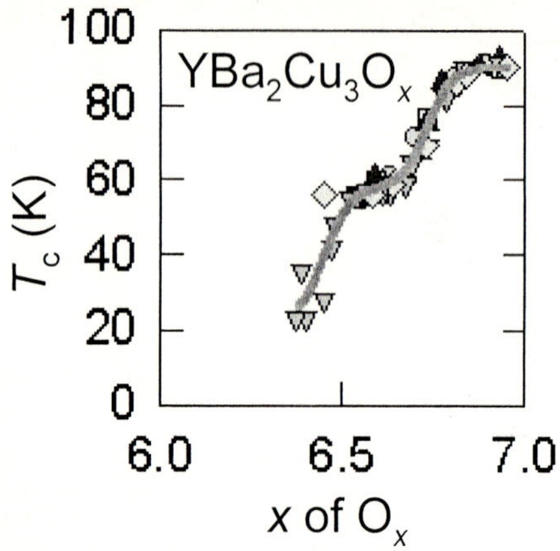

Fig. 4.64. Dependence of the critical temperature on oxygen content (From NIST Ceramic Division, Property Data Summaries for Advanced Materials http://www.ceramics.nist.gov/srd/summary/advmatdb.htm)

Many rare-earth metals such as La, Nd, Sm, Ho, Er, Tm, and Yb have been successfully substituted for Y in 1-2-3 compounds, leading sometimes to better properties. Appendix 12 compares important properties of various (RE)$Ba_2Cu_3O_{7-\delta}$ compounds and other superconductors.

4.4.1.5.2. Bismuth and Thallium Compounds

These compounds can be represented by the general formula $a_xb_2Ca_{n-1}Cu_nO_{4+2n}$ where a is Tl, Bi, Bi+Pb, Bi+Al, Bi+Tl, RE or Tl + Pb, b is Ba or Sr, x=1 or 2 for Tl, x=2 for Bi, and n is the number of $[CuO_2]^{2-}$ layers. All of these compounds have primitive tetragonal unit cells. Unlike 1-2-3 compounds, oxygen stoichiometry, twins, and domains do not play crucial roles in superconductive properties.

A significant advantage of Bi compounds is their moderate ductility due to a flaky structure similar to mica. Bi compounds are resistant to chemical attack and moisture whereas 1-2-3 compounds are generally moisture sensitive and brittle. A comparative study of the chemical stability of $Bi_2Sr_2CaCu_2O_{8+y}$ {2212} and $YBa_2Cu_3O_{6.5+y}$ {123} showed that generally {2212} is more stable in acidic solutions and water but less stable in strongly basic solutions due to the amphoteric nature of Bi [4.276].

It was shown by a process involving partial melting, that composite conductors consisting of $Bi_2Sr_2CaCu_2O_{8-\delta}$ (Bi-2212) can be made to carry high current densities in high magnetic fields at T < 20K [4.267]. This was followed by reports also showing high J_c in a powder-in-tube conductor containing the higher T_c (110 K) phase $(Bi,Pb)_2Sr_2Ca_2Cu_3O_{8-\delta}$ (Bi-2223). According to these results, grain boundary or weak-link problems of 1-2-3 compounds are much less severe in Bi-based compounds. However, at temperatures higher than 40 K, the intragranular (intrinsic) J_c's are still relatively low. There is evidence that (Tl,Pb)–(Ba,Sr)–Ca–Cu–O compounds are useful superconductors at temperatures higher than Bi-based compounds with less severe weak-link problems than Y1-2-3 compounds.

Tl-based compounds form another group of high T_c superconductors. Three compounds in this system can be represented by the general formula $Tl_2Ca_{n-1}Ba_2Cu_nO_{4+2n}$ here n=1, 2, 3 and T_c=80, 105, and 125 K, respectively [4.277]. The structures of these compounds are shown schematically in Fig. 4.65. Figure 4.61 shows the resistance-temperature variation for a $Tl_{2.2}Ca_2Ba_2Cu_3O_{10.3+x}$ sample. It is very important to remember that Tl is an extremely toxic element and should be handled with care. Some important properties of Bi and Tl superconductors are listed in Appendix 12.

4.4.1.5.3. Mercury-Based Compounds

The new Hg-based compounds exhibit the highest T_c's so far [4.265]. The first superconducting compound discovered in this family is $HgBa_2Cu_3O_{4+\delta}$, that has a maximum T_c of 95 K at δ=0.06. As the oxygen content changes, the unit cell dimensions a and c also change. As these dimensions decrease within certain limits, T_c increases. The new sister compounds that contain two or three CuO_2 layers have a T_c of 133 K. Their structure contains a slight excess of oxygen. Under extremely high pressure, T_c values of 150 K were registered [4.265].

Measurement

Superconducting temperature and magnetic susceptibility values are studied using SQUID magnetometers. These are very sensitive detectors of magnetic fields, that themselves employ superconducting materials [4.274]. Electric property characterization usually consists of a simple resistance–temperature measurement of the sample. Another indicator of superconductivity is levitation due to the Meissner effect. Properly prepared samples can levitate over a magnet below the critical temperature. This has been a popular method for demonstrating superconductivity in various superconducting alloys discovered in past years.

4.4.2. Polarization

Dielectric materials do not conduct electricity under an applied electric field. However, a shift in the charge distribution occurs, and the material acquires a dipole moment. This phenomenon is known as polarization [4.5,4.253]. There are various polarization processes that can affect materials (Fig. 4.66). *Electronic* or *atomic polarization* is a small displacement of the valence electron cloud in the atom relative to the nucleus and is a reversible process. Ionic materials undergo *ionic polarization* in an electric field where ions shift slightly in the crystal lattice. Dipolar materials such as water, which have asymmetrical molecules, can also become polarized, and the molecules become oriented with the positive side toward the negative electrode and vice versa. *Dipolar* or *orientation polarization* can be described as a perturbation of the thermal motion of ionic or molecular dipoles. *Space charge polarization* or *interfacial polarization* involves obstructing

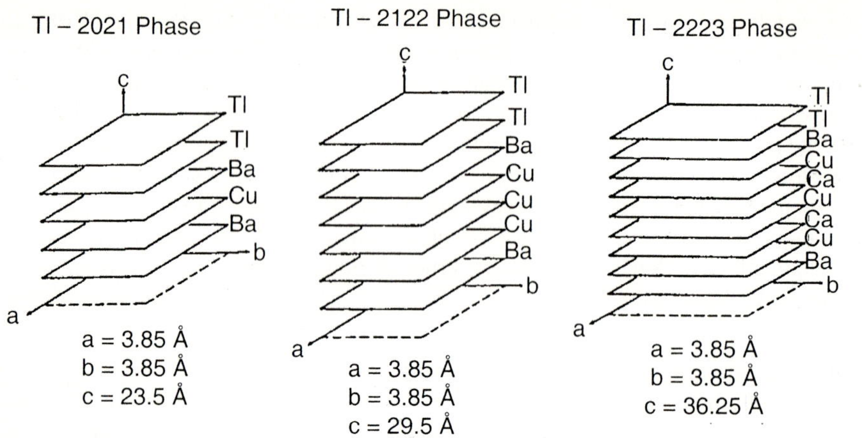

Fig. 4.65. Schematic arrangements of the Tl-based superconducting phases 2021, 2122, and 2223 (Reprinted from A.M. Hermann and Z.Z. Sheng: *Tl-Based Cu-O High Temperature Superconductors*. In: *Advances in Superconductivity*. K.Kitazawa and T. Ishiguro (eds.) Copyright 1989 Springer-Verlag GmbH & Co.)

migrating random charges at physical barriers such as grain boundaries or other discontinuities. This contribution results from the movement of ions or electrons over distances of many atomic spacings. It can be ignored in most cases since it occurs at very low frequencies and most electronic circuits operate well above these frequencies [4.278].

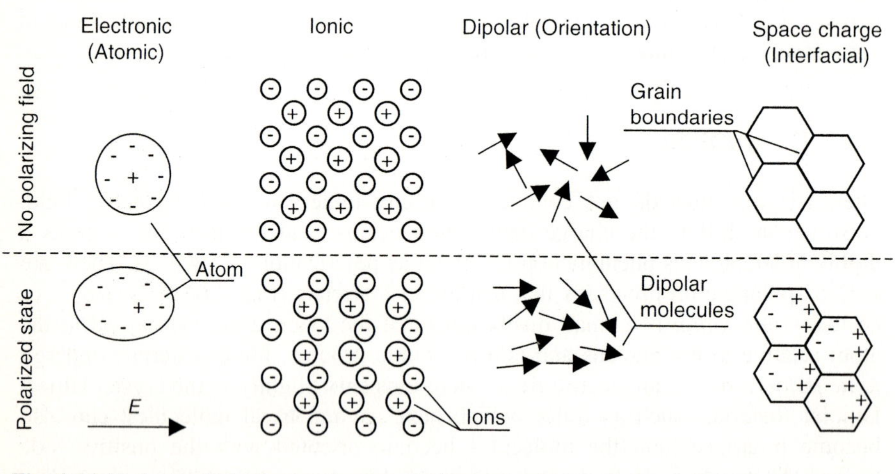

Fig. 4.66. Various polarization mechanisms in the presence of an electric field

Polarization can occur spontaneously or can result from an applied stress. Spontaneous polarization occurs in polar crystals. Of the 32 crystal classes based on symmetry, 12 are centrosymmetric and nonpiezoelectric. The remaining 20 can exhibit piezoelectricity, i.e., they can be polarized under stress. Ten of these 20 are polar (spontaneously polarized), which only undergo a directional change in polarization under stress and a change in the degree of polarization upon heating, termed pyroelectricity. Ferroelectric crystals are in a subclass of pyroelectric crystals in which the direction of polarization is reversed in an electric field. Examples of polar materials are NaCl, TiO_2, SiO_2, and soda glass, and of nonpolar materials, Si, Ge, diamond, and liquid H_2. Polyethylene, paraffin, and poly(tetrafluoroethylene) are examples of weakly polar materials [4.279].

4.4.2.1. Dielectric Constant and Permittivity

The dielectric constant is defined as the degree of polarizability or charge storage capacity of a material in an electric field [4.149,4.234]. The relative dielectric constant or relative permittivity of a material is found by comparing these properties to those of vacuum,

$$K' = K/K_0 = E/E_0 = C/C_0\,, \tag{4.148}$$

where K and E are the material's dielectric constant and permittivity, respectively. C is the capacitance of metal plates separated by the dielectric. The subscript "0" represents vacuum. The permittivity of vacuum is $E_0 = 8.854\times10^{-14}$ Farads/cm. Relative dielectric constants of various ceramics are listed in Appendix 10. In general, dielectrics with low K' values are used as insulators, and those with high K' are used as capacitors.

In most materials, K' does not depend on the frequency of the electric field at ambient temperature. However, at elevated temperatures, this property is frequency-dependent, as shown in the example of an Al_2O_3 crystal (Fig. 4.67). In materials with dominant atomic or ionic polarization, the dielectric constant increases with increasing temperature. Other important factors that affect K' are the relative fractions of crystalline and glassy phases present and the amount of porosity.

Measurement

The dielectric constant (K') of materials and the loss tangent ($\tan\delta$) discussed in the next section, are usually measured together using a disk or tube-shaped specimen with suitable opposite electrodes that form a capacitor. These properties can be calculated from capacitance measurements performed by an impedance analyzer [4.253]. In this method, a sinusoidal voltage is applied to the sample, and the magnitude and phase shift of the output current are measured by electronic equipment [4.244]. Dielectric property characterization with this technique is limited to an upper frequency of ~10 MHz. For higher frequency applications such as microwave

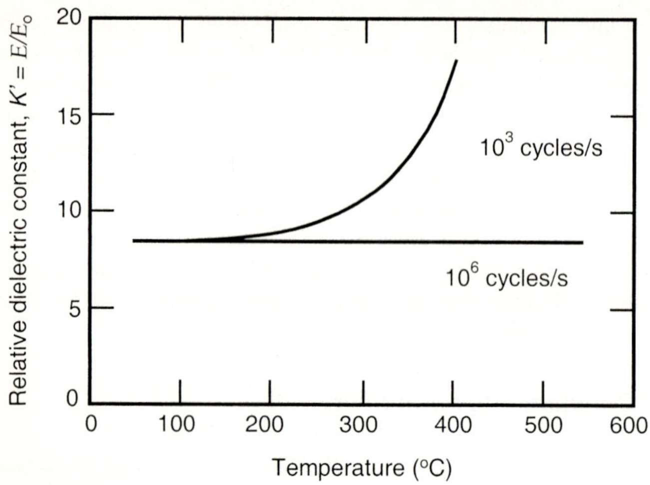

Fig. 4.67. Effect of frequency and temperature on the dielectric constant of an Al_2O_3 crystal with field normal to the c axis (From W.D. Kingery, H.K. Bowen, and D.R. Uhlmann: *Introduction to Ceramics*. Copyright 1976 John Wiley & Sons New York, reprinted by permission)

dielectrics used in satellite and cellular communication systems, a resonance cavity apparatus may be used. Another method uses a calibrated variable capacitor to compare the response of the dielectric to be tested (Fig. 4.68). The capacitance of the parallel plate capacitor is measured in a vacuum and later in the presence of the sample solid. C_0 and C_{solid} are determined by varying the capacitance of the calibrated capacitor to keep the resonance frequency $\omega_0 = [L\,(C_s + C)]^{-1/2}$ constant with vacuum between the metal plates versus solid between the plates. L is the inductance of the coil in Fig. 4.68. The dielectric constant is then calculated using Eq. (4.148) [4.244].

4.4.2.2. Loss Tangent

The logarithm of the time between the maximum electric field and the maximum polarization in an alternating field is represented as the loss angle, δ. The loss angle can also be described as the complement of the angular difference between the applied voltage and the current passing through the material to $\pi/2$ [4.234]. The phase angle Θ (the angle between the current and voltage) becomes $90°-\delta$ in the presence of a dielectric, rather than 90°. The *capacitive loss factor* is defined as [4.235]

$$\mathrm{CLF} = K' \tan \delta \,. \tag{4.149}$$

This property is directly proportional to the energy loss per cycle [4.149],

$$W = \pi E_0 V_0^2 K' \tan\delta\,, \tag{4.150}$$

where V is the maximum of the sinusoidal voltage. Tan δ is also called the dissipation factor [4.255]. As the frequency increases, the dissipation factor also increases. Dissipation factors of various ceramics can be found in Appendix 10. *(Measurement- see preceding section.)*

4.4.2.3. Dielectric Strength

The dielectric strength of an insulating ceramic can be defined as the minimum electric field necessary to cause breakdown. A full understanding of the nature of various breakdown phenomena has not been established so far due to their complexity. Three sorts of breakdown events have been identified nevertheless [4.256]. These are intrinsic, thermal, and discharge breakdown events, explained here.

Measurement

Although the dielectric strength of a material is affected by a large number of factors, testing under specific conditions and using standard geometries is a valuable approach for gathering useful data. A good method is to use thin cross sections rather than thick materials to ensure that breakdown occurs through the material rather than across the specimen surface. Using an insulating oil to prevent ceramic–air contact is also recommended for the same reason. A typical test condition is shown in Fig. 4.69. Commonly encountered ranges of dielectric strengths for a number of ceramics are shown in Appendix 10.

4.4.2.3.1. Intrinsic Breakdown

Under controlled test conditions, breakdown occurs across the insulator under high voltage, typically around 100 MV/m. Electrons ejected from the electrodes or generated from the valence band are accelerated through the crystal. Some interact with phonons and generate heat. Others interact with ions or atoms and generate new electrons. As a result, an avalanche of conducting electrons is generated, and breakdown occurs.

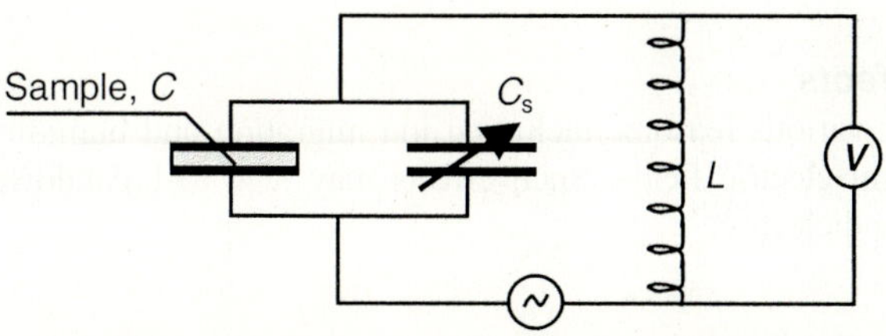

Fig. 4.68. Schematic illustration of dielectric constant and loss tangent measurement setup

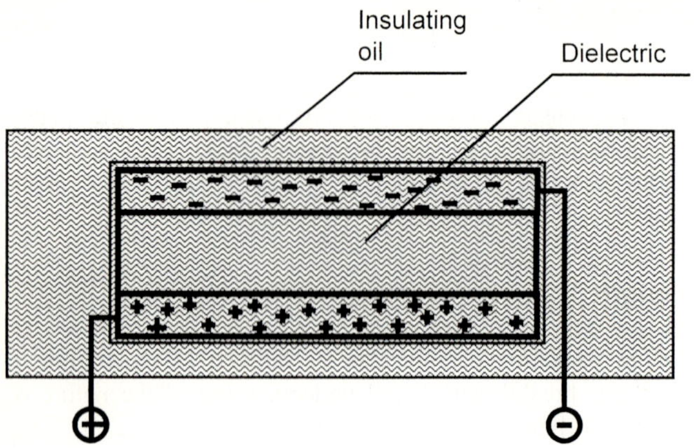

Fig. 4.69. Schematic illustration of dielectric strength measurement setup

4.4.2.3.2. Thermal Breakdown (Thermal Runaway)

Thermal breakdown may occur under conditions where localized heating is generated due to resistive heating or other reasons and where the transmission rate to the surroundings is low. Localized temperature rises can eventually cause melting or evaporation. Thermal stresses may cause failure of the material. An exponential increase in conduction may occur, that leads to breakdown. A typical breakdown time is 10^{-8} s, and this would usually inhibit breakdown at frequencies above 10^8 Hz [4.279]. However, a fast electron may be accelerated upon striking a lattice ion and further by the electric field. For example, breakdown may occur at microwave frequencies of 10^{10} Hz when high power densities are involved.

4.4.2.3.3. Discharge or Ionization Breakdown

Inhomogeneities such as pores can ionize gases locally, leading to heat generation and eventually a condition similar to thermal breakdown. Generally, the dielectric strength of the material decreases as the porosity increases, due to this type of breakdown. Gas ions may damage neighboring surfaces, which may accelerate breakdown. This would typically show itself as intermittent sparking and eventual breakdown as the test field is increased.

4.4.2.3.4. Long-Term Effects

Resistivity may decrease for various reasons, including ion migration and humidity under conditions of long-term electric fields. Such effects may lead to breakdown after prolonged operation of a dielectric.

4.4.2.4. Capacitors and Capacitance

A capacitor is a device consisting of a dielectric sandwiched between two electrodes. Capacitance is the charge storage ability of a capacitor. The capacitance of a parallel plate capacitor (Fig.4.68) is defined as [4.5]

$$C = E_0 K' A / t , \tag{4.151}$$

where A is the area of the electrodes and t is the dielectric thickness. From Eq. (4.151) and other observations, it follows that to achieve high capacitance levels,

1. the dielectric layer thickness can be decreased,
2. materials with large dielectric constants (K') can be used. This can be achieved by controlling the electric susceptibilities and resistivities of grain and grain boundary phases. Internal boundary layer capacitors are a result of such an effort,
3. multilayer capacitors with alternating electrode/dielectric layers can be used,
4. relaxor ferroelectrics (nonlinear dielectrics with high K') can be used, and
5. a combination of all of these alternatives may be used.

Applications of capacitor materials are discussed in Sect. 6.5. *(Measurement- see Sect. 4.4.2.1.)*

4.4.2.5. Piezoelectricity

Piezoelectricity is the generation of electricity (induced polarization) by a crystal upon application of stress. A piezoelectric crystal also undergoes strain when an electric field is applied, termed reverse piezoelectricity. Piezoelectric crystals are polarized under stress, leading to electric potentials when they are isolated. The voltage is proportional to the applied stress [4.251].

The necessary condition for a crystal to be piezoelectric is the absence of a center of symmetry in the crystal structure. Lead–zirconate–titanate (PZT) compounds with the Perovskite structure are currently used for most piezoelectric applications. Other ceramic crystals that exhibit piezoelectricity include ZnO, CdS, AlN, $PbTiO_3$, quartz, $BaTiO_3$, $LiNbO_3$, $SrTiO_3$, and $LiTaO_3$. $BaTiO_3$ has been replaced to a large extent by PZT ceramics in piezoelectric applications [4.280].

The *Curie point*, T_c, is the temperature above which a piezoelectric material behaves as a *paraeletric* (normal dielectric) material. This temperature is one of the important factors that define the useful limits of a piezoelectric material. $La_2Ti_2O_7$ crystals possess the highest Curie point to date (T_c = 1500°C) [4.281].

Macroscopic piezoelectricity cannot be observed in polycrystalline ceramics with random orientation of crystals. However, polarization can be used to align the crystals in one direction. This is typically achieved by poling, i.e., placing the ceramic in a high dc field after all of the processing steps are finished. The direction of polarization in piezoelectric ceramics can be chosen with a considerable degree of freedom. On the other hand, when the electric field is removed, the random orientation of individual crystals produces internal stress and strain. This leads to the lancet shape of the hysteresis loop in polycrystalline materials rather than the roughly rectangular shape seen in single crystals (Fig.4.70). The remanent polarization, P_R, defined as the amount of polarization after the electric field is removed, is much lower in polycrystalline materials. However, the coercive field, E_c, i.e. the field at which polarization changes sign and goes through zero, is somewhat larger. Polycrystalline ferroelectric ceramics may be produced in different shapes and compositions. Monocrystalline piezoelectrics such as quartz, on the other hand, have fixed compositions, and their shapes are restricted by the size of the crystal from which they are cut. Due to these limitations, applications that use monocrystalline elements generally have lower sensitivity and internal capacitance than those with ferroelectric elements [4.282].

In piezoelectric materials, the relationship between induced polarization and the strain, S (caused by applied stress), is given by [4.280]

$$S = \varphi P_s^2, \tag{4.152}$$

where φ is the electrostrictive coefficient and P_s is the spontaneous polarization. The direct piezoelectric response of a material in its simplified form is given by the equation [4.253],

$$D = dT, \tag{4.153}$$

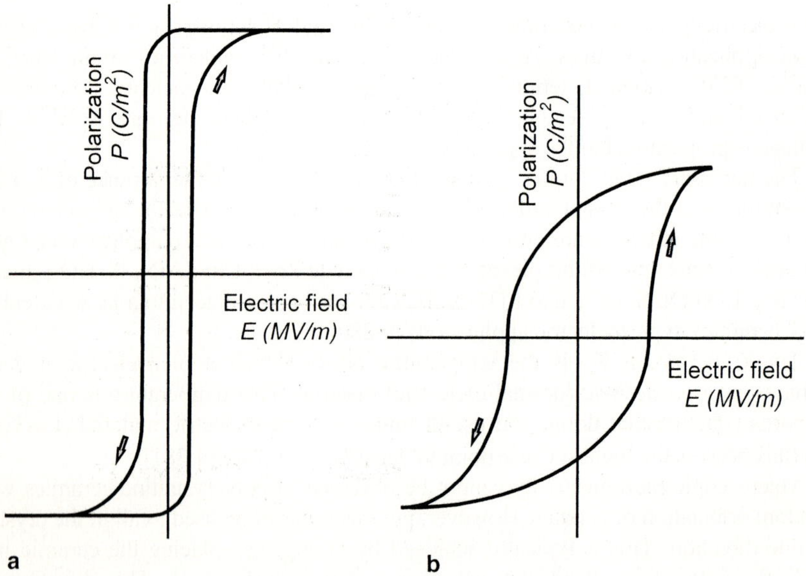

Fig. 4.70. Typical hysteresis loop for a single crystal **a** and polycrystalline piezoelectrics **b**

where D is the dielectric displacement, d is the tensorial piezoelectric coupling coefficient, and T is the stress tensor. Similarly, the converse piezoelectric response, i.e., the development of strain under an applied electric field, is described by

$$S = d^* E , \tag{4.154}$$

where S is the strain, E is the applied electric field, and d^* is the converse piezoelectric coupling coefficient. φ, d, and d^* are all tensor properties, thus they are directional in nature. Instead of the tensor d, the mechanical coupling factors k_{ij} are frequently used in the technical literature. The effective electromechanical coupling coefficient, k_{eff}, is defined by [4.256]

$$k_{eff} = \left(\frac{\text{mechanical energy converted to electrical energy}}{\text{mechanical energy input}} \right)^{1/2} , \tag{4.155}$$

or vice versa for the converse effect. k_{eff} is always less than unity. For example, the planar coupling coefficient ranges from 0.35 for $BaTiO_3$ to 0.72 for PZT. k can be measured along different directions of a certain geometry. For example, k_{33} is the coupling factor in the longitudinal direction of a rod, k_{31} is the lateral coupling factor of plates, and k_p is used only for disks. Appendix 13 lists some of the piezoelectric properties of various ceramics of technical importance discussed. For applications, see Sect. 6.5.

Measurement

The piezoelectric properties of ceramic materials are frequently studied by analyzing the resonance characteristics of a specimen subjected to a varied electric field. Piezoelectric properties of thin films have been measured by laser interferometry [4.253].

4.4.2.6. Pyroelectricity

Pyroelectricity is the electric polarizability of a material caused by a change in temperature. This effect occurs in the family of piezoelectric crystals that have polar axes. Ten of the 20 piezoelectric crystal classes are also pyroelectric [4.251].

For a small change in temperature, ΔT, the components of the polarization vector are given by

$$\Delta P_i = p_i \, \Delta T \qquad i = 1, 2, 3 , \tag{4.156}$$

where p_i is a constant called the pyroelectric coefficient. The inverse effect of pyroelectricity is exhibited by the temperature increase in a crystal when an electric field is applied. The temperature increase is given by

$$\Delta T = (T/\rho\, c_p)\, \Sigma_i\, (p_i\, \Delta E_i)\,, \quad i = 1,2,3 \tag{4.157}$$

where ρ is the density, c_p is the specific heat at constant pressure, T is the absolute temperature, and ΔE is the applied electric field. ΔT in a pyroelectric crystal can approach 0.01°C.

Some ferroelectric materials are also frequently used for their pyroelectric properties. These materials exhibit dramatic polarization variation just below their Curie temperatures. Thus, their highest pyroelectric coefficients are achieved in that temperature range. However in practical applications, compositions are adjusted to operate well below the Curie temperatures, where p_i show higher stability at varying operating temperatures [4.253].

An important parameter in pyroelectric materials, the figure of merit, is given by

$$F_v = P\,/\,(c_v\, K)\,, \tag{4.158}$$

where P is polarization and c_v is the volume specific heat. When significant intrinsic noise exists, the appropriate figure of merit would be

$$F_D = p\,/\,[c_v\, k^{1/2} (\tan\delta)^{1/2}]\,, \tag{4.159}$$

where tan δ is the dissipation factor. Note that figures of merit apply only for given operational conditions and should be used carefully. The properties of some important pyroelectric materials are given in Appendix 14. For applications, see Sect. 6.5.

Measurement

A typical system for pyroelectric property measurement consists of a current amplifier and a heating/cooling system to heat and cool the sample at a controlled rate. In this configuration, the measured current, I_m is approximately equal to the pyroelectric current, I_p, given by [4.253]

$$I_m \approx I_p = pA\, \mathrm{d}T/\mathrm{d}t\,, \tag{4.160}$$

where A is the electrode area, and the pyroelectric coefficient, p, is calculated from the measured current flow and the rate of temperature change dT/dt.

4.4.2.7. Ferroelectricity

Ferroelectric crystals exhibit permanent electric polarization, that can be reversed by an electric field. The degree of spontaneous polarization is called the order parameter of the ferroelectric state [4.251]. Two types of ferroelectric crystals can be distinguished. In one class, spontaneous polarization occurs along one axis. Rochelle salt, $LiNbO_3$, and $PbTiO_3$ belong in this category belong. The second type of ferroelectric material can exhibit spontaneous polarization along several axes. For example, $BaTiO_3$ exhibits spontaneous polarization along <100>, <110>, and <111> in the cubic system [4.283]. Other examples are $BaCoF_4$, HCl, and $NaNO_2$ with two-dimensional polarization [4.5]. Ferroelectrics can also be grouped as of displacive or order-disorder character [4.284]. The recent definition of this distinction is made in terms of the dynamics of the phase transition. In earlier literature, displacive meant that the paraelectric phase was microscopically nonpolar and order-disorder type meant that it was nonpolar only in a macroscopic or thermally averaged sense.

A third type of classification is soft and hard ferroelectrics. Ferroelectrics are considered "soft" when their melting temperature is low, they are mechanically soft, and they are water-soluble. Rochelle salt, GASH (guanidine aluminum sulfate hydrate), some other sulfates, sulfites, tartrates, nitrates, and nitrites belong to this class. Ferroelectrics are considered "hard" if they are mechanically hard and they are not water-soluble. $BaTiO_3$, $KNbO_3$, $PbNb_2O_6$, and $PbTa_2O_6$ are some examples of hard ferroelectrics [4.5].

Ferroelectric crystals are a subgroup of pyroelectric crystals. An important property of ferroelectric materials is hysteresis, which is explained in Sect. 4.4.2.5. in the discussion of piezoelectricity. An ideal ferroelectric material is one that exhibits

1. a high dielectric constant,
2. low dielectric loss,
3. high electrical resistivity,
4. moisture insensitivity, and
5. a high electromechanical coupling coefficient [4.255].

Ferroelectrics with high dielectric constants have been used in ceramic capacitors. Among important ferroelectric materials are PZT, $PbTa_2O_6$, $LiNbO_3$, and ABO_3 *Perovskites* such as $BaTiO_3$ and $PbTiO_3$.

$BaTiO_3$, one of the better known ferroelectric materials, has four crystal structures. At T>132°C, it has a cubic Perovskite structure [4.280], which was schematically shown in Fig. 1.9. Below this temperature, the crystal structure is tetragonal. At about 0°C, an orthorhombic phase is formed. All three phases exhibit ferroelectricity. A rhombohedral phase exists at very low temperatures (below -90°C). During the cubic–tetragonal transformation, spontaneous polarization occurs in the crystal. However, even in a single crystal, polarization does not occur only in one direction. Regions of polarization occur which may point 90° or 180° away from each other. These regions are called *domains*. Domains or domain walls in ferroelectric materials can be observed by a polarizing microscope, SEM, TEM, X-ray topography, or other methods. Figure

4.71 shows wedge-shaped lamellar domains observed by TEM and optical microscope in $BaTiO_3$ single crystals [4.285]. The domains can be oriented in a common direction under applied electric fields. This treatment is called *poling*. In some materials which cannot be poled at ambient temperature, high-temperature poling may be possible. Upon removal of the field, a portion of these domains remains aligned, resulting in remanent polarization (Fig.4.70).

Ferroelectricity, i.e., the reversibility of spontaneous polarization, has only few applications by itself. Ferroelectric recording is such an application where PZT ceramics are used [4.286]. Other applications of ferroelectric materials usually use their piezoelectric or pyroelectric properties.

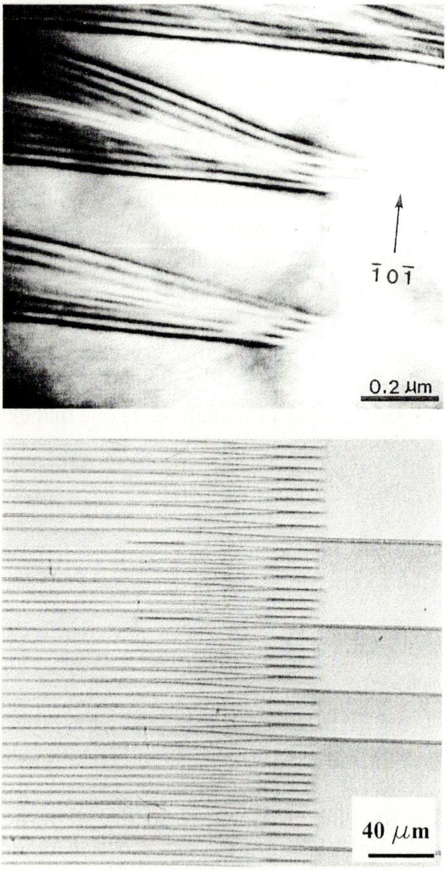

a
b

Fig. 4.71. Ferroelectric domains in $BaTiO_3$ single crystals with wedge-shaped lamellar structure **a** and optical micrograph of a single crystal with a zigzagged boundary **b** (courtesy of Prof. Su Jin Chung, School of Materials Science and Engineering, Seoul National University Seoul Korea)

Measurement

The hysteresis loop behavior of a ferroelectric material is measured by a Sawyer–Tower circuit or a Diamant–Pepinsky bridge [4.287]. In the former method, a variable dc voltage is applied to the horizontal plates of an oscilloscope. The ferroelectric crystal is connected to a linear capacitor in series with the vertical plates. The voltage generated across the linear capacitor is proportional to the polarization of the ferroelectric crystal. Thus, the oscilloscope will show a hysteresis loop [4.244, 4.288].

4.5 Magnetic Properties

Magnetism is caused principally by the interaction of electrons with magnetic fields and with each other [4.289]. Ceramic materials have an important place among magnetic materials due to their complementary properties with magnetic metals. Magnetic ceramics are semiconductors or insulators with respect to their electronic behavior. As a result of their high resistivities, eddy currents produced by alternating electric fields and associated heat losses are significantly reduced in magnetic ceramics. The distinction between hard and soft magnets is based on the size and shape of the hysteresis loop of the material. Hard ferrites usually possess large magnetizations and coercive fields greater than 150 $\mathrm{KAm^{-1}}$. Hard magnets are difficult to demagnetize; a large magnetic field is required to disrupt their magnetic induction. Hexagonal ferrites (Sect. 4.5.4.3) are typically hard magnets. Soft ferrites, on the other hand, display more square B–H hysteresis loops and coercive fields typically below 500 $\mathrm{Am^{-1}}$. Soft magnetic materials are easily magnetized and demagnetized by external magnetic fields. They are suitable for applications requiring switching, i.e., activation and removal of the magnetic field. Soft ferrites crystallize in the spinel structure (Sect. 4.5.4.1).

Other than hard and soft magnets, materials are also categorized according to their magnetic behavior as diamagnetic, paramagnetic, ferromagnetic, antiferromagnetic, and ferrimagnetic. Each of these categories is further discussed in the following sections.

Measurement

A simple experimental arrangement used for measuring the magnetic properties of a solid is shown in Fig. 4.72. The sample is suspended from a sensitive balance and positioned at the center of a shaped pole piece that creates a uniform magnetic field gradient (here shown as a solenoid for clarity). Upon applying a dc current to the solenoid, an axial uniform magnetic field intensity, H, will be created,

$$H = nI\,, \tag{4.161}$$

where n is the number of turns in the solenoid and I is the applied current. The force exerted on the sample along the z axis is given by

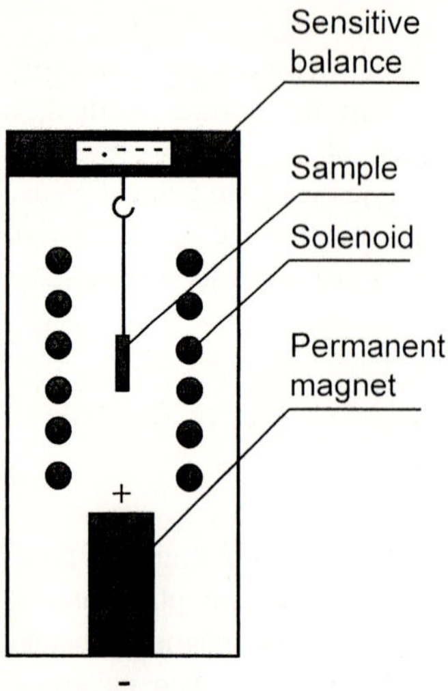

Fig. 4.72. Schematic illustration of simple experimental arrangement used for measuring the magnetic properties of a solid

$$F_z = V \chi_{mag} H \, dB/dz \, , \tag{4.162}$$

where V is the volume of the sample, χ_{mag} is the relative susceptibility, and dB/dz is the magnetic field gradient along the z axis caused by the permanent magnet.

Four possibilities exist in such an experiment:

1. If the weight of the sample diminishes after applying current, the susceptibility is negative, and a repulsive force exists on the sample. Increasing H will linearly increase the repulsive force. This type of material falls into the *diamagnetic material* category.
2. If the weight of the sample increases, the susceptibility is positive, and the sample is weakly attracted to the permanent magnet. Such a material is a *paramagnetic material.* Susceptibility, χ_{mag}, and thus the attractive force will decrease if the temperature is increased. Upon removing the applied current, the weight of the sample returns to the initial condition. Thus, changes caused by the magnetic field are reversible in a paramagnetic material. The same rule also applies to diamagnetic materials.
3. If the susceptibility goes through a maximum with decreasing temperature instead of constantly increasing, the material is an *antiferromagnetic* material.
4. If the sample is strongly attracted to the permanent magnet and the weight change exhibits hysteresis (see Fig. 4.73), the material is *ferromagnetic* or *ferrimagnetic*.

However, it is not possible to distinguish between ferromagnetic and ferrimagnetic materials by this experiment. In Fig. 4.73, B_r is the *remanent induction*, also called *remanence*, and H_c is the coercive field or reverse field required to return B to zero. At low applied field intensities, a weak attraction to the permanent magnet occurs. Increasing the intensity will increase the attraction, and the weight increases until a saturation value. This corresponds to the plateau in the hysteresis curve. After saturation, a further increase in the field intensity leads to a reduction in the weight gain, and the hysteresis loop will be followed. Repeating this experiment at increased temperatures will result in similar behavior until a certain temperature, but a weakening of magnetization occurs. At a certain temperature, the material looses its ferromagnetic behavior and becomes a *paramagnet*.

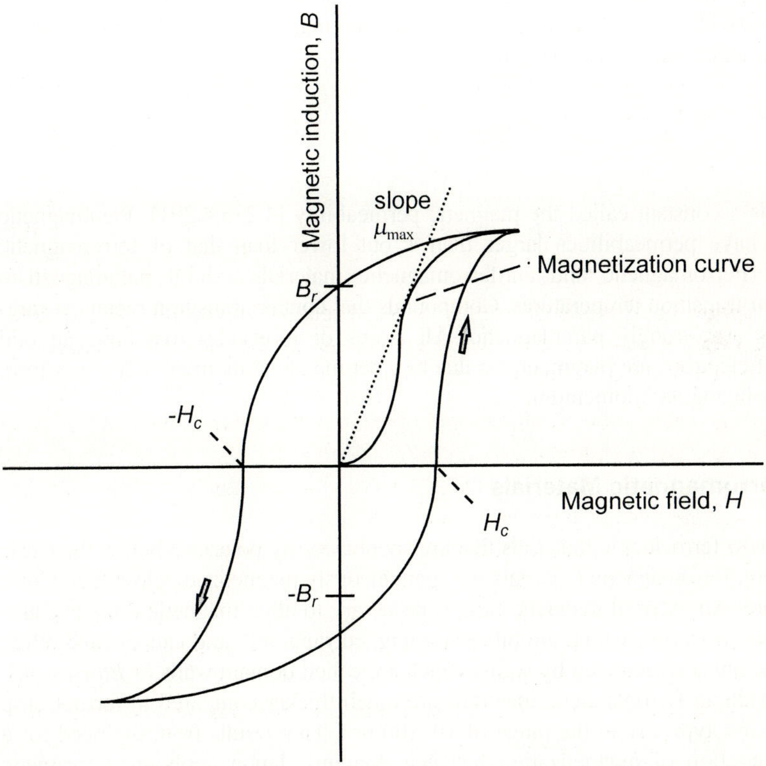

Fig. 4.73. Typical hysteresis curve for a ferromagnetic material

4.5.1. Diamagnetic Materials

Diamagnetic materials have negative magnetic susceptibility, i.e., they magnetize in a direction opposite to that of an applied magnetic field. They have magnetic permeabilities less than 1 and are repelled when placed next to a magnet. Most ceramics fall into this category. In diamagnetic materials, the spin magnetism of electrons is entirely self-canceling and the application of a magnetic field distorts the electron orbits so that the internal magnetic field remains unchanged [4.290]. Superconductors exhibit perfect diamagnetism below their critical temperatures, T_c, as demonstrated by the Meissner effect.

4.5.2. Paramagnetic Materials

Materials in this category exhibit magnetization parallel to a magnetic field. Each atom aligns with the field individually, and the total magnetization (B) is proportional to the magnetic field intensity (H); thus,

$$B/H = \mu \,, \tag{4.163}$$

where μ is a constant called the magnetic permeability [4.235,4.291]. Paramagnetic materials have permeabilities larger than 1 but lower than that of ferromagnetic materials. Ferromagnetic and antiferromagnetic materials exhibit paramagnetism above their transition temperatures. Compounds that contain transition metals or rare-earth ions are strongly paramagnetic. All atoms or molecules that have an odd number of electrons are paramagnetic due to a net magnetic moment that arises from electron spin angular momentum.

4.5.3. Ferromagnetic Materials

In analogy to ferroelectric materials that are spontaneously polarized below the Curie temperature, ferromagnetic materials are spontaneously magnetized below their Curie temperature. An external magnetic field is necessary to align magnetic domains, also called *Weiss domains,* which are otherwise aligned randomly and cancel each other. These domains are separated by walls, which are called domain walls or *Bloch walls*. Domain walls in ferromagnetic materials are much thicker compared to ferroelectric domain walls, typically in the range of 10–100 nm. This results from the need for a gradual variation of magnetization between domains. Upon applying a magnetic field, magnetization increases rapidly to its saturation value. A similar equation to that given by (4.163) can be applied to Fig. 4.73. Upon applying a magnetic field to a ferromagnetic material, beginning from $H= 0$ and $B= 0$, the domain walls begin to move. The movement is reversible in the initial non-linear part, and μ is not constant. In the linear part of the curve, the movement is irreversible, permeability (μ_{max}) is constant, and it attains its maximum value. At the end of this period, called technical saturation, the crystal will contain only a single domain.

Ferromagnetism exhibits itself by the strong attraction of one magnetized body to another. Commonly known "magnets" are hard ferromagnetic materials, i.e., they neither magnetize nor demagnetize easily, and therefore magnetization is permanent.

Above a threshold temperature called the Curie temperature, T_C, ferromagnetic, ferrimagnetic, and antiferromagnetic materials behave as paramagnetic materials since their magnetic domains are oriented randomly due to thermal motion. For ferromagnetic materials in particular, the Curie–Weiss law describes the variation of susceptibility with temperature, T, in the paramagnetic region above T_C [4.159],

$$\chi = C / (T - T_C) = M/H \ , \tag{4.164}$$

where χ is susceptibility, M is magnetization, and C is given by

$$C = nP_m^2 \mu_o / (3k) \ . \tag{4.165}$$

Here, n signifies the total number of dipoles (carriers) per unit volume, P_m is the magnetic dipole moment, μ_o is the permeability of vacuum, and k is Boltzman's constant. Magnetic susceptibility is the proportionality constant between field and magnetization, as Eq. (4.163) indicates.

Ferromagnetism is a relatively rare phenomenon observed in the metals Fe, Co, Ni, Gd, and in certain alloys. CrO_2 is one of the newly developed ferromagnetic oxides.

4.5.4. Ferrimagnetic Materials

Ferrimagnetism occurs when the magnetization of magnetic domains with unequal magnitude are antiparallel to each other, resulting in a net spontaneous magnetization. Ferrites, which are also ferrimagnetic, are among the most important magnetic ceramic materials.

Ferrites are oxide ceramics where Fe_2O_3 is the major component. The term "ferrite" is synonymously used with magnetic ceramics. Common applications of ferrites include waveguides, rectifiers, resonance isolators, junction circulators, and phase shifters. These parts are used in devices that employ microwaves for communication devices (television, computers, telefax machines, and press copy), radar, and heating [4.5]. Typically, ferrites have a crystal structure, which has more than one type of site for cations. Usually, the magnetic moments of metal ions on sites (domains) of one type are parallel to each other and antiparallel to the moments of another type. Thus, ferrites are ferrimagnetic. The three classes of ferrites are spinels, garnets, and hexagonal ferrites.

4.5.4.1. Spinel Ferrites

Ferrites of this type have the common structure characteristic of the mineral spinel, $MgAl_2O_4$ (Fig. 4.74). Their general formula is $MeO{\cdot}Fe_2O_3$ where Me represents divalent ions such as Zn^{2+}, Cd^{2+} , Mg^{2+}, Mn^{2+}, Ni^{2+}, Cu^{2+}, Co^{2+}, and Fe^{2+}. Magnetite, exploited as lodestone for many centuries, is a well-known spinel type ferrite whose composition is $FeO{\cdot}Fe_2O_3$ or Fe_3O_4.

Usually, spinel ferrites have low magnetic anisotropy and are soft magnets, except for those that contain Co^{2+}. Spinel ferrites are commonly used in television, radio, touch-tone telephones, high-frequency welding, high-speed tape and disk recording heads, electronic ignition systems, and transformer cores [4.5]. The properties of selected spinel ferrites are given in Appendix 15 along with other types of ferrites.

4.5.4.2. Garnets

This class of ferrites has the garnet ($3R'O{\cdot}R''_2O_3{\cdot}3SiO_2$, R' and R'' are rare-earth ions with $Z > 61$) structure (Fig. 4.75). Yttrium iron garnet (YIG) is the best known ferrimagnetic garnet because of its applications in microwave devices. It has the composition $Y_3Fe_5O_{12}$ and has 160 atoms in its unit cell. Another important garnet is $Y_3Al_5O_{12}$ doped with Nd^{3+}, known as yttrium aluminum garnet or YAG, that is used in YAG lasers. Other examples are $3CaO{\cdot}Fe_2O_3{\cdot}3SiO_2$ (andradide), $3CaO{\cdot}Al_2O_3{\cdot}3SiO_2$ (grossularite), and $3MnO{\cdot}Al_2O_3{\cdot}3SiO_2$ (spessarite). Appendix 15 lists the properties of important garnet ferrites.

4.5.4.3. Hexagonal Ferrites

A number of compounds exhibit ferrimagnetism in the $BaO–MeO–Fe_2O_3$ system with hexagonal crystal structures. Barium hexaferrite ($BaFe_{12}O_{19}$) is the prototype of

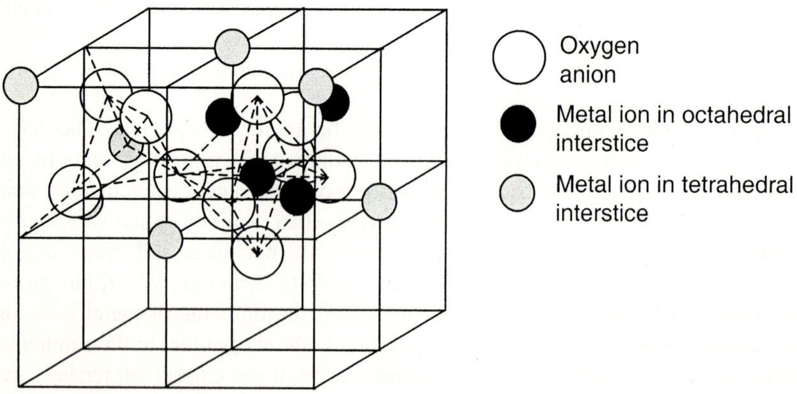

Fig. 4.74. Crystal structure of spinel ($MgAl_2O_4$)

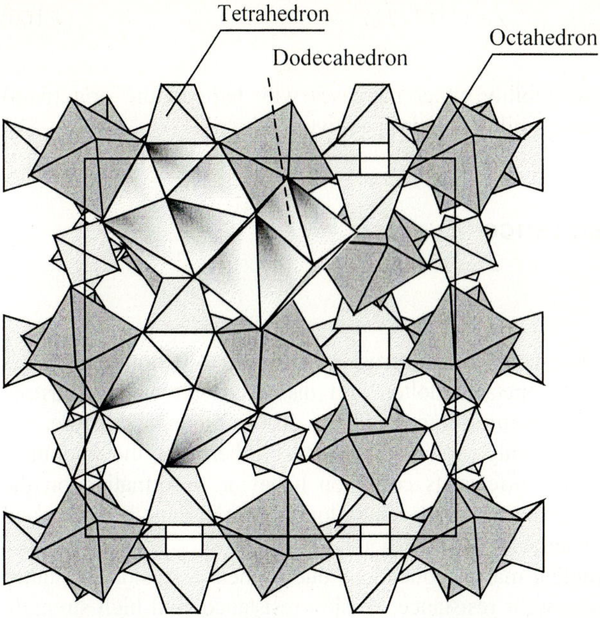

Fig. 4.75. Crystal structure of yttrium iron garnet (YIG): oxygen ions occupy the corners of tetrahedra, octahedra, and dodecahedra. Iron ions occupy the centers of tetrahedra and octahedra. Yttrium ions occupy the centers of the dodecahedra. Only three of the dodecahedra are shown in this view for clarity. They normally fill all the gaps in the structure

M-type hexagonal ferrites and has the same structure as the mineral magnetoplumbite ($PbFe_{7.5}Mn_{3.5}Al_{0.5}Ti_{0.5}O_{19}$). Other hexagonal ferrites include $BaO.2MeO.8Fe_2O_3$, $2(BaO{\cdot}MeO{\cdot}3Fe_2O_3)$ and $3BaO{\cdot}2MeO{\cdot}12Fe_2O_3$, known briefly as W, Y, and Z type hexagonal ferrite families, respectively. Me represents metal ions. For example, Co_2Z represents the compound $3BaO{\cdot}2CoO{\cdot}12Fe_2O_3$. The properties of some hexagonal ferrites are listed in Appendix 15.

4.5.5. Antiferromagnetic Materials

Antiferromagnetic materials can be considered a special case of the ferrimagnetic system. Below the *Neel point*, atomic magnetic moments (or magnetic domains) form an ordered array so that their magnetizations cancel each other, i.e., electron spins of neighboring atoms are opposite to each other. Mn, Cr, and many transition-metal oxides exhibit antiferromagnetism. The spins of an antiferromagnetic material above the Neel point, T_N, lose their antiparallel alignment; they become random, and the substance becomes paramagnetic. Above the Neel point, the susceptibility is given by [4.159]

$$\chi = C / (T + T_N) \,. \tag{4.166}$$

Below the Neel point, susceptibility either increases with temperature or remains constant. In any case, the susceptibility remains positive.

4.6. Corrosion Resistance

4.6.1. Introduction

Ceramic materials are well known for their corrosion resistance to various substances such as acids, alkalis, molten metals, molten and dissolved salts, and industrial exhaust gases. However, each ceramic compound has a different chemical nature, and therefore, resists to corrosion by only certain substances. Furthermore, the amount of impurities in a ceramic greatly affects its corrosion behavior. Information on the corrosion resistance of ceramic materials, especially long-term test data, is scarce, and direct comparison between available data is difficult.

Ceramic materials are finding many applications due to their corrosion resistance combined with refractoriness, wear resistance, erosion resistance, and high strength. Among them are filters for hot gas, particulate, or metal processing; gas turbine components; diesel engine components; crucibles and tubes for handling molten metals; catalyst supports; membranes; and high-pressure sodium lamps, to mention but a few [4.149,4.292,4.293]. The corrosion performance of glasses and ceramics is critical in many applications, including containers for food, beverages and pharmaceuticals; windows for gamma, UV, IR, and visible light protection; dental and bone implants; and nuclear waste encapsulation [4.294]. In spite of the variety of current and potential applications, the fundamental database for the corrosion resistance of ceramics is much smaller than that available for metals. Consequently, extensive testing is required when new applications are considered, or a particular ceramic is selected arbitrarily.

The corrosion resistance of ceramics depends on such factors as the chemical nature of the major and minor phases, surface texture, porosity, whether or not protective layers are formed, and whether mechanical damage occurs together with processes such as erosion and wear. The corrosion resistance of a particular ceramic phase is a function of the atomic bond strength, charge density, and the probability of electron transfer at surface active sites. The major phase may be resistant to corrosion in a certain environment, but the minor phase or impurities generally located at grain boundaries may corrode easily. Thus, corrosion would proceed from locations of low resistance, leading to pitting and porosity. This in turn, might cause strength degradation and eventual failure. The effect of a second phase is particularly apparent in reaction-bonded SiC (RBSC) or Si_3N_4 (RBSN), both of which contain unreacted Si. A comparative electrochemical corrosion study of pressureless sintered α-SiC (no free Si) and reaction-sintered SiC (containing 8–13 wt% free Si) indicated that the corrosion rate of α-SiC in various concentrations of HCl was 127×10^{-6} mm per year and did not vary significantly with HCl concentration [4.295]. The corrosion rates of RBSC, however, were much higher: 3124×10^{-6}, 17323×10^{-6}, and 13437×10^{-6} mm per year in 5%, 18.5%, and 27.8% HCl, respectively. The free Si contained in RBSC is believed responsible for the higher corrosion rates.

Additives and impurities may also decrease corrosion resistance. For example, MgO, used as a sintering aid in Al_2O_3 at typically less than 1 wt%, can lead to intergranular attack in acidic environments in spite of alumina's resistance to most acids [4.149].

Rough surfaces and porous regions typically decrease corrosion resistance due to increased surface area and easy penetration. Elevated temperatures usually accelerate corrosion. However, there are important exceptions to this trend. In some cases, high temperatures may evaporate a liquid and if the corrosion resistance of a ceramic to the gas phase is higher, the corrosion resistance may increase. This behavior is observed in the corrosion of cordierite heat exchangers by sulfuric acid [4.296].

Another exception is the oxidation of porous Si_3N_4. At temperatures above ~1200°C, a continuous protective silica layer forms, which reduces the rate of oxidation. At temperatures ≤1100°C, this layer is not formed, and oxidation occurs rapidly [4.149]. If a reducing atmosphere or low oxygen partial pressure is present, the silica layer will dissociate, allowing oxidation. On the other hand, alkali oxides such as soda (Na_2O) react with protective silica films at high temperatures. When these films are contaminated with Na_2O, lower viscosity and higher oxygen diffusivity result. In turn, surface recession of SiC or Si_3N_4 occurs [4.297]. The formation of protective layers can decrease or totally eliminate corrosion or oxidation. Well-known examples besides Si_3N_4 are SiC and $MoSi_2$. SiC and $MoSi_2$ are used as heating elements that provide temperatures up to 1700°C. Similar to Si_3N_4, a protective SiO_2 layer with low oxygen diffusivity forms on SiC and $MoSi_2$ at elevated temperatures and oxidizing atmospheres.

The stress state of a material is another factor that determines the corrosion resistance. In many ceramics, even very low stress levels may lead to crack growth in corrosive environments. This phenomenon is called *subcritical crack growth*, *stress corrosion*, or *static fatigue*. Experimental and theoretical studies [4.32] indicate that species such as water and ammonia promote slow crack growth of vitreous SiO_2 by attacking strained Si–O bonds, as shown in Fig. 4.76 [4.298]. Chemicals that enhance the crack growth rate in SiO_2 are also active toward single crystal Al_2O_3. This phenomenon is attributed to dissociative chemisorption of electron donor and proton donor species at Al–O crack tip bonds. However, in the case of MgF_2, a totally ionic material which exhibits weak acid/base surface properties, an ion solvation model was proposed to explain environmentally controlled fracture effects [4.198]. Strength deterioration in stress corrosion can be quite severe. For example, high-strength glass filaments exhibit an 86% strength decrease after sustained loading for several days in a moist ambient atmosphere [4.299].

Single-phase ceramics, it was shown, are susceptible to stress corrosion cracking at elevated temperatures in the presence of an amorphous deposit (Figs. 4.77 and 4.78). Silica or silicate-based amorphous phases can penetrate grain boundaries under stress, resulting in localized creep embrittlement. The process is more likely to occur

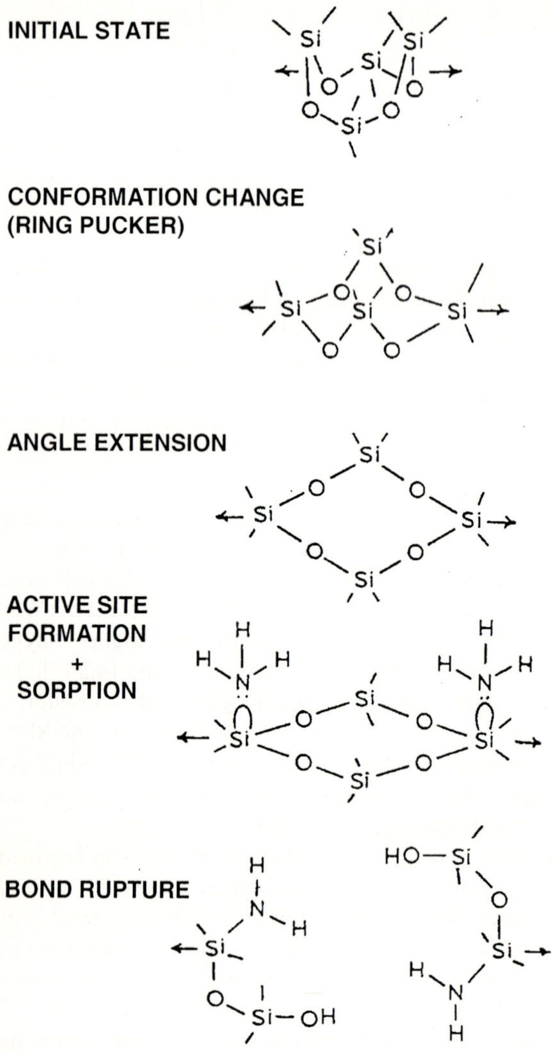

Fig. 4.76. Static fatigue in glasses by attack on Si–O bonds. (1) initial ring structure, (2) conformation change (ring pucker), (3) Si–O–Si angle extension, (4) O–Si–O angle pinching plus adsorption, (5) bond rupture (From T.A. Michalske, B.C. Bunker, and S.W. Freiman: *Stress Corrosion of Ionic and Mixed Ionic/Covalent Solids*. J. Am. Ceram. Soc. **69 [10]**, 721–724 (1986), copyright 1986, American Ceramic Society, reprinted with permission)

between their fatigue behavior [4.301]. The cyclic fatigue lives of MgO–PSZ, Al_2O_3, and Si_3N_4 were significantly lower in air than those in vacuum or inert atmospheres. These results indicate that this difference derives from the stress corrosion effect of air or water vapor. It was shown that Al_2O_3 is more susceptible to stress corrosion cracking or static fatigue in air than Si_3N_4.

Corrosion and degradation of ceramics are accelerated in the presence of wear or erosion. These processes can damage protective layers, induce fresh surfaces for corrosion, or create surface cracks from which corrosion-assisted crack propagation and failure may originate. Such processes may be important in combinations of high temperature and corrosive environments, in which sliding friction or considerable particle impact occurs. The corrosion resistance of some industrially important ceramics is further discussed in the following sections.

The structure of a material on the atomic scale also plays a role in corrosion. Crystalline ceramics and amorphous materials of the same composition often exhibit quite different corrosion behavior [4.294]. This difference was demonstrated in $NaAlSi_3O_8$ which occurs naturally as high-quality crystals of the mineral albite and which can be produced as an aluminosilicate glass. Figure 4.79 shows the corrosion behavior of Na, Al, and Si in corresponding structures. Silicon shows similar corrosion behavior in both phases, but sodium and aluminum corrode at completely different rates in glassy and crystalline phases. The removal of Na from the crystalline phase yields a Na/Si removal rate of 1:3 as required by the formula, thus indicating congruent dissolution. Na removal from the glass phase is much faster, suggesting that it is leached from a depth in the glass in addition to surface dissolution. Al removal also shows different dissolution kinetics in the two phases, and the same considerations are valid in this case, as well. Preferential extraction of leachable ions from glass creates a leach layer that acts as a diffusion barrier to more extraction. Precipitation of network-forming ions such as Si, Al, and B form a gel layer on glass. This may also act as a diffusion barrier, and it serves as a substrate in which the precipitating metal ions are adsorbed. Both of these layers are transient phenomena; ultimately the dissolution of glass is controlled by a surface reaction. Phosphate glasses are usually less resistant to corrosion. Studies of alkali–calcium–phosphate glasses indicate that no leached layer is formed; instead, uniform dissolution occurs by surface reaction.

4.6.2. Liquid–Ceramic Reactions

The chemical mechanisms for the corrosion of ceramics and glasses in liquid environments may be categorized as follows [4.294]:

4.6.2.1. Congruent Dissolution by Simple Dissociation

The ratios of constituent elements in solution are the same as those in the solid that is dissolving. For instance, MgF_2, used as an optical window material, dissolves in water by a simple dissociation reaction:

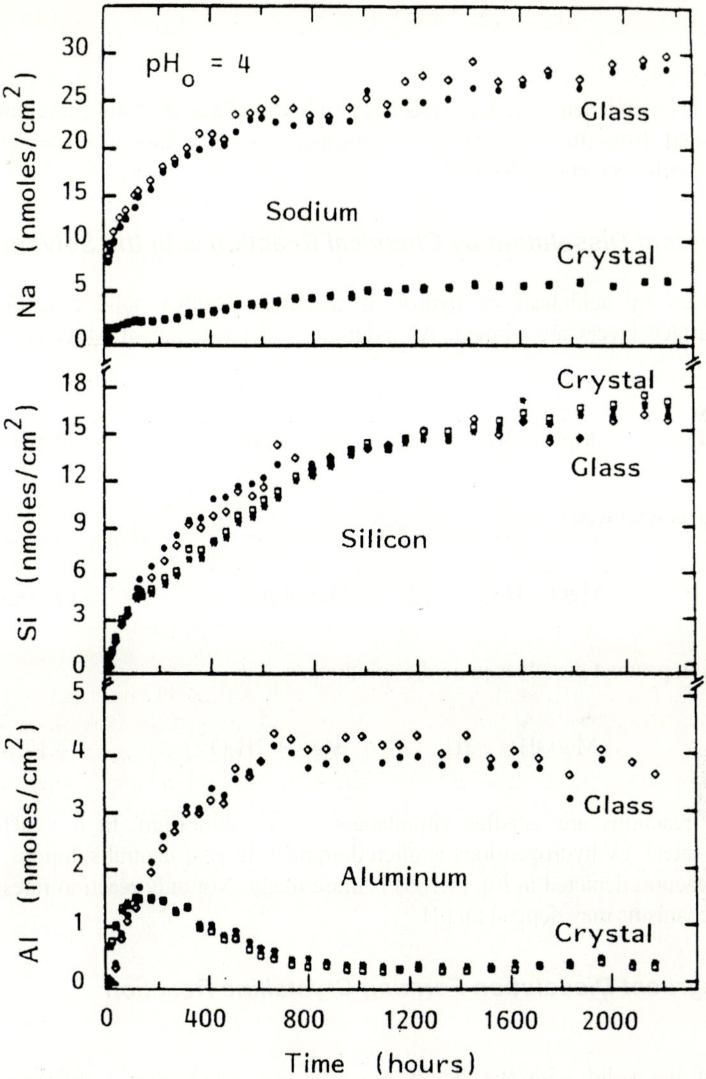

Fig. 4.79. Corrosion behavior of Na, Al, and Si in glasses and crystals that have the albite ($NaAlSi_3O_8$) composition. Solid and open symbols show reproducibility of duplicate runs (From: L.A. Zellmer and W.B. White: *The Comparative Dissolution Kinetics of Albite Crystals and Albite Glass in Aqueous Solutions at 70°C*. In: *Fifth International Symposium on Water Rock Interactions*, Reykjaviqk, Iceland, 1986. Extended Abstracts, pp. 652–655 and L.A. Zellmer: *Dissolution Kinetics of Crystalline and Amorphous Albite*. M.S. Thesis in Geochemistry and Mineralogy, The Pennsylvania State University, 1986, courtesy of Prof. William B. White, Materials Research Laboratory, The Pennsylvania State University)

$$MgF_2 \rightleftarrows Mg^{2+} + 2F^- \tag{4.167}$$

The reaction is independent of pH and reaches a saturation limit at equilibrium that may be calculated from the free energy of formation of the chemical species involved. No protective layers are formed.

4.6.2.2. Congruent Dissolution by Chemical Reaction with the Solvent

Dissolution occurs by acid–base or hydrolysis reactions. Neither solid reaction products nor surface layers are formed. An example is the reaction of MgO with acidic solutions:

$$MgO + 2H^+ \rightleftarrows Mg^{2+} + H_2O \tag{4.168}$$

MgO also reacts with water:

$$MgO + H_2O \rightleftarrows Mg(OH)_2 \tag{4.169}$$

and the reaction product dissolves in acidic solutions:

$$Mg(OH)_2 + 2H^+ \rightleftarrows Mg^{2+} + 2H_2O \tag{4.170}$$

All of these reactions are satisfied simultaneously at equilibrium. In low pH solutions, direct attack by hydrogen ions is often dominant. In near neutral solutions, the hydrolysis reaction depicted in Eq. (4.165) is more likely. Not only reaction rates but reaction mechanisms may depend on pH.

4.6.2.3. Incongruent Dissolution Forming Crystalline Reaction Products

The reaction of the solid with the liquid forms a new solid with a different composition. The concentrations of dissolved species do not have the same ratio as the original solid. The reaction product is less soluble and may accumulate on the surface to form a protective layer. For example, $SrTiO_3$ dissolves in aqueous solutions to release Sr ions into solution but Ti remains in the solid as highly insoluble TiO_2:

$$SrTiO_3 + 2H^+ \rightleftarrows Sr^{2+} + TiO_2 + H_2O \tag{4.171}$$

Thus, if the reaction products remain behind, the Sr concentration drops sharply to zero at the reaction interface, but the Ti concentration remains high out to the original surface.

4.6.2.4. Incongruent Dissolution Forming Noncrystalline Layers

Dissolution may also form a noncrystalline gel layer on the solid–solvent interface. Depending on the service conditions, the gel layer may remain intact, and it may even advance rather than retreat. A typical example would be the dissolution of sodium aluminosilicate:

$$NaAlSi_3O_8 + 8H^+ \rightleftarrows Na^+ + Al_2Si_2O_5(OH)_4 + H_4SiO_4 \qquad (4.172)$$

Although the reaction product is written as the kaolinite composition, it may form as an amorphous gel layer.

4.6.2.5. Ion Exchange

In this type of reaction, mobile ions are leached away from the solid phase, but the matrix remains intact. Sodium aluminosilicate may again be used as an example:

$$NaAlSi_3O_8 + H_3O^+ \rightleftarrows (H_3O)AlSi_3O_8 + Na^+ \qquad (4.173)$$

A continuous loss of Na occurs through the leached layer, whereas the original surface is preserved more or less with the same Si/Al ratio. Usually, some congruent dissolution occurs as an additional reaction. In albite, the following reaction occurs:

$$NaAlSi_3O_8 + 4H_3O^+ \rightleftarrows Na^+ + Al^{3+} + 3H_4SiO_4 \qquad (4.174)$$

The actual corrosion rate is the sum of the rates of the various reactions.

4.6.2.6. Kinetics

In a transport-controlled reaction, the dissolution rate can be calculated by

$$\mathrm{d}C/\mathrm{d}t = k_T(C_s - C), \qquad (4.175)$$

where $\mathrm{d}C/\mathrm{d}t$ is the concentration gradient, C_s is the saturation concentration, C is the concentration of a given species, and k_T is a mass transfer coefficient. The concentration after time t of corrosion is found by integrating Eq. (4.175):

$$C(t) = C_s(1 - e^{k_T t}). \quad (4.176)$$

Empirical studies indicate that mass transport limited dissolution takes place in substances whose solubilities are greater than 10^{-4} moles/liter. Alkali halide optical materials and water-soluble detector (e.g., triglycine sulfate) and transducer (e.g., potassium dihydrogen phosphate) materials are in this category. When solubilities are less than 10^{-5} moles/liter, as it is the case for most ceramics and glasses of practical use, surface reaction controlled dissolution will govern. Provided that the dissolving species will eventually come into equilibrium with the solution (i.e. saturation occurs) by congruent or incongruent dissolution, the rate equation for surface reaction controlled dissolution is given as

$$\mathrm{d}C/\mathrm{d}t = (\mathrm{SA}/V)\, k_s (C_s - C)^n, \quad (4.177)$$

where SA is the surface area, V is the volume, n is the order of the reaction, and k_s is a reaction rate constant. For $n = 1$, or first-order kinetics, the integration of Eq. (4.177) yields

$$C(t) = (\mathrm{SA}/V)\, C_s(1 - \mathrm{e}^{-k_s t}). \quad (4.178)$$

The dissolution of oxides usually follows the rate equation given by,

$$\mathrm{d}C/\mathrm{d}t = (\mathrm{SA}/V)\, k_+ (a_{\mathrm{H}^+})^m - (\mathrm{SA}'/V)\, k_- (C - C_s), \quad (4.179)$$

where SA' is the effective surface area of precipitates, k_+ is the forward reaction rate constant, k_- is the precipitation constant, $(C - C_s)$ is supersaturation with respect to precipitate phases, and a_{H^+} is the hydrogen ion activity. The exponent m is equal to one for most crystalline materials in acidic solutions and varies between –0.2 to –0.6 in alkaline solutions. Equation (4.179) takes into consideration dissolution reactions far from equilibrium. This is so for many ceramics, glasses, and minerals where a back reaction is not observed and the trend toward metastable equilibrium is impeded by precipitation of new, lower solubility, reaction products.

4.6.3. Gas–Ceramic Reactions

Gas–ceramic reactions can be studied with the aid of volatility diagrams [4.302]. These diagrams are not phase diagrams but are representations of various gas–solid reactions by lines between which fields of various condensed phases can be found.

Isomolar lines are used where only O_2 and metal oxide partial pressures are considered during evaporation of MeO_x. In other words, reactive environments are not taken into account. Isobaric lines are used to include the effect of reactive gases, e.g., H_2O/H_2 or CO_2/CO mixtures. In nonreactive environments, the maximum vapor pressure of the evaporating species is given by the intersection of the isomolar line with the maximum equilibrium vapor pressure line at a given temperature. For example, in the Si–O system at 1700 K, the vapor pressure, which is an indication of weight loss, is about 10^{-3} Pa, which would possibly be measurable only by thermogravimetric analysis (TGA). However, in a reactive environment such as H_2O/H_2 at a ratio of 10^{-3}, the equilibrium vapor pressure of SiO_x will be 10^2 Pa at 1700 K, resulting in considerable corrosion and weight loss.

Active/passive oxidation is an important phenomenon in Si and Si-containing ceramics or intermetallics, is. *Active oxidation* implies weight loss through the formation of gaseous oxide species. *Passive oxidation*, on the other hand, means that a protective oxide layer, that prevents further oxidation forms on the solid. This is accompanied by a weight gain. The formation of a passive SiO_2 layer is usually very advantageous for protection in service. At elevated temperatures and in reducing atmospheres, active oxidation typically predominates, given by the reaction [4.303],

$$2Si_3N_4 + 3O_2 \rightarrow 6SiO(gas) + 4N_2 \tag{4.180}$$

The formation of SiO leads to a rapid process of destructive oxidation. Passive oxidation occurs in typical combustion atmospheres (except oxygen) and leads to the formation of a protective SiO_2 layer,

$$Si_3N_4 + 3O_2 \rightarrow 3SiO_2(s, l) + 2N_2 \tag{4.181}$$

Under most practical conditions, the passive ∏ active transition for Si, SiC, and Si_3N_4 is the same because the only relevant reaction is the evaporation of the SiO_2 film. The same does not hold for the active ∏ passive transition. At 1700 K and very low p_{O2}, for example $p_{O2} = 10^{-15}$, the volatility diagram suggests that $p_{SiO,\,equil.} = 10^3$ Pa at the Si (l) interface. Here, SiO_2 (l) is unstable, and active oxidation occurs.

In inert environments, passive ∏ active transformation may proceed slowly due to low SiO vapor pressure and slow evaporation of the SiO_2 scale. Reactive gases such as dry H_2 or CO will promote passive ∏ active transformation by increasing the evaporation rate of the SiO_2 scale. Active ∏ passive transformation of Si_3N_4 and SiC involves similar phenomena, but the oxidation products will differ. Whether passive or active oxidation takes place depends on p_{N2} and p_{CO}, respectively.

The corrosion behavior of Si_3N_4 in a reducing atmosphere, described in Sect. 4.6.6, can also be analyzed by a volatility map of the Si–S–O system [4.303]. For example, at 1300°C, three major possibilities exist:

1. If $p_{SiO} \geq p_{SiS}$, oxidizing conditions with the formation of SiO predominate and corrosion proceeds quickly by active oxidation.

2. If $p_{SiS} \geq p_{SiO}$, and p_{O2} is so low that SiO_2 can not form thermodynamically, the sulfurizing effect predominates and catastrophic corrosion occurs by sulfurization.
3. Finally, if again $p_{SiS} \geq p_{SiO}$, but p_{O2} is high enough for SiO_2 formation, a layer of SiO_2 forms first that will then be reduced to SiO, and the formation of SiS is prevented. Thus, only in this case, corrosion may be sufficiently contained.

The data on a volatility diagram can also be represented by an Ellingham-type diagram [4.304]. Commonly, *Ellingham diagrams* are plots of ΔG versus T for reactions that have the form,

$$2\alpha/\beta\,\mathrm{Me} + \mathrm{O_2} \rightarrow 2/\beta\,\mathrm{Me}_\alpha\mathrm{O}_\beta\,, \tag{4.182}$$

which represents the free energy released when 1 mol of O_2 is used to produce $Me_\alpha O_\beta$. Vapor pressure data on an Ellingham diagram are represented by plotting p_{MeOy} isobars on a diagram with log p_{O2}, log H_2O/H_2, and ΔG as the abscissa and T(°C) as the ordinate. Although Ellingham type representation is convenient, volatility diagrams have two important advantages. First, the vapor pressure of interest, p_{MeOy}, is taken as one of the axes. Secondly, volatility diagrams are usually more convenient than Ellingham-type diagrams when other condensed phases such as carbides, nitrides, or oxynitrides can form.

4.6.4. Corrosion Resistance of Some Important Ceramics

4.6.4.1. Al_2O_3

Al_2O_3 offers good corrosion resistance to many substances including inorganic and organic acids, molten and dissolved salts, weak alkali solutions, anhydrous ammonia, hydrogen sulfide, hydrocarbons, organic and inorganic sulfides, molten Sr, Ba, Na, Be, Fe, Co, P, As, Sb, and Bi, and free molecular hydrogen [4.197,4.206]. On the other hand, hydrofluoric acid, molten or strong hot alkalies, molten Al, Ti, Zr, Si, and fluorides (e.g. NaF or Na_3AlF_6), and some hydrides may corrode Al_2O_3 rapidly. Alumina reacts with steam, C, and Mg, at elevated temperatures. The reaction with free C is given by

$$2\mathrm{Al_2O_3} + 9\mathrm{C} \rightarrow \mathrm{Al_4C_3} + 6\mathrm{CO} \tag{4.183}$$

that has a free energy of reaction of 443.8 cal/mole. Similarly, Al_2O_3 can be reduced by CaC_2, ZrC, and TiC, but not with SiC, at least up to 1850°C. The grain boundary composition of alumina may significantly affect its corrosion resistance. SEM studies showed that the attack of acids or caustic soda solution starts at grain boundaries [4.304]. The presence of SiO_2 reduces the corrosion resistance of Al_2O_3. The amount of SiO_2 is important. Even a minor difference such as 100 ppm versus 290 ppm of

SiO_2 can produce a measurable increase in corrosion. Cracks and open porosity accelerate corrosion.

4.6.4.2. MgO

Magnesia is slightly basic. Thus, it is chemically resistant to alkalis and alkali carbonates. Most acids readily attack MgO. Phosphoric, boric, hydrofluoric, and silicic acids form layers of salt that are not soluble in water. Such layers, e.g., MgF_2, protect MgO from further corrosion at ambient temperature. At elevated temperatures, SiO_2 and many acidic silicates attack MgO. MgO is resistant to molten alkali metals, Fe, Zn, Pb, Sn, Cu, Ni, Pt, U, Zr, and Th. It reacts with Si and forms forsterite:

$$4MgO + Si \rightarrow Mg_2SiO_4 + 2Mg \quad (4.184)$$

Similarly, Mn and MnO attack MgO. MgO is easily reduced at high temperatures in reducing atmospheres, for example, in resistively heated furnaces using graphite heating elements. The good resistance of high purity MgO to molten metals as well as alkali and lead vapor makes it favorable for melting and heat treatment applications up to 2200°C [4.292].

4.6.4.3. ZrO_2

Zirconia is an amphoteric oxide that behaves like an acid with strong bases and like a base with strong mineral acids. Due to the deleterious monoclinic–tetragonal–cubic phase transformation upon heating, ZrO_2 is stabilized in either the cubic, tetragonal, or monoclinic+tetragonal form by the addition of Y_2O_3, CaO, or MgO.

Metals of the fifth, sixth, and seventh groups of the periodic table do not react with ZrO_2. ZrO_2 is fairly stable toward acidic and basic glasses because of its limited solubility in them. Therefore, it can be used for handling molten glass, provided that the glass does not contain excess free alkali. Molten neutral salts, in general, do not significantly corrode ZrO_2, yet molten alkali silicates strongly attack it. Zirconia reacts with alkaline-earth metals, including Be and Mg, and becomes reduced. Al also reduces ZrO_2, but Si and Ti metals do not react with it up to relatively high temperatures. Si reduces ZrO_2 at above 1800°C and in vacuum, whereas molten Ti becomes partially oxidized in ZrO_2 crucibles. At elevated temperatures, ZrO_2 reacts with C and forms ZrC,

$$ZrO_2 + 3C \rightarrow ZrC + 2CO \quad (4.185)$$

4.6.4.4. AlN

AlN is used mostly as an electronic substrate, but it also exhibits excellent corrosion resistance to a variety of substances and can be used in applications that require corrosion resistance. It is fairly stable in all mineral acids and HF. AlN can be wetted with molten Al without reacting with it. Therefore, it is used in special refractory parts for processing molten Al, such as thermowells. Metals and alloys that are reportedly nonreactive toward AlN include Cu, Ag, Sn, Pb, Bi, Ni, Au, Li, U, and superalloys based on Ni–Co, Ni–Cr, and ferrous alloys [4.292]. AlN is also known to be stable in contact with molten GaAs, various molten carbonate eutectic melts under a CO_2 atmosphere, and corrosive salts such as cryolite, K_2CO_3, and K_2SO_4 [4.305]. AlN reacts with TiO_2 and ZrO_2, forming the corresponding nitride and alumina [4.306]. A major drawback of AlN is its low oxidation resistance. It oxidizes in air at above 1100°C, but dense AlN bodies are stable in inert atmospheres up to 1800°C [4.306]. The oxidation rate accelerates above 1200°C [4.307]. The kinetics of oxidation by air are promoted by water vapor.

4.6.4.5. BN

Hexagonal α-BN (h-BN) parts are being used in various fields of chemistry and metallurgy. α-BN is inert to C and CO up to 1800°C [4.308]. It is corrosion resistant to molten glass; molten Si, B, nonoxidizing slags; molten salts such as borax and cryolite; molten metals including Al, Fe, Cu, Zn, Na, Mn, Sn, and Au:, as well as molten semimetals, e.g., Si, Ge, Ga, In, and As. The stability of h-BN to various molten media is attributed to its nonwetting characteristic. Similar to AlN, α-BN is readily oxidized at elevated temperatures. It starts to oxidize at 700°C, but the rate is low up to 1000°C, after which oxidation becomes a major problem. α-BN also reacts with molten alkali carbonates and hydroxides and strong acids. Hot-pressed BN usually contains a few wt% B_2O_3, metal borates, or SiO_2. These densification aids may decrease its chemical stability to various media. In some cases, articles fabricated by CVD or HIP'ing may be preferred due to the absence of densification aids.

4.6.4.6. Si_3N_4

Si_3N_4 has good corrosion resistance to acids but may be attacked by strongly alkaline substances and HF. No reaction occurs with boiling HNO_3 or HCl. High temperature strengths of hot-pressed Si_3N_4 are reduced when exposed to Ca-rich silicate melts. Although most metals, including Al, Zn, Au, Ag, Cd, Bi, Sn, brasses, and bronzes, do not react with Si_3N_4, reaction with metal oxides or slags is possible. For example, Cu reportedly attacks Si_3N_4, probably through the oxide [4.297]. Si_3N_4 is incompatible with molten Mg, Cu, Mn, Fe, Co, V, and Pt metals, lead oxide, fluorite salts, and molten borates [4.309].

As previously described, the formation of a continuous SiO_2 layer is important for the corrosion resistance of Si_3N_4 and SiC. Oxidation resistance is determined by the presence and characteristic of the SiO_2 layer. The oxidation rates of most non-oxide ceramics are too high for high-temperature structural applications [4.307]. Only silica scale-forming ceramics such as SiC and Si_3N_4 meet the oxidation rate requirements of 10 μm^2/h or less at 1600°C. The corrosion rate of both of these ceramics, however, is fast in basic salts due to deterioration of the protective layer, which is acidic. Si_3N_4 also reportedly exhibits good corrosion resistance to water up to 100°C, to salt water at ambient temperature, and to hot NaOH and KOH solutions [4.293]. Si_3N_4 is oxidized by water vapor above ~1150°C and forms SiO_2 on the surface. It reacts with molten K_2CO_3 and forms a protective film upon contact with molten K_2SO_4 [4.310].

It was suggested that the stress corrosion of hot pressed Si_3N_4 in humid environments occurs by ion exchange through glassy grain boundary phases. Tests in acidic environments showed that the hydrogen ion concentration is a controlling factor in the stress corrosion of Si_3N_4. Its strength was decreased by moisture in the environment [4.311].

In reducing atmospheres such as H_2S–H_2O and H_2, typical of many industrial processes such as coal gasification, the corrosion behavior of Si_3N_4 is determined by the partial pressures of oxygen and sulfur. Since the partial oxygen pressure is low in such environments, silicon monoxide usually forms, according to the reaction previously given [4.303]:

$$2Si_3N_4 + 3O_2 \rightarrow 6SiO\ (g) + 4N_2 \tag{4.186}$$

Direct attack of sulfur on Si_3N_4 yields sulfur silicide,

$$Si_3N_4 + 3S \rightarrow 3SiS + 2N_2 \tag{4.187}$$

None of the scales that result from these reactions is resistant to further reaction. Thus, the corrosion resistance of Si_3N_4 is low in reducing environments [4.303].

4.6.4.7. SiC

Silicon carbide is resistant to alkali solutions and to many acids, even concentrated HF, but less resistant to hot NaOH and KOH solutions [4.293]. SiC has good oxidation resistance up to 1500°C due to the formation of a thin, continuous, silica surface layer that allows using it as a heating element in furnaces with oxidizing environments. The presence of free Si, which depends on fabrication conditions, is an important factor in determining corrosion resistance. Silicon-free, hot-pressed or sintered bodies are preferable for applications that require corrosion resistance. Reaction-bonded SiC, which generally contains some unreacted Si, is more readily attacked. High-temperature oxidation obeys similar rules to that of Si_3N_4. Under conditions of low oxygen partial pressure, active oxidation takes place [4.303]:

$$SiC + O_2 \rightarrow SiO\ (gas) + CO \tag{4.188}$$

In contrast, under conditions of high oxygen partial pressure, passive oxidation takes place, forming a protective SiO_2 layer,

$$SiC + 2O_2 \rightarrow SiO_2\,(s, l) + CO_2 \tag{4.189}$$

SiC dissolves slowly in K_2CO_3 melts [4.310] and reacts with $NaCO_3$ at 1000°C [4.312] to form Na_2SiO_4 glass and SiO_2 platelets. At 900°C, it is inert to H_2, H_2S, and pure N_2, but corrosion occurs in the presence of Na_2O as well as Na_2SO_4 [4.313].

At high temperatures, hydrogen reacts with SiC and removes free carbon from it. Weight loss and strength decrease occur in as little as 50 h at 1000°C due to grain boundary corrosion in hydrogen with 25 ppm water. Water vapor in hydrogen greatly accelerates the corrosion rate of SiC since H_2O oxidizes SiC at a faster rate than reduction with hydrogen [4.314].

In reducing H_2/H_2S atmospheres, catastrophic corrosion occurs that is described by the reaction,

$$SiC + H_2S \rightarrow SiS\,(gas) + C + H_2 \tag{4.190}$$

where free C is formed and Si is transported in the gas phase [4.303]. SiC corrodes quickly in the presence of fused basic salts that contain SiO_3 ions, even though a SiO_2 layer may form. High corrosion rates also prevail in the presence of salts that absorb oxygen, preventing the protective SiO_2 layer from forming.

4.6.4.8. Glasses

Glasses have excellent corrosion resistance to most mineral acids and alkaline solutions [4.217]. Acids that can corrode glasses include hydrofluoric acid and hot phosphoric acid. Generally, acidic or basic solutions are much more corrosive to glasses at temperatures above 100°C than at ambient temperature. Similarly, glasses are inert to most organic solutions at ambient temperature, but the resistance may decrease at temperatures above 250°C. Alkali modifiers, used to lower the softening temperature of glass, decrease the resistance of glass to corrosion by aqueous solutions. Silicate glasses with more than 16 mol% alkali oxides are especially susceptible to corrosion. The reaction rate increases rapidly at $pH>9$. Sodium and potassium glasses undergo leaching and changes in surface structure. Reaction with wet carbon dioxide has also been observed in them. A balance between corrosion resistance to aqueous solutions and workability has been found by glassmakers with formulations of soda–lime–silica and potash–lime–silica glasses [4.315]. At high temperatures, the attack on glass by water is due to alkali species in glass. Dissolution of the alkali in glass will raise the alkalinity of the water so that other glass constituents will be attacked. This may be neutralized if acidic species such as boric oxide are present in glass [4.316].

4.6.5. Corrosion Testing

The measurement of the corrosion resistance of ceramic materials under different chemical, thermal, and stress states is a broad subject. Standards for corrosion testing of engineering ceramics have not, so far, been developed. Some standard testing methods exist for conventional ceramics and glasses. ASTM standard C225-85 [4.317] is designed to test the resistance of glass containers to chemical attack. C724-81 is the standard test method for acid resistance of ceramic decorations on architectural type glass [4.318]. The acid resistance test of ceramic tile to chemical substances is described in C650-83 [4.319]. The test conditions, including temperature, exposure time, concentration and pressure of the medium, and whether the medium is static or flowing, of fixed volume or continually replenished, should be considered in reporting corrosion test results or before an application is planned. Tests in which the specimen is exposed to humidity and reactive gases are referred to as weathering tests [4.320]. These tests should be designed such that removal of corrosion products from the surface is minimized.

Two types of testing can be used: laboratory tests and field tests. Laboratory tests have the advantages of being less expensive and providing better control of exposure conditions. In field tests, the sample is placed in the environment of service, or a close simulation thereof. The performance of a material may be better predicted from field test results, but the results are empirical. Thus, a combination of field tests and laboratory tests is the best way to obtain a comprehensive picture of corrosion mechanisms [4.320].

One of the simpler methods for corrosion rate measurement is periodic weight loss measurement after immersion of the material in an electrolyte at a certain temperature. The corrosion rate is then determined as weight loss per unit area and time, for example, as mg/cm^2 year. As the time unit suggests, prolonged test periods are required (over 200 h) for acceptable accuracy. The applicability of this method is limited in some less corrosive media.

Divakar et al. [4.295] showed that standard electrochemical test methods can be applied successfully to ceramic materials. These are easier and faster compared to weight loss measurements. Among several electrochemical techniques used for measuring the corrosion current density I_{corr} in A/cm^2, broadly categorized as those based on direct (dc polarization) or alternating current (ac impedance) conditions, Divakar et al. adapted the dc cathodic polarization method. Testing is done in a standard corrosion cell with an auxiliary high-density graphite electrode and a calomel reference electrode. Polished samples are dipped in the desired electrolyte. The electrode potential is changed while the current is monitored. A typical plot of applied potential versus current density for the polarized specimen in a corrosive medium (*Tafel plot*) is shown in Fig. 4.80. This test yields I_{corr}, by which the corrosion rate in g/s can be calculated from

$$CR = I_{corr}\, A\, (\mathrm{EW})\, /F \qquad (4.191)$$

where A is the area of the corroding surface in cm^2, I_{corr} is the corrosion current density in A/cm^2, F is Faraday's constant, and EW is the equivalent weight of species S in the electrochemical reaction of the form,

$$S = S^{n+} + ne^{-} . \tag{4.192}$$

This equation describes the situation where a species S of valence n undergoes corrosion by giving up n electrons.

Crushed powders or bulk samples are used for corrosion testing. Bulk samples are required when surface analyses are to be performed after the test. Infrared reflection spectroscopy (IRS) and secondary ion mass spectroscopy (SIMS) are routinely used characterization methods are. Nuclear magnetic resonance (NMR), Raman spectroscopy, X-ray photoelectron spectroscopy (XPS), and extended X-ray absorption fine edge structure (EXAFS) are other important characterization methods used in corrosion studies. Solution measurements are necessary to determine the concentration of elements leached from the ceramic into solution. Calorimetric methods have been replaced by inductively coupled plasma spectroscopy (ICP), which is a more practical and accurate technique for solution analysis. Characterization methods are treated in detail in Chap. 5.

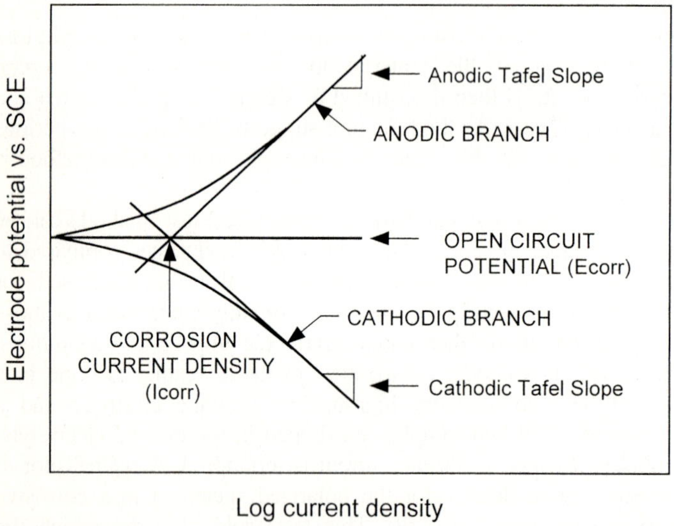

Fig. 4.80. Typical Tafel plot from a polarization experiment (Reprinted with permission from the Annual Book of ASTM Standards, copyright 1985 American Society for Testing and Materials, 100 Barr Harbor Drive, West Conshohocken PA 19428)

5. Characterization

The term characterization has been defined as the analysis of composition and structural features (including defects) of a material that are significant for a particular preparation, study of properties, or use, and suffice for reproducing the material [5.1]. The properties of materials are strongly dependent on the composition and microstructure, as indicated in detail in previous chapters. To ascertain the suitability of a material for a certain application or to be able to understand any property associated with a material, appropriate characterization techniques need to be used.

Incorrect characterization may lead an experimenter to unwarranted assumptions in microstructure–property correlations and probably to incorrect conclusions. Loss of time and money are of consequence. To eliminate such possibilities, one has to carefully select the practices involved in the characterization process. Important characterization techniques relevant to ceramic materials are discussed in this chapter.

5.1. Density and Porosity Measurement

Among the most dependable methods for density measurement are the *Archimedes method* and whenever possible, direct measurement. For theoretical density measurements, X-ray methods and pycnometry are preferred usually.

Numerous variations of the Archimedes method exist for different purposes. One of the relatively easy methods for determining bulk density involves measuring the dry weight (D) of the sample, boiling it in water for five hours, cooling it in water for 24 hours, and measuring the suspended weight in water (S) and wet weight in air (W) (Fig. 5.1). The bulk density is then found from [5.2,5.3],

$$B = D/V = D\,/\,(W - S)\,, \tag{5.1}$$

where V represents the volume. The apparent porosity is calculated from

$$P = (W - D)\,/\,(W - S)\,. \tag{5.2}$$

The details of this method are given in ASTM C373 [5.2].

For specimens with no surface-connected porosity, direct measurement of dry and suspended weights yields the bulk density,

$$B = D/(D - S)\,, \tag{5.3}$$

since the volume of an impermeable material is given by

$$V = D - S\,. \tag{5.4}$$

All of the calculations assume that the density of water is equal to unity. When using a liquid other than water, the density of the liquid has to be incorporated into Eqs. (5.1) – (5.3).

The bulk density of a material can be derived directly from geometrical measurements and the dry weight if the specimen has a well-defined geometrical shape. A micrometer or calipers can be used for precision measurements. An optical microscope with a measuring eyepiece can also be used for surface area measurements.

The theoretical density of a crystalline material can be determined from the volume of the unit crystal lattice determined by X-ray and electron diffraction techniques and from the theoretical weight of each unit lattice. Theoretical densities of various ceramic materials are listed in Appendix 2.

Pycnometry is a particularly useful method for density measurement in amorphous materials, where crystallographic methods can not be used. This method involves pulverizing a sample until particles with no closed porosity are obtained. The powder is placed in a pycnometer bottle with a known volume and weight, and both are weighed together. Then a liquid is added, and the bottle is heated to remove air trapped between particles or adsorbed on surfaces. This results in a mixture of liquid and solid. Weighing of the bottle again and subtracting the initial weight gives the weight of the liquid. If the density of the liquid is known, its volume can be calculated. The difference between the bottle and liquid volumes yields the powder volume:

$$W_{\mathrm{t,1}} = W_{\mathrm{p}} + W_{\mathrm{b}}\,, \tag{5.5}$$

$$W_{\mathrm{t,2}} = W_{\mathrm{p}} + W_{\mathrm{b}} + W_{\mathrm{l}}\,, \tag{5.6}$$

$$W_{\mathrm{l}} = W_{\mathrm{t,2}} - W_{\mathrm{t,1}}\,, \tag{5.7}$$

$$V_{\mathrm{l}} = V_{\mathrm{b}} - V_{\mathrm{p}} = W_{\mathrm{l}}/\rho_{\mathrm{l}}\,, \tag{5.8}$$

$$\rho_{\mathrm{p}} = W_{\mathrm{p}}/V_{\mathrm{p}} = \rho_{\mathrm{th}}\,, \tag{5.9}$$

where subscripts p, b, l, t, and th represent the powder, bottle, liquid, total, and theoretical, respectively.

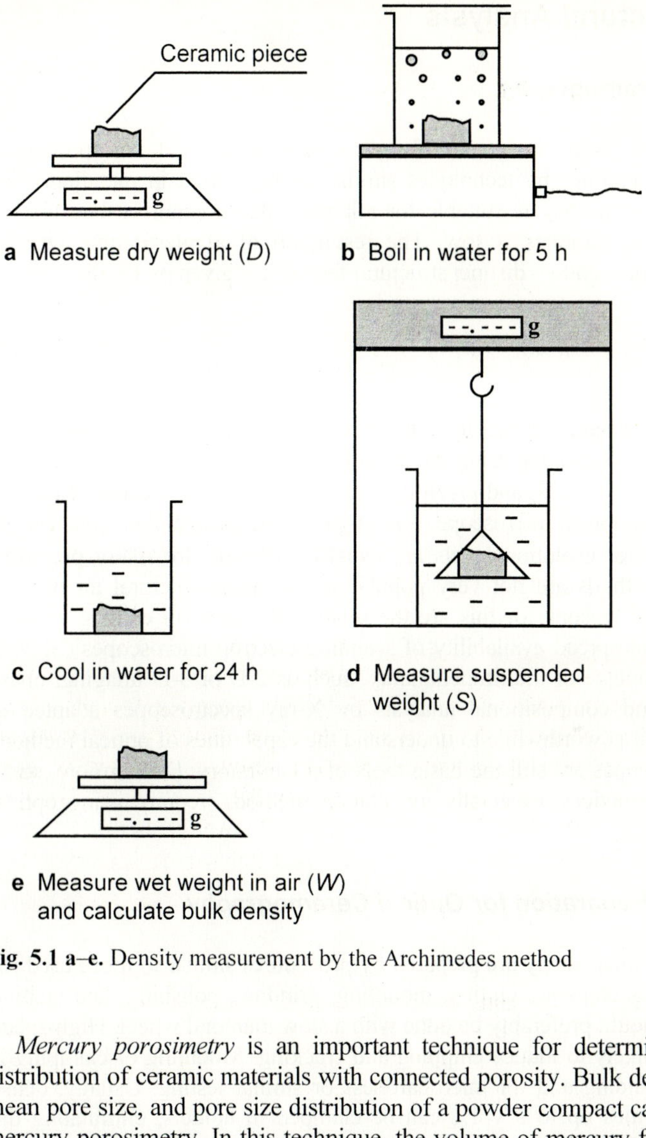

Fig. 5.1 a–e. Density measurement by the Archimedes method

Mercury porosimetry is an important technique for determining the pore size distribution of ceramic materials with connected porosity. Bulk density, pore volume, mean pore size, and pore size distribution of a powder compact can be determined by mercury porosimetry. In this technique, the volume of mercury forced into a sample is measured as a function of pressure. The pore size is determined from the surface tension of mercury and the applied pressure. This method is not suitable for materials with closed porosity or even open porosity with narrow channels since all pores must be filled with mercury at a certain pressure for correct measurements [5.4].

5.2. Microstructural Analysis

5.2.1. Optical Ceramography

Optical microscopes can be used for microstructural and morphological characterization of ceramics by techniques similar to those used in metallography. Generally, optical microscopy is suitable for relatively dense ceramics because the reflectivity of porous ceramics is low. The resolution of a microscope, i.e., the minimum distance between two distinct structural features, is given by [5.5]

$$R = \lambda / 2N, \tag{5.10}$$

where λ is the wavelength of the light beam employed and N is the numerical aperture. Under ideal conditions, using the largest numerical aperture obtained by an oil immersion objective (~1.25) and green light ($\lambda \approx 0.5$ μm), a resolution of 0.2 μm can be achieved in an optical microscope. For a typical microscope, this value can go up to ~0.5 μm. These resolution levels are usually sufficient for many purposes. However, optical methods are not very popular in the microstructural analysis of ceramics. Among the reasons for this are the relative difficulty in etching ceramic materials and the widespread availability of scanning electron microscopes (SEMs). SEMs offer the advantage of multifunctionality, such as 2-D or 3-D imaging, phase contrast imaging, and compositional analysis by X-ray spectroscopes adapted to them. Nevertheless, it is worthwhile to understand the capabilities of optical methods since optical microscopes are still the basic tools of a laboratory. Furthermore, some mechanical testing practices, especially indentation methods, require using optical ceramography.

5.2.1.1. Sample Preparation for Optical Ceramography

Ceramics for optical microscopy are prepared by procedures similar to those used for metals. The necessary steps are cutting, mounting, grinding, polishing, and etching (Fig. 5.2). Cutting should preferably be done with a slow diamond wheel. High-speed abrasive cutters are likely to induce chipping and cracking. Mounting or clamping is required for easy grinding and for microhardness or similar testing. Usually, dense ceramics with porosities up to 5 vol% can be clamped in holders, eliminating the need for mounting [5.6]. If mounting is required, thermosetting polymers such as epoxies, acrylics, and polyesters can be used. These polymers are two component systems; one portion is a resin, and the second is a hardener. The use of compression molding is not recommended for ceramic materials since pressure and heat, which may lead to cracking, are involved.

Grinding is the first step required to remove surface irregularities and damage introduced during cutting. SiC paper, regularly used for metals, can also be used for grinding most ceramics. Ceramics with hardness values comparable to SiC can

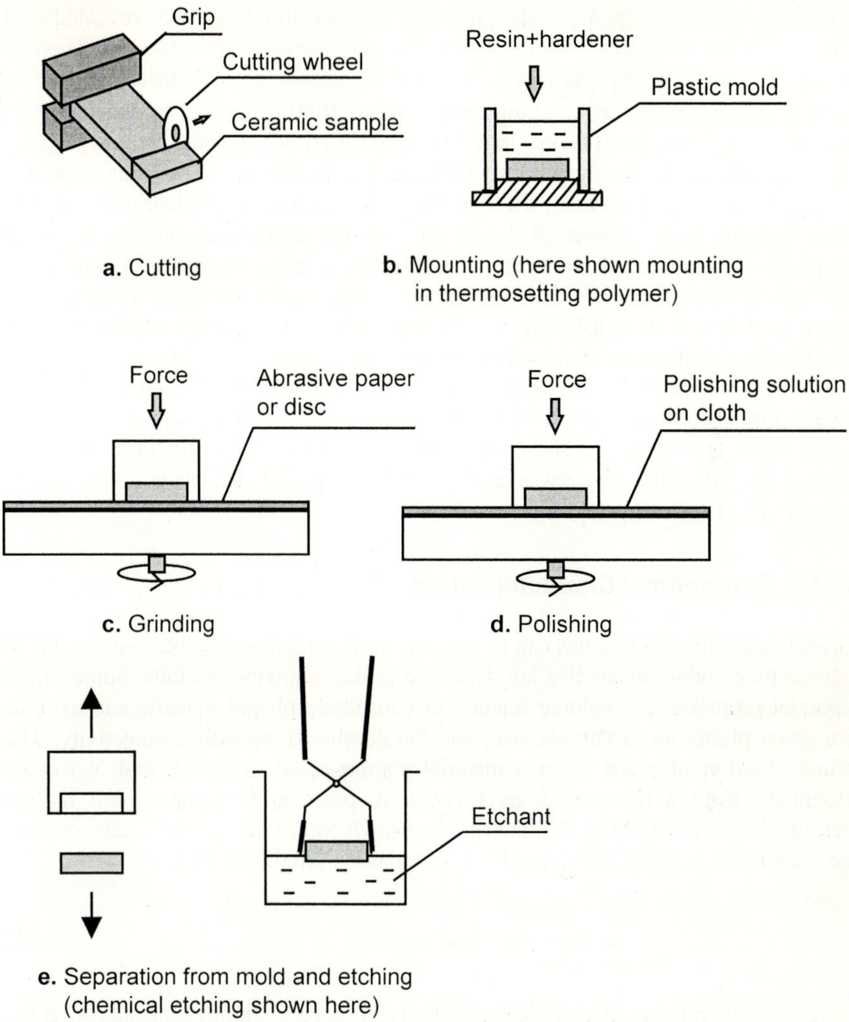

Fig. 5.2 a – e. Steps for preparing ceramic samples for ceramography

be ground with diamond-impregnated plastic or metal lapping disks. Resin-bonded diamond grinding disks yield the highest material removal rate on ceramics but also create the most structural damage [5.6]. Surfaces cut with a fine diamond cutoff wheel usually do not require a grinding step.

Polishing is the step required to achieve a smooth, reflective surface. Alumina slurries on silk cloth provide good results for softer ceramics. For hard ceramics, diamond paste is used on a soft cloth with an oil-based extender. The grain size of the diamond abrasive has to be small enough so that no new grain pullout occurs. Diamond pastes with particle size as small as 0.1 μm can be found commercially.

Etching is a necessary step if the grain size and structure have to be revealed or if different phases have to be distinguished. Various methods are available. Thermal etching in a suitable environment is achieved by exposing the polished surface or whole body to temperatures ranging from 1200 to 1650°C for 1–2 h, depending on the specific ceramic. This practice may lead to grain growth in some ceramics, which may cause some complications if the grain size is to be measured. Another method is ion or plasma etching. Usually, the specimen rests on a cathode plate in a vacuum. Upon applying high voltages (3–4 kV) between the anode and cathode, a plasma appropriate for etching is generated. Etching times range from 10-30 min [5.7]. Although this technique is suitable for many ceramics, the necessary equipment is expensive. Electrolytic polishing is usually not suitable for ceramic materials because good electrical contact and conductivity are hard to achieve.

Although ceramics are generally resistant to chemical attack, certain chemical reagents and procedures can be used for different ceramics. Some examples are given in Appendix 16. Figure 5.3 shows an example of the use of different etchants for polymorphic SiC. The sample was etched with two types of etchants to reveal only α and both $\alpha + \beta$ polymorphs [5.8].

5.2.1.2. Geometrical Characterization

Various geometrical properties can be derived from a micrograph obtained by optical microscopy or other means that are described in the following sections. Some of the important properties are volume fraction of constituent phases, specific surface area of a given phase, mean particle size, and the number of specific connectivity. The volume fraction of phase X in a material composed of phases X and Y can be determined from a micrograph as follows. A point grid is superposed on the micrograph as shown in Fig. 5.4 and the points, which fall into phase X, are counted. The volume fraction of each phase, V_V, is calculated from the equation [5.9]

$$V_V = N_P = N_0/P \,, \tag{5.11}$$

where N_P is the number of specific connectivity in space R^o (point analysis), N_o is the number of connectivity in space R^o (number of points which fall into a given phase), and P is the total number of points in the grid [5.9]. To measure the specific area of a given phase, S_V, an ensemble of lines is superposed on a micrograph. After defining a sense of analysis, the number of transitions X–Y is determined along these lines. From the micrograph shown in Fig. 5.5, the specific area is derived using the following equation:

$$S_V = 4N_L = 4N_1/L \,, \tag{5.12}$$

where N_L is the number of specific connectivity in space R_1 (line analysis), N_1 is the number of connectivity in R_1, and L is the total length of lines superposed.

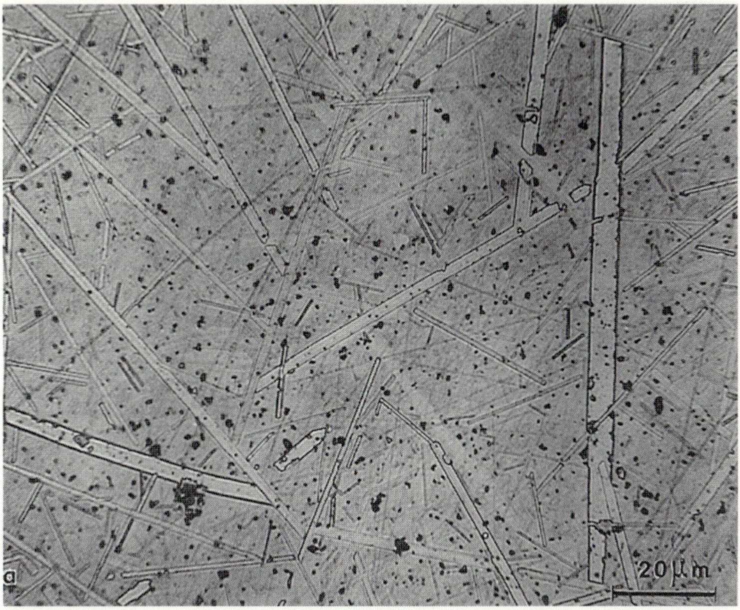

a

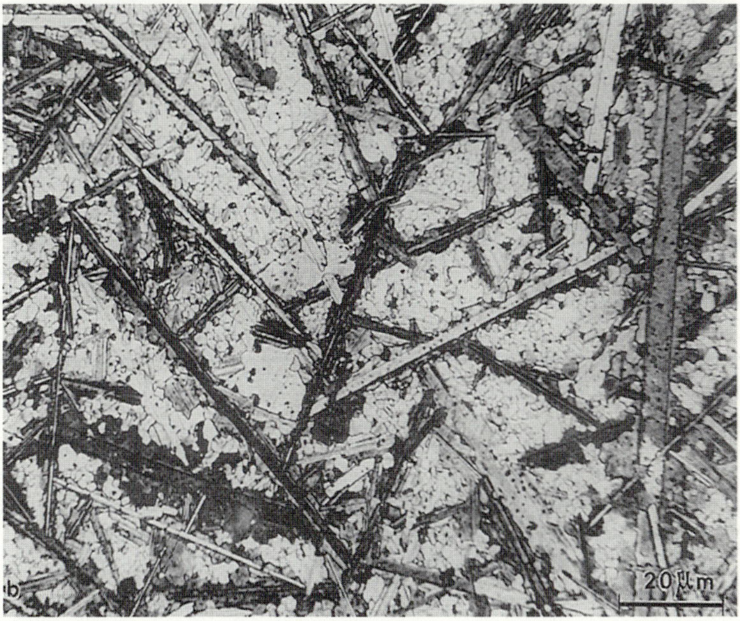

b

Fig. 5.3. Optical micrograph of SiC etched with two types of etchants to reveal only α **a** and both α+β polymorphs **b** (From A.H. Heuer, V. Lou, L. Ogbugji, and T.E. Mitchell: *Lattice Resolution Studies of Engineering Ceramics.* J. Microsc. et Spectr. Electr. **2 [6]**, 475–80 (1977), reprinted with permission of Société Française de Microscopie Paris France)

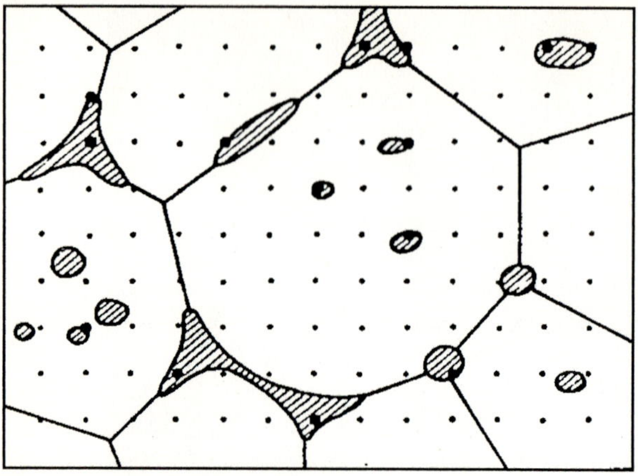

Fig. 5.4. Determination of the volume fraction of inclusions (From J.P. Jernot and J.L. Chermant: *Mesures de Base en Analyse D'Image.* In: *Caractérisation des Poudres et des Céramiques.* J.L. Chermant (ed) (Forceram Hermes France, 1991), pp. 40–56, reprinted with permission of Éditions Hermes Paris France)

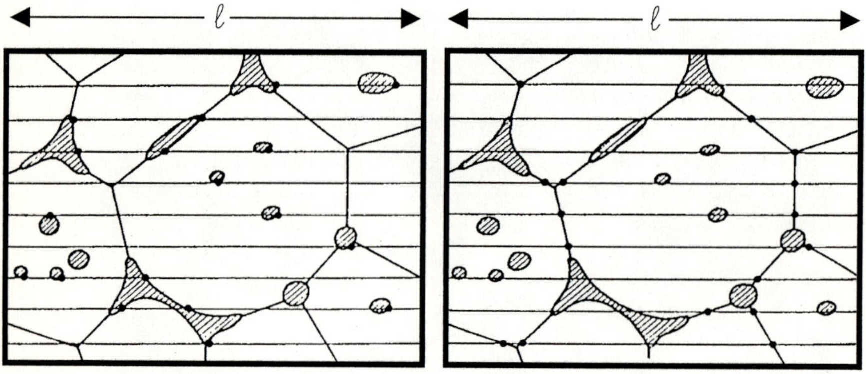

a **b**

Fig. 5.5. Determination of specific surfaces of inclusion/grain **a** and grain/grain interfaces **b** (From J.P. Jernot and J.L. Chermant: *Mesures de Base en Analyse D'Image.* In: *Caractérisation des Poudres et des Céramiques.* J.L. Chermant (ed.) (Forceram Hermes France, 1991), pp. 40-56, reprinted with permission of Éditions Hermes Paris France)

5.2.2. Scanning Electron Microscopy

Scanning electron microscopy is a very popular tool used for microstructural analysis at high magnification. The scanning electron microscope (SEM) uses an electron beam produced by a W-hairpin filament, a LaB_6 tip, or a field emission gun. Electrons are accelerated through a high voltage field (1 to 50 kV) in vacuum [5.10]. The beam is focused onto the specimen by electromagnetic lenses to a spot size of about 1 to 2 nm and scanned on the surface by scanning coils (Fig. 5.6). A lateral resolution of 1–50 nm is achieved in the secondary electron mode. Electrons from the electron source either scatter elastically and produce backscattered electrons, or they interact with the solid. These interactions can produce X rays as well as new electrons, called secondary electrons. All of these emissions are collected by detectors. Upon processing, various types of information can be learned about the material. Specifically, backscattered electrons yield topographic information. For example, surface asperities, grain boundaries, cracks, pores, or for composite materials, fibers or whiskers pulled out of the fracture surface can be observed with superior depth of field compared to optical microscopy. Figure 5.7 shows an SEM image of spray-dried Al_2O_3 powders. An optical microscope would not yield any useful depth of field from such rough surfaces.

Secondary electrons provide atomic number contrast, which can be used to distinguish phases that contain elements with substantially different atomic numbers. This feature can be useful, for example, in yielding a good contrast between ZrO_2 and Al_2O_3 grains in ZTA composites. Many modern SEMs also have X-ray spectroscopic analysis features. This method is discussed in Sect. 5.4.3.

Sample preparation is usually minimal in SEM studies. Many ceramic materials undergo intergranular fracture upon failure. This is a useful feature for analyzing grain structures and sizes. Thus, small (maximum dimension ~100 mm) fragments obtained by fracturing a ceramic are usually useful for fracture surface analysis. Flat, polished surfaces provide useful information only if the sample is etched, or alternatively, if the atomic number contrast can be used. For any insulating ceramic sample, coating with a thin (10 nm) layer of gold, silver, copper, or carbon, is necessary to eliminate problems associated with charging [5.11]. A conductive path is also necessary between the sample and the sample holder, typically made from Al. Usually, a carbon suspension or a piece of Al tape can be used for this purpose. Samples should be cleaned with a suitable solvent to remove grease and dust particles. Samples should be kept in vacuum for a few hours to prevent outgassing during SEM analysis.

A recent development in SEM imaging is the environmental SEM which makes possible sample imaging at 5–15 torr in water vapor at incident beam energies up to 30 kV [5.12]. This microscope can be used for imaging specimens in their natural state without the need for coating or other preparation. The instrument can be used to view samples in almost any gas environment, and even liquid water or oil droplets can be observed with the aid of saturated water vapor conditions. Thus, dynamic processes such as wetting, drying, mixing, crystallization, or curing can be observed and videotaped in real time. Another development in SEM is the use of computers for data collection, analysis, and processing images [5.11].

Digitally stored information can easily be processed by computers for image analysis, e.g., for particle size and area determination, as well as contrast adjustment or elimination of specific undesirable features.

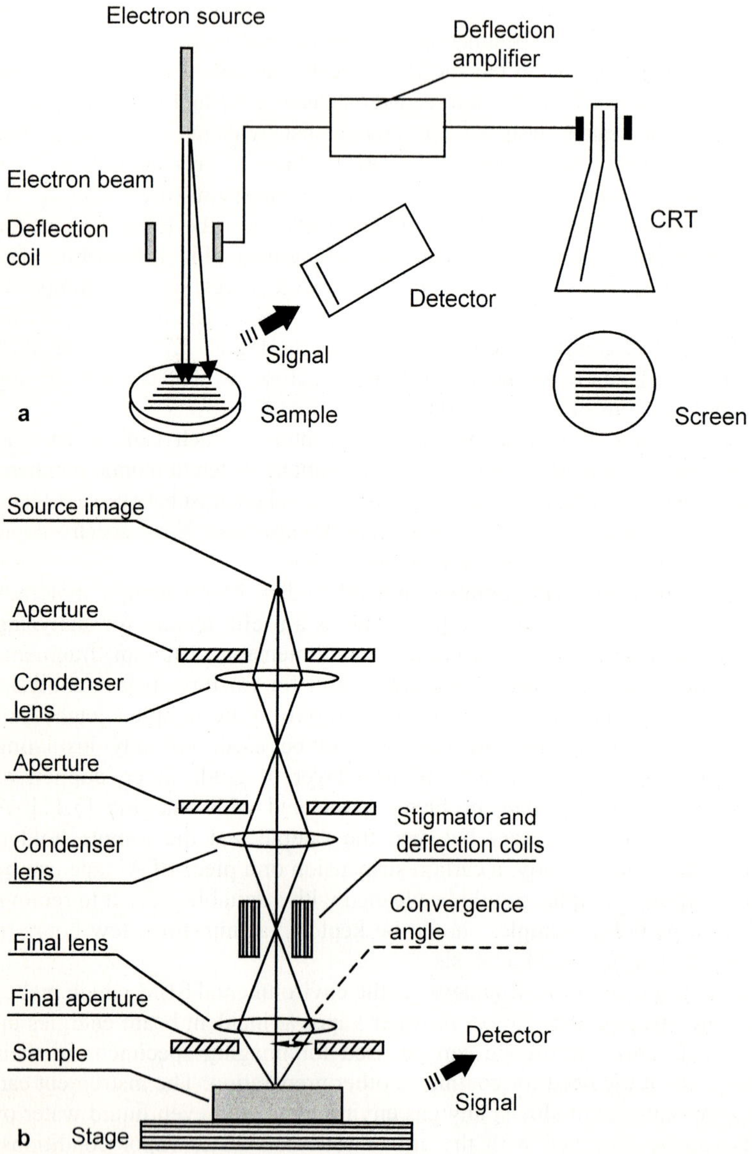

Fig. 5.6. Schematic illustration of SEM operation **a** and detail of typical electron optics **b** (From J.B. Bindell: *Elements of Scanning Electron Microscopy*. Adv. Mater. & Proc. **143 [3]**, 20–27 (1993), adapted with permission of ASM International Materials Park OH)

Fig. 5.7. SEM of spray-dried Al_2O_3 powders (From M. Bengisu, Ph.D. Thesis, New Mexico Institute of Mining and Technology Socorro NM, 1992)

5.2.3. Transmission Electron Microscopy

This technique is indispensable for gaining microstructural information from a material's interior. Such features as vacancy loops, dislocations, stacking faults, microcracks, intergranular phases, etc. can be directly observed in electron transparent specimens by transmission electron microscopy (TEM). For example, Fig. 5.8 shows the internal structure of α and β polymorphs in a hot-pressed SiC sample [5.8]. Crystal structures can be analyzed by various diffraction techniques discussed in Sect. 5.4.1. The TEM uses an electron beam produced in the same fashion as described for SEM. In this case, however, the sample positioned between the electron beam source and the screen is often a small (2–3 mm diameter) disk with a hole in it. An ideal sample should contain a large electron transparent area around the hole. The image is obtained through various contrast mechanisms including diffraction, absorption, and scattering of electrons (Fig. 5.9) [5.8]. The TEM is a unique and convenient tool since it allows switching with a single push-button, from electron microscopy (real space) to electron diffraction (reciprocal space). A fluorescent screen sensitive to electrons is used to observe the image or diffraction pattern. Usually, this screen is movable to allow photography by the camera placed just below it.

The *scanning transmission electron microscope (STEM)* utilizes principles very much like TEM. In this technique, a very fine electron probe obtained from a W-field emission gun or LaB_6 thermoionic field emitter is scanned across a specimen.

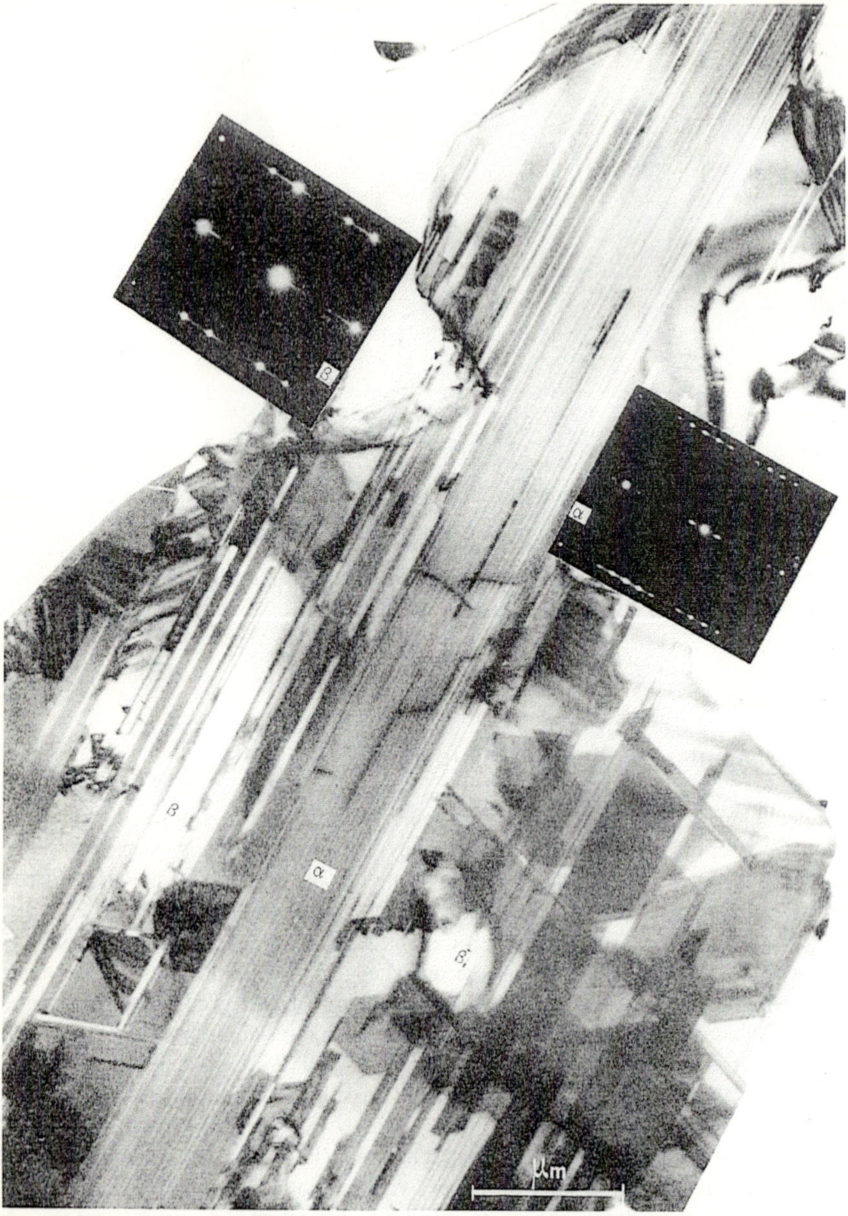

Fig. 5.8. Internal structure of α and β polymorphs in a hot-pressed SiC sample (From A.H. Heuer, V. Lou, L. Ogbugji, and T.E. Mitchell: *Lattice Resolution Studies of Engineering Ceramics.* J. Microsc. et Spectr. Electr. **2 [6]**, 475–480 (1977), reprinted by permission of Société Française de Microscopie Paris France)

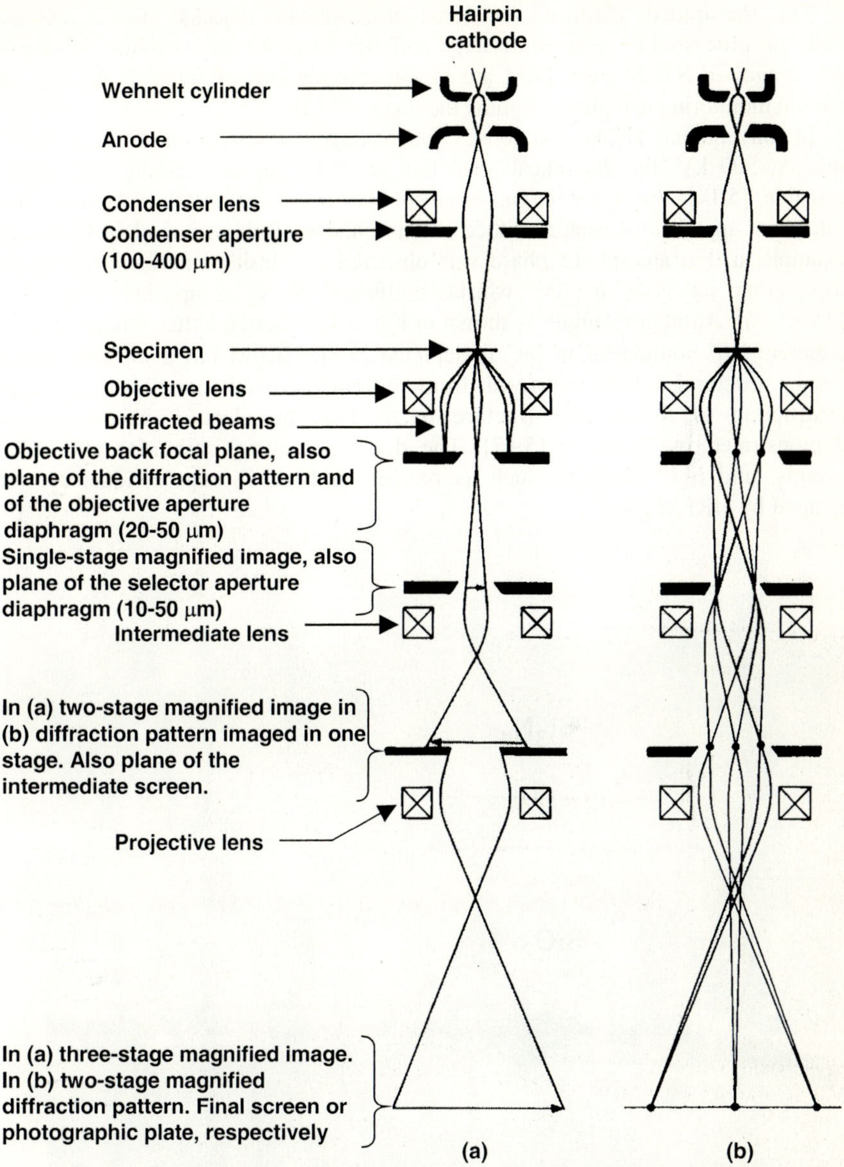

Fig. 5.9. Schematic illustration of TEM techniques. Ray path for bright field imaging **a** and for selected area diffraction (SAD) **b** (From M. von Heimendahl: *Electron Microscopy of Materials, an Introduction.* (Academic Press San Diego CA, 1980) reprinted with permission of Academic Press Orlando FL)

The transmitted, elastically scattered electrons are detected with an annular detector, processed by a signal detector, and viewed on a video monitor. Among the advantages of STEM over TEM are various possibilities of signal processing and ease in monitoring and photographing the image [5.13].

In conventional TEMs with 100-kV bias voltage, the resolution limit is about 0.3 nm. At 500 kV, the theoretical resolution is ~0.15 nm, and atomic resolution is possible [5.14]. *High resolution electron microscopy (HREM)* techniques have important applications such as lattice imaging and analysis of boundary layers. For example, a thin amorphous phase was observed at whisker–matrix interfaces and triple-grain junctions in SiC whisker-reinforced Si_3N_4 composites (Fig. 5.10) [5.15,5.16]. Another example is shown in Fig. 5.11 where a lattice fringe image of coherent α/β boundaries in SiC is depicted [5.8]. HREM has also been used to image structural defects that result from sintering or irradiation. The progressive destruction of the lattice structure in β-Al_2O_3 by electron irradiation was demonstrated by Horiuchi [5.17]. The defect structure, carbonate and sulfite groups, and light elements such as oxygen in high T_c superconductors were imaged by HREM [5.18].

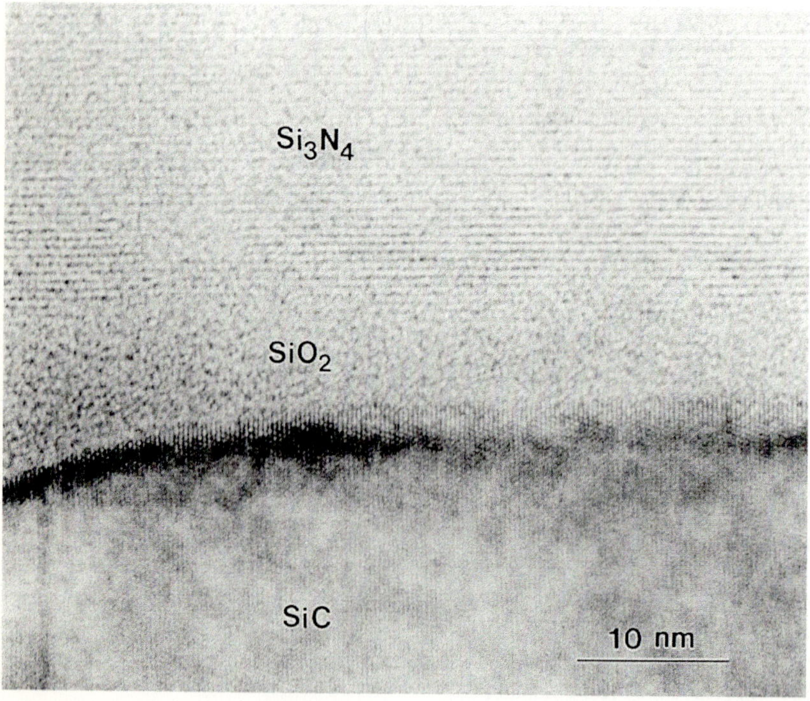

Fig. 5.10. HREM of amorphous SiO_2 phase along a Si_3N_4/SiC interface in the composite (From O. Ünal, J.J. Petrovic, and T.E. Mitchell: *Mechanical Properties of Hot Isostatically Pressed Si_3N_4 and Si_3N_4/SiC Composites*. J. Mater. Res. 8 [3], 626–634 (1993), reprinted with permission of Materials Research Society Pittsburgh PA)

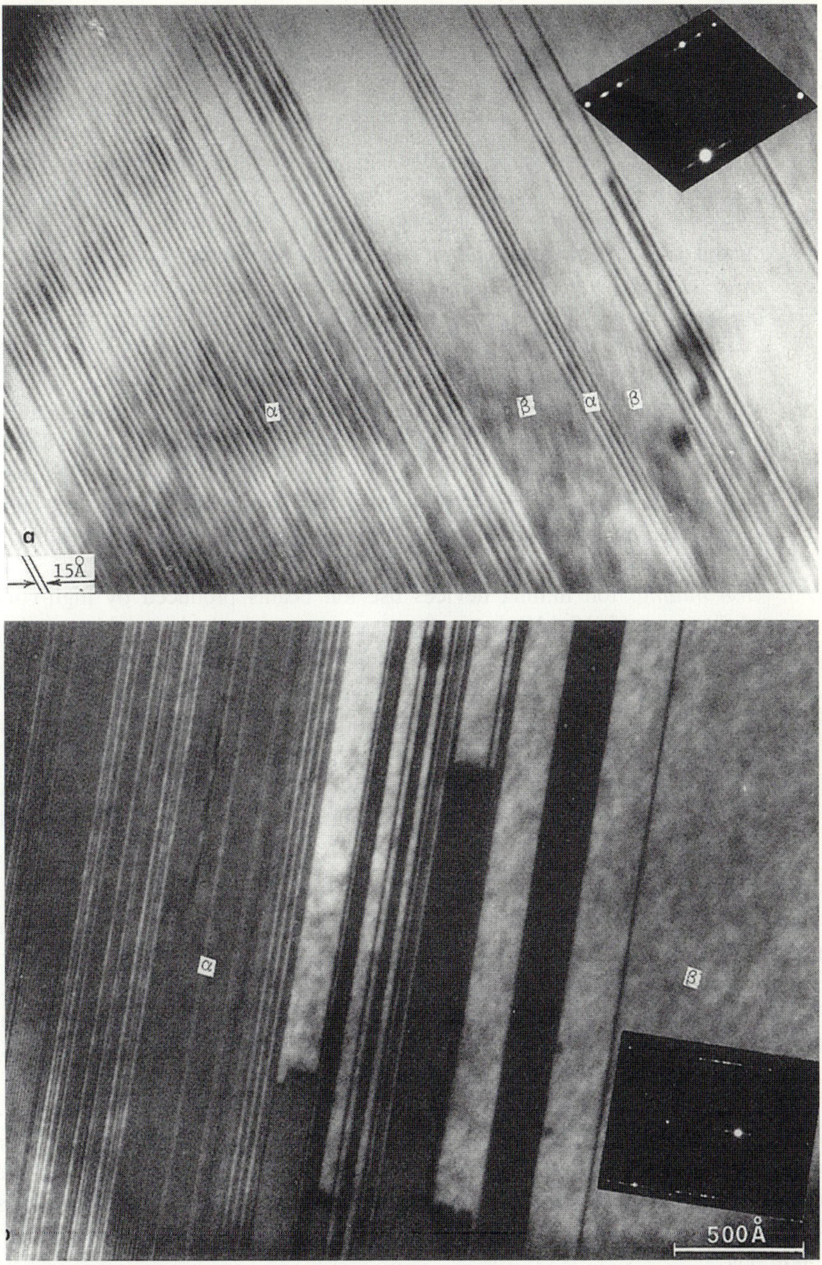

a

b

Fig. 5.11. Internal structure of α and β polymorphs in a hot-pressed SiC sample. Lattice fringe image of coherent α/β boundaries in SiC **a**. and superperiodicity in the β phase **b** (From A.H. Heuer, V. Lou, L. Ogbugji, and T.E. Mitchell: *Lattice Resolution Studies of Engineering Ceramics.* J. Microsc. et Spectr. Electr. **2 [6]**, 475–480 (1977), reprinted with permission of Société Française de Microscopie Paris France)

TEM sample preparation of ceramic materials requires considerable effort and time. The difficulty is greater for composite ceramics, especially if the hardness values of the constituents differ significantly from each other. For most ceramics, however, the following steps provide satisfactory samples [5.19]:

1. *Cutting*: Bulk samples can be cut by a slow diamond blade to obtain slices as thin as 0.5 mm.
2. *Mechanical thinning*: The slice is glued to a metal block for easy handling and it is ground to about 0.1 mm using a diamond lapping disk.
3. *Disk cutting*: Disks are cut out of the thin slice by ultrasonic or vibrating tips using diamond paste. Another method is to use the flat cutting tip on a drill with little pressure and diamond paste in between. Diamond paste with coarse (5 –10 μm) diamond particles is suitable for disk cutting.
4. *Mechanical thinning*: The disks are further thinned to about 20 μm by dimple grinding using diamond paste. A dimple grinder is made of a metal wheel, which thins the middle of a disk and a rotating table, which turns the sample during grinding (Fig. 5.12). If the sample is placed correctly, the dimple grinder produces a sample whose periphery is thick enough for handling and whose interior region is thin enough for ion milling.
5. *Ion milling*: Ion milling devices use ion beams produced by high-voltage differences (5–10 kV) between an anode and cathode to thin specimens slowly and uniformly. The ion current is kept at about 50 – 100 μA. The sample is usually tilted at an inclination of 0 – 30°. Ion bombardment of the surface removes material without significant internal damage and artifacts. Milling times can range from 5 to 20 h until a visible hole appears at the center, which may be detected visually or automatically in modern devices. For better results, insulating ceramics may be coated by a thin conductive layer.

Since sample preparation is time-consuming and requires special equipment, the TEM technique is usually used for analyzing features that are typically not observable under an SEM. Such features include dislocations, vacancy loops, grain boundaries, small impurities, and in HREM, direct information on lattice structures or interfaces. TEM is a reliable tool that can be used for characterizing ceramics throughout different processing steps, from the original powders to the consolidated state and furthermore, at any stage during service. Such a detailed study was performed, for example, on Si_3N_4 ceramics, to understand the influence of sintering additives on microstructural development and mechanical performance [5.20]. Special stages have been developed to image samples at high temperature, low temperature, under stress, in reactive environments, or in other conditions. These contribute to the in situ analysis of structural mutations and defect generation in materials under different conditions.

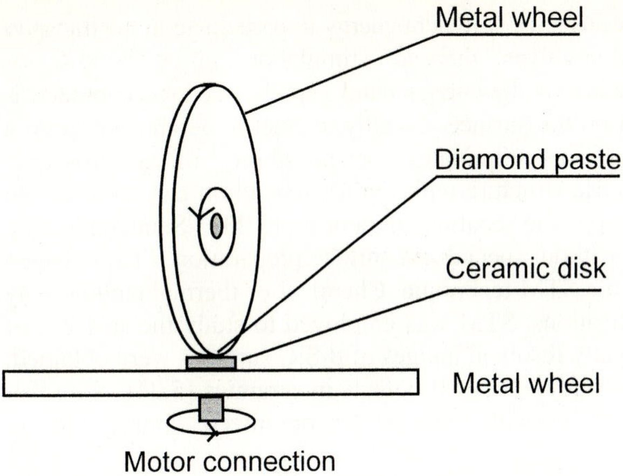

Fig. 5.12. Schematic illustration of TEM sample preparation on a dimple grinder

5.2.4. Scanning Probe Microscopy

Scanning probe microscopy (SPM) is the common name of many related techniques ranging from atomic force microscopy (AFM) to scanning tunneling microscopy (STM). These techniques provide high-resolution, three-dimensional images of solids with lateral spatial resolution as low as 0.1 nm (atomic resolution) and depth resolution better than 0.1 nm. SPM images are digitally stored in a computer in three dimensions, allowing examination from different angles and measurement of exact dimensions along any line [5.21].

Scanning tunneling microscopy is a technique invented in 1981 [5.22], that is still under development. Conducting, semiconducting, and insulating ceramics can be studied by using STM. It does not require electron, photon, or ion beams, eliminating the necessity for lenses. The tip of a fine metal probe is pushed onto a sample surface until the electron clouds of both sides come into contact. When an electric potential is applied between the sample and the tip, a number of electrons tunnel through the potential barrier between the tip and the surface [5.23]. The tunneling current can be related to the electronic structure of the sample and tip materials. In the constant-current imaging mode, the tip is scanned across the surface and yields a map of surface charge density, hence surface topography. In an alternative mode, the bias voltage is changed, and the tunneling current is monitored. The tunneling current variation can be used for determining the surface density of states. This mode of operation is called *scanning tunneling spectroscopy (STS)* and it provides information about the electronic structure of the surface.

STM analysis of ceramic materials with band gaps larger than 1.5 eV and low conductivity exceeds the limits of applicability. Two approaches, however, can be used to image insulating ceramics. One way is to increase the energy of the system so that a higher number of electrons have sufficient energy to participate in the transport processes. This approach involves thermal stimulation up to 200°C or photoexcitation of electrons across the energy band gap. The second approach is to deposit a thin metal film on the surface. Usually, dielectric ceramics require a highly conductive coating. This eliminates the possibility of spectroscopic analysis. The surface geometric structure with limited resolution can be observed where resolution depends on the coating thickness [5.23]. Semiconducting ceramics may be examined without specialized surface preparation. Flat, polished surfaces are best suited for the STM technique. Chemical or thermal etching may be required for certain applications. STM was employed to study the surfaces of SiC and ZnO [5.23]. Atomically resolved images of β-SiC surfaces were obtained. Another application was the study of crystal defects in ceramics [5.24]. Note that atomic resolution images are possible only under optimized sample and tip conditions [5.25].

Atomic force microscopy (AFM) is another SPM technique invented in 1986 [5.26]. It is related to STM, but it measures atomic forces between the tip and the sample surface rather than the current [5.27]. AFM measures topography by moving a sharp probe across the sample surface to sense the contours of the surface similar to the way a phonograph stylus traces the grooves of a record. Unlike STM, samples can be imaged in air or in a liquid medium. In addition to conducting and semiconducting surfaces, stereography of insulating surfaces can also be made in AFM. Applications of AFM include analysis of the full-growth morphology of superconductors, observation of substances that are chemically reactive, nanoprofile analysis of nonconductive materials, and observation of oxide defects in silicon. AFM was used to compare repulsive and van der Waals attractive forces on alkali silicate glass to describe corrosion behavior in different environments [5.28]. An example of an AFM ceramic image is shown in Fig. 5.13.

Numerous new SPM techniques evolved from STM and AFM including magnetic force microscopy, electrostatic force microscopy, lateral-force microscopy, scanning ion conductance microscopy, and laser force microscopy [5.21]. These techniques have specific applications and are still under development.

5.2.5. Scanning Ion Microscopy

This is a technique similar to SEM; the difference is that an ion beam rather than an electron beam is used for imaging. In scanning ion microscopy (SIM), various contrast mechanisms can be used for microstructural analysis. An argon ion beam is preferred since it does not change the chemistry of the sample surface. Insulating ceramics can be analyzed by placing them on a conducting metal grid or by coating them with a very fine conductive film [5.13]. Figure 5.14 shows a schematic diagram of SIM [5.29]. Secondary electrons produced by the ion beam–specimen interactions are preferred over secondary ions for imaging since the electron yield is much greater and microstructural information can be obtained more rapidly and with less damage to the specimen. An important advantage of SIM over SEM is the possibility of in-situ microsectioning and examination of subsurface features by prolonged exposure

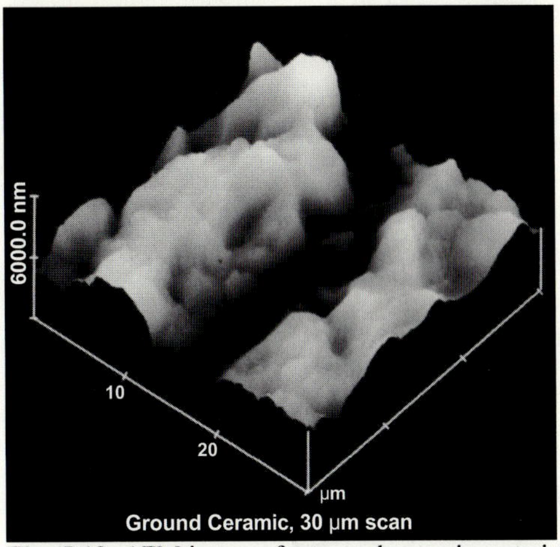

Fig. 5.13. AFM image of a ground ceramic material (courtesy of Digital Instruments. Image generated with a NanoScope® Scanning Probe Microscope, manufactured by Digital Instruments Santa Barbara CA)

and repeated scanning of the focused ion beam on the desired area. Thus, a cross section of the specimen can be obtained. Similarly, cleaning of the surface and material removal to expose the subsurface are possible. Such features are used, for example, in examining microelectronic devices [5.30]. However, the feature that really sets ion microscopy apart from SEM is the ability to provide specific ion or elemental concentration maps of the surface [5.13]. In addition, the method can detect hydrogen, isotopes, and molecular fragments [5.31].

5.3. Diffraction Techniques

5.3.1. Electron Diffraction

Electron diffraction studies normally involve the electron microscope. Selected area diffraction (SAD) is the usual diffraction technique used in a TEM. Convergent electron beam diffraction (CBED) and Kikuchi diffraction are important techniques used in combination with SAD. Reflection electron diffraction (RED) and low energy electron diffraction (LEED) are other important diffraction techniques.

In the *selected area diffraction* mode, the desired electron transparent area is selected by inserting a diaphragm, also called a field-limiting aperture, with a suitable diameter. The fine, parallel beam of electrons, which was described in Sect. 5.2.3, will produce two diffracted beams at position D and the image at position I (Fig. 5.15) upon inserting the diaphragm. By adjusting the power of the intermediate and projector lens, it is possible to focus either the image at position I (plane of the diaphragm), or the diffraction pattern at position D. Thus, it is possible to correlate a

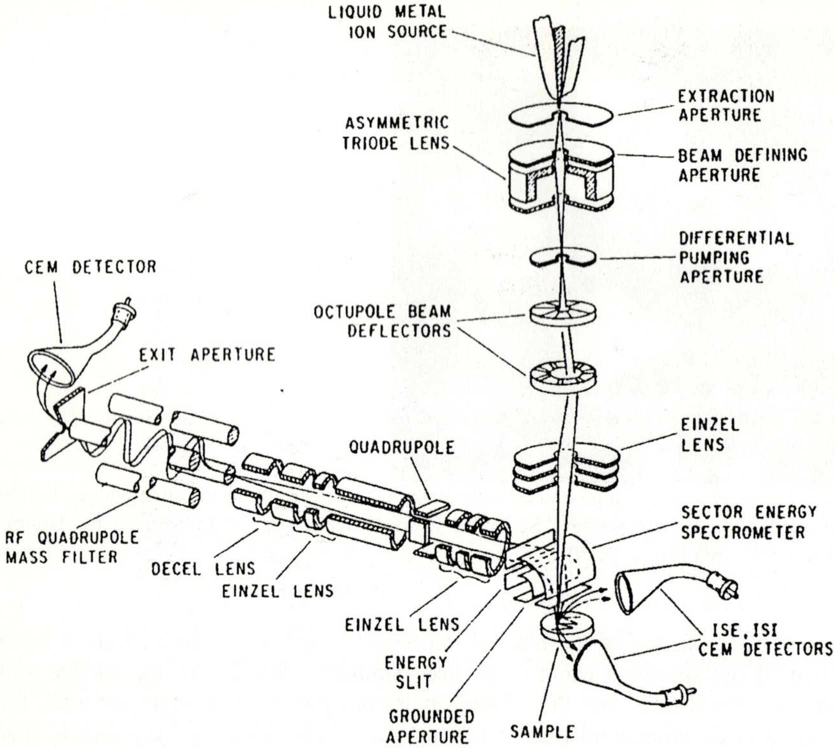

Fig. 5.14. Exploded schematic view of SIM (From R. Levi-Setti, J.M. Chabala, and Y.L. Wang: *Aspects of High Resolution Imaging with a Scanning Ion Microscope.* Ultramiscroscopy **24,** 97–114 (1988), with permission from Elsevier Science)

diffraction pattern with an electron micrograph, i.e., image, from the same region. The SAD technique also allows isolating the diffracted beams. Image signal information specific to those beams can then be observed in the dark field mode [5.26]. The same fluorescent screen and photographic facilities that are used for the image are used for diffraction studies. The main limitations of SAD are spherical aberration of the objective lens and the difficulty of ensuring that the image is situated precisely at position I. Due to spherical aberration, peripheral regions out of the field limiting aperture also contribute to the diffraction pattern, which may complicate indexing.

The diffraction pattern represents the reciprocal of the real crystal lattice of the selected area. Thus, using certain procedures well developed today, it is possible to reconstruct the unit cell and atomic arrangements. Because one diffraction pattern represents only one projection of the reciprocal lattice, at least two diffraction patterns or projections from the same region are required for a three-dimensional reconstruction. This is usually achieved by tilting the sample positioned on a tilt stage.

The high resolution electron diffraction mode can be used for accurate measurement of lattice parameters. In this mode, the specimen is placed just above the projector lens, and very accurate calibration is thus possible.

When the electron beam is sharply focused on a sample surface, diffracted spots become large disks. The sharp focus of the electron beam (0.5 μm and smaller probe diameters) is achieved by using a high brightness field emission gun. Another method is to focus the beam using either condenser lenses or one condenser and the objective lens. The intermediate lens is adjusted so that all real space detail and the shadow image of the sample are eliminated, and only the disk pattern known as *convergent electron beam diffraction (CBED)* pattern detail is left. Within the broad disks, it is possible to observe a regular array of fine higher order Laue zones (HOLZ lines) which arise due to three-dimensional diffraction effects. HOLZ lines can be used as a source of information on specimen orientation, lattice parameter and its variation within a small area, local strain fields, and crystal symmetry. CBED patterns have many applications such as identification of precipitate phases, support for EDS (see Sect. 5.5.1.3) or EELS techniques used in AEM (see Sect. 5.5.7), and measuring local strain fields such as those found between coherent precipitates and the matrix. Other information that can be obtained includes specimen thickness, crystal potential, structure factor, and crystallographic space group.

When single crystals or large grains, both with sufficient thickness, are analyzed by the diffraction mode, the diffraction spots are accompanied by sets of parallel lines. One of them is bright and the other one dark and has a definite association with

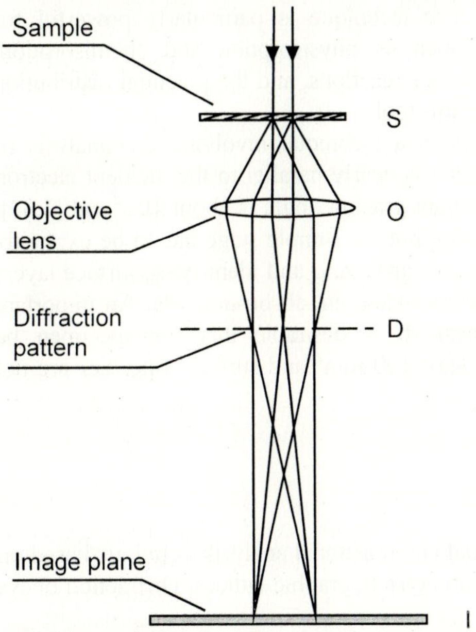

Fig. 5.15. Schematic diagram of the electron microscope in the selected area diffraction (SAD) mode.

the spots. These lines occur due to diffuse scattering of diffracted and incident electron beams, and they are called *Kikuchi lines* or *Kikuchi diffraction patterns*. To aid the indexing of Kikuchi diffraction patterns, standard maps called Kikuchi maps have been developed. These are projections of reciprocal space that correspond to the total composition of Kikuchi diffraction over a finite angle of tilt from one zone. The formation of Kikuchi patterns requires precise crystal symmetry and uniformity. Thus, their formation is an indication of crystal perfection and their absence is due to defects, which enhance scattering. Kikuchi patterns are used mainly as an aid in analyzing diffraction patterns and determining crystallographic uniqueness.

Low energy electron diffraction (LEED) involves only electron acceleration potentials of several hundred volts, whereas all other diffraction methods require potentials higher than 50 kV. A special camera is usually used for LEED studies where the electron beam is directed at the surface of a flat (but not necessarily electron transparent) specimen at a normal angle. The reflected diffraction pattern is observed on the fluorescent screen at the same side from which the beam enters the camera. The pattern on the screen can be observed and photographed through a window from outside the camera [5.32]. The LEED technique can also be applied in RED simply by lowering the acceleration potential and observing the surface structure at a grazing angle. Due to the low potentials, refraction, i.e., deviation from the actual electron path becomes appreciable in LEED. The mean free path for inelastic scattering is typically 0.5 nm. Thus penetration of low-energy electrons is limited to the first few atomic layers at the sample surface [5.33]. LEED has become an important tool for detailed surface structure studies, especially for ceramics, since they cannot be analyzed by field ion microscopy. For example, the (010) surface of V_2O_5 was studied by LEED [5.13]. The technique is particularly powerful for analyzing surface-related phenomena such as physisorption and chemisorption. Absorption of unexpected impurities, surface reactions, and the potential distribution at the surface are complicating factors of this tool.

Reflection electron diffraction (RED) is a technique involving the analysis of diffracted spots from a sample surface that is nearly parallel to the incident electron beam of 10–100 keV [5.33]. The minimum grazing angle is about 10^{-2} rad [5.32]. Thus, the sample does not have to be thin, but the sample stage has to be explicitly designed for this purpose. RED is useful for analyzing and identifying surface layers such as coatings and those that result from oxidation, absorption, etc. An important requirement in RED is surface flatness. It is desirable that the specimen be microscopically flat over an area of at least 100 mm^2 and surface asperities are not thicker than 10 nm.

5.3.2. X-Ray Diffraction

X-ray diffraction (XRD) is a conventional compositional analysis technique based on diffraction of collimated X rays by certain crystallographic lattices. Diffraction obeys the *Bragg law* [5.34],

$$n\lambda = 2d \sin \theta \tag{5.13}$$

where n is the order of the Bragg reflection, typically taken as unity (i.e., first-order reflection) in most calculations, λ is the wavelength of the X rays employed (for Cu–Kα radiation, typically used for routine analysis, λ = 1.544 Å), d is the interplanar distance, and θ is the angle of inclination of the incident X-ray beam.

Diffraction data can be collected in two ways:

1. cameras
2. diffractometers

Various cameras have been developed for specific purposes. The *Laue method* uses a well-formed single crystal of the compound of interest. The Bragg equation is satisfied by using continuous (white) X radiation, whereby a continuous range of λ values is applied at the same time. The crystal is stationary, and the spots of diffraction are formed on a film after proper exposure. The crystal system, unit-cell dimensions, and the space group, at least partially, can be determined from the diffraction pattern [5.35]. The main application of the Laue method is determining crystal orientation and quality. Two types of Laue cameras exist and both are simple to construct. The *transmission camera* (Fig. 5.16a) is made of a collimator, a device that produces a narrow incident beam, a single-crystal specimen holder, and a lighttight film holder (cassette). When the white radiation hits the sample, a part of it is transmitted, and diffraction occurs at the same time. The film intersects diffracted beams at different distances from the crystal, resulting in elliptical spots of different sizes [5.34]. To obtain good results with this method, the sample has to be thin and of low absorption. The *back-reflection camera* is similar to the transmission camera, except that the cassette carries both the film and the collimator (Fig. 5.16b). This method requires longer exposures for diffraction patterns than the former. The diffraction spots obtained from back-reflected beams and intersected by the film, are more or less circular in this case. The back-reflection Laue method is the more widely used between the two Laue methods because it does not require any special preparation of the specimen and the sample thickness is of no concern.

The powder camera is a cylindrical enclosure with a collimator at one side, a beam trap at the contrary side, and a strip of film wrapped around the inner side of the camera. The specimen is usually prepared as a cylinder from fine powders in a plastic or thin-walled glass tube or using an adhesive. The collimated monochromatic X-ray beam falls onto the specimen, and any powder particle (each a tiny crystal) positioned in a diffracting position will diffract the beam according to Bragg's law. The locus of all such diffracted beams is a cone of semiangle 2θ whose apex is at the specimen. The intersection of various diffraction cones, corresponding to various Miller indexes (h,k,l) with the film, are symmetrical curves from which d spacings can be calculated. Most of the crystalline materials studied so far have the major features of their X-ray powder patterns listed in a set of cards that is known as the ASTM (American Society for Testing of Materials) index. From a comparison of d values calculated and listed in this index, one can usually identify an unknown material [5.36].

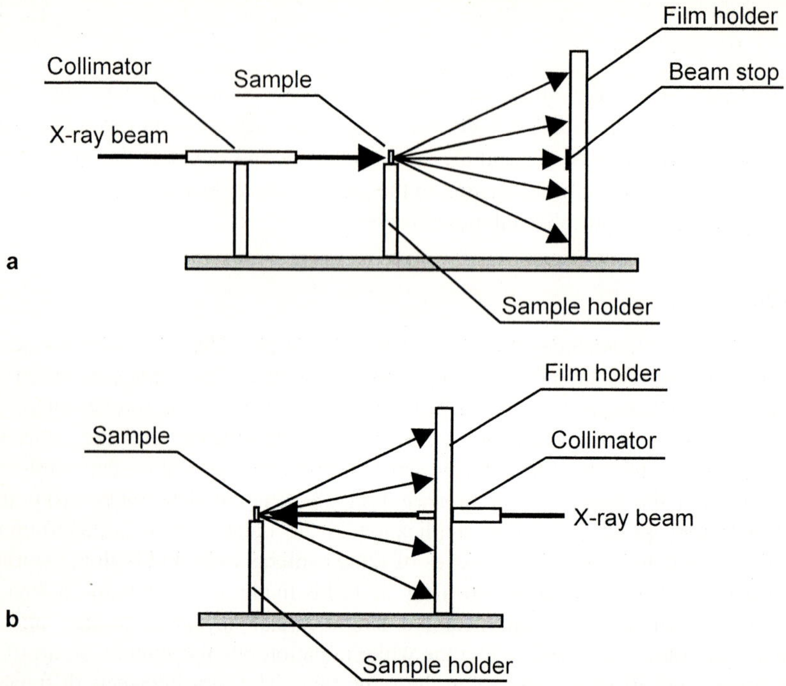

Fig. 5.16. Schematic illustrations of Laue cameras: the transmission camera **a** and the back reflection camera **b**

The *oscillation camera* is a somewhat sophisticated version of a Laue camera. In this case, λ is kept constant by using a filter, and θ is varied by oscillating or rotating the single crystal about a crystallographic axis. Various diffraction cones are formed, and these are recorded by a film wrapped inside a metal camera. From the diffraction lines, unit-cell spacing can be calculated with reasonable accuracy [5.35,5.36]. The main use of the oscillation camera is in determining unknown crystal stuctures [5.34].

The *Weissenberg camera,* which enables a whole layer of information to be collected at one time and which makes indexing quite straightforward, is a more convenient instrument. In this method, the film holder is mechanically coupled with the rotation of the sample crystal and translated parallel to the oscillatory axis. This prevents the overlapping problem found in other cameras.

The *precession camera* produces an undistorted picture of the reciprocal lattice by ensuring that the film is always perpendicular to the crystal axis. Although the camera is quite complex, the resulting photograph has the advantage of being symmetry-true and enabling measurements to be made more easily.

It is possible to measure intensities of diffracted spots by photographic means, but diffractometers are far more practical for this purpose. The *Geiger–Müller counter* records individual X-ray photons. The *proportional counter*, which is similar to the

Geiger-Müller counter, is another type of counter; the difference is that the voltage pulses produced are small (~1 mV) and require amplification. The pulses produced are proportional to the energy of the X-ray photon. Thus, unwanted photons (e.g., second-order reflections) can be detected and eliminated. The counters may be integrated into the various cameras discussed so far, but the typical instruments used for automatic intensity measurements are four-circle diffractometers for single crystals and powder diffractometers for powder samples. In the *four-circle diffractometer*, three types of rotational motion are associated with the crystal and a fourth is provided by the counter, thus the name four-circle. In *powder diffractometers*, an assembly of fine powders is placed on a holder and only the counter moves, thus the design is much simpler. In both of these diffractometers, a computer is usually used to collect data within the required angular range, without any intervention by the user. The results can be printed on chart paper, directly typed in numbers, or collected for data processing at a later time. Figure 5.17 shows an example of a diffraction pattern (graphite) from a powder diffractometer. Various software exist now to process such data, e.g., to directly convert 2θ readings into d_{hkl} measurements, to match these with ASTM powder files, to apply the required correction factors, etc.

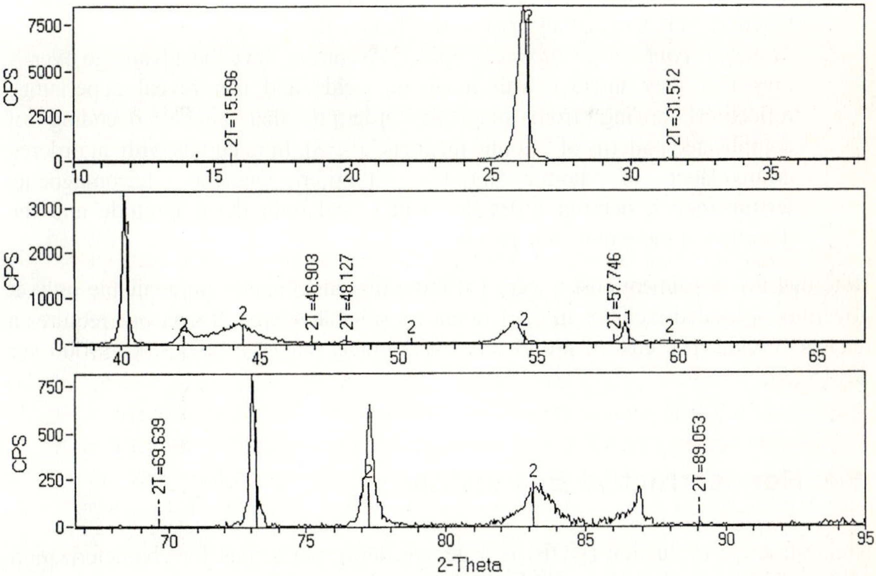

Fig. 5.17. X-ray diffraction pattern of graphite (2), illustrating the presence of a coated material (2T) and W contamination (1) (From: O.T. Inal, E. Bengü, and M. Bengisu, *Low-Voltage Plasma-Assisted Chemical Vapor Deposition.* Invited paper in: *Int. Symp. on Plasma Polymerization/Deposition: Fundamental and Applied Aspects*. Las Vegas Nevada, Nov. 1994)

5.3.3. Neutron Diffraction

In principle, neutron diffraction is similar to X-ray diffraction; the major difference is that neutrons have much greater penetrating capability. Their direct interaction with atomic nuclei of samples leads to the use of this technique in atomic structure determination or structure refinement of phases that contain atoms of widely varying atomic numbers in bulk materials. Neutron diffraction has been one of the most important characterization methods for analyzing glass structures due to its better capability for defining interatomic distances [5.37]. Two special cases exist where neutron diffraction has a definite advantage over X-ray diffraction:

1. *Materials containing hydrogen or neighboring elements in the periodic table*: Scattering or diffraction of X rays occurs due to electron clouds associated with atoms in a structure, but in the case of neutrons, the scattering interaction occurs with nuclei of atoms. Thus, scattering is no longer proportional, as with X rays. As a result, many light elements have scattering power similar to or greater than "heavier" atoms. Hydrogen, in particular, has scattering power similar to many other elements, which makes it possible to locate its nuclei with high accuracy. Neighboring elements in the periodic table are usually easily detected due to their widely different scattering power. For example, neutron diffraction shows strong superlattice lines from ordered FeCo, whereas with X rays, they are practically invisible [5.34].
2. *Materials containing "atomic magnets"*: Neutrons have the advantage over X rays that they interact with magnetic fields and can reveal superlattice reflections arising from magnetic ordering, that enable decoding of complicated patterns of "atomic magnets" [5.38]. In materials with an ordered arrangement of atomic moments (antiferromagnetic, ferromagnetic, ferrimagnetic), neutron diffraction can reveal both the magnitude and the direction of these moments [5.34].

Note that the instrument cost is very high and that instruments are available only at government-funded facilities in the US since a suitable source of neutrons requires a nuclear reactor. The cost of proprietary experiments ranges from \$1000–\$9000 per day [5.39].

5.4. Nondestructive Evaluation

Nondestructive evaluation (NDE) is a generic term that stands for characterization without destroying the material. Many of the NDE techniques can be used for green or densified components without the need for sample preparation, and some of them are developed to image only flat (but not necessarily polished) surfaces. NDE is used for process control as well as quality control of fabricated components.

Destructive evaluation techniques, for example, tensile testing, require samples of a given dimension specially prepared for testing. Sample preparation from ceramic end products is cumbersome and very costly. Destructive evaluation inhibits the further use of a tested sample. Nondestructive techniques, on the contrary, do not harm samples and may allow 100% control of production, if necessary. Another advantage of NDE is that it allows testing a single specimen several times, thus making it possible for example, to monitor the effect of a processing method progressively at various stages.

5.4.1. X-Ray Imaging Techniques

5.4.1.1. X-Ray Radiography and X-Ray Microscopy (Microradiography)

Conventional X-ray radiography is a method based on differential absorption of X rays by different materials or media. Metallic inclusions down to 25 μm in ceramics can be detected readily by this method since the X-ray absorption of dense ceramics is significantly lower than that of metals. Pores, cracks, and undensified regions result in a smaller difference in absorption; the lower limit for such defects is about 100 μm [5.40]. In radiography, the whole section is viewed at once on a film or screen. High resolution radiography, or microradiography, employs microfocus X-ray radiation with an X-ray source $\leq$ 100 mm in size [5.41] to project portions of the specimen onto a film and can provide resolution better than 1 μm [5.42]. Microradiographic contrast occurs upon differential absorption of X rays within a sample, as with other X-ray radiographic techniques. The resolution of the radiographic technique has been enhanced by the development of a hard X-ray microscope [5.43]. This microscope can detect line objects as fine as 0.6 μm wide in transmission images. An important application is the detection of internal flaws in materials. The cross section of a fiber-reinforced composite was successfully analyzed by this method.

5.4.1.2. X-Ray Computed Tomography

In this technique, a thin slice of radiation is used for imaging, as shown in Fig. 5.18. The beam passes through the material, and the transmitted intensity is measured by detectors. The part is rotated, and views are obtained from all directions. This information is collected by a computer, and a map of the density or thickness of the part is reconstructed. Figure 5.19 shows an image obtained by the CT technique in a ceramic part.

Computed tomography (CT) can be used to measure density or phase variations with reasonable sensitivity (0.1 to 1%). However, sharp density variations pose some difficulties, and very high density materials such as tungsten are not suitable for analysis [5.44]. Figure 5.20 is an example of CT capabilities employed in CMCs. An important advantage over conventional radiography is the superior large area, low-contrast image resolution, which is an indication of the ability to detect differences in density [5.45]. Dimensional measurement is also common with this technique; accuracies of ~0.25 mm can be achieved in components up to 0.5 m thick. An example of possible applications is the control of extruded ceramic parts. For

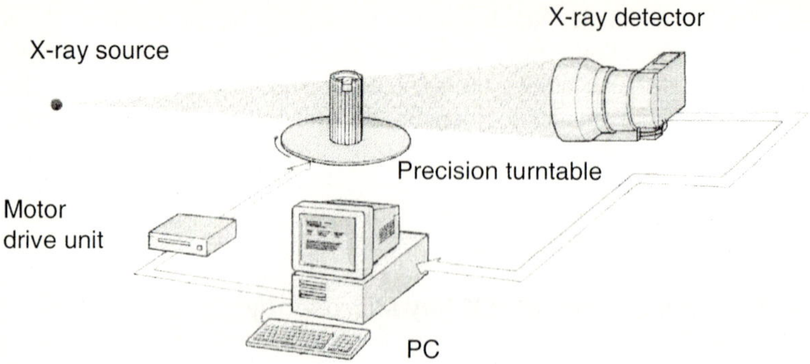

Fig. 5.18. Schematic view of X-ray computed tomography (figure courtesy of AEA Technology plc Oxon UK)

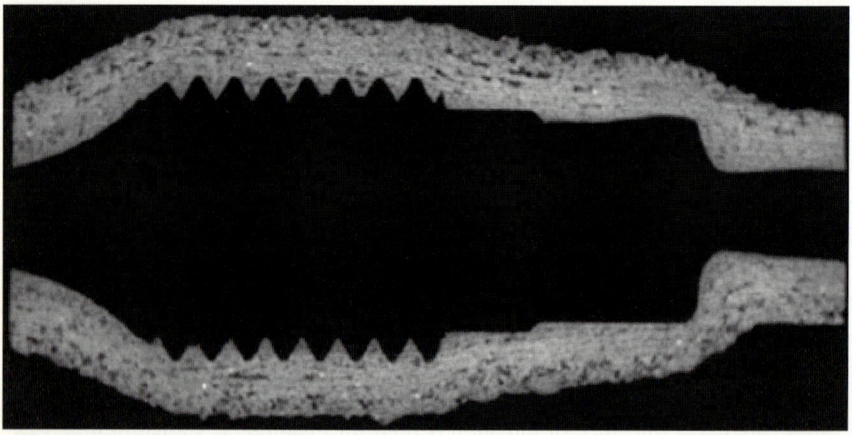

Fig. 5.19. Example of CT imaging in a ceramic mold (figure courtesy of AEA Technology plc Oxon UK)

instance, extruded CMC cutting tool bits often contain cracks or extended voids, which can be detected by computed tomography [5.43]. Note, however, that computed tomography is an extremely slow technique and cannot be used for routine testing in most production settings [5.40].

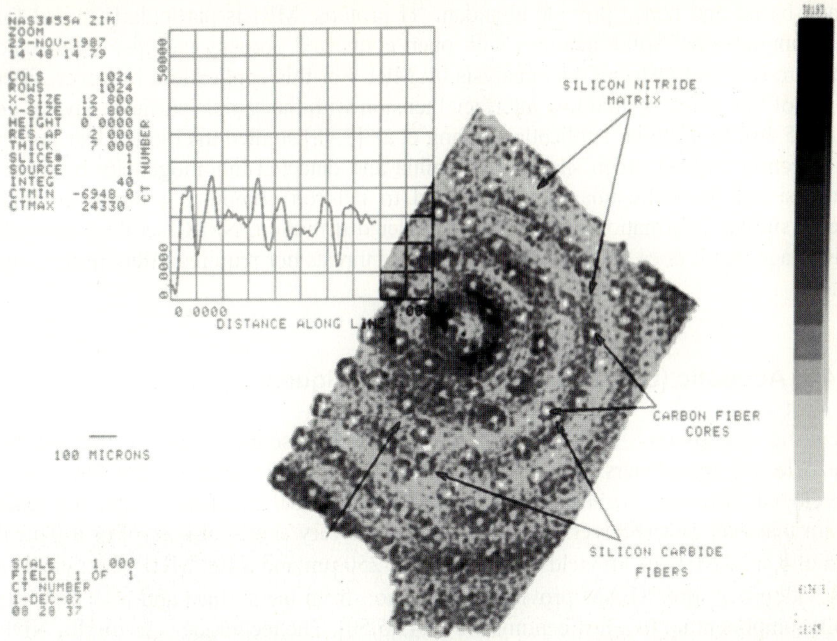

Fig. 5.20. Example of CT capabilities employed in CMCs. The image clearly differentiates high-density SiC fibers and their low-density carbon cores from the Si_3N_4 matrix (courtesy of Advanced Research and Applications Corporation-ARACOR Sunnyvale CA)

5.4.2. Nuclear Magnetic Resonance Imaging

Also called *magnetic resonance imaging (MRI)*, this is a useful technique for monitoring green forming operations such as slip casting and injection molding. Nuclear magnetic resonance (NMR) is a quantum mechanical phenomenon exhibited by atomic nuclei that have an odd number of photons or neutrons. NMR exploits the interaction of nuclear magnetic moments with electromagnetic waves in the radio-frequency region. The sample is placed in a magnetic field and radio-frequency electromagnetic radiation is applied [5.37]. Upon application of an external magnetic field and a smaller pulsed electromagnetic field, particular nuclei absorb this radiation at resonance frequencies which are modified by neighboring atoms, causing an NMR signal to be generated. The signal is spatially encoded to form an image. In MRI, sectional images are obtained that are similar to X-ray computed tomography images.

MRI has been successfully used for near real-time observation of the stages that occur during the slip casting of Al_2O_3 [5.46] and for the NDE of slip cast fiber and whisker-reinforced Al_2O_3 green parts [5.47]. Flaws such as porous regions and fiber or whisker bundles can be observed by this technique. Since most ceramic slips are water-based and hence provide abundant ^{1}H protons, MRI is particularly suited for such applications. Solid ceramics with open connected porosity can also be imaged for pore-size and distributional analysis by MRI. For this application, however, filler fluids of low viscosity and low interfacial tension with the sample surfaces have to be used [5.46]. In a similar application, Wang et al. [5.48] studied the binder distribution in injection molded green shapes of alumina and detected inhomogeneity by three-dimensional ^{1}H-NMR imaging. Compared to neutron computed tomography that yields similar information on water and binder distribution, NMR has the important advantage that it does not pose radiation hazard threats nor requires safety restrictions [5.49].

5.4.3. Acoustic (Ultrasonic) Imaging Techniques

Acoustic microscopy techniques employ ultrasonic sound waves generated by piezoelectric transducers. At least four types of acoustic microscopes have been developed. *Scanning laser acoustic microscopy (SLAM)* involves a scanning laser beam that acts as a receiver. It can be focused to very fine spot sizes. For a typical ceramic, a 10 MHz beam yields a resolution of 250 μm and a 100 MHz beam yields a resolution of 5 μm. SLAM provides information from the surface and also from the bulk component up to a few centimeters [5.42,5.50]. The technique is useful for NDE of surface and internal defects or inclusions.

Scanning acoustic microscopy (SAM) is another technique that is based on the reflection of surface ultrasonic waves to provide information about surface flaws. *C-SAM* is a subsurface ultrasonic wave reflection-based technique for detecting flaws in surfaces and in thin ceramic components. A complete ultrasonic scan is possible with scanning systems using C-scan displays. The smallest detectable defect using focusing probes has a diameter of 20 μm [5.51]. *Scanning photoacoustic microscopy (SPAM)* is a technique similar to SLAM where a laser beam is focused on the specimen and scanned. In SPAM, however, the detection of signals is carried out by a miniature microphone and an amplifier. This technique can be applied to complex-shaped components. For example, internal flaws 20–30 μm in size were detected in ceramic turbine blades by SPAM [5.52]. SAM has been used for estimating Young's modulus of SiC fibers from Rayleigh wave fringe patterns observed within subregions [5.53]. Another application worth mentioning is the examination of Vickers indentation-induced damage in Y–TZP [5.54]. Acoustic imaging techniques are normally applicable to flat surfaces and parts with constant cross section [5.3]. Complex shapes and rough surfaces represent problems in application. Another limitation is attenuation, i.e., scattering of waves as they pass through the material. Features such as pores and flaws that cause scattering increase attenuation. Attenuation limits the thickness that can be effectively imaged.

5.4.4. Birefringence

5.4.4.1. Optical Birefringence

When optically isotropic and transparent materials such as glass are subjected to mechanical stress, they become optically anisotropic due to the fact that the velocity of light varies in the stressed regions (see also Sect. 4.3.3.). This behavior, called birefringence, is easily observed by placing the material under stress between polarized plates so that the polarization planes are perpendicular [5.40]. Thus, light polarized in one direction is not allowed to pass through the other polarized plate. As a result, stressed areas of the material in between become visible: the color of the stressed region changes according to the stress level. This method can be used for detecting inclusions or cracks by their associated strain field and residual stress as little as 55 MPa (in Pyrex™ glass).

5.4.4.2. Ultrasonic Birefringence

Application of a stress state with unequal principal (i.e. nonhydrostatic) stresses results in introducing anisotropic acoustic behavior into otherwise isotropic materials. This effect can be used to detect discontinuous or residual stress. This method can be used in optically opaque materials, unlike the optical birefringence method, as well as in optically transparent materials. Ultrasonic birefringence can be measured in three different ways. Time of flight of ultrasonic waves measured as a function of propagating direction yields stress states of the entire sample. Shear wave velocities measured as a function of the shear wave oscillation plane yield information on smaller sections. Surface stresses can be measured using the velocity of Rayleigh waves as a function of direction [5.40].

5.4.5. Penetrant Techniques

Fluorescent or high-contrast dyes can be used to detect surface flaws (cracks or pores) of ceramics. The method simply involves wetting the part with the penetrant and observing the surface. Certain penetrants are highlighted by infrared illumination in special observation rooms. Special opaque penetrants can also be used in X-ray radiography to increase the contrast and resolution of cracks and pores exposed to the surface [5.40]. The method is not suitable for porous ceramics since absorption of the penetrant by the pores would mask other types of surface flaws [5.49]. Nevertheless, it can be used for routine inspection of nonporous ceramics.

5.5. Chemical and Compositional Analysis

A large number of analytical techniques exist, and new ones are continuously added to the list. This section deals with important analytical techniques especially suitable for ceramic materials and processes.

5.5.1. X-Ray Emission Spectroscopy

If elements in a material emit characteristic X rays upon electron or X-ray bombardment, these X rays can be used to identify the elements. Two such techniques are called *electron probe X-ray microanalysis (EPMA)* and *X-ray fluorescence (XRF)*, each of which are discussed here. Two spectroscopic methods for characterizing emitted X rays are *wavelength dispersive spectroscopy (WDS)* and *energy dispersive spectroscopy (EDS)*, which are explained in Sect. 5.5.1.3. Either spectroscopic method can be used in the two X-ray evaluative techniques mentioned.

5.5.1.1.Electron Probe X-Ray Microanalysis (EPMA)

EPMA is a microanalytical technique that couples electron-beam methods and X-ray microanalysis by spectrometers. Many SEM's are equipped with a WDS or EDS for chemical analysis, using X rays generated by the electron beam–sample interaction. Figure 5.21 shows X-ray maps of Al and Zr atoms in an Al_2O_3–10 vol%ZrO_2 composite obtained by EPMA (with an EDS coupled to an SEM). The distribution of Zr (white spots in Fig. 5.21b and dark spots in Fig. 5.21a) indicates the position of ZrO_2 grains in the Al_2O_3 matrix. Analytical electron microscopes (AEMs) are similar to SEMs: the important difference is the use of an electron transparent sample. AEMs are also coupled with emission spectrometers for chemical analysis of small regions, 2–20 nm in diameter [5.55]. In addition to EDS, energy loss electron spectroscopy (EELS) is usually also available in most AEMs [5.22].

5.5.1.2. X-Ray Fluorescence (XRF)

The interaction of X rays with a sample produces X rays as well as photoelectrons. X-ray detectors (EDS and WDS) can be used to analyze X rays produced by incident X-ray beams. Low atomic weight elements (H, He, Li) cannot be detected with XRF, whereas all other elements are detectable. Although the diameter the beam is quite large (150 μm to 1 mm), this technique has the advantage of eliminating charging problems encountered in X-ray emission spectroscopy. The beam can also be focused to achieve a microbeam, yielding a lateral resolution as low as 10 μm. XRF is useful for bulk compositional analysis of green or sintered compacts [5.56].

5.5.1.3. Spectrometers for X-Ray Emission Spectroscopy

WDS analyzes elements in a sample by determining the angles at which Bragg's law is satisfied, as the whole angular range is scanned. Detection limits are 100 ppm in WDS and 1000 ppm in EDS. The resolution of WDS is typically 5 – 10 eV, which is much better than the resolution of EDS (150 eV). Therefore, more peaks for certain elements can be distinguished by WDS. Furthermore, elements with atomic numbers as low as that of Be (Z=4), can be detected by WDS, whereas EDS spectrometers can detect elements with atomic numbers as low as 5 only if windowless detectors and high quality vacuum systems are used. The main disadvantage of a WDS system is that chemical analysis is slow, usually a few hours, compared to a few seconds required in EDS.

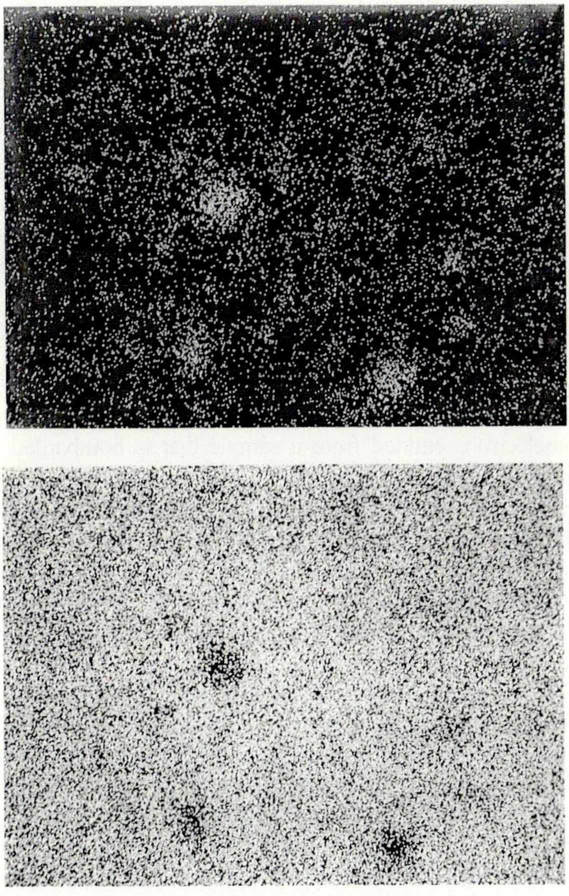

a

b

Fig. 5.21. EDS X-ray maps: Al map **a** and Zr map **b** in Al_2O_3–10%ZrO_2 composite (From M. Bengisu, Ph.D. Thesis, New Mexico Institute of Mining and Technology Socorro NM, 1992)

EDS analysis has the advantage of detecting the total spectrum of interest from 0.75 to 20 keV without the need for scanning. Most EDS instruments use Be windows to separate the spectrometer from the vacuum system of the SEM or AEM. Since this window absorbs low energy X rays, it limits the detectable atomic number to 10 and higher. Windowless EDS detectors are also available which allow the detection of B and heavier elements. However, at energies below 1 keV, which are required for windowless systems, efficiencies of X-ray generation and EDS detection are limited. The lateral resolution of EDS in SEM is 1–2 μm, but in STEM of electron transparent samples, this becomes as small as 1 nm [5.39]. The spatial resolution is ~1 μm due to electron beam spreading effects within the specimen. A 300-kV analytical TEM with a field emission gun was constructed to carry out high spatial resolution analysis on a nanometer scale. The field emission microscope yields a fine electron probe of about 0.5 nm in diameter. This instrument was used to study advanced ceramic materials, including Si_3N_4, $InFeO_3(ZnO)_{13}$, and cubic B–C–N compounds [5.57].

Some quantitative analytical techniques exist which can be used along with elemental analysis. Many of these techniques have been incorporated in microcomputer-based X-ray analysis units. These techniques allow only calculation of the relative amounts of individual elements, not phases. This limits their application to ceramics that are multielemental materials (except for C), unlike metals. Furthermore, all ceramics contain one or more element with $Z<10$, which cannot be detected with most EDS systems. Nevertheless, from relative amounts of metallic species and assumption of equilibrium compounds, one can, for example, determine the relative fraction of different metal oxides in a glass [5.58].

5.5.2. X-Ray Photoelectron Spectroscopy (XPS)

This technique uses photoelectrons emitted from a sample that is bombarded with low-energy (soft) monoenergetic X rays, such as the K-α line of Al or Mg. It is also known as *electron spectroscopy for chemical analysis (ESCA)*. Figure 5.22 illustrates the formation of photoelectrons and Auger electrons (see Sect. 5.5.3) in a material from electron or X-ray bombardment. Most of the useful information is obtained from the top one to ten monolayers of the specimen. Therefore, XPS is mainly a surface analytical technique. Detection and processing of electrons are similar to Auger electron spectroscopy (AES). In XPS, the energy distribution of photoelectrons is measured. The resultant spectra give the kinetic energies of emitted photoelectrons that are compared to specific binding energies of the electrons in the atom. This technique is especially valuable due to its ability to detect all elements except H and He without the need for standards. Lateral resolution ranges from 5 mm to 75 μm and can be improved down to 5 μm in special instruments [5.39]. One of its advantages over AES is the absence of charging from X-ray bombardment of insulating ceramics, unlike the possibility of charging during electron bombardment, a limiting factor in AES.

A typical application of XPS is the analysis of oxidation on fiber, whisker, and powder surfaces, which have significant implications in processing and composite properties. For example, Homeny et al. [5.59] and Campbell et al. [5.60] studied the chemical nature of SiC whisker surfaces by XPS and their effect on composite properties. Figure 5.23 shows XPS spectra of whisker surfaces for two types of whiskers and the detection of SiO_2 in one of them. Holloway [5.61] studied the effect of chemical species on the optical properties of naturally occurring oxide and sulfate minerals. Surface characterization of silicon nitride and silicon carbide powders was undertaken by XPS by Rahaman et al. [5.62]. XPS has also been used to study the electronic and chemical structure of electronic ceramics [5.63,5.64].

5.5.3. Auger Electron Spectroscopy

In this method, an electron beam (1–10 keV) is used to generate Auger electrons from a shallow surface region, usually thinner than 5 nm. Due to this shallow interactive depth, there is little possibility for beam spreading. Thus, the spatial resolution is a direct function of the beam size, which can be as small as 500 nm [5.64]. By combining ion beam sputtering with Auger electron spectroscopy (AES), thicker layers can be analyzed. Elements with less than three electrons (H and He) cannot emit Auger electrons, and thus cannot be detected. However, light elements, including O, C, and N are easy to detect by AES. The lateral resolution of AES is about 300 Å. Auger electrons are characteristic of the atom from which they are emitted. Thus, identification of elements is possible by detecting electrons according to their energies. Insulating ceramics are prone to charging, which may limit the use of AES. Nevertheless, AES is a valuable tool in surface analysis of ceramics. Very thin oxide layers can be detected by this technique. For example, surfaces of as-received whiskers and those incorporated into glass–ceramics were analyzed by AES, revealing C-enrichment as a result of composite fabrication [5.65]. AES was also used to detect grain boundary impurities in MgO, Al_2O_3, and $MgAl_2O_4$ ceramics [5.66]. The corrosion of glass surfaces was studied by AES and ion beam sputter profiling [5.67,5.68].

5.5.4. Secondary Ion Mass Spectrometry

The bombardment of a surface with a 1–25 keV ion beam sputters off the surface layers and produces various entities, including neutral atoms and molecules, secondary electrons, photons, and ions. In secondary ion mass spectroscopy (SIMS), secondary ions are detected via mass spectrometers. Most of the secondary ions are emitted from the top two atomic layers. SIMS can detect every element in the periodic table, as well as their isotopes. This feature makes it a complicated technique. The capabilities of SIMS are strongly related to the type of related equipment [5.69]. Trace impurities can be detected with certain mass spectrometers. Insulating surfaces are subject to charging, and the technique is locally destructive. SIMS has been used for diffusional analysis of Ti in $LiNbO_3$ and characterization of fiber–matrix interfaces in CMCs [5.70,5.71]. Figure 5.24 shows the SIMS analysis of a SnO_2–coated PRD fiber.

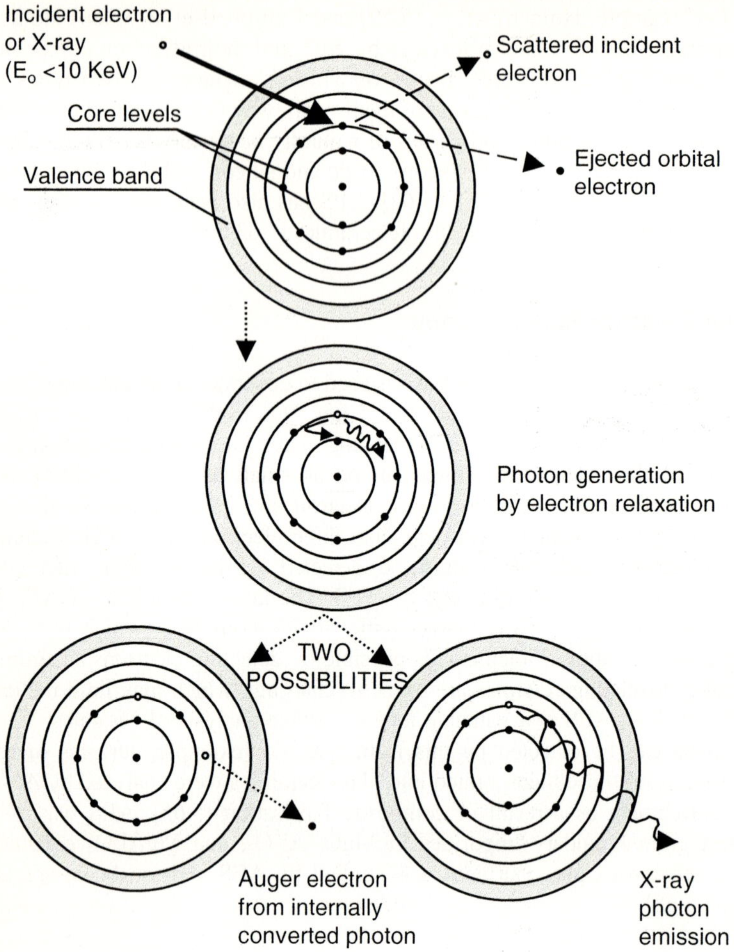

Fig. 5.22 Schematic illustration of X-ray photon and Auger electron emission during deexcitation after initial ionization by X-ray or electron bombardment

5.5.5. Laser Ionization Mass Spectroscopy (LIMS)

This technique is based on material removal and ion emission from surfaces by a laser beam. The beam diameter is usually about 1–2 μm and penetrates several monolayers. All elements from H to U can be detected by LIMS. However, the detection limits may vary significantly from one element to another. For example, the

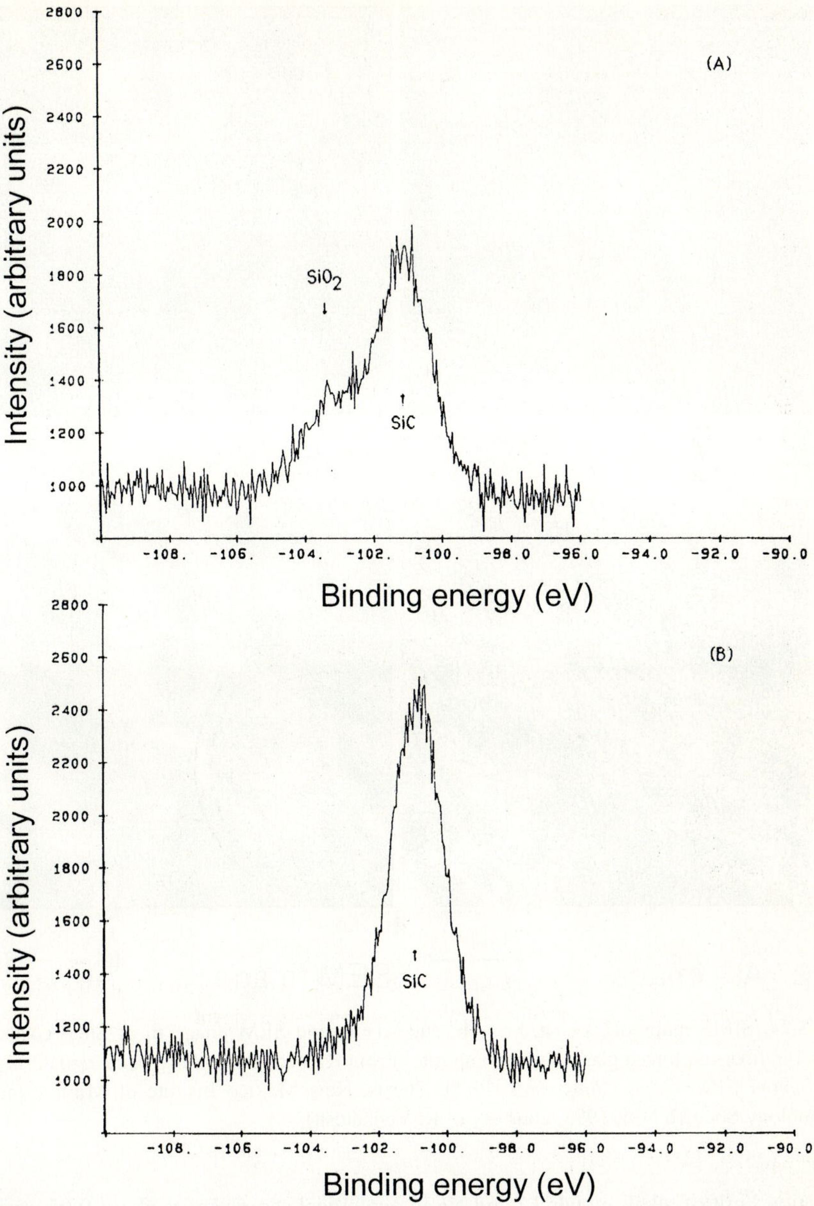

Fig. 5.23. XPS (ESCA) of two SiC whisker surfaces. Si (2p) line spectrum for Silar-SC-9 whisker surface **a** and Tateho-SCW-1-S whisker surface **b** showing the presence of SiO_2 in the former (From J. Homeny and W.L. Vaughn: *Whisker-Reinforced Ceramic Matrix Composites*. MRS Bulletin, 12 [7] 66–71 (1987), reprinted with permission of the Materials Research Society Warrendale PA)

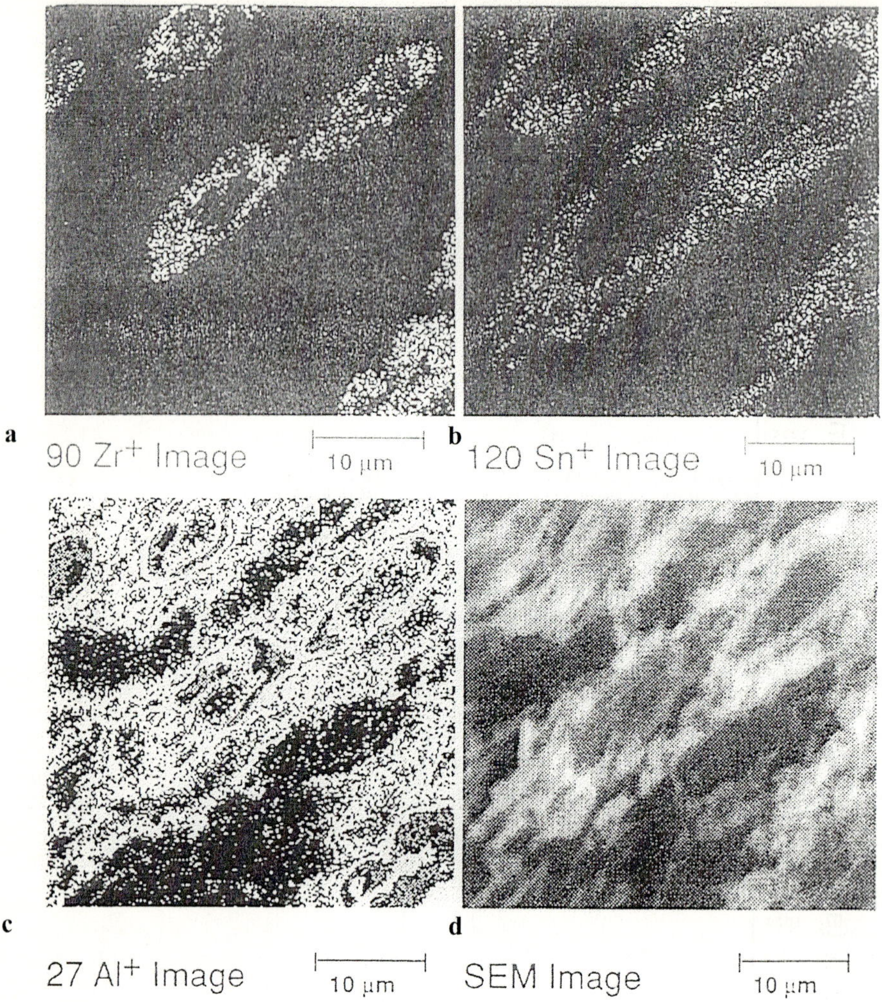

Fig. 5.24. SIMS maps of Zr+ **a**, Sn+ **b**, and Al+ **c**, and SEM image **d**, of SnO_2 coated PRD-166 fiber-reinforced glass matrix composite (From R. Venkatesh: *Interface Engineering of Alumina Fiber/Glass Composites*. Ph.D. Thesis. New Mexico Insitute of Mining and Technology Socorro NM, 1993, courtesy of R. Venkatesh)

detection limit of alkali metals Li and Na in geological samples was about 0.05 ppm in Scott et al.'s studies [5.72], but those of Si and Ti were 20,000 and 50,000 ppm, respectively. The use of a graphite binder in pelletized samples decreased these limits to 200 ppm in both cases. LIMS is one of the few techniques that can analyze adsorbed species on surfaces. For this purpose, the laser beam has to be operated below the ablation threshold [5.66].

5.5.6. Rutherford Backscattering Spectroscopy (RBS)

RBS is an analytical tool that employs MeV level He^{2+} or He^{+} ion acceleration. Ions penetrate into the target where they collide with the surrounding atoms and are then backscattered and collected by surface barrier detectors (Fig. 5.25). Information from a layer up to 10 μm thick can be achieved with a depth resolution of 2–30 nm and a lateral resolution of 1–4 mm (1 μm in specialized equipment) [5.39]. Carbon and heavier elements can be detected [5.73], although the technique is much more sensitive to heavy elements (~100 ppm) compared to light elements (1–20 at.%) [5.69]. Energy loss of helium ions occurs through electronic interactions with solid atoms and small-angle collisions with the target nuclei [5.74]. The technique has the ability of standardless quantitative elemental analysis and depth profiling. Both insulating and conducting ceramics can be analyzed by RBS. Some application examples include studies of glass corrosion [5.75] and chemical composition and annealing studies of superconductor ceramics [5.69,5.75].

5.5.7. Electron Energy Loss Spectroscopy (EELS)

This technique is used in analytical electron microscopes (AEMs). A magnetic prism spectrometer is positioned on the viewing screen of the AEM, which sorts incoming electrons according to their energies. Some electrons of the incident electron beam lose energy due to impact with adsorbed molecules or atoms on the surface, causing vibrational and electronic transitions (ionization) in these molecules or atoms [5.33]. The most common EELS technique is based on detecting electrons, that lose energy by this mechanism. The kinetic energy loss at the other side of the electron transparent specimen is then measured. A lateral resolution of 1 nm to 10 μm is achievable, depending on sample thickness and incident beam diameter. The minimum number of detectable atoms increases as the atomic number decreases. Elements with $Z \geq 3$ (Li) can be detected by EELS. Elemental mapping is also possible by this technique [5.76]. Whisker-matrix interfaces of Al_2O_3–SiC_w composites were analyzed by EELS to investigate the possibility of light element (O, C, N) segregation [5.15].

5.5.8. Fourier Transform Infrared Spectroscopy (FTIR)

FTIR, a branch of infrared spectroscopy, has the important advantage that it does not require vacuum or energetic surface bombardment. FTIR involves measuring the intensity of a beam of infrared radiation before (I_0) and after (I) it interacts with the sample, as a function of light frequency (w_i). From the plot of I/I_0 versus w_i, and comparison with reference standards, the identities, surrounding environments, and concentrations of chemical bonds that are present can be determined. Chemical bonding of matter vibrates with frequencies in the infrared regime. Oscillations induced by certain vibrational modes make it possible to couple with the incident infrared beam and to exchange energy with it when their frequencies are in

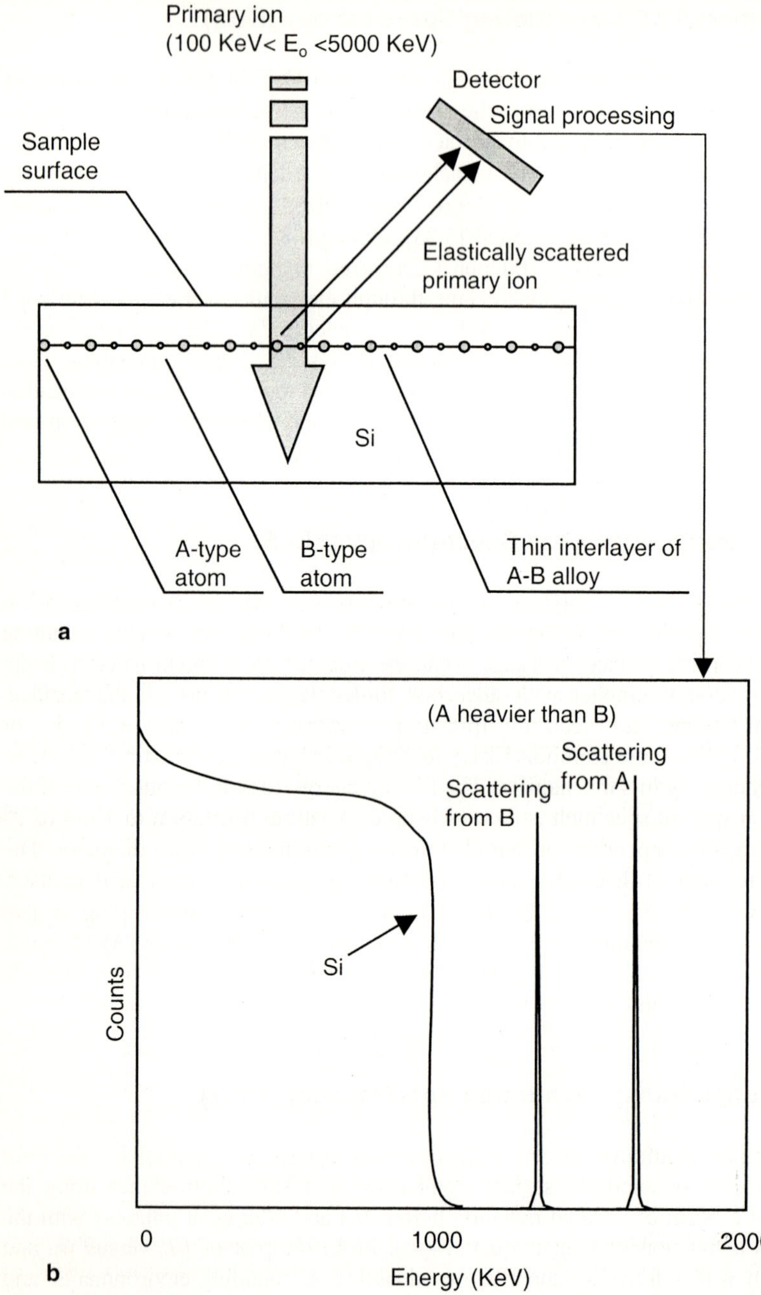

Fig. 5.25. Schematic illustration of Rutherford backscattering spectroscopy **a** and typical RBS signal from A and B type elements embedded in a Si matrix **b**

resonance. Identification of functional groups of chemical bonding is possible by the FTIR method. Solid, liquid, or gas samples can be analyzed. A lateral resolution of 5 mm to 20 μm is possible, depending on the sample [5.39]. Qualitative and quantitative determination of chemical species in solid or thin film samples are the main applications. Figure 5.26 shows the IR spectrum of pyrolytic products of polycarbosilane powders [5.77]. FTIR is suitable for analyzing the products of binder burnout processes before sintering of ceramic components and for better defining the burnout process for specific binders [5.56]. In situ observations of sample–environment reactions can be made. A typical application is the corrosion of glass surfaces [5.78]. Other examples of the application of FTIR are surface species analysis of ceramic powders [5.79] and measurement of absolute infrared reflectivities of ceramic materials [5.80].

5.5.9. Raman Spectroscopy

Raman spectroscopy involves a single line of a continuous gas laser or a suitable light source for exciting vibrations in molecular and crystalline materials and measuring the inelastic light scattering that results from this excitation. The incident light exchanges a quantum of energy with the crystal by creating or annihilating phonons (crystal vibrations). The scattered light loses or gains an energy quantum, respectively. Raman spectroscopy measures the energy shift of the scattered light. The information gathered by this technique is about characteristic values of vibrational frequencies of chemical bonds. Identification of unknown compounds in solid and liquid phases, characterization of structural order, and phase transitions are the main applications. A lateral resolution of 1 μm can be achieved by this technique. A recent review [5.81] demonstrates the use of Raman spectroscopy in characterizing the band configuration, stress state, and thermal stability of wide-band gap semiconductors, as well as in obtaining information about impurities and structural defects in a crystal matrix.

Typically, Raman spectroscopy is used only for qualitative analysis [5.39]. An important application is the identification of coexisting diamond and graphitic phases in diamond films. Bulk structural characterization of laser-synthesized BN powders was carried out by Raman spectroscopy in addition to other techniques [5.79].

Polarized Ar-laser light was used for Raman spectroscopy of AlN single crystals as well as polycrystalline samples. The connections between vibrational spectra and thermal properties were studied [5.82]. In another study, Raman spectroscopy was used to characterize laser-synthesized nanometric BN powders. Figure 5.27 shows the Raman spectrum of such powders, where h-BN peaks are detected [5.79]. Raman spectroscopy was also used to characterize binary and ternary ZrO_2 alloys [5.83]. No difference was observed in the Raman peaks as a function of increasing Y_2O_3 content in ZrO_2. In ternary systems, however, the Raman line corresponding to the stretching modes of two sets of cation–oxygen bonds shifted to greater wavenumbers as Nb_2O_5 or Ta_2O_5 content increased in

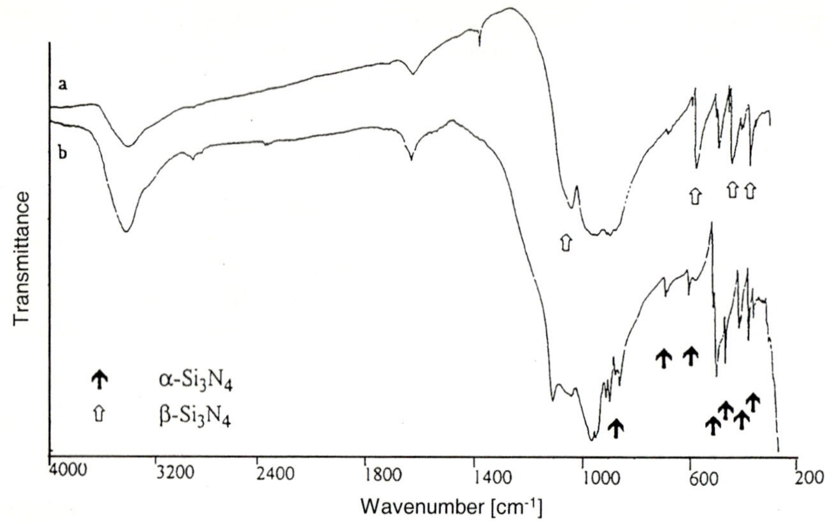

Fig. 5.26. IR spectrum of pyrolytic products of polycarbosilane powders. Presintered at 1700°C **a** and pyrolyzed at 1000°C **b** (From A. Greiner: *Darstellung von Si_3N_4/SiC Verbundkeramiken mit Nanokristallinen SiC-Einschlussen in Mikrokristallinen Si_3N_4-Kornern.* M.S. Thesis. Faculty of Chemistry University of Stuttgart Germany, 1993, courtesy of A. Greiner)

Y_2O_3-stabilized ZrO_2. The difference was attributed to local bonding environments of the pentavalent cations, which were supposed to occupy tetrahedral sites in t-ZrO_2 solid solutions.

5.5.10. Nuclear Magnetic Resonance (NMR) Spectroscopy

NMR imaging, which was discussed in Sect. 5.4.2, and NMR spectroscopy use the same source of nuclear information. They both exploit the interaction of nuclear magnetic moments with electromagnetic radio waves. The solid sample is placed in a strong magnetic field (normally 1–14 tesla) and is exposed to intense radio pulses. The radio pulses are varied over a frequency range necessary to excite a certain atomic nucleus from its ground magnetic (spin) state to a higher state. When the excited nucleus returns to its ground state, it releases a radio signal at the excitation frequency. This signal is detected by electromagnetic induction and transformed into intensity versus frequency information. Sample sizes may be as small as 10 mg in powder form, single crystal, or film. Applications include structural characterization of disordered states, phase identification, and quantification [5.40]. For example, Wang et al. [5.84] studied the binder–ceramic reaction behavior in injection molded green compacts of alumina. They used ^1H nuclear spin echo signals at 400 MHz and measured the nuclear spin–spin relaxation times, T_2, of protons in the binder components for this purpose. NMR spectroscopy was used to analyze the chemical structure of organic precursors used for the synthesis of SiC [5.77]. ^{1}H, ^{13}C{^{1}H}, and ^{29}Si{^{1}H}-NMR spectra were used for this purpose. Another application of NMR is the exploration of coordination numbers and angular distributions in a variety of glasses [5.37].

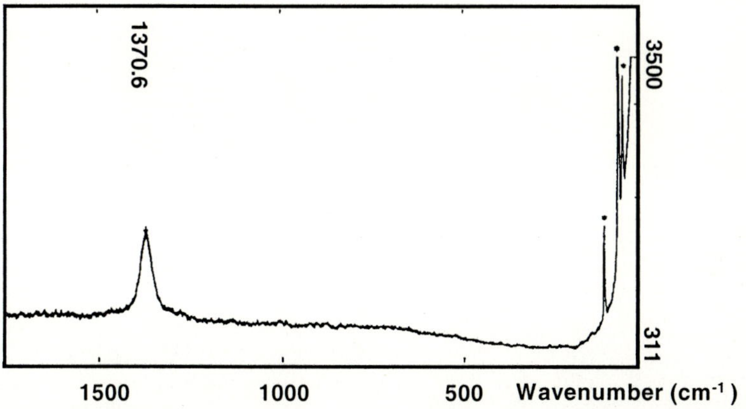

Fig. 5.27. Raman spectrum of nanometric BN powders from laser synthesis (* plasma lines) (reprinted from M.I. Baraton, L. Boulanger, M. Cauchetier, V. Lorenzelli, M. Luce, T. Merle, P. Quintard, and Y.H. Zhou: *Nanometric Boron Nitride Powders: Laser Synthesis, Characterization, and FT-IR Surface Study*. J. Eur. Ceram. Soc. **13 [4],** 371–378 (1994), with permission from Elsevier Science)

6. Applications of Ceramic Materials

The use of ceramic materials in science and industry is becoming increasingly widespread. As discussed in Chap. 4, ceramic materials have important advantages over metals and polymers in electronic devices at high temperatures or in corrosive environments. This chapter will briefly review structural, refractory, energy production, biotechnological, and electronic applications of ceramic materials. Research and development efforts will continue to expand the applications of ceramic materials. Thus, it should be noted that many applications that are termed "potential" here may soon become actual applications, and many new ones will be added to the current list of applications.

6.1. Structural Applications

Due to their brittle nature, monolithic ceramics are sensitive to defects that act as stress concentrators. Therefore, structural applications of monolithic ceramics are limited to parts that are subjected to compressive loading or limited tensile or multiaxial loading. Another important factor to consider is conditions that can lead to impact loading and thermal shock. Most monolithic ceramics are not suitable for these conditions. A well-known example is glass and porcelain ware commonly used in our daily lives. Dropping a glass or porcelain cup to the floor is enough to break it. Filling a glass with hot water is sometimes sufficient to cause fracture. Some special glasses with low thermal expansion have been developed for improved thermal shock resistance. Porous ceramics typically used as refractories and some dense, monolithic ceramics such as Si_3N_4 and SiC have good thermal shock properties. Ceramic matrix composites provide some tolerance to defects and dynamic or thermal loading.

Some examples of structural applications of ceramic materials are bearings, seals, armors, liners, nozzles, and cutting tools. Due to their current high cost, ceramic bearings and journals are used only for precision systems. Si_3N_4 balls are used in spindle bearings for cutting tools, turbomolecular pumps, dental drills, and specialty instrumentation bearings [6.1]. Boron carbide and single-crystal sapphire are used in bearings and seals [6.2]. Silicon nitride and sialons are being considered for gas-turbine bearings [6.3,6.4]. Advantages of such ceramics over conventional materials, e.g., steel, are their lower densities, which reduce the centrifugal load on the balls, higher resistance to wear, and superior high-temperature properties. Slide bearings made from siliconized SiC have been mass-produced since the 1980s [6.5].

A combination of high dielectric strength, corrosion resistance, and mechanical strength makes ceramic materials the choice for many applications, including electrical controls, lasers, radar systems, accelerators, high-power broadcast devices, fuses, night vision systems, high-voltage insulators, and special lamp housings (Figs. 6.1 and 6.2). Ceramic components used in various types of lasers such as excimer lasers, CO_2 lasers, and Nd-YAG lasers include different parts. Pump chambers, feed-throughs, insulators, laser waveguides, and ion laser tubes are some of them. Materials used for these applications include high-purity alumina and porous ceramics. Ceramics are preferred in such applications because of their electrical resistivity, corrosion resistance, thermal conductivity, high strength and hardness, and dimensional stability [6.6].

Ceramic materials such as alumina and silicon carbide are employed as faucet washers and counterfaces in mechanical seals for pumps and blowers (Figs. 6.3 and 6.4). The advantages of ceramic materials in such applications are their corrosion resistance, high hardness, low coefficients of friction, and minimum deposition. The seal is formed by a pair of rings, one stationary and one rotating, that work at a certain gap. The rings are elastically supported by the casing and the shaft. Sb- or synthetic resin-impregnated carbon is used as the wearing partner in hard/soft sliding material combinations. Mechanical seals for boiler feed pumps originally employed Ni-bonded tungsten carbide and Sb-impregnated carbon tribocouples. However, selective corrosion and accelerated abrasion resulted in high leakage rates. To solve this problem, the tungsten carbide rings were replaced with SiC–Si rings and Sb- impregnated carbon with synthetic-resin-impregnated

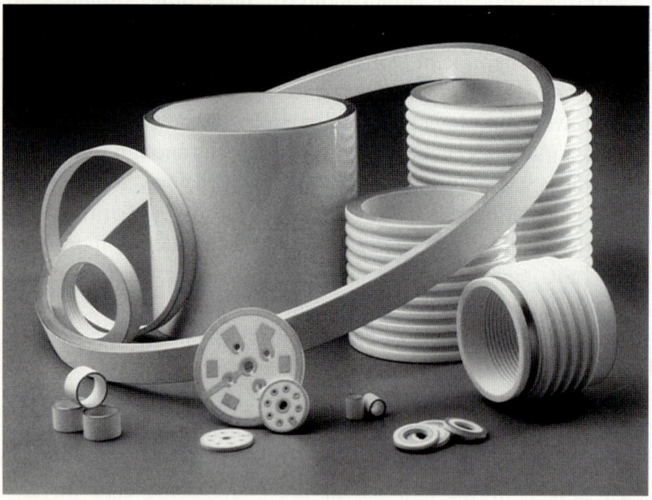

Fig. 6.1. Components for industrial electrical control: Feed-through insulators, brazed windows, absorbers, stand-off insulators (courtesy of Coors Ceramics Company Golden CO)

a

b

Fig. 6.2. Components for electrotechnical applications: Insulators for precision fuses, high-voltage fuses and spark chamber chips. Silica ceramic lamp bases and connectors for lighting systems. Ceramic components for precision engineering and instrumentation, environmental/heating technology as well as thermal controllers **a** and high-voltage insulators for transmission and distribution lines, traction lines, high-voltage apparatus and substations **b** (courtesy of CeramTec AG, Innovative Ceramic Engineering Plochingen Germany)

carbographite that contains a high-strength base carbon [6.7]. This combination enabled sliding speeds up to 52 m/s. In some HR-type seals, the steel casing ring was substituted by ZrO_2 to prevent corrosion and erosion [6.7]. Ceramic materials are preferred in high-performance valve applications for handling corrosive and erosive media such as brine, coal slurries, and drilling muds [6.1].

Fig. 6.3. Products for the fluid handling industry: Mechanical seal faces, rotary unions and faces, pump shaft sleeves, linings, and plungers, valve trim, meter bearings, fine bubble diffusers, faucet valve plates (courtesy of Coors Ceramics Company Golden CO)

Fig. 6.4. Ceramic parts for fuel and water pumps, valves (e.g., inlet, outlet, fuel injection, or exhaust gas control valves), valve drives (e.g., shims), bearings (e.g., bypass exhaust gas valves). Ceramic sealing and control discs for sanitary faucets and taps, industrial fittings, valves (e.g., espresso machines), dishwasher pumps, special designs (courtesy of CeramTec AG, Innovative Ceramic Engineering Plochingen Germany)

Hot pressing and forming dies are well-developed applications. Graphite dies are used almost exclusively under inert conditions because they are easy to machine and they have excellent properties including structural–mechanical stability up to 2000°C, very low sliding resistance that provides ease in part–die separation, and inductive-coupling capability. Al_2O_3, BN, and SiC are used as rolling components in steel plants [6.8]. Tube- and wire-drawing dies are made from Al_2O_3 and PSZ (Figs. 6.5 and 6.6) [6.2]. Al_2O_3 guides are used successfully in the textile industry [6.8] (Fig. 6.7).

Some applications of ceramic materials emerged especially because of their resistance to *corrosive environments*. For example, Si_3N_4 has been successfully substituted in stellite molds used to press carbon dollies for zinc–carbon batteries. The hoop stresses in the mold prevented the use of Si_3N_4 alone, so a titanium outer sleeve with the original external dimensions was used with the ceramic insert. The substitution of Si_3N_4 for stellite increased the life of the mold from 10,000 to 50,000 pressings. Further refinements increased this number to about 100,000, thus proving it feasible [6.10]. Corrosive and abrasive environments occur in the pulp and paper industry. Figures 6.6 and 6.8 show some ceramic components used in this industry. Ceramics are also frequently used for molten metal handling. Graphite and carbon are very resistant to wetting and corrosion, except with metals that form carbides. Al_2O_3, BN, Si_3N_4, and MgO are commonly employed for handling molten metal and glass. The specific chemical characteristic of each of these materials was discussed in Sect. 4.6. A relatively new application in this category is in beverage dispensers used by major soft drink companies and fast food restaurants. Ceramic materials are used in

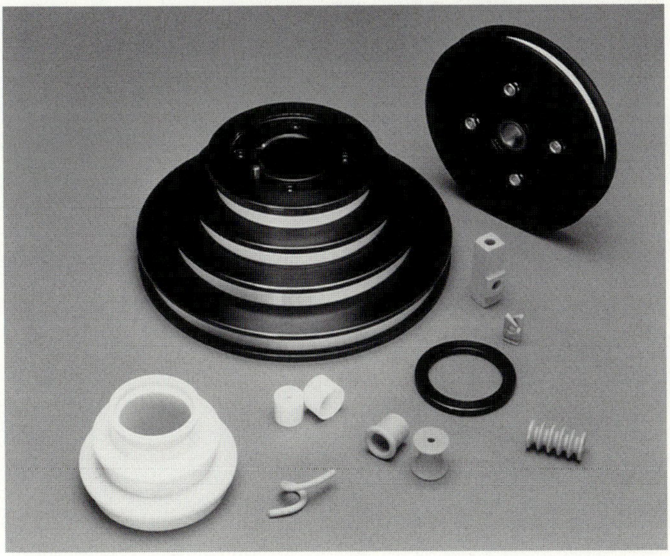

Fig. 6.5. Products for the wire manufacturing industry: Capstans and drawblocks, pulleys and sheaves, guides, rolls, dies (courtesy of Coors Ceramics Company Golden CO)

Fig. 6.6. Ultra-wear-resistant forming tools for wire-drawing machines such as cones and capstans from oxide ceramics/steel composite, wire guide elements. Tools for tube drawing. Parts and components for machinery and apparatus in general: ultra-wear-resistant rollers, ceramics/steel composites, precision parts. Porous silica ceramics for local reinforcements in lightweight structures. Linings for the paper industry (courtesy of CeramTec AG, Innovative Ceramic Engineering Plochingen Germany)

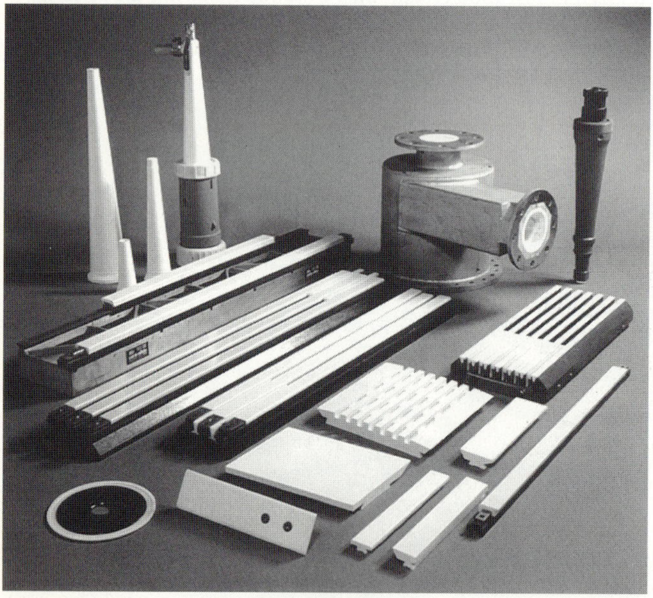

Fig. 6.7. Applications in the textile industry: Friction texturing, intermingling, ring spinning, rotor spinning, thread guides, cutting blades (courtesy of CeramTec AG, Innovative Ceramic Engineering Plochingen Germany)

Fig. 6.8. Products for the pulp and paper industry: Segmented and continuous forming boards, foils, suction box covers, and felt cleaning devices. Stock preparation equipment. Dewatering element support structures. Slitting and sizing knives (courtesy of Coors Ceramics Company Golden CO)

pistons and cylinders of soft drink beverage valves. These valves, which have to withstand both corrosion and erosion, mix CO_2 and syrup [6.6].

Nozzles used for welding and oil burners have to withstand high temperatures and thermal shock. Such nozzles have been produced from reaction-bonded silicon nitride [6.5]. Ceramics are also preferred over other materials due to their hardness, corrosion resistance, wear resistance, and electrical insulation in various mineral processing applications. Examples are handling of ores, slurries, pulverized coal, cement clinker, and flue gas neutralizing compounds (Fig. 6.9).

Ceramic membranes are used in the field of separation. Advantages over other techniques such as distillation, centrifugation, and sedimentation include

1. continuous separation capability,
2. relatively low energy consumption,
3. separation capability at ambient or process temperature,
4. possibility of adding membrane units to an existing process or to each other.

Ceramic membranes are more expensive and less fracture resistant than their organic counterparts, but they have advantages when chemical and physical demands prevail. Examples of such areas are the dairy industries; fruit juice, beer, and wine processing; and chemical and biotechnical applications. To separate oil from water, especially when contaminated with corrosive chemicals or abrasive materials, the best choice is ceramic membranes. Usually, alumina membranes are used for these microfiltration applications [6.11].

Fig. 6.9. Products for the mineral processing industry: Pipe linings, cyclone linings, pump components, burner components, electrostatic precipitator insulators, spray nozzles for flue gas desulfurization, grinding media, fan blade composites, combining tubes, wear-resistant panels, small media mill impellers, waste treatment diffusers, exhauster linings, transfer chute linings (courtesy of Coors Ceramics Company Golden CO)

6.2. Military Applications

High-performance ceramics are an integral part of modern weapons and defense systems. Electronic and optical ceramics are used in missile guidance systems, aircraft, and military ground vehicles. Most military radar communication systems use ceramics. For example, the radar in the Patriot missile system uses ceramic components as sensors. Ceramics are increasingly being applied in military aircraft. The NATO C-130 transport aircraft that flew over Bosnia employ Kevlar/glass ceramic cockpit armor. Ceramics have also been used in equipment for military maneuvers, for example, in Apache helicopters. Lightweight ceramics are used as armors in many modern military attack helicopters [6.12]. C–C composite applications include components for missiles and military aircraft. Important applications are reentry bodies, rocket nozzles, and exit cones for strategic missiles; and brake disks for military and commercial aircraft [6.13]. Advantages of C–C composites are high specific strength, high-temperature strength, high toughness, superior thermal shock, ablation, and high-speed friction properties.

Ceramic materials have been used in *armor applications* since World War I. Probably the first appearance of ceramics in armor was due to the Germans who applied hard-faced enamel coatings to tanks to protect against small arms and shell fragments in World War I. In the World War II period, patents were granted for ceramic armor for aircraft, and materials such as doron (a material made of exceedingly tough spun glass fibers woven into fabric and impregnated with a hard plastic) were tested. During the Korean War, siliceous core armor (steel armor cast around a silicate core) was developed to protect tanks against cannonballs. During the Vietnam War, Al_2O_3 or B_4C coupled with composites such as doron and glass-reinforced plastic were used for protection against small arms, including armor-piercing ammunition, in battlefield helicopters, and personnel aircraft [6.14]. During the 1980s, ceramic tiles were developed to protect ground vehicles against various kinetic and chemical energy threats, such as bullets, kinetic energy penetrators, and shape charges. Ceramic armor is also employed by specialist units in antiterrorist, counterrevolutionary operations, where snipers using high-powered rifles may be encountered.

Modern day experimental or manufactured ceramic armor usually contains ceramic matrix composites such as Al_2O_3–SiC_w, borosilicate glass reinforced with SiC or B_4C particles, LANXIDE™, and improved monolithic ceramics including Al_2O_3, B_4C, SiC, TiB_2, or AlN [6.15,6.16]. Ceramic materials are used in armor applications mainly because of their superior ballistic performance and low weight. Figure 6.10 shows the characteristic performance of a ceramic armor during penetrator impact. Typically, armor ceramics are used in combination with metal or polymer composites as backing plates. For example, a company produces ceramics plates that are used as the outer layer of lightweight composite armor for protecting police and military personnel. The hard ceramic plates can break up armor-piercing bullet, and the woven backing (fiberglass, Kevlar™ or Spectra™) catches the fragments. The assembly is approximately half the weight of steel, and it performs just as well [6.14]. The market for advanced ceramic armor is specialized and performance-driven due to high cost.

6.3. Cutting Tools and Abrasives

6.3.1. Cutting Tools

Widespread application of ceramic materials is in cutting tools. Ceramic cutting tools have high hot-hardness and do not react with the work piece materials. Advantages include long tool lives, machining at very high speeds, and very high metal removal rates. Some applications use ceramics bonded with a suitable binder, but many cutting tools consist of bulk ceramics. Examples of the former type are resin-bonded

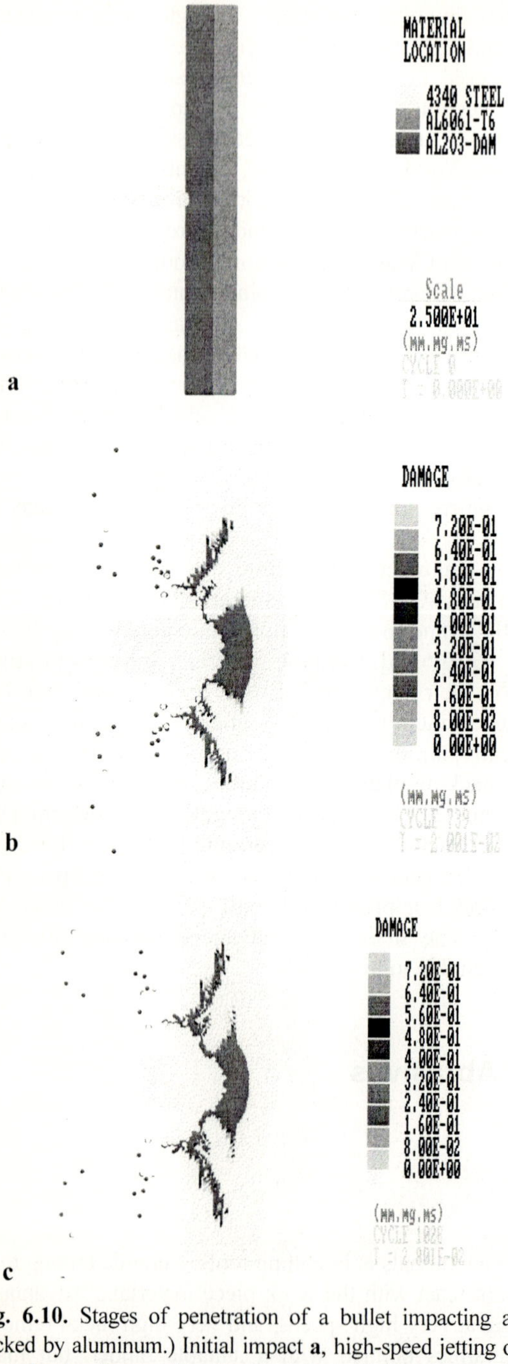

Fig. 6.10. Stages of penetration of a bullet impacting armor (composed of alumina plate backed by aluminum.) Initial impact **a**, high-speed jetting of ceramic debris and high levels of comminution below the penetrator **b**, and discrete cone crack formation in the ceramic **c**. (courtesy of Century Dynamics Limited West Sussex England)

diamond and SiC cutting and grinding wheels and cutting tips made from cemented carbides. Cemented carbides are hard carbides such as WC, TiC, TaC, and NbC, cemented together by a binder, typically Co. Additional coating by TiN, TiC, Al_2O_3 or multiple coatings may be used to improve cutting performance. Cemented carbides have a very broad application range. They can be used to machine the majority of steels, cast iron, Ni-based alloys, soft materials such as Al, or hard materials such as chilled cast iron and Ti [6.17]. Cutting tool inserts made from Al_2O_3 and Al_2O_3–TiC have been in industrial use for many years. Operations include machining of cast iron, hardened steels, and specialized machining of heat-resistant materials. In some cases, grinding can be eliminated after a turning operation. In recent years, many additional ceramic cutting tools have been developed. Among these are ZTA, SiC whisker-reinforced Al_2O_3, Si_3N_4, sialons, and PSZ.

The machining performance of Ce–TZP as a cutting tool was evaluated. Enhanced wedge retention and cutting performance was observed during high-speed turning of spheroidal graphite iron, which is attributed to transformation toughening and cyclic transformation of the ZrO_2 phase [6.18]. Figure 6.11 displays some ceramic and cermet cutting tool bits for machining steel and cast iron. The latter have found an important market in machining Ni-based superalloys in the gas-turbine industry [6.1].

Silicon nitride is the number one choice for machining gray cast iron. It has higher thermal shock resistance and toughness than unreinforced oxide ceramics. Polycrystalline diamond and cubic boron nitride (c-BN) cutting tools prepared by cladding a thin layer of ceramic onto a WC–Co substrate are the hardest tools available, but they are relatively expensive [6.19]. c-BN cutting tools are primarily intended for finishing operations on hard workpiece materials whose Rockwell-C

Fig. 6.11. Examples of cutting tools for steel and cast iron machining: SPK high-performance ceramic inserts, cermets and WURBON ® (c-BN) as well as customized tooling systems for turning, milling and boring (courtesy of CeramTec AG, Innovative Ceramic Engineering Plochingen Germany)

hardness (HRc) is over 48. Soft materials generate excessive wear. Excellent surface textures can be achieved with c-BN edges [6.17]. Diamond-clad (polycrystalline diamond; PCD) or diamond-coated cutting tools offer enhanced productivity [6.20], particularly in machining of Nomex honeycombs, SiC/Al and C–C composites, and composites containing Kevlar fibers [6.1]. An important application is turning and milling abrasive Si–Al alloys. Because of the very brittle nature of PCD, very stable conditions, rigid tools and machines, and high speeds are necessary for machining with them. PCDs are unsuited for ferrous metals. They should be considered for nonferrous metals and nonmetallic materials. Table 6.1 lists a summary of specific applications of various ceramic cutting tools.

6.3.2. Abrasives

Abrasives are hard materials used for cutting or abrading other materials. They are usually employed in grinding wheels and grinding tools with special shapes. Grinding wheels come in different standard sizes and shapes (Fig. 6.12). Abrasive grits are commonly bonded by phenolic resins, sodium silicates, shellac, rubber, vitrified bonding, and metallic bonding. Cloth, paper, or polymer substrates coated with abrasives are also available in belt or disk form. Some abrasive processes

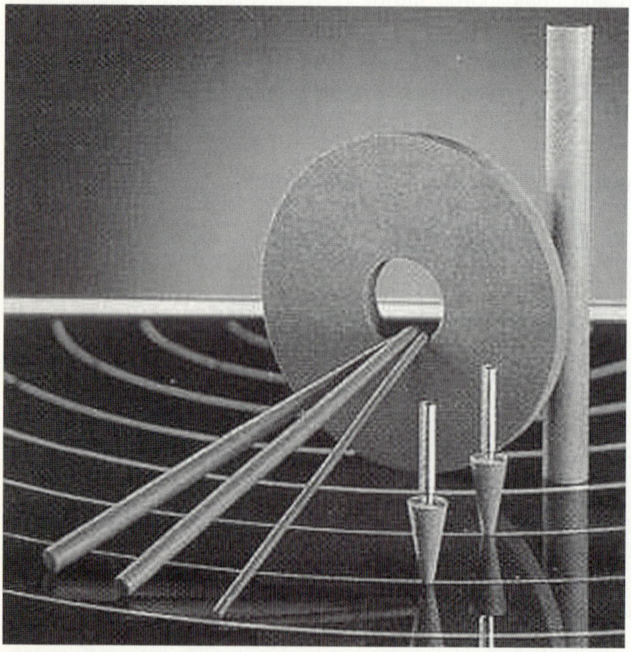

Fig. 6.12. Examples of FRIALIT® DEGUSSIT® grinding tools with different shapes and sizes (Courtesy of Friatec AG, Division FRIALIT® DEGUSSIT®/F310 Mannheim Germany)

Table 6.1. Specific Applications of Ceramic Cutting Tools

Material	Application	Ref. [a]
Al_2O_3	General turning, boring, and grooving of cast iron, tube scarfing	[1]
Al_2O_3–SiCw	Machining of high-temperature alloys	[2]
	Roughing and semifinishing of superalloys, steel, and cast iron, interrupted cuts	[3]
Al_2O_3–TiC	Turning hard cast iron and heat-treated steels hardened to 64 HRc to a very fine surface finish	[2]
	Small automotive gear parts to large form rolls for steel mills	[2]
	General turning, boring, and grooving of cast iron, turning of hard materials (under 65 HRc)	[1]
	Finishing of heat resistant superalloys, cast iron, and steel	[4]
Cubic BN	Turning of hardened steels (over 48 HRc)	[1,3]
	High-speed finish turning and milling of gray cast iron	[1,3]
	Finishing of hardened materials, cast iron, and heat-resistant materials, machining of Co– and Fe–based sintered metals, forming rolls, pearlitic cast iron, martensitic stainless steel	[4]
Carbide cermets	Turning of steel with coolant, machining of bearings, and milling	[3]
Cermets	Finishing nonferrous workpieces	[5]
Coated carbide tools	General turning, grooving, and boring of cast iron and steel	[1,3]
Diamond-coated tools	SiC/Al and C–C composites, composites containing Kevlar fibers and Nomex honeycombs	[6]
Diamond clad tools	Abrasive Al–Si alloys (for fine surface finish), nonmetallic composites, resins, rubber, carbon, presintered ceramics, sintered carbide, Cu, babbit metal, bronze, brass, Mg alloys, Zn alloys, Pb alloys	[3]
Micrograin carbide tools	Low speed machining of steel and stainless steel, milling of cast iron and ductile cast iron	[1]

(Continued)

Table 6.1. *(continued)*

Material	Application	Ref. [a]
Sialons	Rough cutting of Ni-based superalloys for aircraft turbine engines and Ni–based equipment used for corrosive applications in the oil and gas industry	[1]
	Cuts requiring high feed, speed, or depth capabilities; cast iron at high speed	[1]
Si_3N_4–based tools	Rough turning and milling of cast iron under severe conditions up to 13 m/s	[2,6]
	Turning of high-temperature Ni–based alloys (steels are unsuitable)	[2,6]
TiC	Semifinish and finish turning of ductile cast iron and high-speed turning of cast iron with coolant	[1]
TiC–TiN cermets	Finish turning and boring of steel and cast iron, turning and grooving of iron–type sintered metals	[1]
TiN cermets	General turning, boring, and grooving of steel, tube scarfing, machining of bearings	[1]
ZTA (Al_2O_3+ZrO_2)	Turning of gray, nodular, and malleable cast iron hardened to 300 HB at cutting speeds to 15 m/s	[2]
	Turning of carbon steel, alloy steel, tool steel hardened to 38 HRc at cutting speeds to 18 m/s	[2]

[a] References:

1. NTK Technical Ceramics Co. (manufacturer's data)
2. W.W.Gruss, Am. Ceram. Soc. Bull. 67 [6], 993–996 (1988)
3. Sandvik Co: *Modern Metal Cutting* (1995)
4. Sandvik Coromant (manufacturer's data)
5. *Advanced Ceramic Materials*, Charles River Assoc. Inc., International Trade Admin., and National Materials Advisory Board (eds.) (Noyes Park Ridge NJ, 1985)
6. M.R. Pascucci and R.N. Katz: Interceram, 42 [2], 71–78 (1993)

employ abrasive slurries. Lapping or polishing operations of various materials such as glass, marble, metals, and ceramics are performed with slurries of free abrasive grains such as SiC, Al_2O_3, or diamond [6.21]. The slurry is used on a suitable cloth attached to a revolving or vibrating plate. Table 6.2 lists abrasive machining processes, types of particle mounting, and resulting surface roughness values. General requirements of abrasive materials are high hardness, wear resistance, toughness of individual grains, and friability (the ability to crumble) of the abrasive tool [6.22]. Abrasives of unusual hardness are called superabrasives [6.23]. Cubic BN and diamond are two ceramics that fall into this category. Other abrasives have much lower hardness and wear resistance.

The first abrasives used in industry were natural hard materials such as emery (about 50% Al_2O_3 with other oxides, principally magnetite–Fe_3O_4), corundum (natural Al_2O_3), and diamond. Synthetic abrasives were first developed by the Acheson process in 1891 [6.23]. The process, described in Sect. 2.1.1.3, yields two grades of SiC. The one with a lower amount of impurities has a green color. The black grade, which has about the same hardness as the green grade, is used mostly as an abrasive since it costs less. The commercial importance of natural abrasives significantly decreased after the introduction of various synthetic abrasives. Among the limited natural abrasives still employed are diamond, quartz sand, and garnet. Quartz sand is used for abrasive coatings and sand blasting. Today's synthetic abrasives possess uniform properties that permit abrasive processes to become precision manufacturing processes [6.24]. Figure 6.13 shows an example of industrial abrasive grains, in this case a sol–gel derived Al_2O_3.

Common ceramic materials used for various abrasive machining processes include Al_2O_3, SiC, c-BN, and diamond. Al_2O_3 is the most commonly used abrasive, followed by SiC. Al_2O_3 is used for grinding steel and other high-strength ferrous alloys. SiC is suitable for grinding a wide range of metals, both ductile and brittle. Commercially available c-BN has the trade name *Borazon*, and is produced by the General Electric Company. It is employed mostly for grinding hard materials such as hardened tool steels and intermetallic alloys. Cubic BN has considerably better thermal resistance than diamond. It performs better at high grinding speeds and high operating temperatures. Diamond is the hardest material. Therefore, diamond wheels are used to shape other hard and abrasive materials such as ceramics and cemented carbides [6.22,6.24].

The surface morphology and grain size of an abrasive particle (grit) are important parameters in abrasive machining. Grits interact with the surface by cutting, plowing, and rubbing. Grits with large negative rake angles or rounded cutting edges interact by plowing instead of cutting. Controlled breakdown of the abrasive or the matrix is a requirement in abrasive wheels to provide fresh, sharp cutting surfaces. *Dressing* the wheel with porous SiC aids the formation of fresh cutting surfaces. The grain size largely determines surface finish and material removal rates. Small grit sizes provide a better surface finish whereas large grit sizes produce higher material removal rates [6.22]. Hard workpieces require smaller grain sizes for efficient cutting and vice versa. Grain size ranges from 8 (very coarse) to 250 (very fine).

6.4. Automotive and Aerospace Applications

Automotive applications of ceramics include turbocharger rotors, exhaust port liners, honeycombs for catalytic converters, spark plugs, glow plugs, and sensors

Table 6.2. Abrasive machining processes [a]

Process	Type of Particle Mounting	Surface Roughness (ceramic particle size)
Grinding	Bonded	0.2–1.6 microns
Abrasive machining	Bonded	>1.6 microns
Snagging	Bonded belted	>1.6 microns
Honing	Bonded	0.1–0.8 microns
Lapping	Free	0.02–0.4 microns
Superfinishing	Free	0.01–0.2 microns
Polishing	Glued to polishing wheel	0.02–0.8 microns
Buffing	Free in buffing compound	0.01–0.4 microns
Abrasive waterjet	Free in jet	–

[a] References:

1. E.P. DeGarmo, J.T. Black, and R.A. Kohser: *Materials and Processes in Manufacturing* (Prentice-Hall Upper Saddle River NJ, 1997)
2. M.P. Groover: *Fundamentals of Modern Manufacturing* (Prentice-Hall Upper Saddle River NJ, 1996)

Fig. 6.13. Microstructure of Sapphire Blue® ceramic grain for abrasive applications. The microcrystalline structure of sintered aluminum oxide made by sol–gel technology permits microfracturing the abrasive grain. This ensures constant generation of new cutting edges for long tool life, good surface quality, and low surface roughness (courtesy of Hermes Schleifmittel GmbH & Co Hamburg Germany)

[6.25,6.26]. Turbocharger rotors made of Si_3N_4 have been used in some Nissan models since 1987 and by Toyota since 1989. Kyocera in Japan produced more than 30,000 Si_3N_4 turbocharger rotors per month for Toyota Celica and MR-2 models in 1993 [6.27].

Various types of ceramic filters exist for emission control on diesel engines. These filters are required to filter solid particles, which are composed of solid carbon on which uncombusted hydrocarbons and inorganic substances such as Na, Ca, Zn, Fe, and V are adsorbed. Available ceramic filter types include

1. extruded monolithic filters with triangular, hexagonal, or square channels with 35–50% porosity,
2. fiber mats wrapped on a metallic core where the fiber diameter is 3 μm average and porosities are up to 90%,
3. ceramic foams, usually fabricated by impregnating a polymeric foam, followed by heat treatment to burn off the polymer and sintering. The porosity ranges between 55 and 92%.

Two important requirements of filter are high capacity for filtering solid particles and the ability to regenerate (combustion) and eliminate these particles periodically to prevent plugging. Ceramic materials usually employed in diesel filters include SiC, mullite, cordierite, and titanium aluminate ($TiAl_2O_5$). Among these materials, $TiAl_2O_5$ has the lowest CTE, which favors resistance to thermal shock. On the other hand, SiC has significantly higher thermal conductivity (see Appendix 7) which is important at the regeneration stage [6.28]. The ceramic substrate is coated with an oxide coating including Pt or Pd as a catalytically active agent. Diesel oxidation catalysts have been used successfully in mass production for automobile diesel engines since 1989 [6.29]. Oxidation type catalysts reduce hydrocarbon and CO emissions with good efficiency, but they have little or no effect on N–O emissions.

Catalytic converters, together with *lambda sensors* have been used for emission control in passenger cars since the late 70s. Typically, a catalytic converter is composed of one or two extruded monolithic cordierite bodies with a honeycomb structure (see Fig. 3.29). It has a washcoat of alumina mixed with precious metals (Pt,Rh,Pd), wrapped with a vibration damping material and is enclosed in a steel casing [6.30,6.31]. The ceramic material usually employed is cordierite ($2MgO.2Al_2O_3.5SiO_2$). Mullite–aluminum titanate ceramics have been developed for temperatures higher than 200°C [6.30]. Heated by exhaust gases and auxiliary devices when necessary, the three-way catalytic system aids the reaction of CO and various hydrocarbons to produce harmless CO_2 and H_2O and the simultaneous reduction of NO_x to N_2 and O_2.

A typical lambda sensor used in emission control has a ceramic body and a porous ZrO_2 electrode (see Fig. 4.58). The lambda sensor is an oxygen sensor, which provides information for the electronic fuel injection system or electronic carburetor. One surface of the electrode is in contact with air, and the other is in contact with the exhaust gas. This electrode allows the diffusion of oxygen, as explained in Sect. 4.4.1.4.1. When the oxygen content on both sides is different, an electric signal is generated by the sensor. This signal is used to keep the air to gasoline ratio at $\lambda =$

1.00 which is the optimum ratio for the catalytic converter to function and to reduce harmful emissions [6.32].

Other types of *sensors* used in automotive applications are knock sensors, seat pressure sensors, temperature sensors, oil sensors, impact sensors, and road surface sensors which employ piezoelectric ceramics. Knock sensors and seat pressure sensors are used to control engine knock. Temperature sensors are used for emission control. Oil level sensors are used for controlling the oil level in the engine. Road surface sensors monitor road surface conditions and provide input for shock absorbers for improved ride and handling.

Port liners used for thermal insulation of automotive cylinder heads are made from aluminum titanate. These portliners are cast into aluminum or steel cylinder heads. Other automotive parts employing ceramics include spark plug insulation and riser heaters made from barium titanate [6.33,6.34]. Figure 6.14 shows some automotive ceramic components, including exhaust port liners, valve guides, headface plates, wear surface inserts, piston caps, bearings, bushings, and intake manifold liners. Seal rings of 85% Al_2O_3 have been standard in automotive water pumps for more than 20 years, and SiC seals entered this market recently [6.1]. Si_3N_4 is used in diesel glow plugs and precombustion chambers by Isuzu, Mitsubishi, Toyota, and Mazda.

Engineered ceramic fiber paper and felt, made from high alumina or aluminosilicate fibers, are finding applications as heat shields in the exhaust system, floor pan, plastic fuel tanks, and fuel lines. Ceramic fiber paper is a preferred

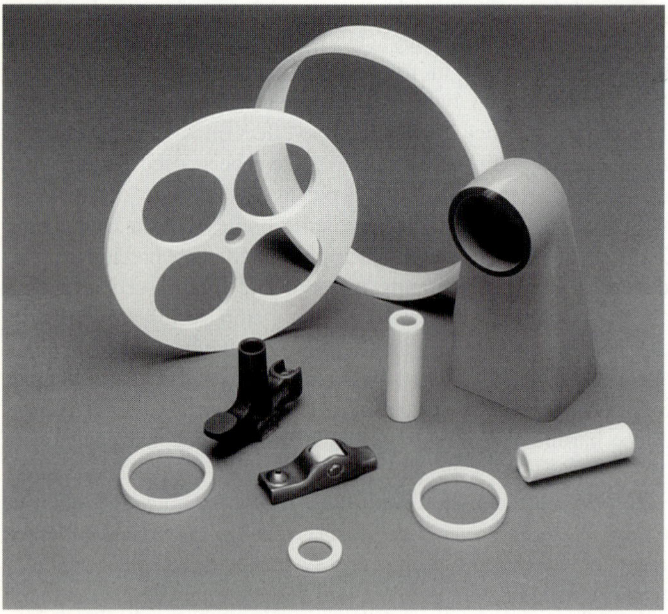

Fig. 6.14. Examples of automotive ceramic components (see text for explanation) (courtesy of Coors Ceramics Company Golden CO)

filtration medium for pyrotechnic air bag inflators due to its uniform permeability, high-temperature corrosion resistance, and its cooling ability. New brake and friction materials are also being developed using engineered ceramic fibers. Significantly higher friction stability over a wide temperature range was demonstrated without sacrificing disk or pad wear [6.35].

Scientists at Daimler–Benz have been investigating the viability of engineering ceramics in passenger car engines since 1984. An intensive program by the company aims to use silicon nitride valves in modern engines [6.36]. For this purpose, 1700 test engines were fitted with ceramic valves. If the test results meet the company's stringent quality requirements, ceramic valves will soon be used in standard production. The main advantage lies in the light weight of silicon nitride. A Mercedes intake valve made from silicon nitride weighs merely 26 grams. The corresponding steel valve, on the other hand, weighs 61 grams. This corresponds to a weight saving of around 57%. A substantial energy gain is achieved due to reduced friction, a direct result of reduced valve weight. Three to six percent savings are expected in fuel consumption in a four-cylinder engine due to reduced friction. The results of bench tests and road tests carried out by Mercedes–Benz to date have been very positive. Seventeen engines have undergone 8,500 bench-test hours and almost 300,000 road-test kilometers without any significant problems. Assuming that the ceramic valves pass the necessary tests, the only question to be resolved by Mercedes engineers and their suppliers concerns cost. Ceramic valves are currently twice as expensive as their steel counterparts.

C fibers are being used commercially in low-volume production cars, prototype cars, and racing cars due to their low weight, high toughness, and high specific strength. McLaren Cars Limited used C fibers for the complete body and chassis integral frame of its high-performance F1 road car. Similarly, the Ettore Bugatti company employed C fibers for the chassis of the EB110 supercar [6.37].

The Honda company developed a process employing Al_2O_3 fibers in engine sleeves for mass production. In this process, a preform made of Al_2O_3 fibers is infiltrated with molten Al. The result is a light and very rigid engine block with wear-resistant sleeves [6.38].

Radomes are an important structural and radio electric application of ceramics. These are hollow cones carried outside aircraft, missiles, and spacecraft as a protective cover and window for electronic guidance and detection equipment. At supersonic speed, considerable skin heating occurs from atmospheric friction, necessitating the use of refractory materials [6.39]. Ceramic materials used for this application are Al_2O_3, MgO, fused SiO_2, ZnS, ZnSe, CdTe, and MgF_2 [6.21]. Corning Pyroceram™ has been used for more than 30 years for a wide range of radar-guided missiles.

Ceramic glass matrices of borosilicate, lithium aluminosilicate, or calcium aluminosilicate glass reinforced with C or SiC fibers are used in various aerospace applications. Including compressor cylinders, valves, brake systems, turbine blades, and supports for satellite mirrors [6.40].

6.5. Refractory Applications

Ceramic materials are well known for their refractory properties. Refractory ceramics have been used for lining boilers, ladles, kilns, and other vessels. Several layers of refractory are used in these applications. The inner layers, where contact with the most severe environment occurs, employ refractories with higher density. These layers must endure corrosive and erosive media such as molten metal, slag, fluidized particles, high velocity corrosive gases, and corrosive waste. The outer layers must provide thermal insulation. Low thermal conductivity, high melting or decomposition points, and low thermal expansion are critical properties. The outer linings usually do not carry high stresses, nor are they exposed to erosive and corrosive media. Porous refractories are better fit for this purpose because they are less costly, lighter, and have better insulation capability [6.21].

Ceramic linings are also employed in waste incinerators. Incineration is the controlled burning of municipal waste. Linings for combustion chambers have to resist wear caused by abrasion and thermal cycling at temperatures up to 1300°C. The heating surfaces of boilers are subject to corrosion and erosion due to polluting gases such as HCl in combination with (heavy) metal-containing fly ash. Ceramic bricks preferred for combustion chamber linings are SiC bricks with 80–90% SiC; silimanite and andalusite bricks with up to 60% Al_2O_3; alumina bricks with 85–90% Al_2O_3; SiC patching mixes for boiler tubes with 70–90% SiC; unshaped high alumina silicate refractories; thermal shockproof castables; chemically bonded moldables and gunning mixtures; and plastic, chemically bonded corundum base moldables. Linings for regions of high thermal stress and corrosive attack are made of SiC bricks that contain glass forming additives (BaO, CaO, metallic oxides) for higher oxidation resistance. The combustion chamber may be lined with chamotte bricks, silimanite/andalusite bricks, or with 85% Al_2O_3 bricks in regions exposed to elevated temperatures. The endurance of ceramic linings can be enhanced by certain combustion and process engineering measures [6.41].

Figure 6.15 shows some refractory ceramics, including compact heat exchangers, heat exchanger tubes, immersion tubes, radiant tubes, furnace components, insulators, thermocouple protection tubes, burner nozzles, atmosphere inlet tubes, muffle tubes, high temperature exhaust stream filters, and kiln furniture. Many of these applications require high-temperature stability and resistance to chemical attack, molten metal and slag erosion, and thermal shock. Conventional refractories are based on single and multiphase oxides and various forms of graphite. Many refractories are produced from naturally occurring minerals, and they contain relatively high amounts of impurities. Examples of conventional refractories include silica, alumina–silica, and basic refractories such as dolomite, magnesite, calcite, and forsterite, zirconia, zircon, and spinel (see Sect. 7.4. for more detail). Appendix 17 lists some important properties of traditional and special refractories. These products are commercially available in a variety of forms such as brick, castables, sheet, cloth, tape, and formable shapes. About 50% of the refractories used in the US are consumed by the iron and steel industry [6.42]. Other refractory consumers include the ceramic and glass

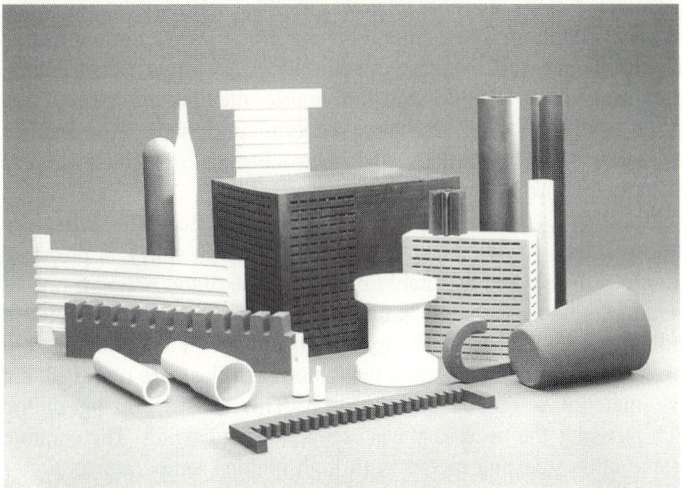

Fig. 6.15. Products for the thermal processing industry (see text for explanation) (courtesy of Coors Ceramics Company Golden CO)

industries, mineral processing, the chemical and petrochemical industries, and nonferrous metal production.

Most refractories are porous and relatively impure, but many refractory applications involve dense, high technology ceramics, including SiC, Si_3N_4, BN, AlN, and pure oxides. SiC and other monolithic ceramics are being developed for heat exchanger components. Ceramic heat exchangers are being used industrially in low-pressure applications such as recuperators and air preheaters. The main advantage over metal heat exchangers is the potential for higher temperature service and longer life. Refractory SiC such as clay-, oxide-, oxynitride-, or nitride-bonded SiC can be used in low-pressure systems where $\Delta P \leq 1$ atm. High pressure ($\Delta P > 1$ atm.) systems require fully dense materials such as siliconized SiC. High thermal shock and cycling resistance are required for these applications [6.43].

SiC is also used in blast furnace liners, shrouding tubes, and immersed nozzles for refining blast furnace iron. One of the oldest applications of SiC is in resistive heating elements. BN, Si_3N_4, and sialons are used as break rings in the continuous casting of steel. These rings determine the meniscus of the metal in horizontal, continuous, casting molds [6.44].

Other examples of refractory applications that use dense ceramics are crucibles, tubes, thermocouple sleeves, and igniters. An area of great consumption of graphite is electric arc furnace electrodes used in pure metal and alloy production. Another established refractory application of ceramic materials is resistive heating elements. Although graphite heaters are preferred under inert atmospheres or vacuum, SiC, $MoSi_2$, and ZrO_2 are used in oxidizing conditions. Typical maximum temperatures of the latter heating elements in air are 1500, 1800, and 2000°C, respectively. Graphite elements can be used to temperatures above 2000°C in inert atmospheres.

Ceramic foams produced from alumina, silicates, mullite, cordierite, or zirconia, are used for molten metal filtration. These filters remove dross, slag, and nonmetallic inclusions, minimize turbulence, and decrease erosion due to hard particles. Ceramic membranes are preferred over metallic and polymeric membranes when high temperatures, abrasion, and corrosion resistance become important. Applications include food processing, pharmaceutical and petrochemical production, gas separation, and molten metal purification [6.45].

Ceramic mold casting is a special technique that benefits from the refractory properties of ceramic powders [6.24]. In this process, a ceramic slurry is prepared from refractory ceramic powders and a suitable binder. In the *Shaw process*, one of the most popular ceramic molding techniques, the slurry is prepared from hydrolyzed ethyl silicate, alcohol, and a gelling agent. The mixture is poured on top of the pattern placed inside a flask. After the mixture sets and the pattern and flask are removed, the resulting rubber-like material is ignited with a torch. This causes crazing of the ceramic mold, which provides the necessary gas escape during casting. The ceramic mold casting technique allows casting metals with high melting temperatures. Other advantages include intricate detail, close tolerances, and smooth surface finish. For metals such as titanium, which react with common mold materials, a mixture of graphite powder, cement, starch, and water is used. After packing and removing the pattern, the mold is fired at 1000°C to consolidate the graphite. Molds prepared by ceramic mold casting are used only once since they are broken to remove the casting after the cast metal solidifies.

6.6. Ceramics for Energy Production

Ceramic materials are currently used in various *nuclear applications*, and many additional ones are foreseen in the future. In *fission reactors*, B_4C is used as control rods and neutron absorbers. Control rods are used in reactor cores to vary the power, depending on demand, and if necessary, to shut down the reactor [6.46]. They are often constituted of stainless steel cylinders in which boron carbide pellets are piled under Ar atmosphere. These rods are used in fast breeder reactors as well as pressurized water reactors. Moderators are used to reduce the neutron energy in reactors, which operate by slow (thermal) neutrons. Graphite and BeO are used as moderators in high-temperature reactors [6.47]. These materials are used to reflect neutrons produced in the fission reactor. B_4C is used in pressurized water reactors, partly substituting for Ag–In–Cd alloys [6.48]. Ceramic sensors are used to monitor oxygen levels in the liquid sodium cooling systems of liquid metal-cooled fast (neutron) breeder reactors (LMFBR). In another application, a miniature electromotive force (emf) cell has been developed and used to measure the variation in oxygen potential with operating (burnup) time of the fuel. The cell consists of an yttria-doped thoria crucible which hosts the fuel specimen and which is pressed in contact with an Fe/FeO or Ni/NiO reference electrode. At a temperature between 700 and 1000°C, the emf is a direct measure of the difference between the oxygen potentials of the reference and the specimen.

The fuel itself, $(U_{0.8}Pu_{0.2})O_{2\,(-x)}$, is also a ceramic material. Other *ceramic fuels* are also considered for fast breeder reactors, including ThO_2 and the so-called advanced nuclear fuels, carbides, nitrides, and carbonitrides of U and Pu. In the *fission process*, a heavy nucleus splits into two smaller fragments. A large amount of energy is released during fission, and this energy is the basis of fission reactors. The nuclear fragments are in highly excited states and emit neutrons and other forms of radiation. The neutrons are slowed down in order to increase the fission probability in the core of the reactor. Therefore, local melting along the track of the fission fragments and a very high level of radiation damage occur. Measurements are very difficult due to the high radiation field in these environments [6.49]. Operating temperatures may reach 2500°C at the center of the fuel pellets, and the impurity concentration grows steadily. Their lifetime is of the order of three years. Typically, a fuel element consists of bundles of fuel rods, which are about 4 m long and 1 cm in diameter. The fuel rod contains stacks of fuel pellets in a metal sheet (a Zr-alloy in the case of water-cooled and moderated reactors).

A variety of applications are predicted for ceramics in *fusion reactors*. Fusion reactors are planned to be deployed in the future, probably around the year 2030, provided that all technological and engineering issues are resolved in time. Some of the potential applications are insulators for magnetic coils, active coils, and divertor coils; windows and dielectrics for radio frequency (RF) heating; structural components; current breaks; and insulators for neutral beam injectors, magnets, and direct converters. Candidate materials for lightly shielded magnetic coil insulators are magnesium aluminate spinel ($MgAl_2O_4$), MgO, and Al_2O_3. The most promising material among these is spinel due to its low swelling upon irradiation. RF heating usually calls for windows, which can be produced from Al_2O_3, BeO, Si_3N_4, SiC, and spinel. Tritium breeding materials are also an important field of application for ceramics. These are Li-based ceramics for solid blankets, which serve to breed and release tritium as fuel to maintain the fusion reaction and at the same time to convert the energy into useful heat. Some examples of these materials are Li_2O, γ-$LiAlO_2$, $LiAl_5O_8{\cdot}Li_2SiO_3{\cdot}Li_4SiO_4$, Li_2TiO_3, and Li_2ZrO_3. These materials are attractive because of their inherent safety advantages [6.50]. Important properties to look for in blanket materials are physical and chemical stability at high temperatures, stability against radiation damage, and compatibility with other blanket components [6.49].

Ceramic materials are also being considered as *nuclear waste storage* materials, especially for corrosive environments. One container concept is composed of two dense Al_2O_3 sections fabricated by HIP and sealed after insertion of the radioactive waste [6.44]. Spent UO_2 fuel may be reprocessed to recover U and Pu by dissolving it in HNO_3 and by separating the minor actinides Np, Am, and Cm. The separated minor actinides form high level waste (HLW) which has to be solidified in a suitable matrix for safe storage and disposal in a dry underground repository [6.49]. Important requirements that have to be satisfied by the container material are high corrosion resistance (also against humidity) and high resistance to radiation damage. Borosilicate glasses are accepted as a suitable material for such applications but other materials are also being considered. Ceramic materials under study often resemble natural minerals that host rare earths such as U and Th. Examples are sphene ($CaTiSiO_5$), monozite orthophosphates (e.g. $(Ce,La)PO_4$), and SYNROC (synthetic rock) consisting of Perovskite ($CaTiO_3$), barium hollandite ($BaAl_2Ti_6O_{16}$), zirconolite ($CaZrTi_2O_7$), and rutile (TiO_2), containing e.g., 10wt% calcined HLW. Considerable (up to 8%) swelling occurs in ceramics along with amorphization. On the other hand, volume changes in glasses are much smaller.

Si_3N_4 and SiC are frequently used in *conventional energy production* since they have higher strength than superalloys above 1100°C (the typical strength of hot pressed Si_3N_4 is 800 MPa at 1100°C and that of a superalloy is 200 MPa). The lower densities of these ceramics (roughly one-third that of superalloys) are also an advantage for rotating and oscillating parts such as turbine blades.

Ceramics are used as insulators and heat exchangers in magnetohydrodynamic (MHD) power generation. Electrical power generation by the MHD system involves expanding a heated, electrically conducting fluid through a magnetic field. Unlike the conventional turbine generator system, MHD does not require a mechanical rotor. The accelerated fluid interacts with an intense transverse magnetic field, inducing an electric field within the fluid. The current is collected by electrodes and electric power is supplied to an external load. The advantage of ceramics in this application is the ability to work at higher temperatures and thus to increase the power output. Toxic emissions such as SO_x and NO_x are also reduced.

Conventional ceramics are used in fluidized bed gasification below 1300°C with dry ash. Ceramics are also candidates for high-temperature (1500–1600°C) gasifiers with liquid ash running down the reactor walls and for temperatures above 1700°C in the burning zone. These applications call for resistance to corrosion, erosion, and high temperature. Si_3N_4 and sialons are promising candidates, especially for the most extreme conditions.

Si_3N_4 and reaction bonded SiC are used in advanced heat exchangers necessary in energy production. These applications require resistance to temperatures up to 1000°C, high corrosion resistance to aggressive liquid or gas media, and thermal shock resistance [6.49].

Oxygen ion conductor ceramic materials are employed in high-temperature fuel cells still under development. Stabilized zirconia is the material of choice due to its high oxygen ion conductivity (see Sect. 4.4.1.4). This material is also used for water vapor electrolysis at high temperatures. Applications in Na- or Li-based batteries were also discussed in Sect. 4.4.1.4. Porous inert ceramic matrices are currently used in high-energy batteries for immobilizing the liquid electrolyte [6.49].

6.7. Biotechnological Applications

Various biotechnological fields, ranging from dental and bone implants to supports for immobilizing enzymes and microbes, employ ceramic materials. Bioceramic materials are used for repair and replacement of human hips, knees, shoulders, elbows, fingers, and wrists [6.51]. Three types of ceramics used as *bone implants* are bioinert, resorbable, and bioactive ceramics. The *first type* of ceramic is microporous Al_2O_3 or a hydroxyapatite (HA) coating on porous metals that enables ingrowth of tissue into the pores and provides good mechanical bonding. As a result of their inertness, the remodeling of bony tissue adjacent to alumina or other bioinert implants is not disturbed by immune reactions or biochemically, i.e., by ions or other matter going into solution [6.52].

In the *second category* are materials such as calcium sulfate (plaster of Paris), tricalcium phosphate, and calcium phosphate salts. These implants degrade gradually and become replaced by the surrounding natural tissue.

The *third type* of ceramics is bioactive, i.e., they interact with the bone and form chemical bonds. Examples of such ceramics are HA, bioactive glasses, and bioactive glass–ceramics. Various sources and forms of HA are available. These include solid HA derived synthetically, coralline HA derived from coral, and phycogene HA derived from calcifying marine algae. Of all of the HA, the algae-derived biomaterial is very similar in its physicochemical properties to those of bone [6.53]. The inner and particulate structure of algae-derived HA is shown in Fig. 6.16 together with various types of algae types used as the raw material. Applications of HA include filling bone defects in dentistry, traumatology, and orthopedics, augmenting hard tissue defects in plastic surgery, and bridging distances between bones in traumatology and orthopedics [6.54]. Permanent implants of HA ceramics have been widely tested in orthognatic surgery for correcting facial bone deformations. They also found large-scale applications in oral surgery as filling material for periosteal defects and alveolar ridge augmentation. HA and bioactive glasses have been applied successfully in middle ear surgery for many years. HA coatings might facilitate the initial healing process because of their bioactivity, but most experts recently agreed that they might disintegrate after 4 to 6 years of service. Thus, they are not recommended for permanent applications [6.52]. Biomedical bone implants for knees and hips involve glass matrix composites of Ca Na phosphosilicate with fibers of Ti or stainless steel or particles of Ti and Ag. In such glasses, the principal components are SiO_2 ($\leq$60%), Na_2O, CaO, P_2O_5, MgO, and K_2O. A high percentage of Na_2O and CaO and a high ratio of CaO/P_2O_5 are required [6.55]. Alumina and stabilized zirconia ceramics are used in orthopedic load-bearing applications. Figure 6.17 shows some implants made from Al_2O_3, ZrO_2, and HA. A hip system with an alumina socket and ball attached to a stem made of TiAlFe alloy achieved a high success rate according to long term follow-up studies [6.56].

Ceramics were first used in *dentistry* in 1774 to substitute for tooth-colored replacements such as ivory, bone, wood, animal teeth, or extracted teeth from human "donors" [6.57]. Ceramics solved the problems of stained, decayed, and malodorous dentures, but they failed to faithfully replicate the optical characteristics of natural teeth. The first ceramics used in dentistry were made of triaxial whiteware porcelain formulations (see Sect. 7.2). Aesthetic properties were improved by minimizing or eliminating clay and quartz from these compositions. Aesthetic porcelains remain mainly alkali-modified aluminosilicate glasses produced from feldspathic minerals with a small amount of crystalline filler. Leucite ($KAlSi_2O_6$)-containing high expansion porcelains are an exception. High contents (40–55 mass%) of leucite are being used as a dispersion strengthening phase. Leucite has the advantage of being able to retain acceptable translucency, important for aesthetic reasons. It also has an index of refraction

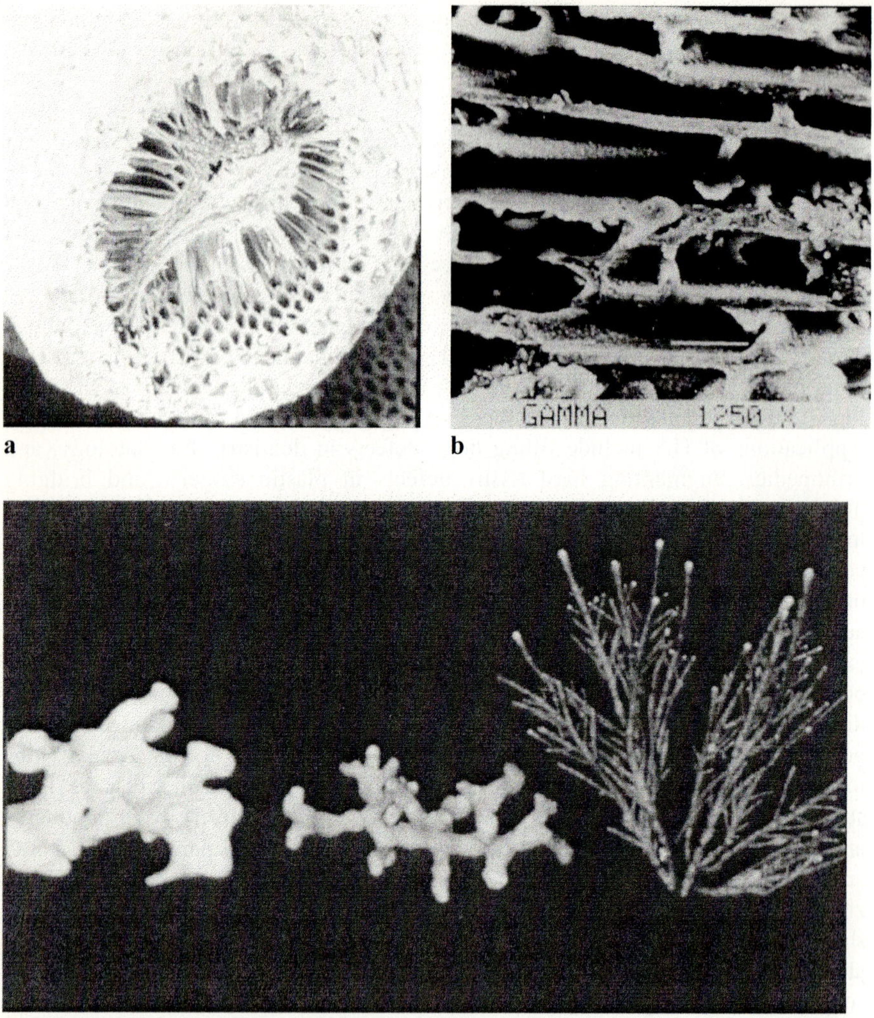

a b c

Fig. 6.16. Algae-derived HA showing 10 μm wide channels whose structure is similar to a honeycomb **a** interconnecting pores of the algae skeleton **b** and various types of algae used for this purpose **c** (courtesy of Friatec AG Medical Technology Division Mannheim Germany)

close to feldspathic glasses and raises the coefficient of thermal expansion (CTE) of the composite. Leucite is also used to increase the fracture toughness of dental porcelains over that of pure glasses.

Most aesthetic ceramics have low strength and are not suitable for applications where functional demands are high, e.g., crowns on molar teeth and fixed partial dentures. Higher strength ceramics are used as core materials instead of cast metal for

Fig. 6.17. Ceramic components for medical applications: alumina ceramic ball heads and inserts (BIOLOX®forte) and composites (BIOLOX® delta) for hip arthroplasty, hydroxylapatite ceramic (BIOLOX®osprovit) bioactive coating or granules as well as molded parts for special surgical applications (courtesy of CeramTec AG Innovative Ceramic Engineering Plochingen Germany)

improved optical results, due to the partial translucency of the former. Initially, dispersion-strengthened, Al_2O_3-filled glass compositions were used as core materials.

Another core material developed for improved strength and toughness is lanthanum aluminosilicate glass-infiltrated alumina (70–85 mass%) that is produced by two alternative methods. One method involves slip casting the alumina body, sintering, and glass infiltration. The second method involves CAD/CAM machining of presintered blocks of porous Al_2O_3, followed by glass infiltration [6.57].

Dental implants are made from alumina, HA, and bioactive glasses. Figure 6.18 depicts various types of dental implants. Initially these were made as one-piece alumina bodies (Fig. 6.18 center). Long-term results revealed some drawbacks related to the fracture resistance of alumina. This has led to the development of second-generation implants that consist of commercially pure titanium coated with hydroxyapatite (Fig. 6.18 left) or left without a coating (Fig. 6.18 right) [6.58]. Alternatively, a stepped Ti cylinder root is joined with a ceramic abutment (Fig. 6.19).

Unreinforced alumina implants are suitable where mechanical properties are not critical. These applications include middle ear implants to replace the ossicular chain partially or totally, a trachea supporting ring for tumor surgery, keratoprostheses (consisting of a corundum single crystal as its optical part and an alumina holding ring), orbital plates, and septum supports [6.56]. A different type of application is the use of yttria-aluminosilicate glass microspheres for the therapeutic treatment of cancer. These microspheres have been used to deliver radiation to cancerous parts of the body [6.59]. Due to their higher fracture

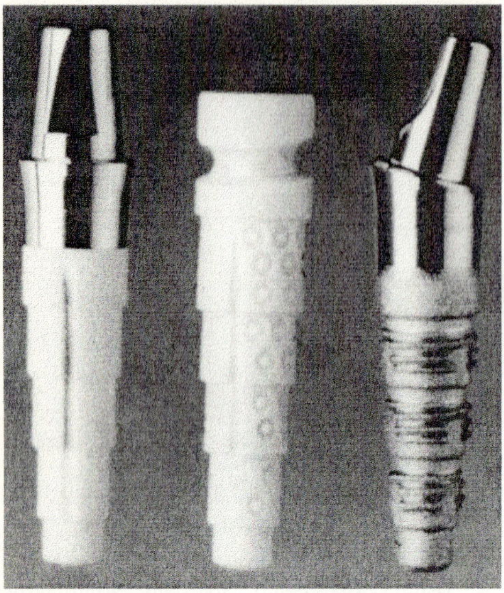

Fig. 6.18. Two versions of the Frialit®-2 Ti-implant (left and right) and the original Frialit®-1 alumina implant (center) (courtesy of Friatec AG Medical Technology Division Mannheim Germany)

a 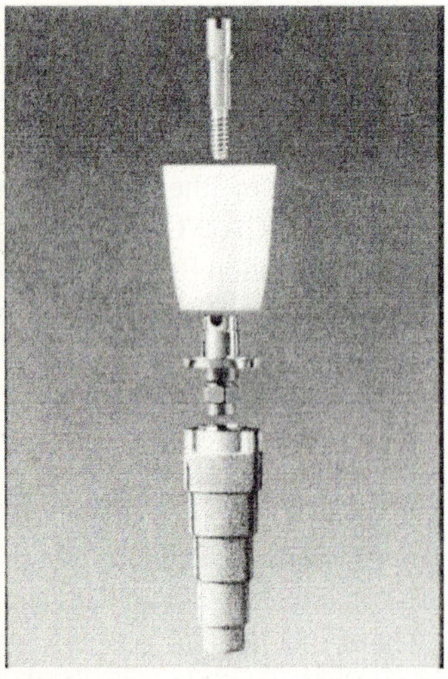 b 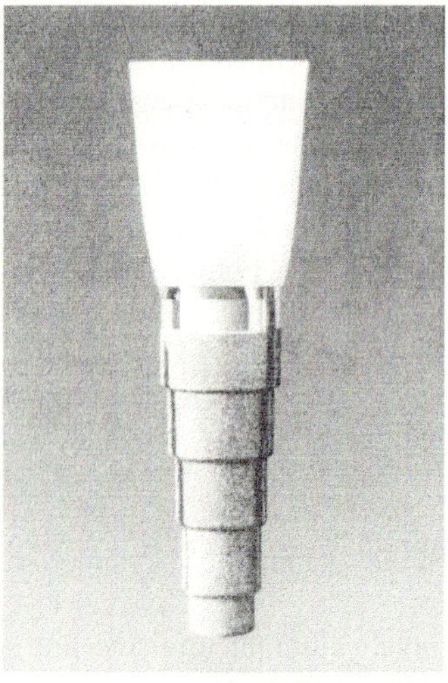

Fig. 6.19. Frialit®-2 implant stepped cylinder with Cerabase abutment, screw and titanium insert before **a** and after assembly **b** (courtesy of Friatec AG Medical Technology Division Mannheim Germany)

toughness, MgO–PSZ and Y_2O_3–PSZ are studied as alternatives to pure alumina in joint replacements. The mechanical performance of zirconia-based ceramics were compared to surgical grade alumina and the former proved superior. Trial implants of Y_2O_3–PSZ were performed in rats and rabbits up to one year, with reference to alumina. Results have shown similarly good biocompatibility. However, long-term biocompatibility information is unavailable so far [6.60]. Carbon has been used in cardiovascular devices, e.g., as an artificial heart valve and in orthopedic devices such as a bone plate and hip cup [6.61].

Porous ceramics are used as supports for immobilizing enzymes, antibodies, antigens, and microbes because of their good chemical, thermal, and biological resistance. Controlled porosity glass, porous sintered glass, cordierite, alumina, silica, and charcoal are commonly used biocatalytic carriers. Other carrier applications include bioaffinity supports for diagnostic reagents and clinical treatment, adsorbents for proteins and chemicals, purification or biotransformation of chemical compounds, and microfiltration [6.62]. Ceramics and glasses are also used in such biotechnological applications as eyeglasses, thermometers, tissue culture flasks, and fiber optics for endoscopy.

Custom-engineered injection molded ceramic shapes are used in medical and diagnostic applications such as shear valves for blood and analysis equipment, pistons, cylinders used to meter precise amounts of fluids or oxygen, and disposable surgical components. Tight dimensional tolerances, highly polished surface finish, mechanical strength, and high resistance to wear are some key properties in these applications [6.6].

6.8. Electrical, Electronic, and Magnetic Applications

Electronic ceramics constitute the major market share of advanced ceramics today. This is due to the diverse range of properties available in a variety of ceramic materials and also due to the absence of significant stresses, allowing adaptation of ceramics for such applications. Ceramics are chemically inert and resistant to high temperatures. Because of these properties, they provide a suitable environment for electronic circuitry [6.63]. Technologically important electronic ceramic classes include insulators, dielectrics, semiconductors, fast ion conductors, varistors, thermistors, piezoelectrics, magnetic ceramics, electro-optic ceramics, and high-temperature superconductors.

Most ceramics have high resistance to electronic mobility. The main application of *insulating ceramics* is in substrates for electronic circuitry. These applications require not only high electrical resistance but at the same time high thermal conductivity, mechanical strength, chemical stability, and CTEs matching those of metals used for conductivity. Al_2O_3 is commonly employed for such applications, but BeO and AlN are considered alternatives due to their better thermal conductivity [6.64]. Higher thermal conductivities are required since the packing density of hybrid integrated circuits and the integration density of semiconductor integrated circuits are being improved gradually. The disadvantages of BeO are the toxicity of Be metal and the high CTE of BeO (nearly twice as large as semiconductor silicon) [6.65]. One important disadvantage of AlN, in fine powder form, is its reactivity with water or humidity [6.66]. Other common areas of application are insulators for power transmission lines (mostly porcelain) and spark plugs.

Dielectric ceramics play a key role in the miniaturization of electronic circuits and devices. Ceramics with higher dielectric constants allow the use of capacitors with smaller surface area. Among the various functions of capacitors are charge storage used in welding and flashlights, blocking dc currents and coupling ac currents, and filtering used for separating ac currents of different frequencies [6.21,6.67]. Various forms are thin-film, thick-film, single-layer, and multilayer capacitors. $BaTiO_3$ is usually used as a ceramic capacitor material due to its high relative dielectric constant (1200–1500 in pure form and 5000–9500 in alloyed form). Other materials used for capacitor applications include $SrTiO_3$, PZT, and PLZT. Thin-film capacitors are used for high capacitance applications such as computers, military equipment, and telecommunication devices. Among thin-film ceramics are Al_2O_3 on Al metal, Ta_2O_5 on Ta metal, $SrTiO_3$, and $PbTiO_3$ [6.21,6.68]. Ferroelectricity in such thin films has been observed at thicknesses as low as 0.04 μm [6.69].

Lead-based relaxor ferroelectrics are recently preferred as capacitor materials due to their high dielectric constants (K'>20,000). Typical compounds are lead magnesium niobate, $Pb(Mg_{1/3},Nb_{2/3})O_3$ and lead zinc niobate, $Pb(Zn_{1/3},Nb_{2/3})O_3$. With the help of additional substituents, these ceramics can be densified at low temperatures (~900°C), permitting the use of relatively inexpensive cofired electrode materials such as silver [6.70]. Another family of ceramic capacitors is internal boundary layer capacitors (IBLCs), based on polycrystalline ceramics that possess phases with different resistivities and dielectric constants at the grain interior and grain boundary. IBLCs usually have dielectric constants between 20,000 and 60,000. Their dielectric properties are the result of electric field splitting between the grain and grain boundary phases, due to their different electrical properties [6.70].

Piezoelectric materials are used in transducers (vibrators, actuators), sensors (accelerometers, microphones, phonograph pick-ups, keyboards), and signal devices (filters, traps, resonators) [6.71]. The phonograph pickup is an example of applications where mechanical force is converted to an electric signal [6.21]. When the record turns on the turntable, a stylus picks up vibrations caused by the contours in the record groove. They are then converted to electric signals by the piezoelectric ceramic. Amplification of the electric signals converts them to audible sound waves. Specially cut and shaped Rochelle salt crystals were used for this application in early commercial phonographs. Although the use of phonographs decreased substantially in the last few years, those that are still being produced use polycrystalline piezoelectrics such as lead zirconate titanate. Microphones and loudspeakers are other examples of the use of piezoelectrics. In a microphone, the vibration created on a diaphragm is transferred to the piezoelectric element and transformed into an electric signal. Loudspeakers and headphones work in the opposite fashion. The piezoelectric element creates vibrations as a result of electric signals received. These vibrations are amplified and transmitted to a diaphragm that transforms them into audible sound waves. Figure 6.20 shows some piezoelectric components used in hydroacoustics, sonar applications, medical technology, and automotive engineering.

The most important application of *pyroelectric ceramics* is infrared radiation detectors. Infrared pyrometers, pollutant control devices, intruder alarms, and smoke detectors are examples of devices that use infrared radiation detection, and in turn, pyroelectric ceramics. Among various pyroelectric materials are triglycine sulfate (TGS), $LiTaO_3$, $Pb(Zr,Ti)O_3$, $SrTiO_3$, as well as cane sugar [6.67,6.71,6.72]. ABO_3 Perovskite ferroelectrics are also frequently used for pyroelectric applications. They include PST (lead scandium tantalate) and PZFN (lead zirconate iron niobate) [6.70].

Semiconductors have important applications as *varistors* and *thermistors* (see also Sect. 4.4.1.2.). Well-known applications of varistors are heating elements (e.g. SiC and $MoSi_2$) and rectifiers (Cu_2O). Rectifiers are devices that convert alternating current to direct current [6.21]. Another important application is surge protectors. These are devices that protect electronic instruments, such as computers, from sudden voltage increases. ZnO is one of the most important materials used as a varistor for such applications. Examples of thermistor applications include temperature sensors, temperature compensators, infrared sensors, switches, and heater systems. Semiconductor ceramics are also used in gas sensors [6.73]. Specific applications are combustible gas sensors and humidity sensors. SnO_2- or ZnO_2-based ceramics are used for gas sensing. The resistivity of these materials decreases, when exposed to combustible gases, due to electron transfer caused by a chemical reaction between the gases and chemisorbed oxygen

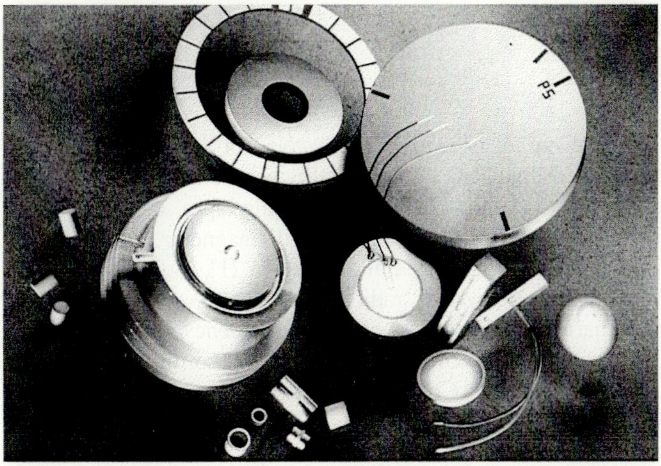

Fig. 6.20. Piezoceramic sensors for hydroacoustics and sonar applications, air-borne noise transmitters, and receivers. Parts and components for ultrasonic apparatus in medical technology for lithotripsy, removing dental scale, and inhalation apparatus. Piezoceramics for use in automotive engineering: back-up sensors, antiknock sensors, vibration and acceleration sensors, sensors for volume and flow measurement. Piezoceramic actuators for injection systems and vibration damping. Metallized alumina ceramics for surge arrestors, high energy rectifier and thyristor housings, vacuum components (courtesy of CeramTec AG, Innovative Ceramic Engineering Plochingen Germany)

on the sensor surface. Two types of humidity sensors are those that depend on electronic and ionic conduction. In humidity sensors, the conductivity usually increases by physisorption or capillary condensation of water. Materials of the ionic class include $MgCr_2O_4$–TiO_2, TiO_2–V_2O_5, and $ZnCr_2O_4$–$LiZnVO_4$. Electronic conduction type sensor materials include Perovskite type oxides and ZrO_2–MgO.

Applications of electronic and ionic conductors as well as superconductors were discussed in Sects. 4.4.1.3–4.4.1.5 and will not be repeated here.

Scintillators are an important group of electro-optic ceramics. Scintillators are radiation sensors that emit light or "scintillate" when struck by high-energy photons such as X rays, γ rays, and α or β particles. In many applications such as medical diagnostics, industrial component inspection, dosimetry (photon counting), and high-energy physics, the scintillator is the primary radiation sensor [6.74]. Visible light emitted by the scintillator is directed onto photodiodes or photomultiplier tubes to be converted to an electric signal. The first commercial application of ceramic scintillators was in a high-performance X-ray computed tomography (CT) device. Other scintillator applications, including high resolution industrial CT, security inspection, baggage scanners in airports, high energy physics calorimeters, and medical positron emission tomography (PET) scanning, are expected to use ceramics as the material of choice in the future due to their superior performance. Ceramic materials developed for scintillator applications include

1. $(Y,Gd)_2O_3$:Eu,Pr,
2. Gd_2O_2S:Pr,Ce,F,
3. $Gd_3Ga_5O_{12}$:Cr,Ce, and
4. $BaHfO_3$:Ce.

Ferrites are predominantly used as *magnetic ceramics* for various applications including "hard" (or permanent) magnets in high-fidelity speakers and small electric motors, magnetic field detectors, audio and video recording tapes, computer disks, generators, and video recorders. Examples of hard magnetic materials are magnetoplumbites such as $BaFe_{12}O_{19}$ and $PbFe_{12}O_{10}$ [6.21]. Soft magnets are employed in telecommunications (television, radio, touch-tone telephones, filters for telecommunication, submarine communication), high-frequency welding, high-frequency fluorescent lighting, low-power transformers, recording heads, and magnetic recording media [6.21,6.70]. Commonly used soft magnets include Mn–Zn– and Ni–Zn-spinel ferrites [6.21].

Figures 6.21 and 6.22 show some ceramic parts used in the electronic industry. Some examples of their uses are listed in Table 6.3. As seen from this table, electronic ceramics are part of commonly used devices such as computers and advanced applications such as nuclear fusion reactors. With further improvement in properties and new developments, many more applications for electronic, magnetic, and electro-optic ceramics will be discovered in the future.

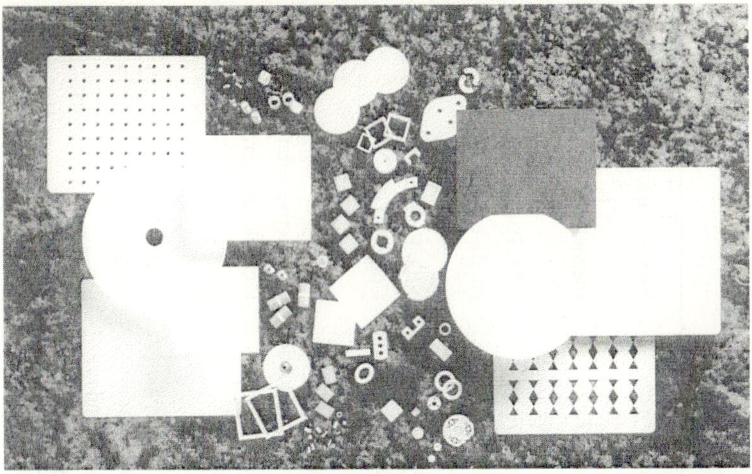

Fig. 6.21. Ceramic components for the electronic industry: Thin-film substrates, thick-film substrates, dry-press multihole substrates, metallized products, laser machined parts, advanced ceramic multilayer packages, and hermetic metal packages (courtesy of Coors Ceramics Company Golden CO)

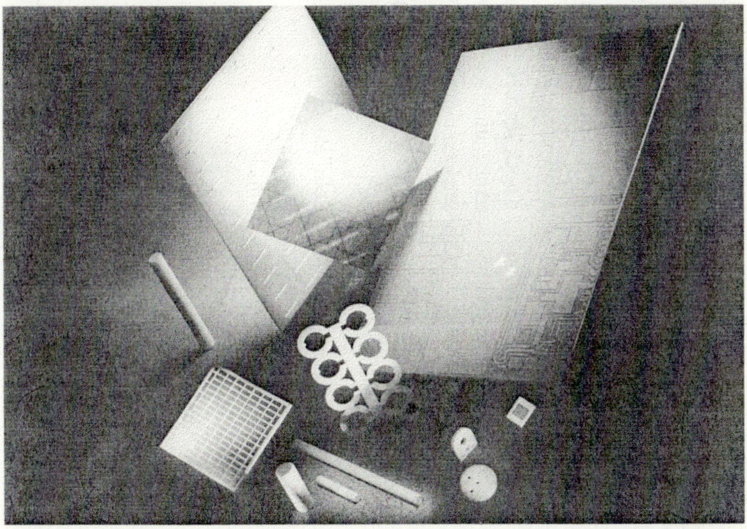

Fig. 6.22. Applications of ceramics in the electronic industry: Rubalit® and Alunit® substrates for hybrid and microelectronics technologies and power electronics. Thick- and thin-film technology. Metallized substrates for SMD inductivities. Thomit ® and Rubalit ® resistor cores for carbon film, metal-coated and metal oxide resistors, with emphasis on high-performance and precision resistors (courtesy of CeramTec AG, Innovative Ceramic Engineering Plochingen Germany)

Table 6.3. Applications of Electronic, Electrooptic, and Magnetic Ceramics

Technology	Application	Component	Function	Ceramic Materials	Ref. [a]
Audio	Alarm devices, audio equipment	Buzzers, speakers, tweeters	Piezoelectricity	Piezoelectric ceramics	[1]
Appliances and industrial technology	Gas igniter products	Spark units, piezoelectric units, gas igniters, lighters	Piezoelectricity	Piezoelectric ceramics	[1]
Automotive and other	Vibrators, oscillators, filters, spark generators, microphones	Knock sensors, pressure sensors, temperature sensors, impact sensors, humidity sensors	Piezoelectricity	PZT, $Pb(Zr,Ti)O_3$, single-crystal SiO_2	[1,2]
Automotive and other	Heat sensors	Thermistor (NTC, PTC)	Electric conduction varied by temperature	NiO, FeO, α-Fe_2O_3, CoO, $BaTiO_3$	[2,3]
Automotive and other	Na-S batteries	Solid electrolyte	Ionic conductivity	β''-Al_2O_3, Nasicon	[4]
		Electrode	Electrical conduction	Conductive ceramics	[5]
Communication	Radios, tranceivers, communication equipment, pagers, telephones	Filters	Piezoelectricity	Piezoelectric ceramics	[1]
Electric energy storage and processing	Computers, consumer electric products, telecommunications	Capacitor	Filtering, bypassing, coupling, tuning, trimming, timing, energy storage	$BaTiO_3$, $SrTiO_3$, PZT, PLZT, Al_2O_3	[2]

(Continued)

Table 6.3. *(continued)*

Technology	Application	Component	Function	Ceramic Materials	Ref. [a]
Environmental, combustion, automotive	Oxygen sensors (λ -sensors for automobiles)	Solid electrolyte	Ionic conduction	ZrO_2	[6]
		Semiconductor	Semiconductivity	TiO_2, $SrTiO_3$	[6]
Industrial equipment and consumer appliances	Ultrasonic equipment	Aerial ultrasonic sensors, immersion transducers, ultrasonic meters, ultrasonic generators, fish-finder transducers	Piezoelectricity	Piezoelectric ceramics	[1]
Information Storage	Computers	Memory core	Magnetic	Ferrites (MnMg–, MnCu–, MnMgZn–Ferrite)	[5]
		Disk	Magnetic	Iron oxide	[7]
		Tape	Magnetic	Chrome oxide, iron oxide	[7]

(Continued)

Table 6.3. *(continued)*

Technology	Application	Component	Function	Ceramic Materials	Ref. [a]
Information recording	Computers	Recording head	Magnetic	Ferrites (MnZn–, NiZn–Ferrite)	[5]
		Recording head substrate	Electric insulation	Al_2O_3, BeO, AlN, SiC	[7]
Information display	Computer, TV, and other monitors		Electric insulation	Lead-borosilicate glass	[7]
		Seal		Lead–zinc–borosilicate Glass	[7]
		Face plate	Optical	Soda–lime glass	[7]
Information printing	Inkjet and electroerosion printers			ZrO_2-containing glass, ceramic–metal composites	[7]
Information transfer	Telecommunication cables	Optical fiber	Electrooptic	SiO_2, B_2O–SiO_2	[7]
Manufacturing, robots, ultra-precision positioning	VTR heads, valves, relays	Bimorphs, piezo-stacks, optical scanners	Piezoelectricity	Piezoelectric ceramics	[1]

(Continued)

Table 6.3. *(continued)*

Technology	Application	Component	Function	Ceramic Materials	Ref. [a]
Medical diagnosis	Nuclear magnetic resonance	Magnet	Superconductivity	$YBa_2Cu_3O_{7-x}$	[8]
	Magnetoencephalography	Squid	Superconductivity	$YBa_2Cu_3O_{7-x}$	[8]
Power conversion		Flyback transformer		MnZn–, NiZn–Ferrites	[5]
		Power transformer		MnZn–Ferrite	[5]
Power generation	Solar cells			CdS	[2]
Research & development	Scanning electron microscopes, transmission electron microscopes	Electron beam cathode	Electron emittance	LaB_6 single crystal	[4]
Safety and other	Infrared pyrometers, smoke detectors, intruder alarms		Pyroelectricity	$LiTaO_3$, PZT, TGS, $Pb(Zr,Ti)O_3$	[6]
Safety, environmental, and other	Gas sensor	Varistor	Semiconductivity	ZnO	[6]
Various	Small motors, precision machinery	Actuator	Piezoelectricity	PZT, $Pb(Zr,Ti)O_3$, Single crystal SiO_2	[9]
Various	Surge protectors		Semiconductivity	ZnO	[10]

(Continued)

Table 6.3. *(continued)*

Technology	Application	Component	Function	Ceramic Materials	Ref. [a]
Various	Light valves, deflectors, displays		Electrooptic	$Pb(Zn_{1/3}Nb_{2/3})O_3$, $(Pb,La)(Zr,Ti)O_3$, $Pb(Mg_{1/3}Nb_{2/3})O_3$–$PbTiO_3$	[11]
Various	Optical memory, optical shutters		Ferroelectricity	PLZT single crystal or polarized	[4]
Various	Cathode luminescence		Luminescence	Y_2O_2S:Eu powder	[4]
	IR Laser		Luminescence	Sintered ThO_2:Nd	[4]
	VL Laser		Luminescence	Al_2O_3:Cr single crystal	[4]
Various	IR detector		Unharmonic oscillation	$Ba_2NaNb_5O_{15}$ single crystal	[4]
Various	Humidity sensors		Ionic conduction	Porous ceramics including $MgCr_2O_4$–TiO_2,	[4]
			Electronic conduction	Perovskite type oxides, ZrO_2–MgO	[4]

[a] References

1. NTK Technical Ceramics Brochure
2. *Charles River Assoc. International Trade Administration and National Materials Advisory Board* (eds) (Noyes Park Ridge NJ, 1985)
3. J.Certo *et al*: J. Eur. Ceram. Soc. **11 [5],** 401-405 (1993)
4. *Fine Ceramics.* S. Saito (ed) (Elsevier Essex England, 1985)
5. *Ceramic Materials for Electronics*. R.C. Buchanan (ed) (Marcel Dekker New York NY, 1986)
6. L. Ketron, Am. Ceram. Soc. Bull. **68 [4],** 860-865 (1989)
7. R. Tummala, Am. Ceram. Soc. Bull. **67 [4],** 752-758 (1988)
8. A.M. Wolsky, R.F. Giese, and E.J. Daniels, Sci. Amer. **Feb. 1989**, 61-69
9. K. Uchino, Key. Eng. Mater. 66-67 (1992) 311-338
10. A.J. Moulson and J.M. Herbert: *Electroceramics* (Chapman&Hall London England, 1990)
11. K. Uchino, Ceram. Intern. **21 [5],** 309-15 (1995)

7. Structure and Properties of Conventional Ceramics

The basic difference that sets engineering ceramics apart from conventional ceramics is that engineering ceramics are composed of a single phase (e.g. Al_2O_3, ZrO_2, Si_3N_4) or only a few phases (e.g., $MgO–Al_2O_3$, $Y_2O_3–ZrO_2$, $Na_2O·5Al_2O_3$), whereas conventional ceramics are typically mixtures of many phases. This difference comes from the origin of both types of ceramics; engineering ceramics are usually products of an artificial process whereas conventional ceramics are made of natural minerals. The crystal structure of an engineering ceramic is usually relatively simple, the microstructure is much finer than conventional ceramics, the impurity level is typically low, and the properties are thus better controlled. Conventional ceramics have complex structures and the constituent phases may vary from one region or time to another due to the diversity of the raw materials used; thus the properties are much more difficult to control. Conventional ceramics include triaxial whiteware compositions, cement and concrete, refractories, structural clay products, glasses, glazes, enamels, and abrasives. To better understand the difference between technical and conventional ceramics, the production and some properties of important conventional ceramic classes are briefly discussed here. The structure of glasses was discussed in Sect. 1.4 and will not be repeated in this chapter. Information on refractory ceramics and abrasives can be found in Sects. 6.2 and 6.3.2, respectively.

Conventional clay-based ceramics are broadly distinguished as unvitrified or vitrified ceramics [7.1]. *Vitrified ceramics* are fired at a high enough temperature to fuse the clay and other components, resulting in a glassy substance. Whiteware compositions are vitrified ceramics. *Unvitrified ceramics*, on the other hand, are low-fired, porous ware such as terra cotta and earthenware. A discussion of clay-based ceramics follows.

7.1. Unvitrified Pottery

7.1.1. Terra-Cotta

Ceramic ware in this group has high porosity (30% or more) and is typically red. It is fired at relatively low temperatures (900°C or below). Terra-cotta represents most unglazed prehistoric pottery. Current applications include flowerpots, roof tiles, bricks, and artware. Terra-cotta is normally not glazed but it may undergo some surface treatment for decoration and improved properties. For example, the surface can be roughened by beating with a paddle that is carved or wrapped with a cord/fabric. The surface roughness increases heat absorption and aids manual transportation. In some cases, the surface is covered with engobe (a liquid solution of fine clays and water) to lower vessel porosity and add aesthetic value such as color and smoothness.

7.1.2. Earthenware

Earthenware is a very common product that is porous and opaque. The usual body formulation is (see Sect. 7.2. for details of each ingredient) [7.2]:

- ball clay: 25%
- china clay: 25%
- flint/sand: 35%
- feldspar: 15%

It is fired at temperatures ranging from 900°C to 1200°C. The pore content is usually 10% to 25%. Glazing may be used on earthenware. The firing temperatures are not sufficient to vitrify the earthenware body but high enough for glazing. Applications of earthenware vary from bricks and tiles to fine ware such as tin-enameled majolica [7.1].

7.2. Whiteware Compositions

Well-known applications of whiteware include hard porcelain for artware, tableware, semivitreous tableware, hotel china, vitreous sanitary ware, vitreous tile ware, electrical porcelain, and dental porcelain [7.1]. Much of this whiteware is produced from three raw ingredients: clay, feldspar, and flint. The role of each of these constituents is as follows [7.2,7.3]:

clays: provide fine particle size and plasticity for easy shaping; give adequate unfired (green) strength to the body for safe handling between the shaping and firing process;

feldspar: forms a flux to aid viscous liquid formation at the firing temperature which is responsible for the strength of the fired product;

flint or silica: acts as a filler at ambient temperature; forms a viscous liquid at the firing temperature; controls the thermal expansion of the fired body.

Figure 7.1 shows areas of triaxial whiteware compositions on the silica (SiO_2)–leucite ($KAlSi_2O_6$)–mullite ($3Al_2O_3 \cdot 2SiO_2$) phase diagram. As the amount of feldspar increases, vitrification is achieved at lower temperatures. Thus, better vitrification and higher translucency are obtained. Increasing the feldspar content increases the vitrification temperature, which in turn makes the firing process more

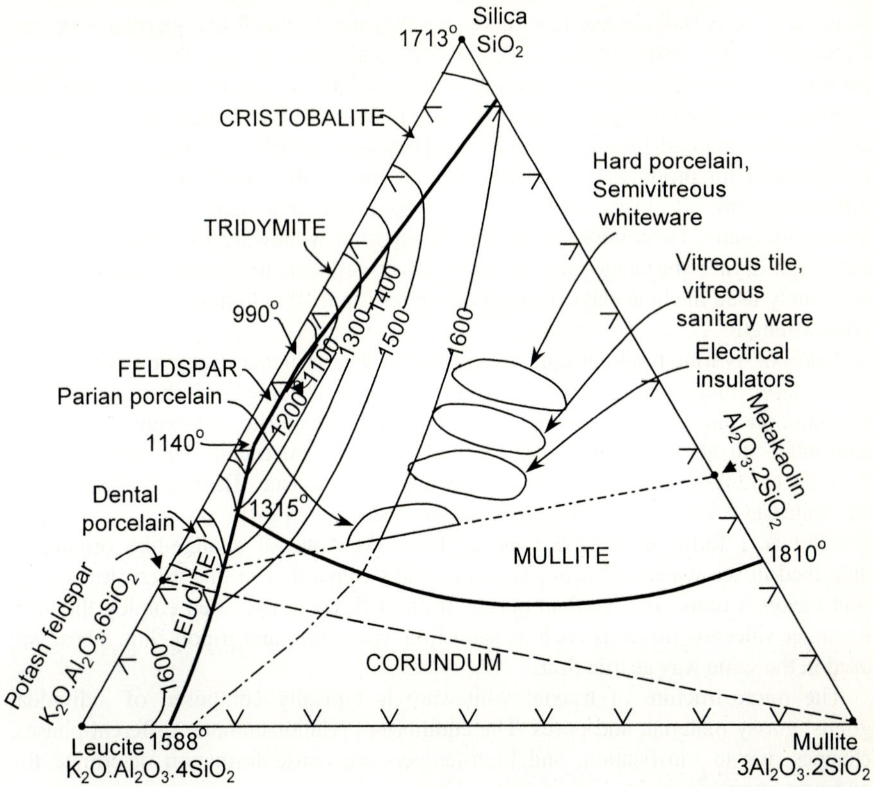

Fig. 7.1. Areas of triaxial whiteware compositions on the silica-leucite-mullite phase diagram. (From W.D. Kingery, H.K. Bowen, and D.R. Uhlmann: *Introduction to Ceramics*. Copyright 1976 John Wiley & Sons New York, reprinted with permission)

difficult and expensive. On the other hand, forming becomes easier, and the resulting mechanical and electrical properties are improved.

The basic form of *clay* found in nature is *kaolinite* ($Al_2O_3 \cdot 2SiO_2 \cdot 2H_2O + K_2O + 4SiO_2$). This is a crystalline material that has flat and hexagonal crystals and a particle size ranging between 5 μm and less than 1 μm. The fine particle size combined with flat particles results in unique properties such as plasticity. Various types of clays exist in nature, depending on the region and its geographical history. *China clays* have relatively large particles but fewer impurities than other types of clay minerals. The impurity level determines the color of the whiteware. The low impurity content of china clays results in the desirable property of firing to a clean white color. As the impurity content increases, which is the case in *secondary clays*, the color turns to off-white, ivory, straw-colored, brown, or red. Among secondary clays are *ball clays* and *fireclays*. The impurity content of ball clays varies greatly from one type to another. The name comes from the "cubes" or "balls" in which the clay used to be cut from the quarry. Ball clays are kaolinitic, but they have a much finer particle size than china clays. This results in high green strength and plasticity. Fireclays have a small particle size, and they contain considerable amounts of fine impurities. They find restricted use in whiteware. Instead they are mainly used in refractories. *Stoneware clays* contain natural fluxes as impurities. They have a buff or brown color and are mainly used for producing stoneware without further flux additions. *Bentonite* is a different form of clay, which is not kaolinitic but contains smectite (or montmorillonite). Bentonites are not used directly in whiteware since they introduce very high drying shrinkage and adversely affect deflocculation characteristics. They are mainly used in china clays in small quantities (up to 1%) to improve plasticity and green strength.

Feldspar is found naturally in several forms. The principal types are potash spar (orthoclase, microcline), soda spar (albite), lime spar (anorthite), barium feldspar (celsian), and lime–soda spar (oligoclase, andesine, labradorite, bytownite). These are aluminum silicates of K, Na, and Ca. For example, microcline has the empirical formula $K_2O{\cdot}Al_2O_3{\cdot}6SiO_2$, albite $Na_2O{\cdot}Al_2O_3{\cdot}6SiO_2$, celsian $BaO{\cdot}Al_2O_3{\cdot}2SiO_2$, and anorthite $CaO{\cdot}Al_2O_3{\cdot}2SiO_2$. None of these are found pure [7.4].

Flint is a form of SiO_2 originating from skeletons of sponge-like organisms dissolved in sea water and deposited as a chalky mineral. The fine microstructure of flint makes it more reactive than quartz. In the US, ceramists often employ the term for other siliceous minerals such as sand flint, rock flint, and tripoli flint. These are used in the same way as true flint.

The microstructure of triaxial whiteware is typically composed of individual grains, glassy material, and pores. The equilibrium relations among different phases, changes due to vitrification, and high-temperature phase formation determine the microstructure.

7.2.1. Tableware

Important categories are porous tableware (earthenware), vitreous bodies (vitreous hotelware, high alumina bodies, and stoneware), and porcelain [7.3]. The constituents are mixed by wet or dry milling and slip cast to obtain the desired shape. Depending on the formulation, the unglazed body, also termed *bisque*, may be fired separately, followed by *glazing* (application of a glass forming mixture for a vitreous layer) and *glost firing* (firing of the glazed body). Some formulations are fired together, such as true porcelain. Colorants and decorations are either applied under the glaze before glost firing, or they are applied on top of the fired glaze and refired.

Vitreous bodies have zero porosity due to a glassy phase. Vitreous hotelware, high alumina bodies and stoneware are opaque vitreous bodies. Porcelain, fine china, and bone china are translucent ceramic ware. *Vitreous hotelware* or *hotel china* is stronger and much tougher than earthenware. The standard body formulation is

- ball clay: 25%
- china clay: 25%

- flint: 30%
- feldspar: 20%

Alumina is introduced into tableware formulations for higher strength and lower thermal expansion. *High alumina bodies* are heavier than earthenware and bone china bodies with densities of 2.7, 2.14, and 2.5 g/cc, respectively. A very white body and strong product can be obtained from the following formulation:

- ball clay: 40%
- alumina: 35%
- feldspar: 25%

Stoneware is a product directly produced from a suitable clay, termed stoneware clay. This type of clay is highly plastic and low in Fe. Generally, other additions are not made. It is fired once with a lead or salt glaze to ~1280°C. Usually, 1% porosity is desired in stoneware, but ovenware, red stoneware and other stoneware products with a suitable body–glaze interface may have up to 10% porosity. Glazed drainpipes, roof tiles, tableware, and artware are application areas of stoneware [7.1].

The term *porcelain* comes from the Italian word *porcellana* (cowrie shell, a white and smooth shell type) used by Marco Polo to define the tableware he saw in China. It took Chinese masters 1500 years to develop porcelain from stoneware. Porcelain emerged during the Tang dynasty (618–907 A.D.) [7.5]. *Chinese* or *hard paste porcelain* is considered true porcelain because it originated in China. It is produced from two minerals, *petuntze*, or china stone, and kaolin, or china clay. When these minerals are mixed in different proportions, both the body and the glaze are formed, without the need to add another ingredient. Petuntze is the fusible part of the two constituents. Mixing it with the body forming paste yields the glaze. The glaze and body are normally fired at the same temperature of about 1350°C. Underglaze decorations used in true porcelain are limited mostly to cobalt blue, since other colors destabilize at the high glazing temperatures. Other colorants are applied over the fired glaze and fired again at a lower temperature. *Artificial* or *soft paste porcelain* was developed in Europe during efforts to imitate Chinese porcelain in the late sixteenth century and beyond. It is obtained by firing a mixture of glass frit and clay at about 1150°C and then the glaze at about 1050°C. The glaze is usually a lead type. Hard paste porcelain is very hard; it is difficult to scratch it with a steel instrument, and it fractures in a conchoidal mode. Soft paste porcelain, however, is easily scratched and usually has a sugary fracture. Hard paste porcelain formulations are generally based on

- china clay: 53%
- quartz sand: 33%
- feldspar: 14%

An actual porcelain body formulation used in production in England until recently is [7.3]

- china clays: 63.6%
- ball clays: 8.7%

- feldspar: 11.8%
- quartz sand: 13.1%
- whiting: 2.7%
- barium carbonate: 0.1%

The last four constituents are ground in a ball mill, and the clays are blunged (mixed with water) separately. Since true porcelain practice is to grind all the materials together in a ball mill at once, the former is a compromise body preparation operation.

Bone china was developed in England by the end of the eighteenth century. The standard formulation is

- bone ash: 50%
- china clay: 25%
- china stone: 25%

Bones are calcined to remove most of the organic matter. On to two percent of residual organic matter is desired for plasticity and strength of the unfired bone china body. After calcination, they are ground so that 70–80% is less than 10 microns in diameter. All of the ingredients are wet mixed in one common blunger. Bisque firing is at 1220°C and glost firing at 1080°C.

Computer-aided design and manufacturing (CAD/CAM) now helps tableware producers to rapidly design and manufacture their products. CAD-CAM saves time in modeling, visualization, changes in the design, and mold preparation. Instead of modeling manually from clay or plaster, the design can be created on the screen, which allows accurate visualization from any angle [7.6].

7.2.2. Sanitary ware

Sanitary ware is usually produced from vitreous china or fireclay. Products such as bidets, wall urinals, water closets, washbasins, and cisterns are examples of sanitary ware. Most sanitary ware is made of vitreous china, whereas large articles are made of fireclay. *Vitreous china* has the following ingredients:

- ball clay: 23–25%
- china clay: 25–27%
- flint or quartz sand: 20–30%
- feldspar or nepheline syenite: 20–30%

The rheology of the slip used for slip casting has to be controlled carefully for good results. The glaze is about three times thicker than used in tableware and is applied by a spray gun. Minor faults detected after firing can be repaired by grinding off the flaws and refiring.

A *Fireclay sanitary ware* body formulation is typically composed of

- fireclay: 70%

- ball clay: 10%
- grog: 19%
- sawdust: 1%

Grog is a naturally occurring calcined fireclay fired by volcanic action. Today, it is produced from new and used refractory rejects such as firebrick, and pottery, and introduced into ceramic ware to improve its physical properties. Another function of grog is to open up the body so that extra thick ware can be dried and fired without excessive warping. Sawdust is used in sanitary ware to open up the slip and ease the casting process. The body is usually covered with at least two layers of engobe and two layers of glaze. Engobe is used to give the body a different texture and color which fires to a more vitreous state than the body. A white engobe is obtained from

- flint: 14%
- feldspar: 32%
- china clay: 42%
- malachite: 12%

The glaze is normally leadless and has the following composition:

- sand: 15%
- feldspar: 25%
- china stone: 34%
- zinc oxide: 10%
- zircon ($ZrSiO_4$): 10%
- whiting ($CaCO_3$ obtained from chalk mineral): 6%

The whole assembly of body, engobe, and glaze is fired in a single step at 1220°C.

7.2.3. Wall Tiles

Wall tile formulations are selected so that shrinkage during firing is very low and does not vary from one location in a kiln to another. An ordinary earthenware body shrinks about 8–10%, but a limestone type body shrinks only about 1%. This is due to the low vitrification obtained from limestone type bodies. Wall tile bodies are composed of:

- ball clay: 22.5–25%
- china clay: 22.5–25%
- flint/quartz sand: 40%
- limestone ($CaCO_3$): 10–15%

The components are mixed in ball mills and pressed into a die cavity to form the tile. After manufacturing the tile under the press, it is transferred to the glazing and drying lines. Tiles that require decorating are transferred to a screen printer. Some designs may require up to three screens [7.7]. Bisque firing is done in a gas-fired tunnel kiln at a maximum temperature of 1100°C for 19–24 h. Glost firing takes place at 1080°C. The resulting product has high porosity (about 36% apparent porosity) and low moisture expansion for craze resistance. Crazing is the phenomenon of polygonized cracking in glazes and enamels that develops from tensile stresses. The already low moisture expansion (0.06%) can be contained by using a glaze with an expansion at 500°C of 0.06–0.08% less than that of the body. Moisture resistance is an important property of wall tiles since they are commonly used in humid areas such as kitchen and bathroom walls and floors.

7.2.4. Glazes

A glaze is a glass layer applied to a ceramic body. The hygiene of sanitary ware is provided by glazes since a glaze has practically zero porosity and therefore, does not host microorganisms. Glazes are also necessary for aesthetic features such as color and gloss. Lead glazes are used for their brilliant appearance. They can be used up to glost firing temperatures of 1150°C, above which they volatilize. Leadless glazes have been developed and properties similar to lead glazes are obtained. For example, Bristol glazes were developed in England to eliminate the toxicity of lead-fluxed glazes. These glazes use zinc as a flux [7.1]. Lead compounds may be soluble in acidic environments, and this may lead to poisoning, for example from a mug filled with orange juice or wine. To prevent such situations, lead compounds are fritted to produce glaze components that are practically insoluble in dilute acids.

Fritting is an operation where the glaze components are heated to the glass formation temperature, cooled, and ground to the desired particle size. Fritting converts soluble compounds into insoluble ones. This is necessary since soluble glaze components would migrate into the pores of the bisque ware and the glaze properties would change radically. Fritting also prevents bubbling and pinholes that may originate from compounds such as calcium carbonate that produces gases upon heating. Raw glazes are glazes that do not undergo any fritting operations. They are generally used at high temperatures, for example, for sanitary ware glaze fired at about 1200°C. Glazes are prepared from oxide mixtures. The raw materials for such oxides may be similar to the raw materials used in the bisque. Typical sources are bauxite or china clay for Al_2O_3, flint or quartz sand for SiO_2; borax or boric acid for B_2O_3; whiting, fluorspar, or colemanite for CaO; lead bisilicate frit ($PbO{\cdot}2SiO_2$) for PbO; pearl ash, potassium nitrate, or potash feldspar for K_2O; and soda ash, borax, or soda feldspar for Na_2O. All ingredients are weighed in accordance with the recipe and are wet ground to pass a 200 mesh sieve. After grinding, the powders are passed over magnets to separate iron-bearing material. After that the viscosity is adjusted, and the glaze is applied by dipping the bisque in the glaze or by spraying the glaze with a gun. Some examples of glaze recipes in mole fraction are given in Table 7.1, where the oxides have been categorized as basic, amphoteric, and acidic, respectively. The sum of basic oxides is equal to one. This convention helps comparing different glaze compositions on an equal basis.

Table 7.1. Compositions of Some Industrial Glazes [a]

Glaze Type	Firing Temperature (°C)	Basic Oxides (Moles)	Amphoteric Oxides (Moles)	Acidic Oxides (Moles)
Lead glaze for earthenware	1000–1150	0.4 PbO 0.3 CaO 0.3 $(Na,K)_2O$	0.25 Al_2O_3	2.5 SiO_2 0.5 B_2O_3
Leadless glaze for earthenware	1000–1150	0.55 CaO 0.45 $(Na,K)_2O$	0.30 Al_2O_3	3.0 SiO_2 0.8 B_2O_3
Sanitary ware glaze	1200–1250	0.6 CaO 0.2 ZnO 0.2 $(Na,K)_2O$	0.35 Al_2O_3	3.0 SiO_2
High-temperature glaze	1465	0.3 $(Na,K)_2O$ 0.7 (Mg,Ca)O	1.1 Al_2O_3	14.7 SiO_2
Wall tile glaze	–	0.35 CaO 0.32 ZnO 0.01 MgO 0.27 Na_2O 0.05 K_2O	0.26 Al_2O_3	0.05 B_2O_3 2.65 SiO_2
Hard paste porcelain glaze	1400	0.7 CaO 0.3 K_2O	1.0 Al_2O_3	10.0 SiO_2
Soft paste porcelain glaze	–	0.7 CaO 0.3 K_2O	0.8 Al_2O_3	7.0 SiO_2

[a] References:

1.W. Ryan and C. Radford: *Whitewares, Production, Testing, and Quality Control*. (Pergamon Press Oxford England, 1987)

2. W.D. Kingery, H.K. Bowen, and D.R. Uhlmann: *Introduction to Ceramics.* (John Wiley & Sons New York, 1976)

3. *Commercial Glasses, Advances in Ceramics,* Vol.18. D.C. Boyd and J.F. McDowell (eds.). (The American Ceramic Society Columbus OH, 1986), pp. 67–72

7.2.5. Colorants

A colorant may be applied to ceramic ware in at least five different ways [7.9]: as a body stain, as an engobe, as an underglaze color, as a colored glaze (in glaze), or as an overglaze color. *Body stains* are added directly to the ceramic body formulation. The main disadvantage of this method is that shading may occur due to temperature differences in the kiln. *Engobes* are colored slips applied to the surface of an unfired body. The engobe must be resistant to the corrosive flux used to adhere itself to the body. It also should be resistant to glost firing while in contact with the molten glaze. Engobes and body stains must be stable up to the bisque firing temperature that usually varies between cone 7 (1225°C) and cone 11 (1300°C). *Underglaze colors* are applied to the body before glazing. Colorants can also be added directly to the glaze, yielding a *colored glaze*. Underglaze colors and colored glazes must be stable to glost firing, which is usually done at cone 06 (1000°C) and cone 4 (1200°C). Pigments that are applied should be resistant to corrosive attack by the molten glaze. *Overglaze colors* are applied to a glazed and fired product. A separate firing step, termed decorating fire, is necessary to stabilize overglaze colors. An overglaze or glass colorant must be stable to the decorating fire applied between cone 020 (625°C) and cone 016 (775°C). The ceramic pigments must be stable to the molten flux used in their application. For example, PbO fluxes common in overglaze decorating are very corrosive.

Colorants are produced by weighing and mixing the ingredients, calcining at temperatures, depending on the particular color involved, wet grinding, and washing [7.3]. Calcination is a critical step for obtaining the right color and for total decomposition of raw materials, so that gas evolution is completed before the ceramic ware is fired. The calcination temperature should be equal or higher than the firing temperature of the decorated ware. Grinding is applied to a mean particle size ranging from 1 to 10 μm, with no residue on a 325 mesh (44 μm) screen.

A ceramic color does not consist only of a coloring oxide. Other components such as modifying agents, diluents or fillers, and fluxes are usually added to the pigment. For example, the colorant *Cobalt Blue* or *Royal Blue* consists of the following components:

- cobalt: the main colorant that gives an intense blue color
- whiting: a modifying agent that gives the blue color a special blue tint and stability
- flint and alumina: act as a diluent or a filler and yield a slightly weaker but more stable color
- feldspar: acts as a flux to aid in sintering all of the components during heat treatment.

Opacifiers are extensively used in the sanitary ware industry while only a limited amount is used in tableware and tiles for special decorative effects. They are added to the glaze. Common opacifiers include zircon (which is generally used with all zirconia and zircon pigments) and titania (which is used with titanium-containing pigments). The compatibility of all these components is a critical requirement for quality and reproducibility. For example, some pigments such as Victoria green garnet, chrome tin orchid cassiterite, and manganese alumina pink corundum are not stable in the presence of ZnO, which may be in a glaze. On the other hand, high ZnO additions are required for the stability of iron brown hematite, chrome alumina pink spinel, or iron chromite brown spinel [7.9].

An important aspect of ceramic colorants is their thermal stability. Pigments have to be thermally stable at the glazing temperatures that range from 800–1400°C. They must also have good saturation since they are frequently mixed with opaque white materials or materials with a high refractive index. Saturation is a critical aspect because it cannot be improved upon and it also determines the amount of pigment needed to achieve a particular color. This in turn, means freedom of formulation and cost. Ceramic colorants should be chemically resistant to acids and alkalis and should have low solubility in glass. Meanwhile, the glass used for glazing should be transparent, stable, and unaggressive so that it is compatible with the pigment applied. The mechanism of color formation and the various pigments used as colorants were discussed in Sect. 4.3.5.

7.3. Cement

Various types of cements exist such as lime, gypsum plaster, natural cement, high alumina cement, pozzolana, masonry cement, magnesium oxychloride cement, calcium aluminate cement, and Portland cement. Among these, Portland cement has the largest share of the world's production [7.10].

Pozzolana cement was widely used before the invention of Portland cement in 1824 [7.11]. The first step in making pozzolana cement is to heat chalk at a relatively low temperature using simple kilns,

$$CaCO_3\ (\text{chalk}) \xrightarrow{heat\ (600^o\,C)} CaO\ (\text{lime}) + CO_2 \tag{7.1}$$

The lime is then mixed with water and volcanic ash. Water reacts with CaO and yields $Ca(OH)_2$ in the bulk,

$$CaO + H_2O \rightarrow Ca(OH)_2 \tag{7.2}$$

Volcanic ashes used to make pozzolana cement have an active surface, and this surface reacts with lime and water, forming what is known as tobomorite gel

$$3CaO + 2SiO_2 + 3H_2O \rightarrow (CaO)_3(SiO_2)_2(H_2O)_3 \tag{7.3}$$

The final product is made up of hydrated lime particles bonded by tobomorite gel.

Portland cement is produced by mixing limestone and clay or other materials containing silica, alumina, and iron oxide, heating the mixture to about 1450°C, and fine grinding the resulting clinker with some gypsum addition [7.10,7.12]. The main reaction, which occurs upon firing limestone and clay is

$$\text{Limestone} + \text{clay} \xrightarrow{heat\ (1450^{o}\text{C})} \text{alite} + \text{belite} + \text{aluminate} + \text{ferrite} \tag{7.4}$$

Clinker is the partially fused, nodular mass obtained by firing. The gypsum addition provides controlled setting, and it may be partly replaced by other calcium sulfate containing materials. A typical Portland cement contains 64–67%CaO, 5–5.5%Al_2O_3, 3%Fe_2O_3, 22%SiO_2, 1.4%MgO, and 2.1%SO_3. Minor amounts of alkali sulfates are also present. The clinker is made up of four major phases: 50-70% alite, 15–30% belite, 5–10% aluminate, and 5–15% ferrite. *Alite* is tricalcium silicate (Ca_3SiO_5) modified by foreign ions such as Mg^{2+}, Al^{3+}, and Fe^{3+}. *Belite* is dicalcium silicate (Ca_2SiO_4) modified largely to the β-polymorph by foreign ions. The *aluminate* phase is tricalcium aluminate ($Ca_3Al_2O_6$) modified by foreign ions such as Si^{4+}, Fe^{3+}, Na^+, and K^+. The *ferrite* phase is tetracalcium aluminoferrite (Ca_2AlFeO_5) modified in composition by changing the Al/Fe ratio and adding foreign ions.

The two basic milling processes are wet and dry processes. In the wet process, the dry ingredients and sufficient water are fed in suitable proportions to the wash mill and milled together to obtain a creamy slurry. Wet milling was preferred over dry milling due to the more accurate control of the cement composition. The modern dry milling processes, however, can control the composition as well as the wet process. In dry milling, the raw materials are crushed, dried, proportioned, and ground by steel balls or other milling media. The mixture is fed into a rotary kiln and burnt at 1300-1500°C. At the lower end of the kiln, the clinker is transferred to coolers. Next, the cooled clinker is ground in grinding mills. Gypsum is added in small quantities during milling. The resulting fine powder is stored in silos and then drawn for packing [7.10].

Cement becomes hard when mixed with water because all of the major phases react with water and form a colloidal gel [$(CaO)_3(SiO_2)_2(H_2O)_3$ – tobomorite gel]. The reactions that produce the tobomorite gel can be simplified as [7.11]

$$2\text{Belite} + 4H_2O \rightarrow (CaO)_3(SiO_2)_2(H_2O)_3 + CaO{\cdot}H_2O + \text{heat} \tag{7.5}$$

$$2\text{Alite} + 6H_2O \rightarrow (CaO)_3(SiO_2)_2(H_2O)_3 + 3CaO{\cdot}H_2O + \text{heat} \tag{7.6}$$

Upon dehydration, the gel becomes a hard solid. Alite reacts relatively quickly with water and in Portland cement. It is the most important component for strength development, especially at ages up to 28 days. Belite, in contrast, reacts slowly with water. During the first 28 days after reaction, its contribution to strength is low. However, at the end of one year, the contributions of alite and belite are about the same. The aluminate phase reacts very quickly with water. This results in undesired rapid setting,

$$\text{Aluminate} + 6H_2O \rightarrow (CaO)_3(Al_2O_3)(H_2O)_6 + \text{heat} \tag{7.7}$$

The setting rate is decreased by a set-controlling agent, usually gypsum. The reaction rate of the ferrite phase may vary, depending on composition and other characteristics but it is usually high initially and intermediate at later stages [7.12]. The most important use of Portland cement is in construction and low stress applications. Portland cement is stronger than pozzolana due to two main factors. One factor is that the gel forms in the bulk of the cement, not just at the surface. The second factor is the formation of spines during hydration. These spines grow and gradually fill the voids between cement grains (Fig. 7.2). They further grow into the porous surfaces of brick, stone, or other structures. This forms an interlocking network of needles consolidated into a rigid structure.

Calcium aluminate cement or *ciment fondu* is prepared by fusing a mixture of limestone and bauxite or other alumina-containing material at 1450-1600°C. The raw materials have to be low in silica for rapid hardening. The molten clinker is cooled, crushed, and ground to yield a surface area of about $300 m^2/kg$. The resulting color is usually gray. Monocalcium aluminate ($CaO{\cdot}Al_2O_3$) develops the main hydraulic activity. Thus it is essential for strength development. It sets normally but hardens rapidly. Useful strength develops in only 6 hours; at 24 hours, the strength level obtained from Portland cement in 28 days can be achieved. Calcium aluminate cements have excellent corrosion resistance and good abrasion resistance. Refractory castables made of calcium aluminate cements and refractory aggregates have found applications in steel and other refractory industries. They have high thermal shock resistance. Cements made from bauxite may be used up to 1350°C. White calcium aluminate cements, low in iron oxide and high in alumina, may be used up to 2000°C [7.12].

Expansive cements are cements that expand slightly on hardening or have no net shrinkage after drying. This is usually achieved by increasing the formation of ettringite ($6CaO{\cdot}Al_2O_3{\cdot}3SO_3{\cdot}32H_2O$) in regular Portland cement. Two categories of expansive cement are shrinkage compensated or self-stressing cements. Regular Portland cement, if restrained, may crack due to tensile stresses that develop during drying shrinkage. *Shrinkage-compensated cement* is a type of cement whose drying shrinkage is balanced to prevent cracking. This balance is achieved by reinforcing the cement. This type of cement has important applications in cases where water leaks, thus cracks, are not tolerated, for example, in multistory car parks and water tanks. *Self-stressing cement* has a net expansion upon setting that results in residual stresses in the structure. This provides a mechanical prestressing effect, which is used for thin and strong articles and structures such as precast pipes. Unreinforced cement has limited use such as wall facing. The majority of cement is used as *concrete,* which is cement mixed with crushed stone and sand as a filler material or as *reinforced concrete*, i.e. concrete reinforced with steel rods.

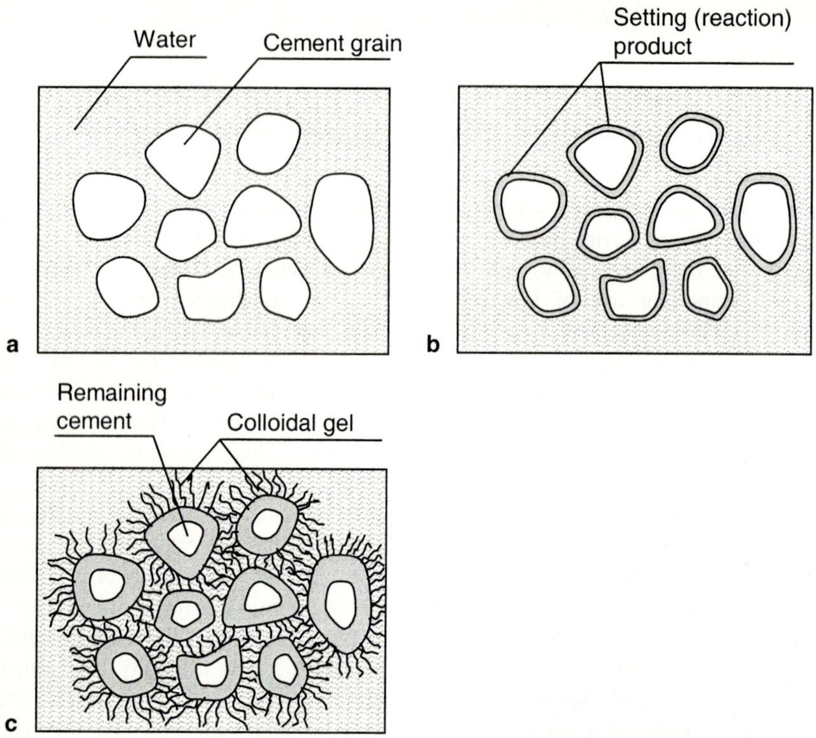

Fig. 7.2. Setting and hardening of Portland cement: Cement grains in water before reaction **a**, weak bonds forming between the grains **b**, and strong bonding formed by interpenetrating spines **c**

The elastic modulus of concrete obeys the rule of mixtures discussed in Sect. 4.1. The fracture toughness of cement is about 0.3 MPa $m^{1/2}$ and that of concrete about 0.4 MPa $m^{1/2}$. The average tensile strength of cement and concrete is around 4 MPa. The fracture toughness of concrete and high-strength concrete increases with increasing compressive strength [7.13].

During compression, the stress–strain curve of cement or concrete exhibits a linear elastic regime where stable crack propagation occurs. A peak forms when the density of cracks reaches a maximum; they link and lead to a general crumbling of the body. The mean crushing strength of 100 mm cubes of concrete is normally around 50 MPa, but this shows a large variation for the same set of materials (between 40 MPa and 60 MPa). Thus, a statistical treatment such as Weibull analysis is necessary to study the strength of concrete. There is also a size effect to consider. Cubes 150 mm wide have about 10% lower strength than those that are 100 mm wide. The strength of cement and concrete can be improved by

1. milling the cement to a fine powder,
2. using the optimum water/cement ratio, which is 0.38,

3. using polymeric lubricants to improve packing, and
4. applying pressure during setting to squeeze out residual porosity.

By using all of these precautions, cement with less than 2% porosity and a tensile strength of 90 MPa can be achieved. There are simpler methods to improve the strength of cement. One is to add polymers, which fill the pores and improve the fracture toughness. Another way is to add glass or steel fibers. Steel-reinforced concrete is a typical fiber-reinforced composite; it carries tensile stresses, which the matrix does not carry and it can be prestressed to keep the concrete in compression. Reinforced and polymer-impregnated concrete are used for special applications due to their relatively high cost [7.11].

In addition to conventional cements, relatively new types of cements and binders have been developed [7.14]. Some of these have found commercial use, and others are still in the developmental stage. They include calcium aluminosilicate glass cements, multiphase high early strength cements (including regulated set Portland cement and alkali-activated aluminosilicate cements), densified systems containing homogeneously arranged ultrafine particles (DSP systems), and macro defect-free cement pastes (MDF). Detailed analysis of these systems can be found in Gartner and Myers' review [7.14].

7.4. Refractories

7.4.1. Compositions and Properties

Classical refractory ceramics are composed of natural materials with high-temperature resistance. Generally they are composed of large refractory grain particles bonded with a fine grained bond. Pores and cracks are common in refractory microstructures [7.3]. The most important industrial use of refractories is in metallurgical furnaces. Therefore, they are commonly classified according to their behavior toward metallurgical slags [7.15]. Acidic refractories are based on SiO_2 and resist acidic slags. Typical refractories in this class include silica, fireclay series, sillimanite, and andalusite. Basic refractories are based on MgO and resist basic slags. These types of refractories include magnesite, dolomite, chrome–magnesite, and magnesite–chrome. Neutral refractories are relatively inert to both acidic and basic slags. Carbon, chromite, and forsterite belong to this class. Appendix 17 lists compositions and properties of various refractories. Industrial refractories are supplied mostly as bricks of standard shapes and sizes. Other commercial forms include tubes, castables, sheet, cloth, tape, and formable shapes.

Among important properties of refractories are coefficient of thermal expansion, thermal conductivity, thermal shock resistance, refractoriness, cold and hot strength, slag resistance, porosity, and apparent density. Low thermal expansion is desired for most cases. This is important in industrial furnaces since the refractories have to accommodate frequent heating–cooling cycles. Low thermal conductivity is obtained by employing highly porous refractory bodies. This property is critical in those zones of a furnace that are designed for thermal insulation. Thermal shock resistance should be high for most applications. This and other thermal properties were discussed in Sects. 4.1.4. and 4.2. Refractoriness is a guide to the quality of a refractory product. It compares the melting temperature of a refractory to the sagging of standard pyrometric cones. Magnesia, one of the most refractory oxides, melts at 2800°C. Cold strength indicates how well a refractory has been fired. Hot strength, on the other hand, has more practical importance because it demonstrates the ability of a refractory to withstand failure under compression at the service temperature. Slag resistance is normally measured qualitatively. Some of the tests try to simulate practical contact conditions with slags. Slag resistance depends on chemical reactions that may occur between the slag and the refractory. Some of the principles of ceramic reactions and corrosion will be found in Sect. 4.6. Porosity and apparent density are directly related properties. Low porosity and therefore, high density are desired in regions of contact with corrosive and abrasive media. In conventional refractories, porosity values typically range from 20–25%. Lower values (10–15%) can be obtained by special processes. High porosity is desirable for better insulation, usually in the outer zones of a furnace.

7.4.2. Production of Refractories

The common steps for producing of conventional refractories are crushing and grinding of raw materials, blending, molding, drying, and firing. Raw materials used include commonly available mineral deposits such as clays, sands, magnesite, dolomite, and spinels. Methods for crushing, grinding, and drying were discussed in Chap. 3. Blending involves mixing different raw materials if the product is not obtained from only one mineral, or it may also involve mixing different powder sizes. Lower porosities can be achieved by using a bimodal size distribution (see Sect. 3.3.1.3.). Blending is done in a paddle mill, which also exerts a kneading action. Water and binders are added at this stage. The plastic mixture is then molded manually or by extrusion followed by pressing for mass production. An alternative method is slip casting that is applicable to colloidal clay suspensions. Figure 7.3 shows examples of refractory products produced by this method. Bricks (or other shapes) are dried on large drying floors or in tunnel kilns. This is followed by firing in a tunnel kiln. The firing temperatures are 1400°C for dolomite, 1450–1500°C for chrome–magnesite and magnesite–chrome, 1000°C for carbon, and 1500°C for silica refractories.

7.5. Glass

7.5.1. Compositions and Properties

Glass is a common term used for noncrystalline solids. The use of glazes on pottery dates back to 12,000 B.C. [7.1], and the earliest samples of solid glass

Fig. 7.3. Slip-cast low cement castables (buff colored) and alumino–silicate refractory products prepared by the ethyl silicate process (light colored) (courtesy of Dyson Precision Ceramics Dewsbury England)

objects date from 7,000 B.C. Both of these discoveries trace back to ancient Egyptian civilizations.

The structure of glass was discussed in Sect. 1.4. Here, some properties and manufacturing methods will be discussed very briefly. Glass is a material, which has hundreds of different applications and corresponding compositions. Some of these applications are window glass, light bulbs, TV/computer display tubes, optics, fiber optics, and cookware. [7.16]. Many books and other forms of literature are dedicated to glass. Thus, it is possible to find extensive literature to analyze the subject further. Important groups of commercial glasses are soda–lime, lead, borosilicate, aluminosilicate, and high-silica glasses. Table 7.2 lists typical compositions and properties of these glass types. Soda-lime glass is the type most commonly employed in daily use. The approximate formula is $Na_2O{\cdot}CaO{\cdot}6SiO_2$. Minor additions of Al_2O_3 are necessary for weathering (corrosion) resistance and devitrification (crystallization) resistance, borates for easier working and lower thermal expansion, and arsenic oxide or antimony oxide for fining (i.e., removal of bubbles) [7.17]. These glasses are inexpensive, and they are used for bottles, electric bulbs, and glazing. Among special purpose glasses are borosilicate, aluminosilicate, lead, and high-silica glasses. Borosilicate glasses (better known by their trade name, Pyrex®) satisfy requirements for corrosion resistance and thermal shock resistance. They are suitable for automobile headlamps, cookware, and laboratory apparatus. Aluminosilicate glasses are used in applications requiring high chemical durability, strength, and resistance to devitrification. Some examples of such applications are cookware, fibers, and seals. Lead glass has a high refractive index, and therefore, high luster. Applications include high quality artware and tableware (so called "crystal" glass), radiation shielding, lamp envelopes, seals, and optical ware. High-silica glasses are resistant to chemical attack. Fused silica is a special engineering glass of more than 99% SiO_2. It is employed in high temperature/thermal cycling applications such as crucibles for Si and Ge melting. Fused quartz, a more expensive form of silica, is used in special optical applications. Due to their high melting temperature, high-silica glasses are very difficult to produce. Suitable viscosities for shaping soda–lime glasses are reached from 700–1000°C, but the corresponding temperature regimes are 800–1200°C for borosilicate glass and 1500–1600°C for fused silica [7.18].

Glass is well known for its superior optical properties compared to those of polymers and other materials. Some optical properties of glasses were discussed in Sect. 4.3. Among important parameters that can influence optical properties are the amount of imperfections, impurity ions, stress in the glass, and surface roughness. The electrical insulating properties of glass are also well exploited. For example, glass is used extensively in electric lighting [7.19]. Similar to other ceramics, glass is susceptible to microflaws. The strength of glass depends on the surface conditions and the quantity of microflaws, among other things. Some applications such as windshields of various vehicles require high strength and toughness. The strength of glass items can be improved by placing their surfaces under compressive stresses. This is achieved by quick and controlled cooling of the glass surface with an air blast (also called tempering) or by substituting smaller alkali cations in the surface with relatively larger alkali cations [7.19].

7.5.2. Production of Glassware

The glass blowing process was invented by the Romans in the first century B.C. [7.17]. A major development in glass production occurred in the mid-nineteenth century. A firm under James Hartley of England devised a new process of rolling glass for manufacturing large window sheets, which made glass a common household commodity [7.20]. In the 1950s, Alistair Pilkington of the Pilkington Glass Company in England developed the floating glass method, which was less expensive and produced higher quality flat glass [7.17,7.20]. Numerous other developments in glass science followed up to modern times.

Traditionally, glass has been melted in a refractory ceramic pot or crucible as a batch. Raw materials required for making the glass are mixed in the desired composition and heated to the melting temperature, generally ranging between 1500 and 1600°C. The principal ingredient in almost all glasses is silica (SiO_2). The primary source of silica is natural quartz sand, which may have a total impurity content of less than 100 ppm [7.17]. Other components added to silica

Table 7.2. Composition and Properties of Commercial Glasses [a]

	Approximate Composition, wt%								Thermal Expansion	Softening Point	Density	Refractive Index	Young's Modulus
Glass Type	SiO_2	Na_2O	K_2O	CaO	PbO	B_2O_3	Al_2O_3	MgO	($\times 10^7$/ °C)	(°C)	(g/cm^3)		(GPa)
Soda–lime	70–75	12–18	0–1	5–14	–	–	0.5–2.5	0–4	87	735	2,5	1,51	69
Lead	53–68	5–10	1–10	0–6	15–40	–	0–2	–	–	–	–	–	–
Borosilicate	73–82	3–10	0.5–1	0–1	1–10	5–20	2–3	–	32	820	2,5	1,51	69
Aluminosilicate	57	1	–	5.5	–	4	20.5	12	–	–	–	–	–
High silica	96	–	–	–	–	3	–	–	5.5	1667	2,2	1,458	69

[a] References:

1. T.D. Taylor: *Structure and Properties of Glasses*. In: *Engineered Materials Handbook,* Vol.4: *Ceramics and Glasses.* (ASM International Materials Park OH, 1991), pp. 564–569
2. R.M. Brick, A.W. Pense, and R.B. Gordon: *Structure and Properties of Engineering Materials.* (McGraw-Hill New York, 1977)
3. R.H. Doremus: *Glass Science.* (John Wiley & Sons New York, 1994)

include soda ash for Na_2O, limestone for CaO, potash for K_2O, alumina, and other minerals to achieve the desired composition [7.21]. Recycled glass is also used extensively in modern practice. Besides being an important aid for environmental preservation, recycled glass facilitates melting. Alkali oxides melt first during heating and dissolve other components such as quartz and alumina, which have higher melting points. Melting is accomplished in special fuel-fired or electric furnaces. Melting, fining, and homogenization of a typical charge takes from 24 to 48 hours.

After melting, glass with the proper viscosity undergoes shaping operations. Three main categories of shaping are discrete processes for piece ware, continuous processes for flat glass, and fiber production processes. Spinning, pressing, and blowing are industrially important processes for piece ware production [7.21].

Spinning involves centrifugal shaping of a gob dropped into a mold. Rotation of the mold spreads the molten glass. The process is used to produce funnel shaped parts such as the back side of cathode ray tubes for TV and computer monitors.

Pressing is used for mass-producing parts such as tableware, bakeware, headlight lenses, TV tube faceplates (front viewing screens), and other flat items. In this process that can be highly automated, a gob of glass is placed inside a mold cavity and pressed into the desired shape by a plunger.

Blowing is a process used for producing hollow shapes. It may consist of a series of blowing operations in different molds that may come after a pressing step. Traditional blowing is a complicated manual process. Modern manufacturing of jars, bottles, light bulb enclosures, drinking glasses, and laboratory glassware involve blowing operations performed by automated equipment.

Flat plate glass can be produced by drawing, rolling, or floating glass processes. In *drawing*, molten glass is pulled from the pool by rollers through a die. Coolers immediately above the mold help solidify the glass. After sufficient hardening, the plate glass is cut into the desired size. *Rolling* is a similar process where molten glass passes through forming rolls, and the resulting plate glass moves over support rolls. These processes have the disadvantage of distortion due to roller–glass contact. In the *floating glass process*, a continuous ribbon of molten glass passes over a bath of molten tin. The surfaces become flat and parallel due to contact with the flat molten tin. The ribbon is cooled down in a cooler region of the tin bath and transferred to an annealing furnace after becoming hard. There are no markings on the glass surface since no roller contact occurs in the viscous state.

Other shapes produced from glass include tubes and fibers. Glass fiber production methods were discussed in Sect. 2.4.1. A lot more detail exists on various manufacturing methods of glassware, but it is beyond the scope of this section. For further information, the reader can consult the references for this chapter.

Appendix

Appendix 1. Typical Hardness and Elastic Modulus Values of Dense Ceramic Materials [a]

Material	Microhardness (GPa)	Elastic Modulus (GPa)
Oxides		
Al_2O_3	18–23 [1]	360–400 [2]
BeO	11–13 [1]	340–400 [1]
CeO_2	–	170 [2]
Cr_2O_3	29 [2]	–
Dy_2O_3	–	160 [4]
Er_2O_3	–	175 [4]
Li_2O	–	140 [5]
Lu_2O_3	–	200 [4]
MgO	9 [12]	240 [12]
Sc_2O_3	–	225 [4]
Sm_2O_3	–	145 [4]
TiO_2	7–11 [2]	280 [2]
Y_2O_3	–	170 [4]
Yb_2O_3	–	190 [4]
c–ZrO2	10–15 [2]	100–210 [2]
PSZ	10–12 [2]	200 [2]
ZTA (6 vol%ZrO_2)	–	340 [6]
ZTA (10 vol%ZrO_2)	–	320 [6]
Y–TZP	10–12 [13]	140–200 [13]
Ce–TZP	7–10 [13]	140–200 [13]
Cordierite	–	120 [1]
Mullite	8 [2]	145–230 [3]
Soda Lime Glass	5.6 [12]	60–80 [2],70 [12]
Glass ceramics	6–7 [2]	80–140 [2]
Fused quartz (SiO_2)	10 [8]	70–75 [8]

(Continued)

Appendix.1. *(continued)*

Material	Microhardness (GPa)	Elastic Modulus (GPa)
Nitrides		
AlN	12 [8]	300–320 [8]
h-BN	2 [12]	20–100 [8]
c-BN	50 [10], 30–55 [12], 49–54 [15]	–
β-Sialon	18 [8]	300–400 [8]
Si_3N_4	8–19 [2]	275–315 [9]
h-TaN	31 [12]	–
c-TaN	11 [12]	–
TiN	16–20 [2]	250 [2]
Carbides		
B_4C	25–37 [11], 44 [12]	370–470 [11]
Cr_3C_2	10–18 [2]	370 [2]
Mo_2C	19 [12]	–
SiC	20–30 [2]	350–440 [9]
TaC	16–24 [2]	285 [2]
TiC	28–35 [2]	430 [2]
WC	22 [12]	680 [14]
W_2C	27 [12]	–
Borides		
CrB_2	18 [3]	–
Cr_3B	11 [12]	–
LaB_6	27 [3]	–
MoB_2	12 [3]	–
TaB_2	25 [3]	250 [14]
TiB_2	33–36 [3]	510–575 [2]
ZrB_2	22 [3]	–

(Continued)

Appendix 1. *(continued)*

Material	Microhardness (GPa)	Elastic Modulus (GPa)
Miscellaneous		
Cemented carbides	8–20 [2]	400–650 [2]
Diamond	130 (dry, (100) plane) [12]	1000 [3]
Graphite (C)	35 [12]	14 [12]
Si	9 [12]	168 [12]
B	35 [12]	–

[a] References:

1. *Handbook of Materials Science*, Vol.2. C.T.Lynch (ed.). (CRC Press Boca Raton FL, 1974)
2. W.J. Lackey et al: Adv. Ceram. Mater. **2 [1],** 24–30 (1987)
3. I.J. McColm: *Ceramic Science for MaterialsTechnologists*. (Leonard Hill New York, 1983)
4. K.K. Phani and S.K. Niyogi: J. Am. Ceram. Soc. **70 [12],** C-362–366 (1987)
5. C.Y. Chu et al.: J. Am. Ceram. Soc. **72 [9],** 1643–1648 (1989)
6. G. De Portu and C. Fion: In: *Advances in Ceramics* 24 B. S. Somiya et al. (eds). (The American Ceramic Society Westerville OH, 1988)
7. D.C. Phillips: In: *Fiber Reinforced Ceramics*. A. Kelly et al. (eds.). (North Holland Amsterdam, 1983)
8. Manufacturer's data
9. *Fine Ceramics*. S. Saito (ed.). (Elsevier Essex England, 1985)
10. *Ceramic Matrix Composites*. R. Warren (ed.). (Blackie and Son London England, 1992)
11. F. Thevenot: J. Eur. Ceram. Soc. **6 [4],** 205–225 (1990)
12. I.J. McColm: *Ceramic Hardness*. (Plenum Press New York, 1990)
13. W.E. Lee and W.M. Rainforth: *Ceramic Microstructures*. (Chapman and Hall London England, 1994)
14. *ASM Engineered Materials Handbook*. (ASM International Metals Park OH, 1989)
15. T. Taniguchi, M. Akaishi, and S. Yamaoka, J. Am. Ceram. Soc. **79 [2],** 547–49 (1996)

Appendix 2. Strength and Fracture Toughness of Monolithic Ceramics [a]

Material	Theoretical Density (g/cm^3)	3-Point Flexural Strength (MPa)	4-Point Flexural Strength (MPa)	Tensile Strength (MPa)	Fracture Toughness ($MPa\ m^{1/2}$)
Al_2O_3	3.99 [3]	323 [4]	–	280–300 [1,5]	2.5–4 [6,7]
Cordierite	–	–	100–180 [18,21]	–	1.5–2.2 [18,21]
Cr_2O_3	5.21 [11]	270 [11]	–	–	3.9 [11]
Mullite	3.22 [1]	–	240 [1]	–	2.2 [2]
Si_2N_2O	–	400 [23]	–	–	3.0 [23]
tet-TiO_2	4.25 [11]	70–100 [11]	–	350 [24]	2.5 [11]
Zircon	–	280 [22]	–	–	–
ZrO_2 (TZP)	6.1 [3]	–	1150 [8]	–	6–9 [8,9]
ZrO_2 (PSZ)	5.7 [11]	–	600 [10]	80–145 [25]	4.1 [10]
B_4C	2.52 [14]	300 [14]	500 [14]	350 [24]	3.0–3.5 [14]
Cr_3C_2	6.7 [11]	50 [11]	–	–	–
RBSC	–	175–450 [26]	–	77 [26]	–
SiC	3.21 [11]	–	350–450 [12]	180–280 [12]	2.5–6 [6,11]
SiC (hot-pr.)	–	300–800 [26]	–	160 [28]	3–4 [26]
SiC (sint.)	–	275–535 [26]	–	–	2.5–6.5 [26]
TaC	14.5 [11]	100–290 [11]	–	96–290 [25]	–
TiC	4.92 [11]	240–270 [11]	–	110–275 [25]	4 [3]
WC	–	480–830 [25]	–	344 [25]	–
TiB_2	4.5 [3]	700–1000 [11]	–	–	5–7 [3]
AlN	3.26 [16]	–	350–400 [15]	125 [27]	2.7 [3]
Si_3N_4	3.18 [11]	1000–1200 [13]	800–900 [12]	600–800 [12]	4–10 [6]
Si_3N_4 (hot-pressed)	–	450–1100 [26]	–	–	2.8–6.6 [26]
Si_3N_4 (sint.)	–	275–840 [26]	–	–	3–5.6 [26]
RBSN	–	50–300 [26]	–	70–210 [26]	3.6 [26]

(Continued)

Appendix 2. *(continued)*

Material	Theoretical Density (g/cm^3)	3-Point Flexural Strength (MPa)	4-Point Flexural Strength (MPa)	Tensile Strength (MPa)	Fracture Toughness ($MPa\ m^{1/2}$)
Borosilicate glass	–	–	80 [17]	–	1.0 [17]
Aluminoborosilicate glass	–	–	100 [18]	–	1.0 [18]
Aluminosilicate glass	–	–	–	–	0.8 [19]
Li aluminosilicate ceramic glass	–	103 [20]	–	–	0.9 [20]
Soda–Lime glass	–	–	–	–	0.7 [19]

[a] References:
1. I.J. McColm: *Ceramic Science for Materials Technologists*. (Leonard Hill New York, 1983)
2. I.A. Aksay et al.: J. Am. Ceram. Soc. **74 [10],** 2343–2358 (1991)
3. *Ceramic Matrix Composite*, R. Warren (ed.). (Blackie and Son London England, 1992)
4. J.R. Brockenbrough et al.: J. Am. Ceram. Soc. **69 [8],** 634–637 (1986)
5. *Handbook of Materials Science*, Vol.2, C.T. Lynch (ed.). (CRC Press Boca Raton FL, 1974)
6. P.F. Becher: J. Am. Ceram. Soc. **74 [2],** 255–269 (1991)
7. J. Liu and P.D.Ownby: Ceram. Eng. Sci. Proc., **12,** 1245–1253 (1991)
8. N. Claussen et al.: J. Am. Ceram. Soc., **69,** 288–292 (1986)
9. J.M. Roberts et al.: Ceram. Eng. Sci. Proc., **12,** 1193–1200 (1991)
10. E. Yasuda et al.: In: *Science and Technology of ZrO_2,* III, ed. S.Somiya et al. pp.1101–6. (The American Ceramic Society Westerville OH, 1988)
11. W.J. Lackey et al.: Adv. Ceram. Mater., **2 [1],** 24–30 (1987)
12. N.L. Hecht et al.: Am. Ceram. Soc. Bull, **71 [4],** 653–659 (1992)
13. S. Saito ed., *Fine Ceramics*. (Elsevier Essex England, 1985)
14. F. Thevenot: J. Eur. Ceram. Soc., **6 [4],** 205–225 (1990)
15. K.S. Mazdiyasni et al.: Am. Ceram. Soc. Bull., **64 [8],** 1149–1154 (1985)
16. T.J. Mroz and A.K. Blakely: Mater. Eng., March 1992
17. F.D. Gac et al.: Ceram. Eng. Sci. Proc., **7,** 978–982 (1986)
18. K.P. Gadkaree and K. Chyung: Am. Ceram. Soc. Bull., **65,** 370–376 (1986)
19. P.F. Becher et al.: J. Am. Ceram. Soc., **71,** 1050–1061 (1988)
20. J.J. Brennan and S.R.Nutt: J. Am. Ceram. Soc., **75,** 1205–1211 (1992)
21. N. Claussen and G. Petzow: In: *13th Int.Conf.on Sci.Ceram.*, ed. P.Odier et al. pp. C1-693–702. (Les Editions Physique Cedex France, 1986)
22. R.N.Singh: J. Mater. Sci., **26,** 1839–46 (1991)
23. S. Iio et al.: J. Am. Ceram. Soc., **74,** 296–300 (1991)
24. Manufacturer's data
25. *ASM Engineered Materials Reference Book*. (ASM International Metals Park OH, 1989)
26. N. Hecht et al.: Ceram. Eng. Sci. Proc., **9 [9–10],** 1313–32 (1988)
27. C.K. Unni and D.E. Gordon: J. Mater. Sci. **30,** 1173–1179 (1995) (NIST Database www.ceramics.nist.gov)
28. S.G. Seshadri and K.Y. Chia: J. Am. Ceram. Soc. **70 [10],** C–242–244 (1987)

Appendix 3. Mechanical Properties of Particulate- and Platelet-Reinforced Composites

Matrix Material	Reinforcement Material [a]	Reinforcement Content (vol%)	Density (% theoretical)	Fracture Toughness (MPa $m^{1/2}$)	Flexural Strength (MPa)		Elastic Modulus (GPa)	Microhardness (GPa)	Ref. [b]
					3-Point	4-Point			
Al_2O_3	B_4C	0	100	3–5	–	–	390	19	[1]
		50	100	3.4	–	–	380	20	[1]
Al_2O_3	TiC	30	100	4.0	–	–	400	21	[1]
Mullite	SiC	0	100	1.2	550	–	–	–	[2]
		10	100	2.3	–	–	240	15	[2]
SiC	TiB_2	0	100	3.4	–	–	440	26	[2]
		16	100	4.5	–	–	430	27	[2]
Al_2O_3	SiC (platelet)	0	–	4.4	345	–	355	–	[3]
		10	–	5.3	280	–	370	–	[3]
		20	–	5.4	140	–	265	–	[3]
		30	–	5.7	200	–	335	–	[3]
Al_2O_3	SiC	10	–	4.3	270	–	380	–	[3]
		20	–	5.1	200	–	290	–	[3]
		30	–	5.4	240	–	290	–	[3]

(Continued)

Appendix 3. *(continued)*

Matrix Material	Reinforcement Material	Reinforcement Content (vol%)	Density (% theoretical)	Fracture Toughness (MPa m$^{1/2}$)	Flexural Strength (MPa) 3-Point	Flexural Strength (MPa) 4-Point	Elastic Modulus (GPa)	Microhardness (GPa)	Ref. [b]
Al_2O_3	BN	30	–	5.4	325	–	140	–	[4]
		40	–	–	135	–	80	–	[4]
		50	–	1.7	175	–	80	–	[4]
Si_3N_4	TiN	0	–	10.2	795	–	275	14.5	[5]
		10	–	10.6	785	–	285	13.9	[5]
		20	–	10.8	720	–	290	13.6	[5]
		30	–	10.7	670	–	300	13.3	[5]
		40	–	9.4	650	–	305	13.1	[5]
Si_3N_4	TiC	0	100	4.8	–	–	300	16	[1]
		30	100	4.5	–	–	–	17	[1]

(Continued)

Appendix 3. *(continued)*

Matrix Material	Reinforcement Material	Reinforcement Content (vol%)	Density (% theoretical)	Fracture Toughness (MPa m$^{1/2}$)	Flexural Strength (MPa)		Elastic Modulus (GPa)	Microhardness (GPa)	Ref. [b]
					3- Point	4- Point			
Si_3N_4–5% Y_2O_3		0	–	7.9	590	–	–	–	[2]
–2%Al_2O_3	TiN	5 (wt%)	–	7.8	590	–	–	–	[2]
	HfN	5 (wt%)	–	7.1	550	–	–	–	[2]
	Cr_2N	5 (wt%)	–	10.7	520	–	–	–	[2]
	NbC	5 (wt%)	–	7.8	590	–	–	–	[2]
	Cr_3C_2	5 (wt%)	–	8.3	420	–	–	–	[2]
	SiC	5 (wt%)	–	5.8	590	–	–	–	[2]
	ZrB_2	5 (wt%)	–	8.8	420	–	–	–	[2]
	TiB_2	5 (wt%)	–	8.5	480	–	–	–	[2]
	NbB	5 (wt%)	–	11.8	570	–	–	–	[2]
	$CrSi_2$	5 (wt%)	–	8.0	540	–	–	–	[2]
	$TaSi_2$	5 (wt%)	–	11.0	620	–	–	–	[2]
	$ZrSi_2$	5 (wt%)	–	11.8	550	–	–	–	[2]

[a] all reinforcements are in particulate form except where indicated

[b] References:

1. R. Warren and V.K. Sarin. In: *Ceramic Matrix Composites* , R. Warren (ed.). (Blackie and Son London, England 1992), pp.146–66
2. Y.G. Gogotsi: J. Mater. Sci. **29 [10],** 2541–2556 (1994)
3. Ceramic Transactions Vol.19, *Advanced Composite Materials* . M.D. Sacks (ed.). (The American Ceramic Society Westerville OH, 1991), pp. 773–779
4. D. Lewis, R.P. Ingel, W.J. McDonoughs, and R.W. Rice: Ceram. Eng. Sci. Proc. **2 [7–8],** 719–727 (1981)
5. Ceramic Transactions. Vol.22: *Ceramic Powder Science IV.* S. Hirano et al. (eds.). (The American Ceramic Society Westerville OH, 1991), pp.655–661

Appendix 4. Mechanical Properties of Whisker–Reinforced Composites

Matrix Material	Whisker Reinforcement	Reinforcement Content (vol%)	Density (% theoretical)	Fracture Toughness (MPa $m^{1/2}$)	Flexural Strength (MPa)		Ref. [a]
					3-Point	4-Point	
Al_2O_3	SiC	14	–	4.5	520	–	[1]
		31.5	–	5.8	670	–	
		53.8	–	7.5	650	–	
		63	–	8.1	630	–	
Al_2O_3	B_4C	0	–	4.0	410	–	[2]
		5	77	5.6	500	–	
		10	76	6.7	500	–	
		15	73	7.2	520	–	
		20	70	7.0	500	–	
		30	–	7.1	–	–	
Al_2O_3–30vol% glass	SiC	0	98	2.7	–	235	[3]
		15	95	3.3	–	465	
		20	92	2.4	–	285	
	Si_3N_4	10	97	3.2	–	225	[3]
		15	95	3.4	–	205	
		20	96	3.4	–	295	
Al_2O_3–15vol% t-ZrO_2	SiC	0	–	6.2	–	–	[4]
		20	–	13.5	–	700	
Al_2O_3–40vol% m-ZrO_2	SiC	0	–	2.0	–	–	
		20	–	8.5	–	880	
Al_2O_3–32vol% m-ZrO_2	SiC	20	–	6.3	–	670	[5]

(Continued)

Appendix 4. *(continued)*

Matrix Material	Whisker Reinforcement	Reinforcement Content (vol%)	Density (% theoretical)	Fracture Toughness (MPa $m^{1/2}$)	Flexural Strength (MPa) 3-Point	Flexural Strength (MPa) 4-Point	Ref. [a]
Cordierite	SiC	0	–	1.5	–	100	[6]
		10	–	3.3	–	200	
		15	–	4.2	–	260	[7]
		20	–	4.9	–	290	
		25	–	5.8	–	400	
Mullite	SiC	0	–	2.8	–	250	[4]
		10	–	4.1	–	280	
		20	–	4.5	–	430	
Mullite	SiC	0	–	2.2	–	200	[8]
		10	–	3.6	–	420	
		20	–	4.7	–	425	
Mullite–10 vol%ZrO_2	SiC	0	–	3.5	–	–	[4]
		10	–	5.4	–	580	
		20	–	6.7	–	550	
Mullite	Si_3N_4	0	97.5	1.2	550	–	[9]
		5	99	2.5	600	–	
		10	99.5	3.0	600	–	
Mullite	Mullite	0	–	1.2	550	–	[7]
		5	–	1.7	450	–	
		10	–	2.2	440	–	
PSZ	SiC	0	–	4.1	–	600	[10]
		10	–	5.0	–	750	
		20	–	8.1	–	950	
		30	–	8.0	–	900	

(Continued)

Appendix 4. *(continued)*

Matrix Material	Whisker Reinforcement	Reinforcement Content (vol%)	Density (% theoretical)	Fracture Toughness (MPa $m^{1/2}$)	Flexural Strength (MPa) 3-Point	Flexural Strength (MPa) 4-Point	Ref. [a]
Si_3N_4	SiC	0	99.5	5.0	–	700	[11]
		20	99	7.5	–	580	
		30	94	5.5	–	400	
Si_3N_4	SiC	0	–	4.7	–	780	[12]
		10	–	4.4	–	740	
		20	–	4.8	–	720	
		30	–	6.4	–	980	
Si_3N_4	Si_3N_4	0	–	6.1	–	850	[13]
		10	–	6.4	–	680	
		20	–	6.3	–	510	
Si_2N_2O	SiC	0	–	3.0	400	–	[14]
		10	–	5.7	660	–	
		20	–	4.9	700	–	
		30	–	6.0	750	–	
TZP	SiC	0	–	6.4	–	1150	[15]
		20	–	10.0	–	650	
		30	–	10.4	–	650	

[a] References:

1. S. Iio, M. Watanabe, M. Matsubara, Y. Matsuo: J. Am. Ceram. Soc. **72,** 1880–1884 (1989)
2. J. Liu and P.D. Ownby: Ceram. Eng. Sci. Proc. **12,** 1245–1253 (1991)
3. R. Chaim, L. Baum, and D.G. Brandon: J. Am. Ceram. Soc. **72 [9],** 1636–1642 (1989)
4. N. Claussen and G. Petzow: In: *13th International Conf. Science of Ceramics*. P. Odier, F. Cabanned, and B. Cales (eds.). (Les Editions de Physique Cedex France, 1986), pp. C1–693–702
5. N. Claussen, M.V. Swain, K.L. Weisskopf: In: *Adv. in Ceramics, Science and Technology of ZrO_2–III.* S. Somiya et al. (eds.). (The American Ceramic Society, Westerville OH, 1988), pp.1101–1106
6. K.P. Gadkaree and K. Chyung: Am. Ceram. Soc. Bull. **65,** 370–376 (1986)

7. K.P. Gadkaree and K.C. Chyung: In: *Whisker and Fiber Toughened Ceramics*. R.A. Bradley et al (eds.). (ASM International Metals Park OH, 1989), pp. 109–111
8. T. Tiegs, P. Becher, and P. Angelini. *In: Ceramic Transactions*, S.Somiya, R.F. Davis, and J.A. Pask (eds.). (The American Ceramic Society Westerville OH, 1990), pp. 463–450
9. Y. Hirata, S. Matsushita, Y. Ishihara, and H. Katsuki: J. Am. Ceram. Soc. **74,** 2438–2442 (1991)
10. E. Yasuda, S. Kimura, Y. Katano, and Y. Akimune: In: *Advances in Ceramics, Science and Technology of ZrO2-III.* S. Somiya et al (eds.). (The American Ceramic Society Westerville OH, 1988), pp. 701–708
11. H. Kodama, T. Suzuki, H. Sakamoto, T. Miyashi: J. Am. Ceram. Soc. **73,** 678–683 (1990)
12. S.T. Buljan, J.G. Baldoni, and M.L. Huckabee: Am. Ceram.Soc. Bull. **66,** 347–352 (1987)
13. J. Duzsa and P. Sajgalik: J. Eur. Ceram. Soc. **9,** 9–17 (1992)
14. S. Iio, H. Yokoi, M. Watanabe, Y. Matsuo: J. Am. Ceram. Soc. **74,** 296–300 (1991)
15. N. Claussen, K.L. Weisskopf, and M. Ruhle: J. Am. Ceram. Soc. **69,** 288–292 (1986)

Appendix 5. Mechanical Properties of C-C Composites

Type of Composite	Test Temperature (°C)	Tensile Strength (MPa)		Tensile Modulus (GPa)		Flexural Strength (MPa)	Shear Strength (MPa)	Shear Modulus (GPa)	Compressive Strength (MPa)	Compressive Modulus (MPa)	Density (g/cc)	Ref. [b]
		x [a]	z	x	z		x–y	x–y	x	x		
3-D billets	25	105	–	415	–	100	–	–	–	–	–	[1]
3-D billets	2700	70	–	10	–	100	–	–	–	–	–	
Reticulated vitreous	25	–	–	–	–	–	–	–	10	505	0.5	[2]
C substrate, CVI lamina composite			–		–		–	–				
Two-dimensional CVI composite	25	–	–	42	–	–	–	–	–	–	–	[3]
Glassy carbon–graphite fiber composite (copyrolysis)	25	210–350	–	–	–	–	–	–	–	–	–	[4]
AS4/PEEK (before carbonization)	AT [c]	–	–	150 (F) [d]	–	1800	–	–	–	–	–	[5]
P75/PEEK (before carbonization)	AT [c]	–	–	280 (F) [d]	–	728	–	–	–	–	–	
AS4/PEEK (carbonized)	AT [c]	–	–	190 (F) [d]	–	900	–	–	–	–	–	[5]
P75/PEEK (carbonized)	AT [c]	–	–	300 (F) [d]	–	600	–	–	–	–	–	[5]

(Continued)

Appendix 5. *(continued)*

Type of Composite	Test Temperature (°C)	Tensile Strength (MPa)		Tensile Modulus (GPa)		Flexural Strength (MPa)	Shear Strength (MPa)	Shear Modulus (GPa)	Compressive Strength (MPa)	Compressive Modulus (MPa)	Density (g/cc)	Ref. [b]
		x [a]	z	x	z		x–y	x–y	x	x		
Unidirectional (1–D) composite w. high-performance fibers (65 vol%)	AT [c]	650–1000	2.0	240–280	3.4	–	7–14	4–7	620	250	–	[6]
Fabric laminate (2–D) composite w. isotropic rayon precursor fibers (25 vol%x, 25 vol%y)	AT [c]	41	5.0	14	–	–	–	–	–	–	–	[6]
Fabric laminate (2–D) composite w. high-performance fibers (31 vol%x, 30 vol%y)	AT [c]	300–350	2.8–5.0	110–125	4.1	–	7–14	4–7	150	100	–	[6]
Woven orthogonal (3–D) composite w. high-performance fibers (13 vol%x, 13 vol%y, 21 vol%z)	AT [c]	170	300	55	96.0	–	21–27	1.4–2.1	140	90	–	[6]

[a] Test direction

[b] References:

1. J. Delmonte: *Technology of Carbon and Graphite Fiber Composites*. (Van Nostrand Reinhold New York, 1981)
2. D.A. Cordit, R.D. Veltri, F.S. Galasso, and G.P. Noyes: Am. Ceram. Soc. Bull. **68 [2],** 359–362 (1987)
3. F.K. Ko: Am. Ceram. Soc. Bull. **68 [2],** 401–414 (1989)
4. R.W. Rice: Ceram. Eng. Sci. Proc. **2 [7–8A],** 493–508 (1981)
5. G. Savage: *Carbon–Carbon Composites*. (Chapman & Hall London England, 1993)
6. J.E. Sheehan, K.W. Buesking, and B.J. Sullivan: Ann. Rev. Mater. Sci. **24,** 19–44 (1994)

[c] AT: ambient temperature
[d] F: flexural

Appendix 6. Mechanical Properties of Fiber–Reinforced Glass and Ceramic Matrix Composites

Type of Composite	Lay-up	Tensile Modulus (GPa)	Tensile Strength (MPa)	Strain to Failure (%)	Compressive Modulus (GPa)	Compressive Strength (MPa)	Strain to Failure (%)	Flexural Strength (MPa)	Shear Modulus (GPa)	Poisson's Ratio	CTE $\times 10^{-6}$ (1/K)	Toughness K_c (MPa $m^{1/2}$)	Ref. [a]
Silica–silica	3-D	21.9	27	0.2	21.9	145	1.6	–	1.5	0.09	0.54	–	[1]
	4-D	11.5	23	1.2	8.6	71	1.5	–	4.4	–	0.47	–	
Glass–C_f or SiC_f	1-D	–	700	–	–	–	–	–	–	–	–	10–25	[2]
Glass–Al_2O_{3f}	1-D	–	280	–	–	–	–	–	–	–		–	
Ba sealing glass–46 vol%Al_2O_{3f}	1-D	–	158	–	–	–	–	–	–	–	–	3.0	[3]
Aluminoborosilicate–47 vol%Al_2O_{3f}	1-D	–	311	–	–	–	–	–	–	–	–	3.3	
Borosilicate (Pyrex)–41 vol%Al_2O_{3f}	1-D	–	305	–	–	–	–	–	–	–	–	3.7	
Low expansion sealing glass–45 vol%Al_2O_{3f}	1-D	–	200	–	–	–	–	–	–	–	–	4.0	

(Continued)

Appendix 6. *(continued)*

Type of Composite	Lay-up	Tensile Modulus (GPa)	Tensile Strength (MPa)	Strain to Failure (%)	Compressive Modulus (GPa)	Compressive Strength (MPa)	Strain to Failure (%)	Flexural Strength (MPa)	Shear Modulus (GPa)	Poisson's Ratio	CTE $\times 10^{-6}$ (1/K)	Toughness K_c (MPa $m^{1/2}$)	Ref. [a]
Pyrex–SiC_f	1-D	–	–	–	–	1200	–	1200	–	–	–	26.3	[4]
Cordierite–6%Nb_2O_5–SiC_f	1-D	–	–	–	–	–	–	295	–	–	–	8.7	[5]
	2-D	–	–	–	–	–	–	174	–	–	–	5.4	
CVI SiC–SiC_f	3-D	–	–	–	–	–	–	324	–	–	–	–	[6]
SiC–Nicalon	1-D	–	–	–	–	–	–	343	–	–	–	–	[7]
Borosilicate glass–C_f	1-D	–	462	–	–	–	–	1104	–	–	–	–	[8]
Glass ceramic–Tyranno	1-D	–	707	–	–	–	–	1150	–	–	–	–	
Borosilicate glass–SiC_f	–	–	–	–	–	–	–	–	–	–	–	50.0	[9]
Glass–Al_2O_{3f}	–	–	–	–	–	–	–	–	–	–	–	7.0	

(Continued)

Appendix 6. *(continued)*

Type of Composite	Lay-up	Tensile Modulus (GPa)	Tensile Strength (MPa)	Strain to Failure (%)	Compressive Modulus (GPa)	Compressive Strength (MPa)	Strain to Failure (%)	Flexural Strength (MPa)	Shear Modulus (GPa)	Poisson's Ratio	CTE $\times 10^{-6}$ (1/K)	Toughness K_c (MPa $m^{1/2}$)	Ref. [a]
Glass (LAS–II)–SiC_f (Nicalon)	1-D	120–140	586–686	0.78–1.08	119.9	1026	0.85	–	–	–	–	–	[10]
Glass (Corning 1723)–SiC_f (Nicalon)	1-D 0°	128	734	1.20	–	–	–	–	–	0.24	–	–	
Glass ceramic (Corning CAS–II) –SiC_f	1-D 90°	101	27	0.035	–	–	–	–	–	–	–	–	
Borosilicate glass – 39 vol%	1-D 0°	124	450	0.95	–	1075	–	48	–	–	4.3 (1/°C)	–	
Thornel 300 C_f	1-D	127	711	–	–	–	–	–	–	–	–	–	
Borosilicate glass – 54 vol% HM-C_f	1-D	180	620	–	–	–	–	–	–	–	–	–	
Borosilicate glass – 58 vol%GY70-C_f	1-D	275	683	–	–	–	–	–	–	–	–	–	
Borosilicate glass – 54 vol%P-100-C_f	1-D	332	683	–	–	–	–	–	–	–	–	–	
ZrO_2–TiO_2–SiC_f	1-D	–	–	–	–	–	–	1000	–	–		20	
ZrO_2–SiO_2–SiC_f	1-D	–	–	–	–	–	–	1000	–	–		20	

(Continued)

Appendix 6. *(continued)*

Type of Composite	Lay-up	Tensile Modulus (GPa)	Tensile Strength (MPa)	Strain to Failure (%)	Compressive Modulus (GPa)	Compressive Strength (MPa)	Strain to Failure (%)	Flexural Strength (MPa)	Shear Modulus (GPa)	Poisson's Ratio	CTE $\times 10^{-6}$ (1/K)	Toughness K_c (MPa $m^{1/2}$)	Ref. [a]
Glass ceramic (Corning CAS-II) –SiC_f	1-D 0°	–	70	0.9	–	–	–	–	–	–	–	3.1[b]	[11]
Glass (LAS-III)–SiC_f	1-D 0°	–	–	0.77	–	50	–	–	–	–	–	2.7[b]	
MAS–SiC_f	1-D 0°	–	175	0.36	–	–	–	–	–	–	–	0.57[b]	
Glass (Corning 1723)–SiC_f	1-D 0°	–	70	1.2	–	–	–	–	–	–	–	5.4[b]	
Borosilicate glass–50 vol% SiC_f	1-D	120	840	–	–	–	–	–	–	–	–	18.9	[12]
(Philips) Borosilicate glass–47 vol% SiC_f	1-D	–	1300	–	–	–	–	–	–	–	–	26	
LAS glass–50 vol% SiC_f	1-D	140	600	–	–	–	–	–	–	–	–	17	
Glass ceramic–55 vol% Al_2O_{3f}	1-D	–	250	–	–	–	–	–	–	–	–	7	
LAS glass – $Nicalon_f$	1-D 0°	130	690	–	–	–	–	–	–	–	2,8	16	[13]
LAS glass – $Nicalon_f$	1-D	45	25	–	–	–	–	–	–	–	1,1	–	
LAS glass – $Nicalon_f$	2-D	76	410	–	–	–	–	–	–	–	2,3	8,5	

[a] References:

1. F.K. Ko: Am. Ceram. Soc. Bull. **68 [2],** 401–414 (1989)
2. R.W. Rice: Ceram. Eng. Sci. Proc. **2 [7–8A],** 493–508 (1981)
3. T. Michalske and J. Hellmann: J. Am. Ceram. Soc. **71 [9],** 725–731 (1988)
4. D.M. Dawson, R.F. Preston, and A. Purser: Ceram. Eng. Sci. Proc. **8 [7–8],** 815–821 (1987)
5. R. Chaim, D. Brandon et al: Ceram. Eng. Sci. Proc. **9 [7–8],** 695–704 (1988)
6. A.J. Caputo, D.P. Stinton, R.A. Lowden, and T.M. Besmann: Am. Ceram. Soc. Bull. **66 [2],** 368–372 (1987)
7. *Ceramic Transactions,* Vol.19, *Advanced Composite Materials*. M.D. Sacks (ed.). (The American Ceramic Society Westerville OH, 1991),
8. J.J. Davies, R.F. Preston, R.J. Lee, and K.N. Walls: In: *New Materials and Their Applications*, Conf. Proc. (University of Warwick Inst. of Physics London, 1990)
9. J.E. Bailey: In: *Conf. Ceramic Composites and Coatings.* (London, 1988)
10. K.S. Mazdiyasni (ed.): *Fiber Reinforced Ceramic Composites*. (Noyes Park Ridge NJ, 1990)
11. B.F. Sorensen and R. Talreja: J. Eur. Ceram. Soc. **15 [11],** 1047–1059 (1995)
12. M.C. Willson and R.J. Dover: Metals and Mater. **4 [12],** 752–756 (1988)
13. W.B. Hillig: Ann. Rev. Mater. Sci. **17,** 341–383 (1987)

[b] Toughness values given in MJ/m^3

Appendix 7. Thermal Properties of Ceramic Materials [a]

Ceramic Class	Composition	Coefficient of Thermal Expansion CTE ($\times 10^6$/K)	Temperature Range of CTE (°C)	Thermal Conductivity at Room Temp. (W/mK)	Heat Capacity [J/(Kg K)]	Thermal Diffusivity (mm^2/s)	Melting Point (°C)
Oxides	Al_2O_3 (99.9%)	6.5–8.9 [1]	200–1200	38.9	1088 [6]	–	2050 [1], 2322 [11]
	α-Al_2O_3 (a axis)	8.3 [2]	0–1450	–	–	–	–
	α-Al_2O_3 (c axis)	9.03 [2]	0–1450	–	–	–	–
	BeO	6.8 ($\times 10^{-6}$/°C) [3]	20–200	240–290 [3,8]	–	–	2550 [4], 2725 [11]
	Cordierite	2.5 ($\times 10^{-6}$/°C) [12]	20–1000	2.5 (W/m°C)	–	–	–
	Cr_2O_3	–	–	–	–	–	2275 [1] (volatilizes)
	Enstatite ($MgO{\cdot}SiO_2$)	–	–	–	–	–	1557 [1]
	MgO	–	–	40 [6]	929 [7]		2800 [1], 3098 [11]
	MgO (single crystal, a axis)	13.5 [2]	0–1000	–	–	–	–
	MgO–PSZ	6.8 [1]	0–1000	1–2	–	1.14	3037±83 [4]

(Continued)

Appendix 7. *(continued)*

Ceramic Class	Composition	Coefficient of Thermal Expansion CTE ($\times 10^6$/K)	Temperature Range of CTE (°C)	Thermal Conductivity at Room Temp. (W/mK)	Heat Capacity [J/(Kg K)]	Thermal Diffusivity (mm^2/s)	Melting Point (°C)
	Mullite ($3Al_2O_3 \cdot 2SiO_2$)	4.5 ($\times 10^{-6}$/°C) [12]	20–1000	7 [3], 6 [12]	1046 [2]	–	1810 [1]
		3.1–4.1 // a [1]	300–900	–	–	–	–
		5.6–7.0 // b	300–900	–	–	–	–
		5.6–6.1 // c	300–900	–	–	–	–
	Quartz (SiO_2, a axis)	22.2 [2]	0–500	–	–	–	–
	Quartz (SiO_2, c axis)	12.8 [2]	0–500	–	–	–	–
	Quartz (SiO_2)	–	–	1.46 [5]	670–740 [5]	–	1723 [1], 1978 [11]
	Spinel ($MgO \cdot Al_2O_3$)	–	–	–	–	–	2135 [1]
	Silicate ($2CaO \cdot SiO_2$)	–	–	–	–	–	2130 [1]
	$TiAl_2O_5$	0.8–1.3 ($\times 10^{-6}$/°C) [12]	20–1000	3.4 (W/m°C)	–	–	–
	TiO_2	8.8 [5]	20–800	5 [5]	–	–	1840 [4], 2113 [11]

(Continued)

Appendix 7. *(continued)*

Ceramic Class	Composition	Coefficient of Thermal Expansion CTE ($\times 10^6$/K)	Temperature Range of CTE (°C)	Thermal Conductivity at Room Temp. (W/mK)	Heat Capacity [J/(Kg K)]	Thermal Diffusivity (mm^2/s)	Melting Point (°C)
	Y_2O_3–PSZ	10.23 [1]	0–1000	1–2	–	1.04	3037±83 [4]
	Y_2O_3–TZP	10.6 [1]	0–1000	1–2	–	1.19	3037±83 [4]
	t-ZrO_2 (a axis)	8–10 [2]	0–1000	–	–	–	–
	t-ZrO_2 (c axis)	10.5–13 [2]	0–1000	–	–	–	–
	m-ZrO_2 (a axis)	6.8–8.4 [2]	0–1000	–	–	–	–
	m-ZrO_2 (c axis)	12–14 [2]	0–1000	–	–	–	–
	c-ZrO_2	7.5–13 [2]	0–1000	1.8 [6]	–	–	2677 [4], 3123 [11]
Nitrides	AlN	4.2 ($\times 10^{-6}$/°C) [3]	20–200	100–230 [3,8]	–	–	2475 at 4 atm N_2 (sublimes) [9]
	BN (hot-pressed)	1–6 [5]	20–1000	15–37 [5]	780 [5]	–	2730 [4], 3000 [11]
	β-Sialon	3.04 ($\times 10^{-6}$/°C) [1]	0–1200	22	–	–	–
	%60 α-Sialon	3.2 ($\times 10^{-6}$/°C) [1]	0–1200	10.9	–	–	–
	Si_3N_4 (hot-pressed)	2–3 [1]	0–1000	15–50	680–800 [5]	–	1878 [1] (oxidizes)

(Continued)

Appendix 7. *(continued)*

Ceramic Class	Composition	Coefficient of Thermal Expansion CTE ($\times 10^6$/K)	Temperature Range of CTE (°C)	Thermal Conductivity at Room Temp. (W/mK)	Heat Capacity [J/(Kg K)]	Thermal Diffusivity (mm^2/s)	Melting Point (°C)
	Si_3N_4 (reaction-bonded)	2–3 [1]	0–1000	3–30	690 [5]	–	1878 [1] (oxidizes)
	β-Si_3N_4 (a axis)	3.23 [2]	0–1000	–	–	–	–
	β-Si_3N_4 (c axis)	3.72 [2]	0–1000	–	–	–	–
Carbides	B_4C (hot-pressed)	5.6 [5]	20–1000	27–36 [5]	950 [5]	–	2450 [4], 2720 [11]
	Carbon (graphite, a axis)	1 [2]	0–1000	10	–	–	4000 (oxidizes) [1]
	Carbon (graphite, c axis)	27 [2]	0–1000	2000	–	–	–
	SiC (reaction-bonded)	4.2 [1]	0–1500	–	670–710 [5]	–	–
	SiC (hot-pressed)	3.7–4.5 [1]	0–1500	50 (600°C)	670–710 [5]	–	2500 (oxidizes) [1]
	SiC (sintered)	4.9 [1]	0–1500	55 (600°C)	–	–	–
	TaC	–	–	32 [6]	167 [6]	–	3880 [4], 3813 [11]
Other	Diamond	–	–	900 [2]	–	–	–
	TiB_2	–	–	65–127 [6]	632 [6]	–	2980 [10], 3253 [11]
	Si	3.5 [2]	–	125 [8]	–	–	–

[a] References:

1. W.E. Lee and W.M. Rainforth: *Ceramic Microstructures.* (Chapman & Hall London England, 1994)
2. *Handbook of Materials Science,* Vol. 2. C.T. Lynch (ed.). (CRC Press Boca Raton FL, 1974)
3. R.R. Tummala: J. Am. Ceram. Soc. **74 [5]**, 895–908 (1991)
4. *CRC Handbook of Chemistry and Physics*. R.C. Weast (ed.). (CRC Press Boca Raton FL, 1986)
5. Manufacturer's Data
6. R. Morrell: *Handbook of Properties of Technical and Engineering Ceramics Part 1*. (Her Majesty's Stationery Office London, 1985)
7. A.N. Copp: Am. Ceram. Soc. Bull. **68 [5],** 1054–1055 (1989)
8. *Fine Ceramics*. S. Saito (ed.) (Elsevier Essex England, 1985)
9. T.J. Mroz and K.A. Blakely: Mater.Eng. **109,** March 1992
10. I.J. McColm: *Ceramic Science for Ceramic Technologists.* (Leonard Hill New York NY, 1983)
11. *CRC Materials Science and Engineering Handbook.* J.S. Hackelford, W. Alexander, J.S. Park (eds.). (CRC Press Boca Raton FL 1994)
12. J.M. Tulliani, L. Montanaro, C. Borello, and P.P. Demaestri: Ceramurgia, **25 [2],** 81–90 (1995)

Appendix 8. Optical Properties of Crystals, Glasses, and Ceramic Materials [a]

Material	Average Refractive Index [1,2,5] n	Abbe Number [2,5] ν_D	Spectral Emissivity [3] ε_λ	Average Spectral Emissivity on Smooth Metal Surface [3] ε_λ	Transmission Range [4] (μm at 298K)
Al_2O_3	1.76	–	0.22–0.40	0.30	0.15–6.5
$BaTiO_3$	2.40	–	–	–	–
BeO	1.71	–	0.07–0.37	0.35	–
Ce_2O_3	–	–	0.58–0.80	–	–
Cr_2O_3	–	–	0.60–0.80	0.70	–
GeO_2	–	41.2	–	–	–
$LiNbO_3$ (single crystal)	–	–	–	–	0.33–5.2
MgO	1.74	–	0.10–0.43	0.20	0.25–8.5
Mullite ($3Al_2O_3 \cdot 2SiO_2$)	1.64	–	–	–	–
NiO	–	–	0.85–0.96	0.90	–
SiO_2	1.55	67.7	–	–	0.12–4.5
SnO	–	–	0.32–0.60	–	–
TiO_2	2.71	–	–	0.50	0.43–6.2
Y_2O_3	1.92	–	–	0.60	–
ZrO_2	–	–	0.18–0.43	0.40	–

(Continued)

Appendix 8. *(continued)*

Material	Average Refractive Index [1,2,5] n	Abbe Number [2,5] ν_D	Spectral Emissivity [3] ε_λ	Average Spectral Emissivity on Smooth Metal Surface [3] ε_λ	Transmission Range [4] (μm at 298K)
BaF_2 (single crystal)	–	–	–	–	0.25–15.0
CaF_2	1.434	–	–	–	0.13–12.0
LiF	1.392	–	–	–	0.12–9.0
NaF	1.326	–	–	–	0.19–15.0
MgF_2 (film)	–	–	–	–	0.2–5.0
SiC	2.680	–	–	–	–
BN	2.120	–	–	–	–
Diamond	2.380	–	–	–	–
Ultran 20	–	74.25	–	–	–
Corning 7940	1.459	67.80	–	–	–
Corning 9753	–	54.98	–	–	–
Corning 9754	–	46.50	–	–	–
Schott IRG2	–	30.03	–	–	–
Schott IRG11	–	44.21	–	–	–
Ohara HTF-1	–	92.46	–	–	–

(Continued)

Appendix 8. *(continued)*

Material	Average Refractive Index [1,2,5] n	Abbe Number [2,5] ν_D	Spectral Emissivity [3] ε_λ	Average Spectral Emissivity on Smooth Metal Surface [3] ε_λ	Transmission Range [4] (μm at 298K)
Silicate glass					
$20Na_2O·20CaO·60SiO_2$	1.551	55.6	–	–	–
$20Na_2O·20BaO·60SiO_2$	1.577	52.9	–	–	–
$40BaO·60SiO_2$	1.639	53.7	–	–	–
Borate glass					
$20BaO·80B_2O_3$	1.559	62.8	–	–	–
$25BaO·75B_2O_3$	1.583	62.5	–	–	–
$30BaO·70B_2O_3$	1.610	62.2	–	–	–
Aluminate glass					
$65CaO·35Al_2O_3$	1.665	45.7	–	–	–
Tellurite glass					
$10Al_2O_3·90TeO_2$	2.048	20.1	–	–	–
$20Tl_2O·80TeO_2$	2.206	14.6	–	–	–

(Continued)

Appendix 8. *(continued)*

Material	Average Refractive Index [1,2,5] n	Abbe Number [2,5] ν_D	Spectral Emissivity [3] ε_λ	Average Spectral Emissivity on Smooth Metal Surface [3] ε_λ	Transmission Range [4] (μm at 298K)
Antimonate glass					
$20Tl_2O \cdot 80Sb_2O_3$	2.181	13.5	–	–	–
$25Tl_2O \cdot 75Sb_2O_3$	2.287	12.7	–	–	–
$30Tl_2O \cdot 70Sb_2O_3$	2.266	11.7	–	–	–
Heavy metal gallate					
$80PbO \cdot 20Ga_2O_3$	2.294	11.1	–	–	–
$60PbO \cdot 20Bi_2O_3 \cdot 20Ga_2O_3$	2.346	10.8	–	–	–
$40PbO \cdot 40Bi_2O_3 \cdot 20Ga_2O_3$	2.383	10.6	–	–	–

[a] References:

1. *ASM Engineered Materials Reference Book.* (ASM International Metals Park OH, 1989)
2. *Handbook of Optics,* Vol. II (McGraw Hill New York, 1995)
3. *Handbook of Optics.* W.G. Driscoll and W. Vaughan (eds.). (McGraw Hill New York, 1978)
4. *CRC Materials Science and Engineering Handbook*. J.F. Shackelford, W. Alexander, and J.S. Park (eds.). (CRC Press Boca Raton FL, 1994)
5. S. Fujino, H. Takebe, and K. Morinaga: J.Am.Ceram.Soc. **78 [5],** 1179–1184 (1995)

Appendix 9. Ceramic Pigments Used in Glazes and Paints [a]

Material	Color	Stable at				
Pigments for glazes and body stains		Cone 015 (787–843°C)	Cone 06 (991–1023°C)	Cone 4 (1168–1209°C)	Cone 7 (1215–1264°C)	Cone 11 (1294–1336°C)
$CoAl_2O_4$ spinel	Blue	x	x	x	x	x
Co_2SiO_4 olivine	Blue	x	x	x		
Co Zn silicate phenacite	Blue	x	x	x		
Co Zn aluminate spinel	Blue	x	x	x	x	x
Co Sn alumina spinel	Blue	x	x	x		
Co chromite spinel	Blue–green (turquoise)	x	x	x	x	x
$ZrSiO_4$:V (V-doped zircon)	Turquoise	x	x	x	x	x
Co chromate	Turquoise					
Cr_2O_3	Green					
Zn chromate	Green					
Victoria green garnet	Bright green	x	x	x		
Ni silicate olivine	Green	x	x	x		
Co titanate spinel	Green	x	x	x		
VTiSnO	Yellow	x	x	x	x	x
$PbCrO_4$	Yellow					
$ZrO_2 \cdot V_2O_3$	Lemon to Orange					
CdS	Strong Yellow					
Cr:TiO_2	Orange–Yellow/ Maple					
Pb antimonate pyrochlore	Yellow	x	x			
Ni Sb Ti rutile	Yellow	x				
Ni Nb Ti rutile	Yellow	x				

(Continued)

Appendix 9. *(continued)*

Material	Color	Stable at				
Pigments for glazes and body stains		Cone 015 (787–843°C)	Cone 06 (991–1023°C)	Cone 4 (1168–1209°C)	Cone 7 (1215–1264°C)	Cone 11 (1294–1336°C)
Sn V cassiterite	Yellow	x	x	x	x	x
Fe_2O_3 (hematite)	Brown					
Mn oxides	Dark brown					
Mn Cr Sb Ti rutile	Brown	x				
Mn Nb Ti rutile	Brown	x				
Fe chromite spinel	Brown	x	x	x	x	x
Ni ferrite spinel	Brown	x	x	x	x	x
Zn ferrite spinel	Brown	x	x	x	x	x
Zn Fe chromite	Brown	x	x	x	x	x
Cr Fe Mn spinel	Brown	x	x	x	x	
Cr Mn Zn spinel	Brown	x	x	x	x	x
Zr Fe zircon (corals)	Pink	x	x	x	x	x
CrSnCaZrSi	Pastel pink	x	x	x		
CrZnCaSiPb	Pink	x	x	x		
Cr alumina corundum	Pink	x	x	x	x	x
CrSnCaPb	Rose	x	x	x		
ZrSiCdSeSCrSnCa	Indian red, flame	x	x	x		
CrZnCaSiPb	Maroon	x	x	x		
CrCoSnCaPbSi	Purple	x	x	x		
Cr Sn cassiterite	Orchid (pink with purple and maroon	x	x	x	x	x
Cr Sn sphene	Pink	x	x	x	x	x
CrSn	Violet, mauve	x	x	x		
Cr SnO_2	Purple shaded pink					
Cr sphene	Rose to maroon shaded pink					

(Continued)

Appendix 9. *(continued)*

Material	Color	Stable at				
Pigments for glazes and body stains		Cone 015 (787–843°C)	Cone 06 (991–1023°C)	Cone 4 (1168–1209°C)	Cone 7 (1215–1264°C)	Cone 11 (1294–1336°C)
CrCoSn	Lavender	x	x	x		
Cr alumina spinel	Pink	x	x	x	x	x
PrFeZrSi	Ocher	x	x	x	x	x
PrFeVZrSi	Old gold, nectarine, mustard	x	x	x	x	x
PrVFeZrSi	Khaki, mint green	x	x	x	x	x
PrVZrSi	Paris green, lime	x	x	x	x	x
Cd sulfoselenides	Bright red/orange					
S/Se:1.7:1–1.3:1	Red					
1:1	Maroon					
19.5%Se	Orange					
CdSeSZrSiPr	Mandarin	x	x	x	x	x
CdSeSZrSiPrFe	Deep orange	x	x	x	x	x
CdSeSZrSi	Strong red	x	x	x	x	x
PbZrSi	Citrus	x	x	x	x	x
PbSb	Sunshine, orange, egg yellow	x	x			
FeZrSi	Coral	x	x	x	x	x
FeVZrSi	Light brown, gray brown	x	x	x	x	x
Co Mg borate	Red–blue					
Co phosphate	Violet					
Co Li phosphate	Violet					
CoSi	Dark blue	x	x	x	x	x

(Continued)

Appendix 9. *(continued)*

Material	Color	Stable at				
Pigments for glazes and body stains		Cone 015 (787–843°C)	Cone 06 (991–1023°C)	Cone 4 (1168–1209°C)	Cone 7 (1215–1264°C)	Cone 11 (1294–1336°C)
TiO_2	White					
ZrO_2	White					
SnO_2	White					
ZnS	White					
SnSiO	White	x	x	x	x	x
SnSbO	Blue gray	x	x	x	x	x
VSnSbO	Steel gray, brown gray	x	x	x	x	x
MnO_2	Black					
Cu chromite spinel	Black	x				
Fe Co spinel	Black	x	x	x	x	x
Fe Co chromite spinel	Black	x	x	x	x	x
Mn ferrite spinel	Black	x	x	x	x	
Cr Fe Ni spinel	Black	x	x	x	x	x
Cr Sn Ni spinel	Black					
Co Ni periclase	Gray	x	x	x	x	x
Ti V Sb rutile	Gray	x				
Sn Sb cassiterite	Gray	x	x	x	x	x
Cr Sb Ti rutile	Buff	x				
Cr Nb Ti rutile	Buff	x				
Cr W Ti rutile	Buff	x				
Mn Sb Ti rutile	Buff	x				

(Continued)

Appendix 9. *(continued)*

Pigments for organic coatings (paints)	
TiO_2	White
$2PbCO_3{\cdot}Pb(OH)_2$	White
C	Black
$Fe_4[Fe(CN)_6]_3$ (prussian blue)	Blue
Prussian blue + $PbCrO_4$	Green
$MnPO_4$	Purple
$PbCrO_4$	Yellow
CdS	Yellow
Fe $(OH)_3$	Brown–yellow
$PbCrO_4{\cdot}PbO$	Orange
$Pb(Cr,Mo)O_4$	Orange
Pb_3O_4	Orange–red
CdSe	Red

[a] References:
1. R.A. Eppler: Am. Ceram. Soc. Bull. **66 [11],** 1600–1604 (1987)
2. L.H. VanVlack: *Physical Ceramics for Engineers.* (Addison–Wesley Reading MA, 1964)
3. W.D. Kingery, H.K. Bowen, and D.R. Uhlmann: *Introduction to Ceramics*. (John Wiley & Sons New York, 1976)
4. Ceramic Industry **141 [2],** p.183 (1993)
5. G.R. Streatfield: Br. Ceram. Trans. J. **89 [5],** 177–180 (1990)

Appendix 10. Electrical Properties of Ceramic and Other Materials at Ambient Conditions [a]

	Material	Dielectric Constant ε	Dielectric Strength (kV/mm)	Volume Resistivity (Ω cm)	Loss Factor (tan δ) $\times 10^{-4}$ at 50 Hz	1 KHz	1 MHz
Oxides	Al_2O_3	9.4–9.8 [1,2]	13 [3]	10^{12} [3], $>10^{15}$ [4]	2 [3]	5 [3], 50 [4]	10 [3], 4 [4]
	BeO	6.7–6.8 [2,5]	13 [3]	$>10^{14}$ [5]	10 [3]	10	10
	Cordierite (MgO·2Al2O3·5SiO2)	4.1–5.3 [4]	2.4–7.9 [4], 10 [3]	10^{14} [4]	250 [3]	–	70
	MgO	9–65 [17] (102–108 Hz)	–	$>10^{15}$ [8]	–	–	–
	Mullite ($3Al_2O_3 \cdot 2SiO_2$)	6.5–6.7 [1,2]	17 [3]	10 [3]–105 [18]	–	–	–
	Quartz (fused SiO_2)	3.8 [7]	25–40 [9]	1013–1017 [9]	–	–	–
	TiO_2	160–200 [17] (104–107 Hz) 85.8–87.3 [17] (102–104 Hz)	8 [3]	10^{10} [3]	–	65 [3]	20
	Zircon ($ZrSiO_4$)	8–9 [6]	30–40 [6]	1011–1013 [6]	–	–	4–14 [3]
	ZnO	8.8 [13]	50 [13]	–	–	–	–
	ZrO_2	–	–	10^9 [3]	–	–	20 [3]

(Continued)

Appendix 10. *(continued)*

	Material	Dielectric Constant ε	Dielectric Strength (kV/mm)	Volume Resistivity (Ω cm)	Loss Factor (tan δ) × 10^{-4} at 50 Hz	1 KHz	1 MHz
Nitrides	AlN	8.8 [10],10 [11]	5–10 [11]	10^{11}–10^{13} [11]	–	–	–
	BN	4.1 [7]	374 [18]	10^{11}–10^{13} [9]	–	–	14 [18]
	HfN	–	–	33×10^{-6} [19]	–	–	–
	Si_3N_4	7.0 [10],10 [3]	–	10^{12}–10^{15} [9]	–	–	–
	TiN	–	–	22×10^{-6} [16,19]	–	–	–
	ZrN	–	–	13.6×10^{-6} [19]	–	–	–
Carbides	HfC	–	–	109×10^{-6} [19]	–	–	–
	β-SiC	9.7 [12]	400 [12]	150 [12]	–	–	–
	TaC	–	–	10^{-4}–10^{-5} [16]	–	–	–
	TiC	–	–	10^{-4} [16], 180–250 × 10^{-6} [19]	–	–	–
	WC	–	–	10^{-6} [16]	–	–	–
	ZrC	–	–	70×10^{-6} [19]	–	–	–

(Continued)

Appendix 10. *(continued)*

	Material	Dielectric Constant ε	Dielectric Strength (kV/mm)	Volume Resistivity (Ω cm)	Loss Factor (tan δ) × 10^{-4} at 50 Hz	1 KHz	1 MHz
Borides	HfB_2	–	–	$10–12 \times 10^{-6}$ [19]	–	–	–
	TiB_2	10^{16} [18]	$(9–15) \times 10^{16}$ [18]	28.4×10^{-6} [19]	–	–	–
	ZrB_2	–	–	9.2×10^{-6} [19]	–	–	–
Other	Borosilicate glass	4.0 [10]	–	–	–	–	–
	CdS	9.5 [13]	50 [13]	–	–	–	–
	Diamond	5.5 [12]	10^3 [12]	10^{13} [12]	–	–	–
	Ge	16.3 [15]	–	–	–	–	–
	Hf	–	–	$32–50 \times 10^{-6}$ [19]	–	–	–
	Polymers	<4 [14]	–	–	–	–	–
	Si	11.8 [12]	30 [12]	10^3 [12]	–	–	–
	Ti	–	–	91×10^{-6} [19]	–	–	–
	Zr	–	–	43×10^{-6} [19]			

[a] References:
1. I.A. Aksay, D.M. Dabbs, and M. Sarikaya: J. Am. Ceram. Soc. **74 [10],** 2343–2358 (1991)
2. R.R. Tummala: J. Am. Ceram. Soc. **74 [5],** 895–908 (1991)
3. R. Morrell: *Handbook of Properties of Technical and Engineering Ceramics, Part 1*. (Her Majesty's Stationery Office London, 1985)
4. *Handbook of Materials Science,* Vol.2, C.T.Lynch (ed.). (CRC Press Boca Raton FL, 1974), pp. 360–362
5. *Fine Ceramics*, S. Saito (ed.). (Elsevier Essex England, 1985)
6. T. Margaria, M. Rapa: *Materiali e Componenti per L'elettronica*. (Levrotto&Bella Turin Italy, 1986)
7. D. Mattox, J. Olenick et.al.: Ceram. Eng. Sci. Proc. **9 [11–12],** 1567–1578 (1988)
8. A.N. Copp: Am. Ceram. Soc. Bull. **68 [5],** 1054–1055 (1989)
9. Manufacturer's Data
10. L.M. Sheppard: Am. Ceram. Soc. Bull. **70 [9],** 1471 (1991)
11. *Advances in Ceramics,* Vol.26, M.F. Yan et al. (eds.). (The American Ceramic Society Westerville OH, 1989)
12. K.E. Spear: J. Am. Ceram. Soc. **72 [2],** 171–191 (1989)
13. N.D. Patel, P.H. Boldt, and P.S. Nicholson: In: *Ceramic Transactions,* Vol.11. (The American Ceramic Society Westerville OH, 1990), pp.19–34
14. G. Geiger: Am. Ceram. Soc. Bull. **69 [7],** 1131 (1990)
15. Z.D. Jastrzebski: *The Nature and Properties of Engineering Materials*. (John Wiley & Sons New York, 1977)
16. *Introduction to Fine Ceramics*, N. Ichinase (ed.). (John Wiley & Sons New York, 1987)
17. *Handbook of Optics,* W.G. Driscoll and W. Vaughan (eds.). (Mc Graw-Hill New York, 1978)
18. *ASM Engineered Materials Reference Book.* (ASM International Metals Park OH, 1989)
19. C.C. Wang, S.A. Akbar, W. Chen, and V.D. Patton: J. Mater. Sci. **30 [7],** 1627–1641 (1995)

Appendix 11. Properties of Some Fast Ionic Conductors [a]

Compound	Ionic Conductivity ($\Omega^{-1}cm^{-1}$)	at T (°C)	Activation Energy (eV)	Enthalpy (kcal/mole)	Stability Range of High Conduct. Form (°C)	Comments
Cationic conductors						
AgCl	8×10^{-5}	200	0.18	–	150–430	3-D conductivity
AgBr	8×10^{-4}	200	0.15	–	175–400	3-D conductivity
α-AgI	1	150	0.05	1.15	146–555 (m.p.)	BCC iodine skeleton, 3-D cond.
α-Ag_2S	3.8	200	0.05	1.15	>179	3-D conductivity
α-Ag_2Se	3.6	220	0.1	–	>133	3-D conductivity
$AgAl_{11}O_{17}$ (silver β-alumina)	0.1	500	0.17	3.91	Very stable	2-D conductivity
β-alumina	0.35	300	0.01	0.23	–	Currently used in Na–S batteries as solid electrolyte, effective formula: 11 $Al_2O_3 \cdot (1+x) M_2O$
α-CuBr	0.3	470	Small	–	>470	
α-CuS	0.2	400	0.25	5.75	>91	Cu–deficient phase, 3-D cond.

(Continued)

Appendix 11. *(continued)*

Compound	Ionic Conductivity ($\Omega^{-1}cm^{-1}$)	at T (°C)	Activation Energy (eV)	Enthalpy (kcal/ mole)	Stability Range of High Conduct. Form (°C)	Comments
Li_2SO_4	1.04	600	0.4	9.20	575–860 (studied range)	
$Li_2Ti_xO_{2x+1}$ $(x=1,2,3)$	0.1	500	0.5	–	–	
$NaSbO_3$	0.06	300	0.35	8.05	–	Cubic structure stabilized by NaF
$NaAl_{11}O_{17}$	0.1	500	0.17	3.91	Very stable	2-D conductivity
$Na_3Zr_2Si_2PO_{12}$	0.2	25	0.27	6.21	Appears stable up to melting point of Na	Highest conductivity in the system $Na_{1+x}Zr_2P_{3-x}Si_xO_{12}$ $(x=0,1,2,3)$
Anionic conductors						
Bi_2O_3–25%Y_2O_3	0.012	500	0.6	13.80	–	
	0.16	700	–	–	–	
Bi_2O_3–20%SrO	0.006	500	–	–	–	
	0.025	600	–	–	–	
	0.22	700	–	–	–	
CeO_2–11%La_2O_3	0.08	–	0.91	–	–	5% anion vacancies

(Continued)

Appendix 11. *(continued)*

Compound	Ionic Conductivity ($\Omega^{-1}cm^{-1}$)	at T (°C)	Activation Energy (eV)	Enthalpy (kcal/mole)	Stability Range of High Conduct. Form (°C)	Comments
HfO_2–8%Y_2O_3	0.029	–	1.1	–	–	3.7% anion vacancies
La_2O_3–15%CaO	0.024	–	0.88	20.24	–	2.7% anion vacancies
ThO_2–8%Y_2O_3	0.0048	1000	1.1	25.30	–	3.7% anion vacancies
ZrO_2–12%CaO	0.055	1000	1.1	25.30	–	Fluorite FCC structure, 6% anion vacancies
ZrO_2–9%Y_2O_3	0.12	1000	0.8	18.40	–	Fluorite FCC, 4.1% anion vacan.
ZrO_2–8%Yb_2O_3	0.088	1000	0.75	–	–	Fluorite FCC, 3.7% anion vacan.
ZrO_2–10%Sc_2O_3	0.25	1000	0.65	14.95	–	Fluorite FCC, 4.5% anion vacan.
CaF_2	0.01	700	1	23.00	–	
LaF_3–5%SrF_2	0.01	400	1	23.00	–	
PbF_2	0.001	200	0.45	10.35	–	FCC lattice

[a] References:

1. L.L. Hench and J.K. West: *Principles of Electronic Ceramics.* (John Wiley&Sons, New York, 1990)
2. P. McGeehin and A. Hooper: J. Mater. Sci. **12 [1],** 1–27 (1977)

Appendix 12 . Properties of Superconducting Ceramics

Material	Process	No. of CuO Sheets	Critical Onset Temperature $T_c^{ons.}$ (°C)	Critical Temperature T_c (°C)	Critical Current Density J_c (A/cm^2)	Estimated Critical Magnetic Field Strength H_c (kOe)	Ref. [a]
YBaCuO compounds							
$YBa_2Cu_3O_7$	–	1	125	83	–	–	[1]
$Y_{0.4}Ba_{0.6}CuO_{2.22}$	–	1	91.7	88	–	–	[1]
$YBa_2Cu_3O_{6.9}$	–	1	93.5	91	–	–	[1]
$YBa_2Cu_3O_{6.8}$	–	1	94	92	–	–	[1]
$YBaCu_2O_{6-y}$	–	1	93	92	–	–	[1]
$YBa_2Cu_3O_{7-y}$	–	–	–	90	–	1860	[2]
$YBa_2Cu_3O_{7-y}$	Polycrystalline film	–	–	42	1.2×10^7	–	[3]
	Single-crystal film	–	–	77	2.5×10^6	–	
	Polycrystalline bulk material	–	–	77	4×10^3	–	
$YBa_2Cu_3O_{7-y}$	Pulsed laser deposition/ablation on $SrTiO_3$	–	–	92	1.5×10^6–5×10^6	–	[4]

(Continued)

Appendix 12. *(continued)*

Material	Process	No. of CuO Sheets	Critical Onset Temperature $T_c^{ons.}$ (°C)	Critical Temperature T_c (°C)	Critical Current Density J_c (A/cm²)	Estimated Critical Magnetic Field Strength H_c (kOe)	Ref. [a]
$YBa_2Cu_3O_{7-y}$	Metallo-organic CVD at 550–800°C	–	–	90	–	–	[4]
$YBa_2Cu_3O_{7-y}$	Plasma-based molecular beam epitaxy at 650°C	–	–	89	–	–	[4]
$YBa_2Cu_3O_{7-y}$	Hot-pressed at 500 psi, 900–960°C	–	–	99	1.4×10^3	–	[5]
(RE) BaCuO compounds							
$DyBa_2Cu_3O_{7-y}$	–	–	–	91.8–92.0	–	1714	[2]
$ErBa_2Cu_3O_{7-y}$	–	–	–	89.5–91.4	–	1261	[2]
$EuBa_2Cu_3O_{7-y}$	–	–	–	93.5–93.9	–	1296	[2]
$GdBa_2Cu_3O_{7-y}$	–	–	–	93.5–93.6	–	1873	[2]
$HoBa_2Cu_3O_{7-y}$	–	–	–	91.7–92.5	–	1659	[2]
$HoBa_2Cu_3O_{7-y}$	Pulsed laser deposition/ablation on MgO/YSZ at 500–650°C	–	–	84–86	10^6	–	[4]
$LaBa_2Cu_3O_{7-y}$	–	–	–	62.5	–	906	[2]
$LaBa_2Cu_3O_{6.6}$	–	1	77	60	–	–	[1]

(Continued)

Appendix 12. *(continued)*

Material	Process	No. of CuO Sheets	Critical Onset Temperature $T_c^{ors.}$ (°C)	Critical Temperature T_c (°C)	Critical Current Density J_c (A/cm^2)	Estimated Critical Magnetic Field Strength H_c (kOe)	Ref. [a]
$LaBa_2Cu_3O_{6+x}$	–	1	91	75	–	–	[1]
$La_{1.85}Sr_{0.15}CuO_4$	–	1	41	24.5	–	–	[1]
$NdBa_2Cu_3O_{7-y}$	–	–	–	88–93.6	–	1290	[2]
$SmBa_2Cu_3O_{7-y}$	–	–	–	93.5–94.2	–	1365	[2]
$TmBa_2Cu_3O_{7-y}$	–	–	–	90.2–91	–	1318	[2]
$YbBa_2Cu_3O_{7-y}$	–	–	–	88.7–89.5	–	1482	[2]
Bi-based compounds							
$Bi_{1.5}CaSrCu_2O_x$	–	–	–	65	–	–	[2]
$Bi_{1.5}Ca_{1.5}SrCu_2O_x$	–	–	–	63	–	–	[2]
$Bi_{0.5}Pb_{0.5}CaSrCu_2O_x$	–	–	–	71	–	–	[2]
$BiAl_{0.1}CaSrCu_2O_x$	–	–	–	73	–	–	[2]
$BiAl_{0.3}CaSrCu_2O_x$	–	–	–	78	–	–	[2]
BiCaSr CuO	–	2	110	–	–	–	[1]

(Continued)

Appendix 12 . *(continued)*

Material	Process	No. of CuO Sheets	Critical Onset Temperature $T_c^{ons.}$ (°C)	Critical Temperature T_c (°C)	Critical Current Density J_c (A/cm^2)	Estimated Critical Magnetic Field Strength H_c (kOe)	Ref. [a]
$BiCaSr_2Cu_2O_{9-y}$	–	2	120	–	–	–	[1]
$Bi_{1.8}Pb_{0.2}Ca_2Sr_2Cu_3O_{10}$	–	2	125	107	–	–	[1]
$BiSrCaCu_2O_x$	Metallo-organic CVD on MgO at 800–850°C	–	–	81	10^4	–	[4]
BiPbCaSr CuO	RF plasma deposition on MgO at 680°C	–	–	90	–	–	[4]
$BiSr_2CaCu_2O_x$ / $BiSr_2Ca_2Cu_3O_x$	Sputter deposition on MgO	–	–	84	10^4	–	[4]
$Bi_2Ca_2Sr_2Cu_2O_8$	Hot-pressed at 850°C	–	–	85	–	–	[5]
Tl-based compounds							
$Tl_2Ba_2CuO_6$	–	–	–	80	–	–	[2]
$Tl_2CaBa_2Cu_2O_8$	–	–	–	105	–	–	[2]
$Tl_2Ca_2Ba_2Cu_3O_{10}$	–	–	–	125	–	–	[2]
Tl_2 $CaBa_2Cu_3O_x$	–	2	125	107	–	–	[1]
$Tl_2Ca_2Ba_2Cu_3O_x$	–	3	–	118	–	–	[1]

(Continued)

Appendix 12. *(continued)*

Material	Process	No. of CuO Sheets	Critical Onset Temperature $T_c^{ons.}$ (°C)	Critical Temperature T_c (°C)	Critical Current Density J_c (A/cm^2)	Estimated Critical Magnetic Field Strength H_c (kOe)	Ref. [a]
Hg-based compounds							
$HgBa_2Ca_2Cu_3O_y$	–	–	–	135	–	–	[6]

[a] References:

1. L.L. Hench and J.K. West: *Principles of Electronic Ceramics*. (John Wiley&Sons New York, 1990)
2. J.D. Doss: *Engineer's Guide to High Temperature Superconductivity*. (John Wiley&Sons New York, 1989)
3. S. Johnson, M. Gusman et al. In: *Ceramic Superconductors II*. M.F. Yan (ed.). (The American Ceramic Society Westerville OH, 1988), pp. 300–307
4. *New Directions in Materials Research.* Am. Ceram. Soc. Bull. **69 [2],** 174 (1990)
5. L.M. Sheppard: Am. Ceram. Soc. Bull. **70 [9],** 1479 (1991)
6. T.P. Sheahen: *Introduction to High Temperature Superconductivity*. (Plenum Press New York, 1994)

Appendix 13. Selected Properties of Various Piezoelectric Materials

Material	Electro-strictive Coefficient φ	Density (gr/cm^3)	Curie temp. (°C)	Relative Dielectric Constant $\varepsilon_{11}^T/\varepsilon_o$	$\varepsilon_{33}^T/\varepsilon_o$	Ref. [a]
α - Quartz (single crystal)	–	2.65	–	–	4.6	[1]
$BaTiO_3$	500	5.70	130	1600	1900	[1]
Sintered PZT 4	450–1750	7.65	–	–	1300	[2,3]
Hot-pressed PZT 4	490	7.99	–	–	–	[2,3]
Sintered PZT 5	70–165	7.73	–	–	3400	[2,3]
Hot-pressed PZT 5	80	8.01	–	–	–	[2,3]
Sintered PZT 7	1025	7.45	–	–	–	[2,3]
Hot-pressed PZT 7	1180	7.97	–	–	–	[2,3]
$LiNbO_3$ (single crystal)	–	4.64	1210	85	29	[1]
$LiTaO_3$ (single crystal)	–	7.46	665	53	43	[1]
$Na_{0.5}K_{0.5}NbO_3$	240	4.50	420	600	400	[1]
$PbNb_2O_6$	11	5.90	560	–	225	[1]
$PbTiO_3$	326	7.12	494	–	203	[1]
PLZT (2La/65Zr/35Ti)	–	–	–	–	–	[4]
PLZT (7La/65Zr/35Ti)	–	–	–	–	–	[4]
PLZT (8La/65Zr/35Ti)	–	–	–	–	–	[4]
PLZT (5La/55Zr/45Ti)	–	–	–	–	–	[4]
PLZT (12La/40Zr/60Ti)	–	–	–	–	–	[4]
PLZT (7La/60Zr/40Ti)	–	–	–	–	–	[4]
Hot-pressed SbSI	–	–	30	–	–	[5]

(Continued)

Appendix 13. *(continued)*

Material	Loss tangent ($\tan \delta$) (%)	Electromechanical coupling factor				Ref. [a]
		K_p	K_{31}	K_{33}	K_{15}	
α - Quartz (single crystal)	–	$K_{11} = 0.1$		$K_{14} = 0.05$		[1]
$BaTiO_3$	0.7	0.38	0.21	0.49	0.44	[1]
Sintered PZT 4	0.4	0.54	0.32	0.70	–	[2,3]
Hot-pressed PZT 4	0.3	0.62	0.37	–	–	[2,3]
Sintered PZT 5	1.8	0.63	0.38	0.75	–	[2,3]
Hot-pressed PZT 5	1.6	0.72	0.43	–	–	[2,3]
Sintered PZT 7	1.5	0.32	0.19	–	–	[2,3]
Hot-pressed PZT 7	0.6	0.38	0.23	–	–	[2,3]
$LiNbO_3$ (single crystal)	–	0.035	0.02	0.17	0.61	[1]
$LiTaO_3$ (single crystal)	–	0.1	0.07	0.14	–	[1]
$Na_{0.5}K_{0.5}NbO_3$	1.0	0.45	0.27	0.53	–	[1]
$PbNb_2O_6$	1.0	0.07	0.045	0.38	–	[1]
$PbTiO_3$	2.2	–	0.052	0.35	0.36	[1]
PLZT (2La/65Zr/35Ti)	–	0.45	–	–	–	[4]
PLZT (7La/65Zr/35Ti)	–	0.62	–	–	–	[4]
PLZT (8La/65Zr/35Ti)	–	0.65	–	–	–	[4]
PLZT (5La/55Zr/45Ti)	–	0.705	–	–	–	[4]
PLZT (12La/40Zr/60Ti)	–	0.47	–	–	–	[4]
PLZT (7La/60Zr/40Ti)	–	0.72	–	–	–	[4]
Hot-pressed SbSI	–	–	–	–	–	[5]

(Continued)

Appendix 13. *(continued)*

Material	Piezoelectric strain constant (pC/N)			Elastic compliance ($\mu m^2/N$)					Ref. [a]
	d_{31}	d_{33}	d_{15}	S_{11}	S_{12}	S_{13}	S_{33}	S_{44}	
α - Quartz (single crystal)	$d_{11} = 2.3$ $d_{14} = 0.67$			12.8	−1.8	−1.2	9.6	20.0	[1]
$BaTiO_3$	−79	190	270	8.6	−2.6	−2.9	9.1	23.0	[1]
Sintered PZT 4	−115	289	–	–	–	–	18.8	–	[2,3]
Hot-pressed PZT 4	−136	–	–	–	–	–	–	–	[2,3]
Sintered PZT 5	−170	593	–	–	–	–	20.7	–	[2,3]
Hot-pressed PZT 5	−230	–	–	–	–	–	–	–	[2,3]
Sintered PZT 7	−30	–	–	–	–	–	–	–	[2,3]
Hot-pressed PZT 7	−40	–	–	–	–	–	–	–	[2,3]
$LiNbO_3$ (single crystal)	−0.85	6	69	5.8	−1.2	−1.42	5.0	17.1	[1]
$LiTaO_3$ (single crystal)	−3.0	5.7	26	4.9	−0.5	−1.28	4.3	10.5	[1]
$Na_{0.5}K_{0.5}NbO_3$	−50	160	–	9.6	–	–	10	–	[1]
$PbNb_2O_6$	−11	80	–	29	–	−5 to −8	25	–	[1]
$PbTiO_3$	−7.4	47	–	11	–	–	11	–	[1]
PLZT (2La/65Zr/35Ti)	–	150[b]	–	–	–	–	–	–	[4]
PLZT (7La/65Zr/35Ti)	–	400[b]	–	–	–	–	–	–	[4]
PLZT (8La/65Zr/35Ti)	–	682[b]	–	–	–	–	–	–	[4]
PLZT (5La/55Zr/45Ti)	–	578[b]	–	–	–	–	–	–	[4]
PLZT (12La/40Zr/60Ti)	–	235[b]	–	–	–	–	–	–	[4]
PLZT (7La/60Zr/40Ti)	–	710[b]	–	–	–	–	–	–	[4]
Hot-pressed SbSI	–	40	−60	–	–	–	–	–	[5]

[a] References:

1. A.J. Moulson and J.M. Herbert: *Electroceramics*. (Chapman & Hall London England, 1990)
2. N.D. Patel and P.S. Nicholson: Am. Ceram. Soc. Bull. **65 [5],** 783–787 (1986)
3. R.W. Whatmore. In: *Fundamentals of Ceramic Engineering*. P. Vincenzini (ed.). (Elsevier Applied Science Essex England, 1991)
4. *Ceramic Materials for Electronics*, R.C. Buchanan (ed.). (Marcel Dekker New York, 1986)
5. K. Yon, M. Kahn, and R.W. Rice: Adv.Ceram.Mater. **1 [1],** 64–67 (1986)

[b] unit: pm/V

Appendix 14. Selected Properties of Various Pyroelectric Materials

Material	Curie Temperature T_c	Pyroelectric Coef. P	Dielectric Properties		Volume Specific Heat C_v	Figures of Merit		Ref. [a]
	(°C)	($\times 10^{-8}$ C/cm^2 K)	Dielectric Constant ε	Loss Tangent ($\tan\delta$)	[J/(cm^3 K)]	F_v (m^2/C)	F_D (μm^3/J)$^{1/2}$	
DTGS (deuterated TGS) single crystal (at 40 °C)	61	5.5	43	2×10^{-2}	2.4	0.60	83	[1]
$LiTaO_3$ single crystal	665	2.3	47	1×10^{-4}	3.2	0.17	350	[1]
TGS (triglycine sulphate) single crystal (at 35°C)	49	2.8	38	1×10^{-2}	2.3	0.36	66	[1]
PVDF (polyvinylidene fluoride) polymer	80	0.27	12	10–2	2.4	0.1	8.8	[1]
$(SrBa)Nb_2O_6$ single crystal	121	5.5	400	3×10^{-3}	2.3	0.07	72	[1]
$LiNbO_3$	1210	30	0.4	–	2.8	0.48[b]	–	[2]
$Pb(Zr,Ti)O_3{\cdot}Pb(Sn_{0.5}Sb_{0.5})O_3$	220	18.0	380	–	2.4	2.0[b]	–	[2]
$PbTiO_3$	470	6.0	200	–	3.2	0.94[b]	–	[2]

(Continued)

Appendix 14. *(continued)*

Material	Curie Temperature T_c	Pyroelectric Coef. P	Dielectric Properties		Volume Specific Heat C_v	Figures of Merit		Ref. [a]
	(°C)	($\times 10^{-8}$ C/cm^2 K)	Dielectric Constant ε	Loss Tangent ($\tan\delta$)	[J/(cm^3 K)]	F_v (m^2/C)	F_D (μm^3/J)$^{1/2}$	
$Sr_{0.48}Ba_{0.52}Nb_2O_6$	115	6.5	380	–	2.1	0.8[b]	–	[2]
$LiTiO_3$ single crystal	660	2.3	54	–	3.15	–	–	[3]
$PbZrO_3$-based ceramic	200	3.5	250	5×10^{-3}	2.6	0.54[b]	1.2[c]	[4]
$PbZrO_3$–(PbZn)NbO_3-based ceramic	235	2.8	300	2×10^{-2}	2.6	0.36[b]	0.44[c]	[4]
PZT–5A polycrystalline	365	4.0	1900	–	3.1	–	–	[3]

[a] References:

1. A.J. Moulson and J.M. Herbert: *Electroceramics*. (Chapman & Hall London, 1990)
2. *Fine Ceramics*. S. Saito (ed.). (Elsevier Essex England, 1985)
3. *Introduction to Fine Ceramics*. N. Ichinose (ed.). (John Wiley & Sons New York, 1987)
4. T. Takenaka, A.S. Bhalla, L.E. Cross, and K. Sakata: J. Am. Ceram. Soc. **72 [6]**, 1016–1023 (1989)

[b] ($\times 10^{-10}$ C cm/J)
[c] ($\times 10^{-8}$ C cm/J)

Appendix 15. Selected Properties of Various Magnetic Materials

Material Composition	Curie Temp. T_c (°C)	Initial Permeability μ	Saturation Magnetization M_s at 0 K (G)	Saturation Magnetization M_s at 20°C (G)	Remanence B_r (G)	Coercivity H_c (Oe)	Resonance Line Width ΔH (Oe)	Dc Resistivity (Ω·cm)	Dielectric Loss ε	Loss Tangent (tanδ)	Ref. [a]
Spinel ferrites											
27 MnO·20 ZnO·53 Fe_2O_3	433	2000	–	–	–	0.2	–	20	–	–	[1]
34 MnO·14 ZnO·52 Fe_2O_3	493	1500	–	–	–	0.25	–	20	–	–	[1]
30 MnO·15 ZnO·55 Fe_2O_3	483	650	–	–	–	0.7	–	150	–	–	[1]
32 NiO·18 ZnO·50 Fe_2O_3	653	110	–	–	–	4.0	–	10^6	–	–	[1]
50 NiO·50 Fe_2O_3	713	20	–	–	–	14.0	–	10^6	–	–	[1]
MgO·Fe_2O_3	898	–	–	1450	–	–	900	10^8	8.5	0.016	[2]
MnO·Fe_2O_3	848	–	–	4800	–	–	600	10^4	9.3	0.051	[2]
NiO·Fe_2O_3	1133	–	–	3200	–	–	500	10^4	8.9	0.017	[2]
(NiO·ZnO)·Fe_2O_3	923	–	5000	–	–	0.9	135	–	12.5	0.001	[2]
(NiO·CoO)·Fe_2O_3	1133	–	3000	–	–	12	350	–	13.0	0.003	[2]
(NiO·Al_2O_3)·Fe_2O_3	1108	–	2000	–	–	4	450	–	12.8	0.001	[2]
(NiO·CoO·Al_2O_3)·Fe_2O_3	1093	–	1800	–	–	36	1000	–	9.0	0.002	[2]
(MgO·MnO)·Fe_2O_3	858–86	–	2150–2	–	–	2.5	500–540	–	13.0	<0.003	[2]
(MgO·MnO·Al_2O_3)·Fe_2O_3	838	–	1760	–	–	2.1	490	–	12.0	0.0005	[2]

(Continued)

Appendix 15. *(continued)*

Material Composition	Curie Temp. T_c (°C)	Initial Permeability μ	Saturation Magnetization M_s at 0 K (G)	Saturation Magnetization M_s at 20°C (G)	Remanence B_r (G)	Coercivity H_c (Oe)	Resonance Line Width ΔH (Oe)	Dc Resistivity (Ω·cm)	Dielectric Loss ε	Loss Tangent (tanδ)	Ref. [a]
Hexagonal ferrites											
$BaO{\cdot}6Fe_2O_3$	450	–	530 [b]	380 [b]	3950	1850	–	–	–	–	[1]
$2(BaO{\cdot}CoO{\cdot}3Fe_2O_3)$	320	–	204 [b]	–	–	–	–	–	–	–	[1]
$BaFe_{18}O_{27}$	445	–	520 [b]	–	–	–	–	–	–	–	[1]
$SrO{\cdot}6Fe_2O_3$	450	–	530 [b]	380 [b]	3425	3300	–	–	–	–	[1]
Garnets											
$Y_3Fe_5O_{12}$ (Y16)	823	–	–	1740	–	–	50	10^{10}	12.0	0.002	[2]
YGd–garnet	813	–	1600	–	–	1.1	60	–	16.0	<0.0003	[2]
YAl–garnet	823	–	1540	–	–	–	50	–	14.8	0.0005	[2]
YAl–garnet	808	–	1200	–	–	0.5	60	–	15.5	<0.0003	[2]
YGdAl–garnet	798	–	1200	–	–	–	55	–	14.8	0.0005	[2]

[a] References:
1. L.L. Hench and J.K. West: *Principles of Electronic Ceramics.* (John Wiley & Sons New York, 1990)
2. *Handbook of Microwave Ferrite Materials.* W.H. von Aulock (ed.). (Academic Press Orlando FL, 1965)

[b] emu/cm^3

Appendix 16. Etching Techniques for Selected Ceramics

Material	Etching Method	Etchant Composition	Etching Conditions	Ref. [a]
Oxides				
Al_2O_3 (>99.9%)	Chemical	Conc H_2SO_4	20 min, 200°C	[1]
Al_2O_3 (>99.9%)			5 min, 275°C	[1]
Al_2O_3 (>99.9%)	Thermal	–	15 min, 1470°C	[1]
Al_2O_3 (>99.7%)	Chemical	H_3PO_4	10 min, boiling	[1]
Al_2O_3 (>99.7%)	Thermal	–	20 min, 1450°C	[1]
Al_2O_3 (>99%)	Chemical	$KHSO_4$	5 min, 500°C	[1]
Al_2O_3 (>99%)	Thermal	–	5 min, 1600°C	[1]
Al_2O_3 (>97.5%)	Chemical	10% HF	10 s, 25°C	[1]
Al_2O_3 (95–96%)	Thermal	–	15 min, 1350°C	[1]
Al_2O_3	Thermal	–	120 min, 1100–1500°C in air	[2]
Al_2O_3, PSZ	Chemical	H_3PO_4	5–16 min, 180–250°C, chemical hood, remove plastic mount before polishing	[3]
Al_2O_3 (90–99%	Chemical	H_2SO_4	5 s–1min, 330°C	[3]
Al_2O_3 (85–90% dense)	Chemical	2 mL HF, 98 mL water	20 °C	[3]
$BaTiO_3$	Chemical	4 mL HNO3, 94 mL H_2O, 2 mL HF	10–40 s, 20°C	[3]
$BaTiO_3$	Chemical	H_3PO_4	120 min, 100°C, distinguishes poling effects	[3]
$BaTiO_3$	Chemical	HNO_3	20°C, distinguishes poling effects	[3]
$BaTiO_3$, $SrTiO_3$	Chemical	1 mL HF, 99 mL H_2O	20 s, 20°C, distinguishes poling effects	[3]

(Continued)

Appendix 16. *(continued)*

Material	Etching Method	Etchant Composition	Etching Conditions	Ref. [a]
MgO	Chemical	H_3PO_4	4 min, 20°C	[3]
MgO	Chemical	50 ml HNO_3, 50 ml H_2O	1–5 min, 20°C	[3]
MgO	Chemical	H_2SO_4	1 min, 55°C	[3]
$PbZrO_3$–$PbTiO_3$	Chemical	5 mL HCl, 95 mL H_2O, 3 drops of HF	3–4 min, 20°C	[3]
SiO_2	Chemical	H_3PO_4		[3]
SiO_2	Chemical	HF	5 s, 20°C	[3]
SnO_2	Thermal	–	4 min, 1100°C in air	[3]
$SrTiO_3$	Chemical	10 mL HF, 20 mL H_2O, 20 mL HNO_3	5 min, 25°C	[3]
TiO_2	Chemical	KOH	8 min, 650°C, no mount	[3]
UO_2	Chemical	50 mL HNO_3, 50 mL H_2O	1–5 min, 20°C	[3]
UO_2	Chemical	18–20 mL hydrogen peroxide, 2 mL H_2SO_4	1 min	[3]
ZrO_2	Chemical	H_3PO_4	3 s–2 min, 250°C	[1]
ZrO_2	Chemical	50% H_2SO_4	1–5 min, boiling	[1]
ZrO_2	Chemical	40% HF	2 s–6 min, room temp.	[1]
ZrO_2	Thermal	–	1–60 min, 1100–1500°C in air (e.g., 10 min, 1400°C)	[1]
ZrO_2	Ion beam	–	1–7 kV, ion current 60–400 mA, 30 min, 5–60°C	[1]
ZrO_2	Chemical	50 mL H_2O, 50 mL H_2SO_4	1–5 min, boiling	[3]

(Continued)

Appendix 16. *(continued)*

Material	Etching Method	Etchant Composition	Etching Conditions	Ref. [a]
Nitrides				
Si_3N_4 /Sialon	Chemical	Molten NaOH (water free)	20 s–3 min, 350°C	[1]
Si_3N_4 (Hot-pressed)	Chemical	Molten KOH	1 min	[1]
Si_3N_4 (Reaction bonded) /Sialons	Chemical	Molten KOH + KNO_3, 1:9	5–6 min	[1]
Si_3N_4/Sialon	Chemical	Concentrated H_3PO_4	5–45min, 250oC	[1]
Si_3N_4/Sialon	Plasma	CF_4 gas + O_2, 2:1	5 min	[1]
Si_3N_4	Thermal	–	5 hr, 1600°C in high-purity N_2	[2]
Si_3N_4	Electrolytic	2–10 g chromic acid, 90 mL water	3 V, 2–10 s, 20°C, stainless steel cathode	[3]
Carbides				
α-SiC	Chemical	Modified Murakami Solution (3 g NaOH, 30 g K3Fe(CN)$_6$, 100 mL distilled water)	15–20 min, boiling	[1]
β-SiC	Chemical	Molten (90 g KOH + 10 g KNO_3)	2–3min, 450°C	[1]
SiC (Reaction-bonded)	Chemical	Murakami solution (10 g NaOH, 10 g $K_3Fe(CN)_6$, 100 mL distilled water)	15 min, boiling	[1]
SiC (Hot-pressed), SiC+B	Thermal	–	30 min, 1200°C in vacuum, 1550°C in Ar	[1]

(Continued)

Appendix 16. *(continued)*

Material	Etching Method	Etchant Composition	Etching Conditions	Ref. [a]
SiC (RB, Sint., HP)	Electrolytic	400 cm^3 glacial acetic acid, 120 g Cr_2O_3, 380 mL water, 30 mL H_2SO_4, 2 mL C_2H_5OH	Sample connected as anode in 6 V dc cell with stainless steel cathode, 10 s	[1]
SiC	Electrolytic	40 g KOH, 160 mL water	60 V, 20 s, stainless steel cathode	[3]
SiC	Chemical	30 g NaF, 60 g KCO_3	10–60 min, 650°C (remove specimen from mounting before etching)	[3]
SiC	Chemical	20 g NaOH, 20 g Na–peroxide	400–600°C	[3]
TaC	Chemical	16 mL HCl, 4 mL HNO_3, 20 mL HF	Chemical hood	[3]
TiC	Electrolytic	5 g Cu–sulfate, 16.5 g Na–hydroxide, 76 mL H_3PO_4, 1 mL H_2SO_4, 320 mL water	3–5 V, 25°C, Cu–cathode, 20–30 s	[3]
TiC, WC	Chemical	10 g K–ferricyanide, 10 g K–hydroxide, 100 mL water	–	[3]
WC	Chemical	20 mL HCl, 20 mL hydrogen peroxide	–	[3]
ZrC	Chemical	27 mL HCl, 9 mL HNO_3, 3 mL HF	Chemical hood	[3]

(Continued)

Appendix 16. *(continued)*

Material	Etching Method	Etchant Composition	Etching Conditions	Ref. [a]
Borides				
HfB, HfB_2, ZrB_2	Hybrid	10–20 g polishing abrasive, 10 g potassium ferricyanide, 10 g potassium hydroxide, 100 mL water	For final polishing (if manual), use rubber gloves	[3]
LaB_6, tantalum borides, titanium borides	Electrolytic	10 g NaOH, 490 mL water	Stainless steel cathode, 20°C	[3]
TiB_2, HfB_2, ZrB_2	Chemical	10 mL HF, 10 mL HNO_3, 30 mL lactic acid	–	[3]

[a] References:

1. W.E. Lee and W.M.Rainforth: *Ceramic Microstructures.* (Chapman & Hall London England, 1994)
2. D.C. Zipperian: Adv. Mater. Proc. **140 [5],** 37–39 (1991)
3. J.H. Richardson: *Optical Microscopy for the Material Sciences.* (Marcel Dekker New York, 1971)

Appendix 17. Selected Properties and Applications of Refractories

Refractory Name	Chemical Composition (%)												Ref. [a]
	Al_2O_3	SiO_2	ZrO_2	Cr_2O_3	CaO	C	MgO	SiC	Fe_2O_3	TiO_2	Alkalies	B_2O_3	
Silica brick[b]	0.9	95.4	–	–	1.87	–	–	–	–	–	–	–	[1]
Alumina–chromia brick[b]	77.2	3.2	3.0	13.8	0.16	–	–	–	–	–	–	–	
High alumina brick[b]	83.5	16.0	–	–	–	–	–	–	–	–	–	–	
Zircon brick[b]	3.7	35.4	60.1	–	–	–	–	–	–	–	–	–	
Magnesite	1–3	2–3	–	–	1–3	–	85–90	–	2–5	–	–	–	[2]
Chromite	15–20	4–6	–	30–40	1–2	–	15–20	–	12–15	–	–	–	
Chrome–magnesite	15–20	4–6	–	28–30	1–2	–	38–40	–	10–15	–	–	–	
Magnesite–chrome	4–6	3–4	–	6–10	1–2	–	65–75	–	4–6	–	–	–	
Dolomite	2–3	12–15	–	–	38–40	–	40–42	–	2–3	–	–	–	
Forsterite	2–3	30–40	–	–	0.5–1.0	–	50–60	–	6–8	–	–	–	
High alumina ladle brick	85.9	11.2	–	–	–	–	–	–	–	–	–	–	[3]
Al_2O_3–C ladle brick	82.0	7.5	–	–	–	6.4	–	–	–	–	–	–	
MgO–C brick	–	–	–	–	–	18.7	76.8	–	–	–	–	–	
Al_2O_3–MgO–C ladle brick	54.8	3.2	–	–	–	6.5	28.2	–	–	–	–	–	

(Continued)

Appendix 17. *(continued)*

Refractory Name	Chemical Composition (%)												Ref. [a]
	Al_2O_3	SiO_2	ZrO_2	Cr_2O_3	CaO	C	MgO	SiC	Fe_2O_3	TiO_2	Alkalies	B_2O_3	
Conventional magnesia–chromite brick	–	1.6	–	15.2	0.7	–	63.7	–	–	–	–	–	[4]
Hybrid magnesia–chromite brick	–	1.1	–	18.0	9.5	–	61.1	–	–	–	–	–	
Special magnesia–chromite brick	–	1.0	–	21.3	0.6	–	60.0	–	–	–	–	–	
Rebonded magnesia–chromite brick	–	1.8	–	19.2	9.6	–	59.8	–	–	–	–	–	
Casting-through materials													
A	78–82	1–3	–	–	–	1–3	–	11–14	–	–	–	–	[5]
B	70–74	1–4	–	–	–	1–3	–	14–17	–	–	–	–	
C	68–72	1–4	–	–	–	3–5	–	14–17	–	–	–	–	
D	62–68	6–9	–	–	–	2–4	–	16–19	–	–	–	–	
E	64–68	6–9	–	–	–	2–4	–	14–17	–	–	–	–	
F	30–33	48–52	–	–	–	2–4	–	7–10	–	–	–	–	
G	55–59	6–9	–	–	–	4–6	–	18–21	–	–	–	–	

(Continued)

Appendix 17. *(continued)*

Refractory Name	Chemical Composition (%)												Ref. [a]
	Al_2O_3	SiO_2	ZrO_2	Cr_2O_3	CaO	C	MgO	SiC	Fe_2O_3	TiO_2	Alkalies	B_2O_3	
Castable refractories in petrochemical applications													
Cast installation or pneumatic gunning													
A	44–52	38–50	–	–	2.8–7.6	–	–	–	0.6–1.2	1.2–1.4	0.2–0.4	–	[6]
B	45–53	41–42	–	–	3.0–8.1	–	–	–	1.1–1.2	0.3–1.1	0.5–1.0	–	
C	57–94	5–39	–	–	1.0–1.2	–	–	–	0–0.2	0–1.8	–	–	
D	54	37	–	–	4.8–5.0	–	–	–	1.8	0.8	1.2	–	
Cast installation by vibration												–	
E	92	6.5	–	–	1.4	–	–	–	<0.1	<0.1	<0.1	–	
F	66	30	–	–	1.0	–	–	–	0.9	1.6	<0.2	–	
Ceramic foam materials													
Phosphate-bonded alumina	–	–	–	–	–	–	–	–	–	–	–	–	[7]
Mullite (3 $Al_2O_3 \cdot 2\ SiO_2$)	–	–	–	–	–	–	–	–	–	–	–	–	

(Continued)

Appendix 17. *(continued)*

Refractory Name	Chemical Composition (%)												Ref. [a]
	Al_2O_3	SiO_2	ZrO_2	Cr_2O_3	CaO	C	MgO	SiC	Fe_2O_3	TiO_2	Alkalies	B_2O_3	
Sintered alumina	>99	–	–	–	–	–	–	–	–	–	–	–	[7]
Partially stabilized zirconia	–	–	–	–	–	–	–	–	–	–	–	–	
Zirconia–alumina	35	–	65	–	–	–	–	–	–	–	–	–	
Fused-cast refractories													
Mullite corundum	73.55	0.08	1.01	–	0.39	–	0.47	–	1.02	4.12	0.87	–	[8]
Alpha A1	99.34	0.60	–	–	0.13	–	–	–	0.06	–	0.39	0.50	
Alpha A2	99.80	1.09	–	–	–	–	–	–	0.09	–	0.03	–	
Alpha-beta alumina	94.81	0.12	–	–	0.28	–	0.15	–	0.06	–	3.59	–	
Beta alumina	94.43	1.77	–	–	0.12	–	0.06	–	0.06	0.03	5.17	–	
Chrome–alumina	60.40	15.29	–	27.26	–	–	6.05	–	4.21	–	0.31	0.15	
33%Zirconia–alumina	49.69	11.34	33.24	–	0.18	–	–	–	0.19	0.18	1.26	0.16	
36%Zirconia–alumina	50.72	10.61	36.37	–	–	–	–	–	0.16	–	1.25	0.15	
41%Zirconia–alumina	47.33	–	40.97	–	–	–	–	–	0.15	–	0.84	–	

(Continued)

Appendix 17. *(continued)*

Refractory Name	Chemical Composition (%)												Ref. [a]
	Al_2O_3	SiO_2	ZrO_2	Cr_2O_3	CaO	C	MgO	SiC	Fe_2O_3	TiO_2	Alkalies	B_2O_3	
Chromic oxide, zircon, and zirconia refractories													[9]
Dense chromic oxide	–	–	–	94.2	–	–	–	–	–	3.8	–	–	
Chrome 20	–	–	–	94.2	–	–	–	–	–	3.8	–	–	
Exp.–HD chromic oxide	–	–	–	93.4	–	–	–	–	–	3.8	–	–	
Dense zircon	0.2	32.2	65.9	–	–	–	–	–	–	1.3	–	–	
Zircon 20	0.3	34.2	64.7	–	–	–	–	–	–	0.7	–	–	
ZMg	–	0.5	97.0	–	–	–	–	–	–	–	–	–	
Sintered refractories													[10]
C–125 (High purity chromic oxide, isopressed or slip cast, corrosion resistant)													[11]
C–1220 (Isostatically pressed chromic oxide, thermal shock resistant)													
C–1222 (Mechanically pressed high-purity chromic oxide, thermal shock resistant)													
Zircon 20 (Dense, high-purity zircon, slip cast or isostatically pressed)													
ZS 1300 (Zirconium silicate, thermal shock resistant)													
ZS 835 (Pure isostatically pressed $ZrSiO_4$ with low blistering potential)													
T–1185/KGT (Tin oxide refractory, highly corrosion resistant)													
BPM 70 (Enriched mullite grade materials for large and complex shapes)													
AKRAL 60 V (High alumina grade materials for large and complex shapes)													

Appendix 17. *(continued)*

Refractory Name	Physical Properties										Ref. [a]
	Bulk Density (g/cm^3)	Apparent Porosity (%)	MOR (at room temp.) (MPa)	MOR (MPa)	(at T) (°C)	Crushing Strength (MPa)	Elastic Modulus (GPa)	Thermal Conductivity ($g \cdot cal \cdot cm/cm^2 \cdot {}^\circ C \cdot s$)	Shear Modulus (GPa)	Electrical Resistivity at 1093°C (Ω·cm)	
Silica brick[b]	1.82	20.6	13	–	–	–	11.6	–	4.5	–	[1]
Alumina–chromia brick[b]	3.20	16.0	25	–	–	–	63.4	–	26.4	–	
High alumina brick[b]	2.78	12.7	30	–	–	–	77.3	–	32.0	–	
Zircon brick[b]	3.77	13.2	28	–	–	–	114.4	–	47.2	–	
Magnesite	2.80	19–24	–	–	–	70.0	–	–	–	–	[2]
Chromite	3.00	22–27	–	–	–	48.3	–	–	–	–	
Chrome–magnesite	3.00	20–25	–	–	–	27.6	–	–	–	–	
Magnesite–chrome	2.90	20–23	–	–	–	27.6	–	–	–	–	
Dolomite	2.50	22–24	–	–	–	55.2	–	–	–	–	
Forsterite	2.60	20–23	–	–	–	27.6	–	–	–	–	

(Continued)

Appendix 17. *(continued)*

Refractory Name	Physical Properties										Ref. [a]
	Bulk Density (g/cm^3)	Apparent Porosity (%)	MOR (at room temp.) (MPa)	MOR (MPa)	(at *T*) (°C)	Crushing Strength (MPa)	Elastic Modulus (GPa)	Thermal Conductivity ($g \cdot cal \cdot cm/cm^2 \cdot {}^oC \cdot s$)	Shear Modulus (GPa)	Electrical Resistivity at 1093°C (Ω·cm)	
High alumina ladle brick	3.00	16.9	–	2.9	1400	88	–	–	–	–	[3]
Al_2O_3–C ladle brick	2.95	5.5	–	5.9	1400	73	–	–	–	–	
MgO–C brick	2.88	4.0	–	10.8	1400	44	–	–	–	–	
Al_2O_3–MgO–C ladle brick	3.00	6.0	–	8.3	1400	69	–	–	–	–	
Conventional magnesia–chromite brick	3.10	16.0	4	12	1260	37	–	–	–	–	[4]
Hybrid magnesia–chromite brick	3.20	16.0	9	12	1260	62	–	–	–	–	
Special magnesia–chromite brick	3.30	13.0	10	14	1260	63	–	–	–	–	
Rebonded magnesia–chromite brick	3.30	13.0	11	16	1260	46	–	–	–	–	
				5	1480		–	–	–	–	

(Continued)

Appendix 17. *(continued)*

Refractory Name	Physical Properties										Ref. [a]
	Bulk Density (g/cm^3)	Apparent Porosity (%)	MOR (at room temp.) (MPa)	MOR (MPa)	(at T) (°C)	Crushing Strength (MPa)	Elastic Modulus (GPa)	Thermal Conductivity (g·cal·cm/cm^2·°C·s)	Shear Modulus (GPa)	Electrical Resistivity at 1093°C (Ω·cm)	
Casting-through materials											
A	2.97	–	2.5–13	–	–	–	–	–	–	–	[5]
B	2.76	–	1–12	–	–	–	–	–	–	–	
C	2.71	–	1.5–10	–	–	–	–	–	–	–	
D	2.39	–	1.5–9	–	–	–	–	–	–	–	
E	2.48	–	1–6	–	–	–	–	–	–	–	
F	2.15	–	1–6	–	–	–	–	–	–	–	
G	2.36	–	1–10	–	–	–	–	–	–	–	

(Continued)

Appendix 17. *(continued)*

Refractory Name	Physical Properties										Ref. [a]
	Bulk Density (g/cm^3)	Apparent Porosity (%)	MOR (at room temp.) (MPa)	MOR (MPa)	(at T) (°C)	Crushing Strength (MPa)	Elastic Modulus (GPa)	Thermal Conductivity ($g \cdot cal \cdot cm/cm^2 \cdot {}^{\circ}C \cdot s$)	Shear Modulus (GPa)	Electrical Resistivity at 1093°C (Ω·cm)	
Castable refractories in petrochemical applications											
Cast installation or pneumatic gunning											[6]
A	2.0–2.2	–	–	–	–	–	–	–	–	–	
B	1.7–1.8	–	–	–	–	–	–	–	–	–	
C	2.5–3.0	–	–	–	–	–	–	–	–	–	
D	1.0–1.2	–	–	–	–	–	–	–	–	–	
Cast installation by vibration											
E	2.8	–	–	17.2	104	–	–	–	–	–	
F	2.6	7.7	–	16.2	104	–	–	–	–	–	
				32.7	1593						

(Continued)

Appendix 17. *(continued)*

Refractory Name	Physical Properties										Ref. [a]
	Bulk Density (g/cm^3)	Apparent Porosity (%)	MOR (at room temp.) (MPa)	MOR (MPa)	(at *T*) (°C)	Crushing Strength (MPa)	Elastic Modulus (GPa)	Thermal Conductivity ($g \cdot cal \cdot cm/cm^2 \cdot {}^\circ C \cdot s$)	Shear Modulus (GPa)	Electrical Resistivity at 1093°C (Ω·cm)	
Ceramic foam materials											
Phosphate-bonded alumina	–	–	0.35	–	0.86	–	–	–	–	–	[7]
Mullite (3 $Al_2O_3 \cdot 2\ SiO_2$)	–	–	0.69	–	2.01	–	–	–	–	–	
Sintered alumina	–	–	0.86	–	3.50	–	–	–	–	–	
Partially stabilized zirconia	–	–	–	–	–	–	–	–	–	–	
Zirconia–alumina	–	–	0.68	–	2.01	–	–	–	–	–	
Fused-cast refractories											
Mullite corundum	3.45	2.55	–	–	–	–	–	0.0120	–	–	[8]
Alpha A1	3.52	1.06	–	–	–	–	–	0.0169	–	715	
Alpha A2	3.56	15	–	–	–	–	–	0.0172	–	–	
Alpha-beta alumina	3.17	1.91	–	–	–	–	–	0.0117	–	18	

(Continued)

Appendix 17. *(continued)*

Refractory Name	Physical Properties										Ref. [a]
	Bulk Density (g/cm^3)	Apparent Porosity (%)	MOR (at room temp.) (MPa)	MOR (MPa)	(at *T*) (°C)	Crushing Strength (MPa)	Elastic Modulus (GPa)	Thermal Conductivity ($g \cdot cal \cdot cm/cm^2 \cdot {}^\circ C \cdot s$)	Shear Modulus (GPa)	Electrical Resistivity at 1093°C ($\Omega \cdot cm$)	
Beta alumina	2.88	4.09	–	–	–	–	–	0.0083	–	44	
Chrome–alumina	3.45	4.23	–	–	–	–		0.0120	–	82	
33%Zirconia–alumina	3.45	0.45	–	–	–	–		0.0069	–	90	
36%Zirconia–alumina	3.52	0.61	–	–	–	–	–	0.0072	–	114	
41%Zirconia–alumina	3.69	1.34	–	–	–	–	–	0.0079	–	130	
Chromic oxide, zircon, and zirconia refractories											[9]
Dense chromic oxide	4.24	15	–	–	–	–	–	–	–	–	
Chrome 20	4.09	20	–	–	–	–	–	–	–	–	
Exp.–HD chromic oxide	4.85	2.4	–	–	–	–	–	–	–	–	
Dense zircon	4.24	0.5	–	–	–	–	–	–	–	–	
Zircon 20	3.92	12.8	–	–	–	–	–	–	–	–	
ZMg	4.70	17.0	–	–	–	–	–	–	–	–	

(Continued)

Appendix 17. *(continued)*

Refractory Name	Physical Properties										Ref. [a]
	Bulk Density (g/cm^3)	Apparent Porosity (%)	MOR (at room temp.) (MPa)	MOR (MPa)	(at T) (°C)	Crushing Strength (MPa)	Elastic Modulus (GPa)	Thermal Conductivity ($g·cal·cm/cm^2·°C·s$)	Shear Modulus (GPa)	Electrical Resistivity at 1093°C (Ω·cm)	
Sintered refractories											[10]
C–125 (High purity chromic oxide, isopressed or slip cast,	4.24	–	75.8	–	–	–	–	–	–	–	[11]
C–1220 (Isostatically pressed chromic oxide, thermal shock resistant)	3.96	–	22.7	–	–	–	–	–	–	–	
C–1222 (Mechanically pressed high-purity chromic oxide, thermal shock resistant)	3.97	–	40.6	–	–	–	–	–	–	–	
Zircon 20 (Dense, high-purity zircon, slip cast or isostatically pressed)	4.25	–	74.8	–	–	–	–	–	–	–	

(Continued)

Appendix 17. *(continued)*

Refractory Name	Physical Properties										Ref. [a]
	Bulk Density (g/cm^3)	Apparent Porosity (%)	MOR (at room temp.) (MPa)	MOR (MPa)	(at *T*) (°C)	Crushing Strength (MPa)	Elastic Modulus (GPa)	Thermal Conductivity (g·cal·cm/cm^2·°C·s)	Shear Modulus (GPa)	Electrical Resistivity at 1093°C (Ω·cm)	
ZS 1300 (Zirconium silicate, thermal shock resistant)	3.92	–	27.6	–	–	–	–	–	–	–	[10] [11]
ZS 835 (Pure isostatically pressed $ZrSiO_4$ with low blistering potential)	3.93	–	46.8	–	–	–	–	–	–	–	
T–1185/KGT (Tin oxide refractory, highly corrosion resistant)	6.01–6.48	–	66.3	–	–	–	–	–	–	–	
BPM 70 (Enriched mullite grade materials for large and complex shapes)	2.64	–	22.1	–	–	–	–	–	–	–	
AKRAL 60 V (High alumina grade materials for large and complex shapes)	2.55	–	20.7	–	–	–	–	–	–	–	

(Continued)

Appendix 17. *(continued)*

Refractory Name	Applications/Notes	Ref. [a]
Silica brick[b]		[1]
Alumina–chromia brick[b]		
High alumina brick[b]		
Zircon brick[b]		
Magnesite		[2]
Chromite		
Chrome–magnesite		
Magnesite–chrome		
Dolomite		
Forsterite		
High alumina ladle brick	Alumina phosphate bond	[3]
Al_2O_3–C ladle brick	Resin bond	
MgO–C brick	Resin bond	
Al_2O_3–MgO–C ladle brick	Resin bond	

(Continued)

Appendix 17. *(continued)*

Refractory Name	Applications/Notes	Ref. [a]
Conventional magnesia–chromite brick	RH and RH OB degasser, upper and intermediate	[4]
Hybrid magnesia–chromite brick	RH/RH OB, snorkel, throat, bottom, lower intermediate	
Special magnesia–chromite brick	RH/RH OB, snorkel, OB tuyere,throat	
Rebonded magnesia–chromite brick	RH, RH OB lower vessel	
Casting-through materials		
A	Main through, impact zone, shimmer zone	[5]
B	Main through, impact zone, shimmer zone,	
C	Main through, impact zone, metal line	
D	Main through, metal line	
E	Tilting runner	
F	Iron runner	
G	Slag runner	

(Continued)

Appendix 17. *(continued)*

Refractory Name	Applications/Notes	Ref. [a]
Castable refractories in petrochemical applications		
Cast installation or pneumatic gunning		[6]
A	Transfer line	
B	Overhead line	
C	Cyclones	
D	Regenerator, reactor, and gasifier walls	
Cast installation by vibration		
E	Burner blocks (CO boiler)	
F	Sulfur burning and recovery units	
Ceramic foam materials		
Phosphate-bonded alumina	Aluminum and nonferrous filtration	[7]
Mullite ($3\ Al_2O_3 \cdot 2\ SiO_2$)	Iron and nonreactive Ni–base alloy filtration	

(Continued)

Appendix 17. *(continued)*

Refractory Name	Applications/Notes	Ref. [a]
Sintered alumina	Ferrous (esp. Fe) and high melting nonferrous filtration incl. reactive metals	[7]
Partially stabilized zirconia	Superalloy and ferrous filtr.	
Zirconia–alumina	Ferrous filtration	
Fused-cast refractories		
Mullite corundum		[8]
Alpha A1		
Alpha A2		
Alpha-beta alumina		
Beta alumina		
Chrome–alumina		
33%Zirconia–alumina		
36%Zirconia–alumina		
41%Zirconia–alumina		

(Continued)

Appendix 17. *(continued)*

Refractory Name	Applications/Notes	Ref. [a]
Chromic oxide, zircon, and zirconia refractories		
Dense chromic oxide	Glass contact melter sidewalls, channels, forehearths	[9]
Chrome 20	Flow blocks, backup to dense chrome in forehearth, channels	
Exp.–HD chromic oxide	Metal line melter, dog house area	
Dense zircon	Backup to dense CrO_3 in melter, channels,	
Zircon 20	Bushing blocks	
ZMg	Bushing blocks	

(Continued)

Appendix 17. *(continued)*

Refractory Name	Applications/Notes	Ref. [a]
Sintered refractories		
C–125 (High purity chromic oxide, isopressed or slip cast, corrosion resistant)	Melter and refiner sidewalls, bottom paving, forehearth sidewalls, corner blocks of textile furnaces	[10]
C–1220 (Isostatically pressed chromic oxide, thermal shock resistant)	Flow blocks for textile fiberglass furnaces, insulating wool furnace throats	[11]
C–1222 (Mechanically pressed high-purity chromic oxide, thermal shock resistant)	Backup refractories in melters, channels, forehearths in textile fiberglas furnaces	
Zircon 20 (Dense, high-purity zircon, slip cast or isostatically pressed)	Contact areas in furnaces for textile fiberglass and low alkali glass	

(Continued)

Appendix 17. *(continued)*

Refractory Name	Applications/Notes	Ref. [a]
ZS 1300 (Zirconium silicate, thermal shock resistant)	Bushing blocks in textile fiberglass furnaces	[10] [11]
ZS 835 (Pure isostatically pressed $ZrSiO_4$ with low blistering potential)	Forehearths, refiner blocks, bushing blocks, feeder parts for borosilicate/opal glass furnaces	
T–1185/KGT (Tin oxide refractory, highly corrosion resistant)	Electrodes for glass furnaces, direct melting or electrical boosting in gas and oil-fired furnaces	
BPM 70 (Enriched mullite grade materials for large and complex shapes)	Forehearths, superstructure textile fiberglass container, specialty glasses	
AKRAL 60 V (High alumina grade materials for large and complex shapes)	Same as above	

[a] References:
1. K. Chen and Y. Ko: Am. Ceram. Soc. Bull. **67 [7],** 1228–1234 (1988)
2. J.D. Gilchrist: *Fuels, Furnaces and Refractories.*(Pergamon Press Oxford England, 1977)
3. *UNITECR '89 Proceedings*, Vol.1, L.J. Trostel (ed.). (American Ceramic Society

Westerville OH, 1989), pp. 586–603
4. Idem, pp.695–707
5. Y. Sakano and H. Takahashi: Am. Ceram. Soc. Bull. **67 [7],** 1164–1175 (1988)
6. S. Banerjee: Ceram. Eng. Sci. Proc. **7 [1–2],** 236–242 (1986)
7. J.W. Brockmeyer and L.S. Aubrey: Ceram. Eng. Sci. Proc. **8 [1–2],** 63–74 (1987)
8. *Alumina Chemicals Science and Technology*, L.D. Hart (ed.). (American Ceramic Society Westerville OH, 1990), pp.393–426
9. Am. Ceram. Soc. Bull. **69 [1],** 61 (1990)
10. Am. Ceram. Soc. Bull. **69 [10],** 1667 (1990)
11. Corhart Refractories Co. Manufacturer's Data

[b] Chemical composition in wt%

References

Chapter 1

[1.1] I. Asimov: *Sternstunden der Forschung Twentieth Century Discovery.* (Deutsche Verlags-Anstalt Stuttgart Germany, 1971)

[1.2] N. Bohr: *On the Constitution of Atoms and Molecules*. Philos. Mag. **26,** 1–25 (1913)

[1.3] S.H. Avner: *Introduction to Physical Metallurgy.* (McGraw–Hill Book Co., New York, 1974)

[1.4] A. Sommerfeld: *Atomic Structure and Spectral Lines*. (Methuen London, 1929)

[1.5] E. Schroedinger: Ann. Phys. **81**, 190 (1936)

[1.6] W. Heisenberg: Zeit. Phys. **43**, 172 (1927)

[1.7] L.H. Van Vlack: *Physical Ceramics for Engineers*. (Addison–Wesley Reading MA, 1964)

[1.8] W.D. Kingery, H.K. Bowen, and D.R. Uhlmann: *Introduction to Ceramics*. (John Wiley & Sons New York, 1976)

[1.9] H.W. King: *Structure of the Pure Metals*. In: *Physical Metallurgy.* R.W. Cahn and P. Haasen (eds.). (North Holland Publishing, Amsterdam, 1983), pp. 37–71

[1.10] Bilim ve Teknik **29 [348],** 29–30 (1996)

[1.11] W.J. Lehman: *Atomic and Molecular Structure*. (John Wiley&Sons New York, 1972)

[1.12] D.W. Richerson: *Modern Ceramic Engineering.* (Marcel Dekker New York, 1982)

[1.13] M.F. Ashby and D.R. Jones: *Engineering Materials 2*. (Pergamon Press Oxford England, 1986)

[1.14] R.E. Reed-Hill: *Physical Metallurgy Principles.* (Van Nostrand New York, 1973)

[1.15] R.W. Hanks: *Materials Engineering Science, An Introduction.* (Harcourt, Brace & World New York, 1970)

[1.16] T.B. Shaffer and A. Goel: *Silicon Nitride*. In: *ART Handbook of Advanced Ceramic Materials*. (Advanced Refractory Technologies Inc., March 1993)

[1.17] H. Salmang and H. Scholze: *Keramik*. (Springer-Verlag Berlin, 1982)

[1.18] R.H. Doremus: *Glass Science.* (John Wiley & Sons0 New York,1994)

[1.19] W.H. Zachariasen: *The Atomic Arrangement in Glass*. J. Am. Chem. Soc. **54,** 3941 (1932)

[1.20] I.J. McColm: *Ceramic Science for Materials Technologists*. (Leonard Hill Glasgow Scotland, 1983)

[1.21] H. Inoue, N. Aoki, and I. Yasui: *Molecular Dynamics Simulation of the Structure of Borate Glasses*. J. Am. Ceram. Soc. **70 [9]**, 622–627 (1987)

[1.22] A.K. Varshneya: *Fundamentals of Inorganic Glasses*. (Academic Press Inc. San Diego CA, 1994)

[1.23] D.R. Uhlmann and H. Yinnon: *Glass: Science and Technology Vol. 1*. (Academic Press Inc. San Diego CA, 1983)

[1.24] Z.E. Erkmen: *Seramik Faz Diyagramlari.* (Anadolu Universitesi Basimevi Eskisehir Turkiye, 1996)

[1.25] M.W. Barsoum: *Fundamentals of Ceramics*. (McGraw-Hill New York, 1997)

[1.26] D.R. Gaskell: *Introduction to Metallurgical Thermodynamics*. (McGraw-Hill New York, 1981)

[1.27] W. Heywang and H. Thormann: *Tailoring of Piezoelectric Ceramics*. Ann. Rev. Mater. Sci. **14,** 27–47 (1987)

[1.28] M.S. Dresselhaus: *Future Directions in Carbon Science*. Ann. Rev. Mater. Sci. **27,** 1–34 (1997)

[1.29] F.P. Bundy, W.A. Bassett, W.S. Weathers, R.J. Hemley, H.K. Mao, and A.F. Goncharov: *The pressure-temperature phase and transformation diagram for carbon, updated through 1994*. Carbon **34 [2],** 141–153 (1996)

[1.30] E.N. Bunting: J. Res. Natl. Bur. Stand. **6 [6]**, 948 (1931)

[1.31] M.L. Keith and J.F. Schairer: J. Geol. **60,** 182 (1952)

Chapter 2

[2.1] D.W. Richerson: *Modern Ceramic Engineering*. (Marcel Dekker New York, 1982)

[2.2] L.H. Van Vlack: *Physical Ceramics for Engineers*, Addison- Wesley Reading MA, 1964)

[2.3] J.S. Reed: *Introduction to the Principles of Ceramic Processing*. (John Wiley&Sons New York, 1988)

[2.4] W.E. Rhine and H.K. Bowen: *An Overview of Chemical and Physical Routes to Advanced Ceramic Powders*. Ceram. Int. **17 [3],** 143–52, (1991)

[2.5] *Materials Handbook.* Ceram. Ind. **136 [1]**, 102 (1991)

[2.6] R. Stevens: *Zirconia and Zirconia Ceramics.* (Magnesia Elektron Publication # 113, 1986)

[2.7] W.E. Lee and W.M. Rainforth: *Ceramic Microstructures.* (Chapman&Hall London, 1994)

[2.8] F.H.Norton: *Fine Ceramics*. (Krieger Malabar FL, 1978)

[2.9] T.B. Shaffer and A. Goel: *Silicon Nitride* in *ART Handbook of Advanced Ceramic Materials*. (Advanced Refractory Technologies, March 1993)

[2.10] S. Saito (ed): *Fine Ceramics.* (Elsevier Essex England, 1985), pp. 205–211

[2.11] D.W. Johnson Jr.: *Innovations in Ceramic Powder Preparation.* In: *Advances in Ceramics,* Vol.21, *Ceramic Powder Science*. G.L. Messing, K.S. Mazdiyasni, J.W. McCauley, and R.A. Haber (eds.). (The American Ceramic Society Westerville OH, 1984), pp. 3–19

[2.12] W.F. Kladnig and J.E. Horn: *Submicron Oxide Powder Preparation by Microwave Processing*. Ceram. Int. **16 [2],** 99–106 (1990)

[2.13] C.L.J. Adkins: *Powder Preparation by Gas-Phase Techniques.* In: *Characterization of Ceramics.* R.E. Loehman (ed.). (Butterworth-Heinemann Stoneham MA, 1993)

[2.14] G.H. Maher, C.E. Hutchins, and S.D. Ross: *Preparation and Characterization of Ceramic Fine Powders Produced by the Emulsion Process*. Am. Ceram. Soc. Bull. **72 [5],** 72–76 (1993)

[2.15] D.L. Segal: *Advanced Methods for the Production of Ceramic Powders.* Ceram. Acta **4 [5–6],** 15–30 (1992)

[2.16] W.J. Dawson: *Hydrothermal Synthesis of Advanced Ceramic Powders.* Am. Ceram. Soc. Bull. **67 [10],** 1673–1678 (1988)

[2.17] S. Komarneni, E. Fregeau, E. Breval, and R. Roy: *Hydrothermal Preparation of Ultrafine Ferrites and Their Sintering*. J. Am. Ceram. Soc. **71 [1],** C-26–28 (1988)

[2.18] L.L. Hench and J.K. West: *The Sol-Gel Process*. Chem. Rev. **90 [1],** 33–72 (1990)

[2.19] E. Matijevich: *Monodispersed Metal (Hydrous) Oxides: A Fascinating Field of Colloidal Science*. Acc. Chem. Res. **14 [1],** 22–29 (1981)

[2.20] E.A. Barringer and H.K. Bowen: *Formation, Packing, and Sintering of Monodisperse TiO_2 Powders*. J. Am. Ceram. Soc. **65 [12],** C199–201 (1982)

[2.21] P. Colomban: *Gel Technology in Ceramics, Glass–Ceramics, and Ceramic–Ceramic Composites*. Ceram. Int. **15 [1],** 23–50 (1989)

[2.22] J. Livage, M. Henry, J.P. Jolivet, and C. Sanchez: *Chemical Synthesis of Fine Powders*. MRS Bull. **15 [1],** 18–25 (1990)

[2.23] J.A. Switzer: *Electrochemical Synthesis of Ceramic Films and Powders.* Am. Ceram. Soc. Bull. **66 [10],** 1521–1524 (1987)

[2.24] Y. Zhou, R.J. Phillips, and J.A. Switzer *Electrochemical Synthesis and Sintering of Nanocrystalline Cerium(IV) Oxide Powders*. J. Am. Ceram. Soc. **78 [4],** 981–985 (1995)

[2.25] I. Zhitomirsky, L. Gal-or, A. Kohn, and H.W. Hennicke: *Electrodeposition of Ceramic Films from Non-Aqeous and Mixed Solutions*. J. Mater. Sci. **30 [20],** 5307–5312 (1995)

[2.26] A. Kumar and R. Roy: *Reactive-Electrode Submerged-Arc Process for Producing Fine Non-Oxide Powders*. J. Am. Ceram. Soc. **72 [2],** 354–356 (1989)

[2.27] T. Onu, P.W. Brown, J. Adair, and P. Ravindranathan: *Synthesis of Ultra-Fine AlN Powders by the Reactive Electrode Submerged-Arc Method (RESA) - abstract*. (University of Florida, Adair Group Website, http://www.mse.ufl.edu/~adair/pubs/ abstract/job41.txt)

[2.28] K.S. Mazdiyasni, C.T. Lynch, and J.S. Smith II: J. Am. Ceram. Soc. **48 [7]**, 372–375 (1965)

[2.29] S.K. Varshney and C.L. Beatty: *Formation of Silicon Nitride and Silicon Carbide by Vapor-Phase Reaction* . Ceram. Sci. Eng. Proc. **3 [9–10]**, 555–564 (1982)

[2.30] C.J. Brinker and G.W. Scherer: *Sol-Gel Science*. (Academic Press San Diego CA, 1990)

[2.31] L.M. Sheppard: *Vapor-Phase Synthesis of Ceramics*. Adv. Mater. Proc. **131 [4]**, 53–58 (1987)

[2.32] D.S. Phillips and G.J. Vogt: *Plasma Synthesis of Ceramic Powders*. MRS Bull. **12 [7]**, 54–57 (1987)

[2.33] C.A. Pickles, J.M. Toguri, and C.J. Simpson: *Plasma Arc Production of Silicon Carbide Crystals*. Br. Ceram. Trans. **94 [3]**, 89–96 (1994)

[2.34] R.W. Chorley and P.W. Lednor: *Synthetic Routes to High Surface Area Non-Oxide Materials*. Adv. Mater. **3 [10]**, 474–485 (1991)

[2.35] J.S. Haggerty: In: *Ultrastructure Processing of Ceramics, Glasses, and Composites*. L.L.Hench and D.R.Ulrich (eds.). (John Wiley&Sons New York, 1984), pp. 353–366

[2.36] M.I. Baraton, L. Boulanger, M. Cauchetier, V. Lorenzelli, M. Luce, T. Merle, P. Quintard, and Y.H. Zhou, *Nanometric Boron Nitride Powders: Laser Synthesis, Characterization, and FT-IR Surface Study*. J. Eur. Ceram. Soc. **13 [4]**, 371–378 (1994)

[2.37] R. Pampuch, L. Stobierski, and J. Lis: *Synthesis of Sinterable β-SiC Powders by a Solid Combustion Method*. J. Am. Ceram. Soc. **72 [8]**, 1434–1435 (1989)

[2.38] O. Yamada, Y. Miyamoto, M. Koizumi: *Self Propagating High Temperature Synthesis (SHS) of SiC Powders and the Properties of the Sintered Compact*. J. Jpn. Soc. Powder Metall. **33 [6]**, 286–290 (1986)

[2.39] J.J. Kingsley and K.C. Patil. *A Novel Combustion Process for the Synthesis of Fine Particle α-Alumina and Related Oxide Materials*. Mater. Lett. **6 [11,12]**, 427–432 (1988)

[2.40] A.G. Merzhanov: *Review-History and Recent Developments in SHS*. Ceram. Intern. **21 [5]**, 371–379 (1995)

[2.41] R.S. DeCarli and J.C. Jamieson: Science **133,** 1821 (1961)

[2.42] R. Liepins, K.P. Staudhammer, K.A. Johnson, and M. Thomson: *Shock-Induced Synthesis, I.Cubic Boron Nitride From Ammonia Borane*, Mater. Lett. **7 [1,2]**, 44–46 (1988)

[2.43] G. LeCaer, P. Matteazzi, E. Bauer Grosse, A. Pianelli, and E. Bouzy: *Mechanically Driven Syntheses of Carbides and Silicides*, J. Mater. Sci. **25 [11]**, 4726–4731 (1990)

[2.44] P. Matteazzi and G. LeCaer: *Room-Temperature Mechanosynthesis of Carbides by Grinding of Elemental Powders*. J. Am. Ceram. Soc. **74 [6]**, 1382–1390 (1991)

[2.45] P.E.D. Morgan: *Structuring Chemical Technology to Produce Cost-Effective Ceramic Products on the Large Scale*. Am. Ceram. Soc. Bull. **72 [7]**, 65–70 (1993)

[2.46] A.W. Weimer, G.A. Cochran, G.E. Eisman, J.P. Henley, B.D. Hook, L.K. Mills, T.A. Guiton, A.K. Knudsen, N.R. Nicholas, J.E. Volmering, and W.G. Moore: *Rapid Process for Manufacturing Aluminum Nitride Powder*. J. Am. Ceram. Soc. **77 [1]**, 3–18 (1994)

[2.47] P. Bracke, H. Shurmans, and J. Verhoest: *Inorganic Fibers and Composite Materials*. (Pergamon Press Oxford England, 1984)

[2.48] A. Levitt (ed.): *Whisker Technology*. (John Wiley & Sons New York, 1970)

[2.49] H. Wada and M.J. Wang: *Ceramic Whisker Synthesis and Phase Stability in the Si–C–N–O System*. In: *Whisker and Fiber Toughened Ceramics*. R.A. Bradley, D.E. Clark, D.C. Larsen, and J.O. Stiegler (eds.). (The American Society for Metals Metals Park OH, 1989), pp. 63–72

[2.50] S. Hashimoto and A. Yamaguchi: *Synthesis of $MgAl_2O_4$ Whiskers by an Oxidation–Reduction Reaction*. J. Am. Ceram. Soc. **79 [2]**, 491–494 (1996)

[2.51] M. Futamoto, I. Yuito, and U. Kawabe: *Hafnium Carbide and Nitride Whisker Growth by Chemical Vapor Deposition*. J. Cryst. Growth **61 [1]**, 69–74 (1983)

[2.52] W.J. Lackey, J.A. Hanigofsky, and G.B. Freeman: *Experimental Whisker Growth and Thermodynamic Study of the Hafnium–Carbon System for Chemical Vapor Deposition Applications*. J. Am. Ceram. Soc. **73 [6]**, 1593–1598 (1990)

[2.53] G. Gorny and R. Pampuch: *Growth of SiC Whiskers and Platelets from the Gas Phase*. In: *Euro Ceramics,* Vol.1, *Processing of Ceramics*. G. DeWith, R.A. Terpstra, and R.Metselaar (eds.). (Elsevier Essex England, 1989), pp. 1.18–1.22

[2.54] J.W. Milewski: *Growth of ß-SiC Whiskers by the VLS Process*. J. Mater. Sci., **20 [4]**, 1160–1166 (1985)

[2.55] P.D. Shalek, D.S. Phillips, D.E. Christiansen, J.D. Katz, W.J. Parkinson, and J.J. Petrovic: *Synthesis and Characterization of VLS-Derived Silicon Carbide Whiskers*. In: *Whisker and Fiber Toughened Ceramics*. R.A. Bradley, D.E. Clark, D.C. Larsen, and J.O. Stiegler (eds.). (The American Society for Metals Metals Park OH, 1989), pp. 53–62

[2.56] J.V. Milevski, F.D. Gac, J.J. Petrovic, and S.R. Skaggs: *Growth of Beta-Silicon Carbide Whiskers by VLS Process*. J. Mater. Sci. **20 [4]**, 1160–1166 (1985)

[2.57] T.F. Cooke: *Inorganic Fibers - A Literature Review*. J. Am. Ceram. Soc. **74 [12]**, 2959–2978 (1991)

[2.58] J.G. Lee and I.B. Cutler: *Formation of SiC from Rice Hulls*. J. Am. Ceram. Soc. **54 [2]**, 195 (1975)

[2.59] A.K. Ray, G. Mahanty, and A. Ghose: *Effect of Catalysts and Temperature on Silicon Carbide Whiskers Formation from Rice Husks*. J. Mater. Sci. Lett. **10 [4]**, 227–229 (1991)

[2.60] R.V. Krishnarao, M.M. Godhinki, P.G.I. Mukunda, and M. Chakraborty: *Direct Pyrolysis of Raw Rice Husks for Maximization of Silicon Carbide Whisker Formation*. J. Am. Ceram. Soc. **74 [11]**, 2869–2875 (1991)

[2.61] A.P. Divecha, S.G. Fishman, and S.D. Karmarkar: *Silicon Carbide Reinforced Aluminum-A Formable Composite*. J. Metals **33 [9]**, 12–17 (1981)

[2.62] K.C. Patil, S.S. Manoharan, and D. Gajapathy: *Preparation of High-Density Ferrites. In: Handbook of Ceramics and Composites, Vol.1, Synthesis and Properties*. (Marcel Dekker New York, 1990)

[2.63] C.E. Bamberger, D.W. Coffey, and T.A. Nolan: *Formation of TiN Whiskers from Oxide Containing Cyanide Melts*. J. Mater. Sci. **25 [12]**, 4992–4996 (1990)

[2.64] M. Yoshimura, T. Hiuga, and S. Somiya: *Dissolution and Reaction of Yttria-Stabilized Zirconia Single Crystals in Hydrothermal Solutions*. J. Am. Ceram. Soc. **69 [7]**, 583–584 (1986)

[2.65] M. Yoshimura, H. Suda, K. Okamoto, and K. Ioku. *Hydrothermal Synthesis of Biocompatible Whiskers*. J. Mater. Sci. **29 [13]**, 3399–3402 (1994)

[2.66] B.D. Cullity: *Elements of X-Ray Diffraction*. (Addison-Wesley Reading MA, 1978)

[2.67] R.W. Cahn and P. Haasen (eds): *Physical Metallurgy*. (North Holland Amsterdam, 1983)

[2.68] E. Jesnitzer: *Verformung Metallischer Werkstoffe*. Ph.D.Thesis. (Institut fur Werkstoffkunde der Technichen Universitat Hannover Germany)

[2.69] E.A. Weaver, H.D. Merchant, and R.P. Poplawsky: *Growth of Fe–Ni Oxide Spinel Single Crystals by the Arc Image Technique*. J. Am. Ceram. Soc. **52 [4]**, 214–215 (1969)

[2.70] F. Schmid, C.P. Khattak, and D.M. Felt: *Producing Large Sapphire for Optical Applications*. Am. Ceram. Soc. Bull. 73 [2], 39–44 (1994)

[2.71] K.K. Chawla: *Composite Materials*. (Springer-Verlag New York, 1987)

[2.72] J.I. Duffy (ed): *Glass Technology*. (Noyes Park Ridge NJ, 1981)

[2.73] J.O. Carlsson: *Review-Techniques for the Preparation of Boron Fibers*. J. Mater. Sci. **14 [2]**, 255–264 (1979)

[2.74] J. Delmonte: *Technology of Carbon and Graphite Fiber Composites*. (Van Nostrand Reinhold New York, 1981)

[2.75] G. Savage: *Carbon–Carbon Composites*. (Chapman & Hall London, 1993)

[2.76] S. Yajima, Y. Hasegawa, J. Hayashi, and M. Imura: *Synthesis of Continuous Silicon Carbide Fibre with High Tensile Strength and High Young's Modulus*. J. Mater. Sci. **13 [12]**, 2569–2576 (1978)

[2.77] L.C. Sawyer, M. Jamieson, D. Brikowski, M.I. Haider, and R.T. Chen. *Strength, Structure, and Fracture Properties of Ceramic Fibers Produced from Polymeric Precursors: I, Base Line Studies*. J. Am. Ceram. Soc. **70 [11]**, 798–810 (1987)

[2.78] A.R. Bunsell: *Ceramic Fibers for Reinforcement*. In: *Ceramic-Matrix Composites*. R.Warren (ed.). (Blackie and Son London, 1992)

[2.79] N.R. Langley, G.E. LeGrow, and J. Lipowitz: *Properties of Ceramic Fibers from Organosilicon Polymers.* In: *Fiber Reinforced Ceramic Composites*. (Noyes Park Ridge NJ, 1990), pp. 63–92

[2.80] J.C. Romine: *New High Temperature Ceramic Fiber*. Ceram. Eng. Sci. Proc. **8,** 755–765 (1987)

[2.81] H.G. Sowman and D.D. Johnson: *Oxide Fibers from Chemical Processes*. In: *Fiber Reinforced Ceramic Composites*. K.S.Mazdiyasni (ed.). (Noyes Park Ridge NJ, 1990), pp. 122–140

[2.82] D. Sporn: *Technology for the Development of Polycrystalline Oxidic Functional Fibers*. Ceram. Forum Intern. **71 [9]**, 547–549 (1994)

[2.83] J.S. Haggerty, K.C. Wills, and J.E. Sheehan: *Growth and Properties of Single Crystal Oxide Fibers*. Ceram. Eng. Sci. Proc. **12 [9–10]**, 1785–1801 (1991)

[2.84] Y.M. Sung, S.A. Dunn, and J.A. Koutsky: *Inviscid Melt Spinning (IMS) and Phase Identification of $CaO–Al_2O_3–MgO$ Fibers*. Ceram. Inter. **20 [5]**, 337–341 (1994)

Chapter 3

[3.1] S.G.Malghan: *Comminution.* In: *Engineered Materials Handbook* Vol.4: *Ceramics and Glasses.* (ASM International Materials Park OH, 1991), pp. 75–82

[3.2] J.S. Reed: *Introduction to the Principles of Ceramic Processing*. (John Wiley & Sons New York, 1988)

[3.3] N. Sarkar and G.K. Greminger, Jr.: *Methylcellulose Polymers as Multifunctional Processing Aids In Ceramics.* Am. Ceram. Soc. Bull. **62 [11]**, 1280–1283 (1983)

[3.4] R.D. Nelson: *Handbook of Powder Technology,* Vol.7: *Dispersing Powders in Liquids*. (Elsevier New York, 1988)

[3.5] D.A. Stanley, L.Y. Sadler III, D.R. Brooks, and M.A. Schwartz: *Attrition Milling of Ceramic Oxides*. Am. Ceram. Soc. Bull. **53 [11]**, 813–829 (1974)

[3.6] N. Claussen and J. Jahn, *Mechanical Properties of Sintered and Hot-Pressed Si_3N_4–ZrO_2 Composites*, J. Am. Ceram. Soc. **61 [1],** 94–95 (1978)

[3.7] S. Prochazka: *Attrition Milling of Hard Substances to Submicrometer Grain Size*. In: *Ceramic Powder Science, Advances in Ceramics,* Vol.21. G.L. Messing, K.S. Mazdiyasni, J.W. McCauley, and R.A. Haber (eds.). (The American Ceramic Society Westerville OH, 1987), pp. 311–320

[3.8] G. von Bernuth: *Sample Preparation in Ceramics Laboratories*. Ceram. Forum Int. **68 [5],** 221–223 (1991)

[3.9] C.L. Prasher: *Crushing and Grinding Process Handbook*. (John Wiley & Sons New York, 1987)

[3.10] D. Rickwood: *The Theory and Practice of Centrifugation*. In: *Centrifugation*. D. Rickwood (ed.). (IRL Press Oxford England, 1984), pp. 1–43

[3.11] M.L. Green: *Rheology of Ceramic Oxide Dispersions*. (Ceramics Processing Research Laboratory Report No.18 MIT Department of Materials Science and Engineering Cambridge MA, 1982)

[3.12] B.J. Derjaguin, N.V. Churaev, and V.M. Muller: *Surface Forces*. (Consultants Bureau New York, 1987)

[3.13] R.G. Horn: *Surface Forces and Their Action in Ceramic Materials*. J. Am. Ceram. Soc. **73 [5],** 1117–135 (1990)

[3.14] J.W.Goodwin: *Rheology of Ceramic Materials*. Am. Ceram. Soc. Bull. **69 [10],** 1694–1698 (1990)

[3.15] F.M. Fowles: *Dispersions of Ceramic Powders in Organic Media*. In: *Advances in Ceramics,* Vol.21: *Ceramic Powder Science*. G.L. Messing, K.S. Mazdiyasni, J.W. McCauley, and R.A. Haber (eds.). (The American Ceramic Society Westerville OH, 1987), pp. 411–421

[3.16] B.V. Derjaguin and L.D. Landau: *Theory of Stability of Highly Charged Lyophobic Sols and Adhesion of Highly Charged Particles in Solutions of Electrolytes*. Acta Physicochim. URSS **14 [6],** 633–662 (1941)

[3.17] E.J.W. Verwey and J.Th.G. Overbeek: *Theory of Stability of Lyophobic Colloids*. (Elsevier Amsterdam, 1941)

[3.18] D.W. Fuersteneau, R. Herrera-Urbina, and J.S. Hanson: *Adsorption of Processing Additives and the Dispersion of Ceramic Powders*. In: *Ceramic Transactions, Ceramic Powder Science II*. G.L. Messing, E. Fuller, Jr., and H. Hausner (eds.). (The American Ceramic Society Westerville OH, 1987), pp. 333–51

[3.19] R. Hogg, T.W. Healy, and D.W. Fuersteneau: *Mutual Coagulation of Colloidal Dispersions*. Trans. Faraday Soc. **62 [522],** 1638–1651 (1966)

[3.20] F.M. Fowkes: *Dispersions of Ceramic Powders in Organic Media*. In: *Advances in Ceramics,* Vol.21: *Ceramic Powder Science*. G.L. Messing, K.S. Mazdiyasni, J.W. McCauley, and R.A. Haber (eds.). (The American Ceramic Society Westerville OH, 1987), pp. 411–421

[3.21] F.V. Shaw: *Spray Drying: A Traditional Process for Advanced Applications*. Am. Ceram. Soc. Bull. **69 [9],** 1484–1489 (1990)

[3.22] A. Hendry: *Processing of Engineering Ceramics.* Powder Met. **31 [1],** 20–22 (1988)

[3.23] A.E. McHale: *Processing Additives.* In: *Engineered Materials Handbook,* Vol.4: *Ceramics and Glasses*. (ASM International Materials Park OH, 1991), pp. 116–121

[3.24] G.C. Robinson: *Additives Assist Fabrication of Ceramics*. Ceram. Ind. **138 [4],** 53–55 (1992)

[3.25] T. Chartier, M. Ferrato, and J.F. Baumard: *Debinding of Ceramics: a Review.* Ceram. Acta **6 [6],** 17–27 (1994)

[3.26] L. Hermansson: *Organics Removal*. In: *Engineered Materials Handbook,* Vol.4*: Ceramics and Glasses*. (ASM International Materials Park OH, 1991), pp. 135–140

[3.27] O.J. Whittemore, Jr: *Particle Compaction.* In: *Ceramic Processing Before Firing*. G.Y. Onoda, Jr. and L.L. Hench (eds.). (John Wiley & Sons New York, 1978), pp. 343–355

[3.28] J.S. Reed and R.B. Runk: *Dry Pressing*. In: *Ceramic Fabrication Processes*. F.F.Y. Wang (ed.). (Academic Press New York, 1976), pp. 71 –92

[3.29] W.M. Price. In: *Proc.1st Conf.Compaction and Consolidation of Particulate Matter.* A.S. Goldberg (ed.). (Brighton England, 1972)

[3.30] R.E. Caligaris, R. Topolevsky, P. Maggi, and F. Brog: *Compaction Behavior of Ceramic Powders.* Powder Tech. **42,** 263–267 (1985)

[3.31] S.J. Lukasiewitz and J.S. Redd: *Character and Compaction Response of Spray Dried Agglomerates*. Am. Ceram. Soc. Bull. **57 [9],** 798–805 (1978)

[3.32] A.R. Cooper, Jr., and L.E. Eaton: *Compaction Behavior of Several Ceramic Powders.* J. Am. Ceram. Soc. **45 [3],** 97–101 (1962)

[3.33] B.J. McEntire: *Dry Pressing.* In: *Engineered Materials Handbook,* Vol.4*: Ceramics and Glasses*. (ASM International Materials Park OH, 1991) pp. 141–46

[3.34] D.B. Quinn, R.E. Bedford, and F.L. Kennard: *Dry Bag Isostatic Pressing and Contour Grinding of Technical Ceramics*. In: *Advances in Ceramics,* Vol.9: *Forming of Ceramics*. J.A. Mangels and G.L. Messing (eds.). (The American Ceramic Society Columbus OH, 1984), pp. 4–15

[3.35] F. Kennard: *Cold Isostatic Pressing.* In: *Engineered Materials Handbook,* Vol.4: *Ceramics and Glasses*. (ASM International Materials Park OH, 1991), pp. 147–152

[3.36] R. Swindells and G. Deplace: *Cold Isostatic Pressing Applications*. In: *Isostatic Pressing Technology.* P.J. James (ed.). (Applied Science Publishers Essex England, 1983), pp. 145–167

[3.37] F.L. Kennard: *Ceramic Component Fabrication*. Ceram. Eng. Sci. Proc. **7 [9–10],** 1095–1111 (1986)

[3.38] J.E. Funk: *Slip Casting and Casters.* In: *Advances in Ceramics,* Vol.9: *Forming of Ceramics.* J.A.Mangels and G.L. Messing (eds.). (The American Ceramic Society Columbus OH, 1984), pp. 76–84

[3.39] M. Bengisu and O.T. Inal: *Sintering of MgO and MgO–TiC Ceramics by Plasma, Microwave, and Conventional Heating.* J. Mater. Sci. **29 [20],** 5475–5480 (1994)

[3.40] R.E. Cowan: *Slip Casting.* In: *Ceramic Fabrication Processes.* F.F.Y. Wang (ed.). (Academic Press New York, 1976), pp. 154–171

[3.41] C.H. Schilling and I.A. Aksay: *Slip Casting.* In: *Engineered Materials Handbook,* Vol.4: *Ceramics and Glasses.* (ASM International Materials Park OH, 1991), pp. 153–60

[3.42] J.H.D. Hampton, S.B. Savage, and R.A.L. Drew: *Experimental Analysis and Modeling of Slip Casting.* J. Am. Ceram. Soc. **71 [12],** 1040–1045 (1988)

[3.43] J.P. Torre and Y. Bigay: *Fabrication of Silicon Nitride Parts by Slip Casting.* Ceram. Eng. Sci. Proc. **7 [7–8],** 893–899 (1986)

[3.44] K. Wilfinger and W.R. Cannon: *Processing of Transformation Toughened Alumina.* Ceram. Eng. Sci. Proc. **7 [9–10],** 1169–1181 (1986)

[3.45] F.F. Lange, B.I. Davis, and I.A. Aksay: *Processing-Related Fracture Origins: III, Differential Sintering of* ZrO_2 *Agglomerates in* Al_2O_3/ZrO_2 *Composite.* J. Am. Ceram. Soc. **66 [6],** 407–408 (1983)

[3.46] C.R. Blanchard-Ardid and R.A. Page: *The Effect of Ultrasonic Treatments on the Particle Size Distribution and Microstructure of a Slip Cast* Al_2O_3. In: *Ceramic Transactions, Ceramic Powder Science II.* G.L. Messing, E. Fuller, Jr., and H. Hausner (eds.). (The American Ceramic Society Westerville OH, 1988), pp. 724–732

[3.47] J.A. Mangels: *The Role of Powder Properties in Ceramic Processing.* Ceram. Eng. Sci. Proc. **7 [9–10],** 1112–1121 (1986)

[3.48] F.L. Kennard: *Ceramic Component Fabrication.* Ceram. Eng. Sci. Proc. **7 [9–10],** 1095–1111 (1986)

[3.49] D.W. Richerson: *Modern Ceramic Engineering.* (Marcel Dekker New York, 1982)

[3.50] F. Kools and O. Fiquet: *Wet Pressing for Forming Advanced Ceramics.* In: *Euro Ceramics,* Vol.1: *Processing of Ceramics.* G. deWith, R.A. Terpstra, and R. Metselaar (eds.). (Elsevier Essex England, 1989), pp. 1.258–1.262

[3.51] O. Lyckfeldt, E. Liden, M. Persson, R. Carlsson, and P. Apell: *Progress in the Fabrication of* Si_3N_4 *Turbine Rotors by Pressure Slip Casting.* J. Eur. Ceram. Soc. **14 [5],** 383–395 (1994)

[3.52] E.G. Blanchard: *Pressure Casting Improves Productivity.* Am. Ceram. Soc. Bull. **67 [10],** 1680–83 (1988)

[3.53] F.F. Lange: *Powder Processing Science and Technology for Increased Reliability.* J. Am. Ceram. Soc. **72 [1],** 3–15 (1989)

[3.54] H.H. Grazzini and D.S. Wilkinson: *Slip Casting Under Pressure*. Ceram. Eng. Sci. Proc. **13 [7–8]**, 528–535 (1992)

[3.55] S.N. Heavens: *Electrophoretic Deposition as a Processing Route for Ceramics*. In: *Advanced Ceramic Processing Technology*, Vol.1. J.G.P. Binner (ed.). (Noyes Park Ridge NJ, 1990)

[3.56] R.W. Powers: *The Electrophoretic Forming of Beta-Alumina Ceramic*. J. Electrochem. Soc. **122 [4]**, 490–500 (1975)

[3.57] R.W. Powers: *Ceramic Aspects of Forming Beta Alumina by Electrophoretic Deposition*. Am. Ceram. Soc. Bull. **65 [9]**, 1270–1277 (1986)

[3.58] H. Wittwer and H.G. Krüger: *Möglichkeiten und Grenzen der Elektrophorese - Possibilities and Limits of Electrophoresis*. Ceram. Forum Int. **72 [9]**, 556–560 (1995)

[3.59] J.A. Mangels: *Injection Molding Ceramics*. Ceram. Eng. Sci. Proc. **3 [9–10]**, 529–537 (1982)

[3.60] B.C. Mutsuddy and R.G. Ford: *Ceramic Injection Molding*. (Chapman & Hall London, 1995)

[3.61] M.J. Edirishinge: *Injection Molding of Ceramics*. Powder Met. **6 [6]**, 367–370 (1990)

[3.62] J.R. Peshek: *Machinery for Injection Molding of Ceramic Shapes*. In: *Advances in Ceramics*, Vol. 9: *Forming of Ceramics*. J.A. Mangels and G.L. Messing (eds.). (The American Ceramic Society Columbus OH, 1984), pp. 234–238

[3.63] B.C. Mutsuddy: *Equipment Selection for Injection Molding*. Am. Ceram. Soc. Bull. **68 [10]**, 1796–1802 (1989)

[3.64] C.L. Quackenbush, K. French, and J.T. Neil: *Fabrication of Sinterable Silicon Nitride by Injection Molding*. Ceram. Eng. Sci. Proc. **3 [1–2]**, 20–34 (1982)

[3.65] B.C. Mutsuddy: *Injection Molding*. In: *Engineered Materials Handbook*, Vol.4: *Ceramics and Glasses*. (ASM International Materials Park OH, 1991), pp. 173–180

[3.66] R.H. Smoak and R.S. Storm: *Iterative Development of Injection Molded Sintered α-SiC Turbine Materials*. ASME paper No.79-GT-77, presented at the *24th Annual Gas Turbine Conference* (San Diego CA, March 12–15, 1979)

[3.67] R.D. Rivers. U.S. Patent 4,113,480 (1978)

[3.68] J.E. Schuetz: *Methylcellulose Polymers as Binders for Extrusion of Ceramics*. Am. Ceram. Soc. Bull. **65 [12]**, 1556–1559 (1986)

[3.69] A.J. Fanelli and R.D. Silvers. U.S.Patent 4,734,237 (1988)

[3.70] A.J. Fanelli, R.D. Silvers, W.S. Frei, J.V. Burlew, and G.B. Marsh: *New Aqueous Injection Molding Process for Ceramic Powders*. J. Am. Ceram. Soc. **72 [10]**, 1833–1836 (1989)

[3.71] I. Ruppel: *Extrusion.* In: *Engineered Materials Handbook,* Vol.4: *Ceramics and Glasses*. (ASM International Materials Park OH, 1991), pp. 166–172

[3.72] J.J. Benbow, S.H. Jazayeri, and J. Bridgewater: *Ceramic Extrusion Mechanics, the Effect of Paste Formulation and Liquid Phase Rheology on Die-Flow Resistance.* In: *Ceramic Transactions, Ceramic Powder Science II.* G.L. Messing, E. Fuller, Jr., and H. Hausner (eds.). (The American Ceramic Society Westerville OH, 1987), pp. 624–634

[3.73] P.A. Haas, F.G. Kills, and H. Beutler: *Preparation of Reactor Fuels by Sol-gel Processes*. Chem. Eng. Progr. Symp. Ser. Nucl. Eng. Part XVIII **63**, 17–23 (1967)

[3.74] K.S. Mazdiyasni, C.T. Lynch, and J.S. Smith: *Cubic Phase Stabilization of Translucent Yttria-Zirconia at Very Low Temperature*. J. Am. Ceram. Soc. **50 [9],** 532–537 (1967)

[3.75] G.S. Snow: *Fabrication of Transparent Electronic PLZT Ceramics by Atmosphere Sintering.* J. Am. Ceram. Soc. **56 [2],** 91–96 (1973)

[3.76] B.E. Yoldas: *Alumina Gels that Form Transparent* Al_2O_3. J. Mater. Sci. **10 [11],** 1856–1860 (1975)

[3.77] B.E. Yoldas: *Preparation of Glasses and Ceramics from Metal-Organic Compounds.* J. Mater. Sci. **12 [6],** 1203–1208 (1977)

[3.78] M. Yamane, A. Shinji, and T. Sakaino. *Preparation of a Gel From Metal Alkoxide and Its Properties as a Precursor of Oxide Glass.* J. Mater. Sci. **13 [4],** 865–870 (1978)

[3.79] C.J. Brinker and G.W. Scherer: *Sol-Gel Science*. (Academic Press San Diego CA, 1990)

[3.80] S. Sakka and K. Kamiya: *Preparation of Shaped Glasses Through Sol-Gel Method.* In: *Emergent Methods for High Technology Ceramics*. T. Davis, H. Palmour, and T. Porter (eds.). (Plenum Press New York, 1984), pp. 83–94

[3.81] B.E. Yoldas: *Effect of Variation in Polymerized Oxides on Sintering and Crystalline Transformations.* J. Am. Ceram. Soc. **65 [8],** 387–393 (1977)

[3.82] L.L. Hench and J.K. West: *The Sol-Gel Process.* Chem. Rev. **90 [1],** 33–72 (1990)

[3.83] I.A. Aksay, W.Y. Shih, and M. Sarikaya: *Colloidal Processing of Ceramics with Ultrafine Particles.* In: *Ultrastructure Processing of Advanced Ceramics*. J.D. Mackenzie and D.R. Ulrich (eds.). (John Wiley & Sons New York, 1988), pp. 393–406

[3.84] N. Shinohara, D.M. Dabbs, and I.A. Aksay: *Infrared Transparent Mullite Through Densification of Monolithic Gels at 1250°C.* Proc. SPIE. **683,** 19 (1986)

[3.85] D.R. Uhlmann, B.J.J. Zelinski, and G.E. Wnek: *The Ceramist as a Chemist-Opportunities for New Materials*. In: *Better Ceramics Through Chemistry.* C.J. Brinker, D.E. Clark, and D.R. Ulrich (eds.). (Elsevier Science New York, 1984), pp. 59–70

[3.86] P. Colomban and L. Magnolles: J. Mater. Sci. **26,** 3503–3510 (1991)

[3.87] M.A. Janney, S.D. Nunn, C.A. Walls, O.O. Omatete, R.B. Ogle, G.H. Kirby, and A.D. McMillan: *Gelcasting.* To be published in: *The Handbook of Ceramic Engineering.* M.N. Rahaman (ed.). (Manuscript from Oak Ridge National Lab. website http://www.ornl.gov/MC-CPG/gelcasting.html)

[3.88] A.C. Young, O.O. Omatete, M.A. Janney, and P.A. Menchhofer: *Gelcasting of Alumina.* J. Am. Ceram. Soc. **74 [3],** 612–618 (1991)

[3.89] I.A. Aksay, J.T. Staley, and R.K. Prud'homme: *Ceramic Processing with Biogenic Additives.* In: *Biomimetic Materials Chemistry.* S. Mann (ed.). (Wiley-VCH New York, 1996), pp. 361–378

[3.90] S.L. Morissette and J.A. Lewis: *Chemorheology of Aqueous-Based Alumina–Poly(vinyl alcohol) Gelcasting Suspensions.* J. Am. Ceram. Soc. **82 [3],** 521–528 (1999)

[3.91] M.A. Huha and J.A. Lewis: *Polymer Effects on the Chemorheological and Drying Behavior of Alumina–Poly(vinyl alcohol) Gelcasting Suspensions.* J. Am. Ceram. Soc. **83 [8],** 1957–1963 (2000)

[3.92] A.J. Millan, M.I. Nieto, and R. Moreno: *Aqueous Gel-Forming of Silicon Nitride Using Carrageenans.* J. Am. Ceram. Soc. **84 [1],** (2001) 62–64

[3.93] G.V. Franks, S.B. Johnson, and D.E. Dunstan: *Novel Gelforming Process for Near Net Shape Ceramic Component Production.* J. Aust. Ceram. Soc. 36 [2], 1–5 (2000)

[3.94] M. Bengisu, E. Yilmaz, and A. Badreddine: *Gelcasting of Ceramic Parts Using Chitin and Chitosan Gels.* To be published in: *Chitosan in Pharmacy and Chemistry,* R.A.A. Muzzarelli (ed.). (ATEC Italy, 2001)

[3.95] A. Roosen: *Basic Requirements for Tape Casting of Ceramic Powders.* In: *Ceramic Transactions, Ceramic Powder Science II.* G.L. Messing, E. Fuller, Jr., and H. Hausner (eds.). (The American Ceramic Society Westerville OH, 1987), pp. 675–92

[3.96] R.R. Landham, P. Nahass, D.K. Leung, M. Ungureit, W.E. Rhine, H.K. Bowen, and P. Calvert: *Potential Use of Polymerizable Solvents and Dispersants for Tape Casting of Ceramics.* Am. Ceram. Soc. Bull. **66 [10],** 1513–1516 (1987)

[3.97] K.P. Plucknett, C.H. Caceres, and D.S. Wilkinson: *Tape Casting of Fine Alumina/Zirconia Powders for Composite Fabrication.* J. Am. Ceram. Soc. **77 [8],** 2137–2144 (1994)

[3.98] E.A. Groat and T.J. Mroz, Jr.: *Aqueous Processing of AlN Powders.* Ceram. Ind. **140 [3],** 34–38 (1993)

[3.99] E. Streicher and T. Chartier. *Influence of Organic Components on Properties of Tape-Cast Aluminum Nitride Substrates.* Ceram. Int. **16 [4],** 247–252 (1990)

[3.100] M. Descamps, G. Ringuet, D. Leger, and B. Thierry: *Tape-Casting: Relationship Between Organic Constituents and the Physical and*

Mechanical Properties of Tapes. J. Eur. Ceram. Soc. **15 [4],** 357–362 (1995)

[3.101] E.P. Hyatt: *Continuous Tape Casting for Small Volumes.* Am. Ceram. Soc. Bull. **68 [4],** 69–70 (1989)

[3.102] G.A. Steinlage, R.K. Roeder, K.P. Trumble, and K.J. Bowman: *Centrifugal Slip Casting of Components.* Am. Ceram. Soc. Bull. **75 [5],** 92–94 (1996)

[3.103] W. Huisman, T. Graule, and L.J. Gauckler: *Alumina of High Reliability by Centrifugal Casting.* J. Eur. Ceram. Soc. **15 [9],** 811–821 (1995)

[3.104] W. Huisman, T. Graule, and L.J. Gauckler: *Centrifugal Slip Casting of Zirconia (TZP).* J. Eur. Ceram. Soc. **13 [1],** 33–39 (1994)

[3.105] K.M. Haunton, J.K. Wright, and J.R.G. Evans: *The Vacuum Forming of Ceramics.* Br. Ceram. Trans. J. **89 [2],** 53–56 (1990)

[3.106] F.F. Lange: *Influence of Particle Arrangement on Sintering: A Thermodynamic Viewpoint.* In: *13th International Conference on Science of Ceramics.* P. Odier, F. Cabannes, and B. Cales (eds.). (Les Editions de Physique Les Unix Cedex France, 1986) pp.C1-205–217

[3.107] M.P. Harmer: *Science of Sintering as Related to Ceramic Powder Processing.* In: *Ceramic Transactions, Ceramic Powder Science II.* G.L. Messing, E. Fuller, Jr., and H. Hausner (eds.). (American Ceramic Society Westerville OH, 1987), pp. 824–839

[3.108] M.P. Anderson, D.J. Srolovitz, G.S. Grest, and P.S. Sahni. *Computer Simulation of Grain-Growth-I.Kinetics.* Acta Metall. **32 [5],** 783–791 (1984)

[3.109] G.N. Hassold, I.W. Chen, and D.J. Srolovitz. *Computer Simulation of Final Stage Sintering: I, Model, Kinetics, and Microstructure.* J. Am. Ceram. Soc. **73 [10],** 2857–2864 (1990)

[3.110] R.T. DeHoff: *A Cell Model for Microstructural Evolution During Sintering.* In: *Materials Science Research,* Vol.16: *Sintering and Heterogeneous Catalysis.* G.C. Kuczynski, A.E. Miller, and G.A. Sargent (eds.). (Plenum Press New York, 1984), pp. 23–24

[3.111] G.C. Kuczynski: *Statistical Approach to the Theory of Sintering.* In: *Materials Science Research, Vol.10.* G.C. Kuczynski (ed.). (Plenum Press New York, 1975), pp. 325–337

[3.112] R.L. Coble: *Effects of Particle-Size Distribution in Initial- Stage Sintering.* J. Am. Ceram. Soc. **56 [9],** 461–466 (1973)

[3.113] D.L. Johnson: *Comment on Temperature-Gradient Driven Diffusion in Rapid-Rate Sintering.* J. Am. Ceram. Soc. **73 [8],** 2576–2578 (1990)

[3.114] G.C. Kuczynski: *Statistical Theory of Pore Shrinkage and Grain Growth During Powder Compact Densification.* In: *Ceramic Microstructures '76.* R.M. Fulrath and J.A. Pask (eds.). (Westview Press Boulder CO, 1977), pp. 233–245

[3.115] T.T. Fang and H. Palmour III: *Useful Extensions of the Statistical Theory of Sintering.* Ceram. Intern. **15 [6],** 329–335 (1989)

[3.116] T.M. Hare: *Statistics of Early Sintering and Rearrangement by Computer Simulation.* In: *Materials Science Research* Vol.13: *Sintering Processes.* G.C. Kuczynski (ed.). (Plenum Press New York, 1980), pp. 77–93

[3.117] D.J. Srolovitz, M.P. Anderson, P.S. Sahni, and G.S. Grest: *Computer Simulation of Grain Growth II. Grain Size Distribution, Topology, and Local Dynamics*. Acta Metall. **32 [5],** 793–802 (1984)

[3.118] D.J. Srolovitz, G.S. Gretz, and M.P. Anderson: *Computer Simulation of Recrystallization, I. Homogeneous Nucleation and Growth.* Acta Metall. **34 [9],** 1833–1845 (1986)

[3.119] D.J. Srolovitz, G.S. Gretz, M.P. Anderson, and A.D. Rollett: *Computer Simulation of Recrystallization, II. Heterogeneous Nucleation and Growth.* Acta Metall. **36 [8],** 2115–2128 (1988)

[3.120] I.W. Chen, G.N. Hassold, and D.J. Srolovitz: *Computer Simulation of Final-Stage Sintering: II, Influence of Initial Pore Size.* J. Am. Ceram. Soc. **73 [10],** 2865–2872 (1990)

[3.121] R.L. Coble: *Sintering Crystalline Solids.* J. Appl. Phys. **32 [5],** 793–799 (1961)

[3.122] R.L. Coble and J.E. Burke: *Sintering in Ceramics.* In: *Progress in Ceramic Science,* Vol.3. J.E. Burke (ed.). (Pergamon Press London, 1963), p. 197

[3.123] G. Rossi and J.E. Burke: *Influence of Additives on the Microstructure of Sintered Al_2O_3.* J. Am. Ceram. Soc. **56 [12],** 654–659 (1973)

[3.124] M.P. Harmer and R.J. Brook: *The Effect of MgO Additions on the Kinetics of Hot Pressing in Al_2O_3.* J. Mater. Sci. **15 [12],** 3017–3024 (1980)

[3.125] J.E. Burke: *Control of Grain Boundary Mobility.* In: *Ceramic Transactions,* Vol.7: *Sintering of Advanced Ceramics.* C.A. Handwerker, J.E. Blendell, and W. Kaysser (eds.). (The American Ceramic Society Westerville OH, 1990), pp. 215–228

[3.126] R.D. Monohan and J.W. Halloran: *Single Crystal Boundary Migration in Hot-Pressed Aluminum Oxide.* J. Am. Ceram. Soc. **62 [11–12],** 564–567 (1979)

[3.127] J.E. Burke, K.W. Lay, and S. Prochazka: *The Effect of MgO on the Mobility of Grain Boundaries and Pores in Aluminum Oxides* In: *Materials Science Research,* Vol.13: *Sintering Processes.* G.C.Kuczynski (ed.). (Plenum Press New York, 1980), pp. 417–425

[3.128] K.A. Berry and M.P. Harmer: *Effect of MgO Solute on Microstructure Development in Alumina.* J. Am. Ceram. Soc. **69 [2],** 143–149 (1986)

[3.129] D.L. Johnson: *Impurity Effects in the Initial Sintering of Oxides.* In: *Sintering and Related Phenomena.* G.C. Kuczynski, N.A. Horton, and C.F. Gibbon (eds.). (Gordon & Breach New York, 1967), p. 393

[3.130] S. Hori and R. Kurita: *Influence of Small ZrO_2 Additions on the Microstructure and Mechanical Properties of Al_2O_3.* In: *Science and Technology of ZrO_2 III.* S. Somiya, N. Yamamoto, and H. Yanagida (eds.). (The American Ceramic Society Westerville OH, 1988), pp. 423–439

[3.131] F.F. Lange and M.M. Hirlinger: *Hindrance of Grain Growth in Al_2O_3 by ZrO_2 Inclusions*. J. Am. Ceram. Soc. **67 [2],** 164–168 (1984)

[3.132] J. Wang and R. Stevens: *Review-Zirconia Toughened Alumina (ZTA) Ceramics.* J. Mater. Sci. **25 [10],** 3421–3440 (1990)

[3.133] F.F. Lange and M.M. Hirlinger: *Grain-Growth in Two-Phase Ceramics: Al_2O_3 Inclusions in ZrO_2.* J. Am. Ceram. Soc. **70 [11],** 827–830 (1987)

[3.134] M. Bengisu and O.T. Inal: *Rapid-Rate Sintering of Particulate Ceramic Matrix Composites.* Ceram. Int. **17 [3],** 187–198 (1991)

[3.135] S. Prochazka and C.F. Bobik: *Sintering of Aluminum Nitride.* In: *Materials Science Research,* Vol.13: *Sintering Processes.* G.C. Kuczynski (ed.). (Plenum Press New York, 1980), pp. 321–331

[3.136] B. Mikijely and O.J. Whittemore: *Grain Cuboidization During Sintering of MgO–$MgCl_2$ (1%).* Am. Ceram. Soc. Bull. **66 [5],** 809–812 (1987)

[3.137] P.H. Duvigneaud and D. Reinhard: *Activated Sintering of Tin Oxide.* In: *Science of Sintering,* Vol.12. P.Vincenzini (ed.). (Ceramurgica Faenza Italy, 1980), p. 287

[3.138] V. Tikare: *Microstructural Evolution of Si_3N_4 During Sintering.* Presented in: *92nd Annual Ceramic Society Meeting* (Dallas TX, 1990).

[3.139] S. Prochazka: *Sintering of Silicon Carbide.* In: *Ceramics for High-Performance Applications.* J.J. Burke, A.E. Gorum, and R.N. Katz (eds.). (Brook Hill Chestnut Hill MA, 1974), pp. 239–252

[3.140] R. Hamminger, G. Grathwohl, and F. Thummler: *Microanalytical Investigation of Sintered SiC, Part 1: Bulk Material and Inclusions.* J. Mater. Sci. **18 [2],** 353–364 (1983)

[3.141] R. Hamminger, G. Grathwohl, and F. Thummler: *Microanalytical Investigation of Sintered SiC, Part 2: Study of the Grain Boundaries of Sintered SiC by High Resolution Auger Electron Spectroscopy.* J. Mater. Sci. **18 [2],** 3154–3160 (1983)

[3.142] T. Mizrah, M. Hoffman, and L. Gauckler: *Pressureless Sintering of α-SiC.* Powder Met. Int. **16 [5],** 217–220 (1984)

[3.143] S.L. Dole and S. Prochazka: *Densification and Microstructure Development in Boron Carbide.* Ceram. Eng. Sci. Proc. **6 [7–8],** 1151–1160 (1985)

[3.144] R. Hamminger, G. Grathwohl, and F. Thummler: *Examination of Carbon Inclusions in Sintered Silicon Carbide.* Int. J. High Tech. Ceram. **3 [2],** 129–141 (1987)

[3.145] M. Rühle and G. Petzow: *Microstructure and Chemical Composition of Grain Boundaries in Ceramics.* In: *Surfaces and Interfaces in Ceramic and Ceramic-Metal Systems,* Vol.14. J. Pask and E. Evans (eds.). (Plenum Press New York, 1981), p. 167

[3.146] G.H. Wroblewska, E. Nold, and F. Thummler: *The Role of Boron and Carbon Additions on the Microstructural Development of Pressureless Sintered Silicon Carbide.* Ceram. Int. **16 [4],** 201–209 (1990)

[3.147] D.P. Birnie III: *A Model for Silicon Self-Diffusion in Silicon Carbide: Anti-Site Defect Motion.* J. Am. Ceram. Soc. **6 [2],** C33–35 (1986)

[3.148] K. Schwetz and G. Vogt: *Process for the Production of Dense Sintered Articles of Polycrystalline Boron Carbide.* (U.S. Patent 4,195,066 May 25, 1980)

[3.149] R.L. Lange, Z.A. Munir, and J.B. Holt: *Sintering Kinetics of Pure and Doped Boron Carbide.* In: *Materials Science Research,* Vol.13: *Sintering Processes.* G.C. Kuczynski (ed.). (Plenum Press New York, 1980), pp. 311–320

[3.150] D.R. Clarke: *Grain Boundaries in Polyphase Ceramics.* In: *International Conference on the Structure and Properties of Internal Interfaces.* M. Ruhle, R.W. Baluffi, H. Fischmeister, and S.L. Sass (eds.). (Les Editions de Physique Les Unix Cedex France, 1985), pp. C4-51–59

[3.151] O. Ünal, J.J. Petrovic, and T.E. Mitchell: *Mechanical Properties of Hot Isostatically Pressed Si_3N_4 and Si_3N_4/SiC Composites.* J. Mater. Res. **8 [3],** 626–634 (1993)

[3.152] P.A. Morris: *Impurities in Ceramics: Processing and Effects on Properties.* In: *Ceramic Transaction,* Vol.7: *Sintering of Advanced Ceramics.* C.A. Handwerker and J.E. Blendell (eds.). (The American Ceramic Society Westerville OH, 1990), pp. 50–85

[3.153] E.A. Barringer and H.K. Bowen: *Formation, Packing, and Sintering of Monodisperse TiO_2 Powders.* J. Am. Ceram. Soc. **65 [12],** C199–201 (1982)

[3.154] C. Han, A. Aksay, and O.J. Whittemore: *Characterization of Microstructural Evolution by Mercury Porosimetry.* In: *Advances in Materials Characterization II.* R.L. Snyder, R.A. Condrate, Sr., and P.F. Johnson (eds.). (Plenum Press New York, 1985), pp. 339–347

[3.155] T.S. Yeh and M.D. Sacks. *Effect of Particle Size Distribution on the Sintering of Alumina.* J. Am. Ceram. Soc. **71 [12],** C-484–487 (1988)

[3.156] T.S. Yeh and M.D. Sacks: *Effect of Green Microstructure on the Sintering of Alumina.* In: *Ceramic Transaction,* Vol.7: *Sintering of Advanced Ceramics.* C.A. Handwerker and J.E. Blendell (eds.). (The American Ceramic Society Westerville OH, 1990), pp. 309–331

[3.157] S.J. Milne, M. Patel, and E. Dickinson: *Experimental Studies of Particle Packing and Sintering Behavior of Monosize and Bimodal Spherical Silica Powders.* J. Eur. Ceram. Soc. **11 [1],** 1–7 (1993)

[3.158] J.M. Ting and R.Y. Lin: *Effect of Particle Size Distribution on Sintering, Part II. Sintering of Alumina.* J. Mater. Sci. **30 [9],** 2382–2389 (1995)

[3.159] M. Bengisu: *Densification and Mechanical Properties of Whisker and/or Zirconia Toughened, Effect of Shock Treatment and Consolidation*

Method. Ph.D. Thesis. (New Mexico Institute of Mining and Technology Socorro NM, 1992)

[3.160] H. Hahn, J. Logas, and R.S. Averback: *Sintering Characteristics of Nanocrystalline* TiO_2. J. Mater. Res. **5 [3],** 609–614 (1990)

[3.161] I.A. Aksay, F.F. Lange, and B.I. Davis: *Uniformity of* Al_2O_3*–*ZrO_2 *Composites by Colloidal Filtration.* J. Am. Ceram. Soc. **10,** C190–192 (1983)

[3.162] Y. Zhou, R.J. Phillips, and J.A. Switzer: *Electrochemical Synthesis and Sintering of Nanocrystalline Cerium(IV) Oxide Powders.* J. Am. Ceram. Soc. **78 [4],** 981–985 (1995)

[3.163] K. Haberko: *Characteristics and Sintering Behavior of* ZrO_2 *Ultrafine Powders.* Ceram. Int. **5 [4],** 148–154 (1979)

[3.164] A. Roosen and H. Hausner: *Sintering Kinetics of* ZrO_2 *Powders.* In: *Science and Technology of Zirconia II.* N. Claussen, M. Ruhle, and A.H. Heuer (eds.). (The American Ceramic Society Columbus OH, 1984), pp. 714–726

[3.165] F.F. Lange and B.I. Davis: *Sinterability of* ZrO_2 *and* Al_2O_3 *Powders, the Role of Pore Coordination Number Distribution.* In: *Science and Technology of Zirconia II.* N. Claussen, M. Ruhle, and A.H. Heuer (eds.). (The American Ceramic Society Columbus OH, 1984), pp. 699–713

[3.166] W.D. Kingery and B. Francois: *Sintering of Crystalline Oxides I: Interaction Between Grain Boundaries and Pores.* In: *Sintering and Related Phenomena.* G.C. Kuczynski, N.A. Horton, and C.F. Gibbon (eds.). (Gordon & Breach New York, 1967), pp. 471–498

[3.167] C.H. Hsueh, A.G. Evans, and R.L. Coble: *Microstructure Development During Final/Intermediate Stage Sintering-I.Pore/Grain Boundary Separation.* Acta Metall. **30 [7],** 1269–1279 (1982)

[3.168] C.A. Handwerker, R.M. Cannon, and R.L. Coble. *Final-Stage Sintering of MgO.* In: *Advances in Ceramics,* Vol.10: *Structure and Properties of MgO and* Al_2O_3 *Ceramics.* W.D.Kingery (ed.). (The American Ceramic Society Columbus OH, 1984), pp. 619–643

[3.169] S. Wu, E. Gilbart, and R.J. Brook: *Solid State Sintering: The Attainment of High Density.* In: *Advances in Ceramics,* Vol.10: *Structure and Properties of MgO and* Al_2O_3 *Ceramics.* W.D. Kingery (ed.). (The American Ceramic Society Columbus OH, 1984), pp. 574–582

[3.170] C.E.G. Bennett, N.A. McKinnon and L.S. Williams: *Sintering in Gas Discharges.* Nature **217 [5135],** 1287–1288 (1968)

[3.171] D.L. Johnson and R.A. Rizzo: *Plasma Sintering of ß"-Alumina.* Am. Ceram. Soc. Bull. **59 [4],** 467–472 (1982)

[3.172] K. Upadhya: *An Innovative Technique for Plasma Sintering of Ceramics and Composite Materials.* Am. Ceram. Soc. Bull. **67 [10],** 1691–1694 (1988)

[3.173] T.T. Meek, R.D. Blake, and J.J. Petrovic: *Microwave Sintering of Al_2O_3 and Al_2O_3–SiC Whisker Composites.* Ceram. Eng. Sci. Proc. **8 [7–8],** 861–871 (1987)

[3.174] J. Wilson and S.M. Kunz: *Microwave Sintering of Partially Stabilized Zirconia.* J. Am. Ceram. Soc. **71 [1],** C-40–41 (1988)

[3.175] A.Morrel and A.Hermosin: *Fast Sintering of Soft Mn–Zn and Ni–Zn Ferrite Pot Cores.* Am. Ceram. Soc. Bull. **59 [6],** 626–629 (1980)

[3.176] D.L. Johnson: *Ultra Rapid Sintering.* In: *Materials Science Research,* Vol.16: *Sintering and Heterogeneous Catalysis.* G.C. Kuczynski, A.E. Miller, and G.A. Sargent (eds.). (Plenum Press NY, 1984), pp. 243–252

[3.177] D.E. Garcia, J. Seidel, R. Janssen, and N. Claussen: *Fast Firing of Alumina.* J. Eur. Ceram. Soc. **15 [10],** 935–938 (1995)

[3.178] J.A. Varela, O.J. Whittemore, and E. Longo: *Pore Size Evolution During Sintering of Ceramic Oxides.* Ceram. Int. **16 [3],** 177–189 (1990)

[3.179] R.L. Coble: *Sintering of Alumina: Effect of Atmospheres.* J. Am. Ceram. Soc. **45 [3],** 123–127 (1962)

[3.180] S. Prochazka, C.A. Johnson, and R.A. Giddings: *Atmosphere Effects in Sintering of Silicon Carbide. In: Proceedings of the International Symposium, Factors in Densification and Sintering of Oxide and Non-Oxide Ceramics.* S. Somiya and S. Saito (eds.). (Association for Science Documents Information Tokyo Institute of Technology Okayama Meguro Japan, 1978), pp. 366–381

[3.181] G.J. Ghorra: *Theory of Fast Firing.* Ceram. Eng. Sci. Proc. **14 [1–2],** 77–115 (1993)

[3.182] F.F. Lange: *Transformation Toughening, Part 4: Fabrication, Fracture Toughness, and Strength of Al_2O_3–ZrO_2 Composites.* J. Mater. Sci. **17 [1],** 247–254 (1982)

[3.183] C.J. McHargue, H. Naramoto, C.W. White, J.M. Williams, P.S. Sklad, and P. Angelini: *Structure of Ceramic Surfaces Modified by Ion Beam Techniques.* In: *Materials Science Research,* Vol.17. R.F. Davis, H. Palmour III, and R.L. Porter (eds.). (Plenum Press New York, 1984), pp. 519–31

[3.184] C.J. McHargue: *Structure and Mechanical Properties of Ion Implanted Ceramics.* Nucl. Inst. Meth. Phys. Res. **B19/20,** 797–804 (1987)

[3.185] B. Manley, J.B. Holt, and Z.A. Munir: *Sintering of Combustion Synthesized Titanium Carbide.* In: *Materials Science Research,* Vol.16: *Sintering and Heterogeneous Catalysis.* G.C. Kuczynski, A.E. Miller, and G.A. Sargent (eds.). (Plenum Press New York, 1984), pp. 303–316

[3.186] M. Ouabdesselam and Z.A. Munir: *The Sintering of Combustion Synthesized Titanium Diboride.* J. Mater. Sci. **22 [5],** 1799–1807 (1987)

[3.187] J.E. Bailey, D. Lewis, Z.M. Librant, and L.J. Porter: *Phase Transformations in Milled Zirconia.* Trans. J. Brit. Ceram. Soc.. **71,** 23–30 (1972)

[3.188] O.R. Bergmann and J. Barrington: *Effects of Shock Waves on Ceramic Materials*. J. Am. Ceram. Soc. **49 [9],** 502–507 (1966)

[3.189] S. Somiya, M. Yoshimura, S. Fujiwara, K.I. Kondo, and A. Sawaoka: *Hot Isostatic Pressing of Shocked Si_3N_4 Powder*. J. Am. Ceram. Soc. **67 [3],** C-51–52 (1984)

[3.190] E.K. Beauchamp: *Shock Activated Sintering.* In: *High Pressure Explosive Processing of Ceramics*. R.A. Graham and A.B. Sawaoka (eds.). (Trans Tech Zürich Switzerland, 1987), pp. 141–174

[3.191] E.K. Beauchamp and M.J. Carr: *Hot Pressing of Shock Modified Powders.* In: *High Pressure Explosive Processing of Ceramics.* R.A. Graham and A.B. Sawaoka (eds.). (Trans Tech Zürich Switzerland, 1987), pp. 174–206

[3.192] C.S. Yust and L.A. Harris: *Observation of Dislocation and Twins in Explosively Compacted Alumina.* In: *Shock Waves and High Strain-Rate Phenomena in Metals; Concepts and Applications*. M.A. Meyers and L.E. Murr (eds.). (Plenum Press New York, 1981), pp. 881–894

[3.193] H. Palmour III, T.M. Hare, A.D. Batchelor, K.Y. Kim, K.L. More, and T.T. Fang: *Sintering of Shock-Conditioned Materials*. In: *Advances in Ceramics,* Vol.10: *Structure and Properties of MgO and Al_2O_3 Ceramics*. W.D. Kingery (ed.). (The American Ceramic Society Columbus OH, 1984), pp. 506–525

[3.194] E.K. Beuchamp, R.E. Loehman, R.A. Graham, B. Morosin, and E.L. Venturini: *Densification Kinetics of Shock-Activated Nitrides.* In: *Emergent Process Methods for High-Technology Ceramics,* Vol.17. R.F. Davis, H. Palmour III, and R.L. Porter. (Plenum Publication New York, 1984), pp. 735–748

[3.195] R. Prummer and G. Ziegler: *Structural Changes in the Explosive Compaction of Alumina Powders.* Ber. Deutsche Keram. Ges. **51,** 343–347 (1974)

[3.196] B. Morosin, R.A. Graham, and E.K. Beauchamp: In: *Final Report of the DARPA Dynamic Materials Synthesis and Consolidation Program, Vol.II Shock-Activated Sintering.* C.F. Cline (ed.). (Lawrence Livermore Laboratory Report UCID-19663–85, June 1985), pp. 138–368

[3.197] T.H. Hare, K.L. More, A.D. Batchelor, and H. Palmour III: *Sintering Behavior of Overcompacted Shock-Conditioned Alumina Powder*. In: *Materials Science Research,* Vol.16: *Sintering and Heterogeneous Catalysis*. G.C. Kuczynski, A.E. Miller, and G.A. Sargent (eds.). (Plenum Press New York, 1984), pp. 265–279

[3.198] W.D. Kingery: *Densification During Sintering in the Presence of a Liquid Phase, I.Theory.* J. App. Phys. **30 [3],** 301–306 (1959)

[3.199] R.M. German: *Liquid Phase Sintering*. (Plenum Press New York, 1985)

[3.200] W.D. Kingery and M.D. Narasimhan: *Densification During Sintering in the Presence of a Liquid Phase. II. Experimental*. J. Appl. Phys. **30 [1],** 307–310 (1950)

[3.201] W.D. Kingery, E. Niki, and M.D. Narasimhan: *Sintering of Oxide and Carbide–Metal Compositions in Presence of a Liquid Phase.* J. Am. Ceram. Soc. **44 [1]**, 29–35 (1961)

[3.202] V.K. Singh: *Sintering of Alumina in the Presence of a Liquid Phase.* Trans. Ind. Ceram. Soc. **37 [2]**, 55–57 (1978)

[3.203] W.J. Huppmann, S. Pejornik, and S.N. Horn: *Rearrangement During Liquid Phase Sintering of Ceramics.* In: *Materials Science Research,* Vol.11: *Processing of Crystalline Ceramics.* H. Palmour , R.F. Davis, and T.M. Hare (eds.). (Plenum Press New York, 1979), p. 233

[3.204] W.J. Huppmann and G. Petzow: *The Elementary Mechanisms of Liquid Phase Sintering.* In: *Materials Science Research,* Vol.13: *Sintering Processes.* G.C. Kuczynski (ed.). (Plenum Press New York, 1980), pp. 189–201

[3.205] K.G. Ewsuk: *Consolidation of Bulk Ceramics.* In: *Characterization of Ceramics.* R.E. Loehman (ed.). (Butterworth-Heinemann Stoneham MA, 1993)

[3.206] O-H. Kwon: *Liquid Phase Sintering.* In: *Engineered Materials Handbook,* Vol.4: *Ceramics and Glasses.* (ASM International Materials Park OH, 1991), pp. 285–290

[3.207] W.D.G. Boecker, R. Hamminger, J. Heinrich, J. Huber, and A. Roosen: *Covalent High-Performance Ceramics.* Adv. Mater. **4 [3]**, 169–178 (1992)

[3.208] F.K. van Dijen and E. Mayer: *Liquid Phase Sintering of Silicon Carbide.* J. Eur. Ceram. Soc. **16 [3]**, 413–420 (1996)

[3.209] E. Dorre and H. Hubner: *Alumina-Processing, Properties, and Applications.* (Springer-Verlag Berlin, 1984)

[3.210] C. Herring, *Diffusional Viscosity of a Polycrystalline Solid*, J. App. Phys., **21** 437–45 (1950)

[3.211] S. Suresh and J.R. Brockenbrough: *A Theory for Creep by Interfacial Flaw Growth in Ceramics and Ceramic Composites.* Acta Metall. **38 [1]**, 55–68 (1990)

[3.212] R.L. Coble: *Diffusion Models for Hot Pressing with Surface Energy and Pressure Effects as Driving Forces.* J. Appl. Phys. **41 [12]**, 4798–4807 (1970)

[3.213] W.D. Kingery, H.K. Bowen, and D.R. Uhlmann: *Introduction to Ceramics.* (John Wiley & Sons New York, 1976)

[3.214] M.R. Notis, R.H. Smoak, and V. Krishnamachari: *Interpretation of Hot Pressing Kinetics by Densification Mapping Techniques.* In: *Materials Science Research,* Vol.10: *Sintering and Catalysis.* G.C. Kuczynski (ed.). (Plenum Press New York, 1975), pp. 493–507

[3.215] M.F. Ashby: *A First Report on Deformation-Mechanism Maps.* Acta Metall. **20 [7]**, 887–897 (1972)

[3.216] H.J. Frost and M.F. Ashby: *Deformation-Mechanism Maps.* (Pergamon Press Oxford England, 1982)

[3.217] K.H. Hardtl: *Gas Isostatic Pressing Without Molds*. Am. Ceram. Soc. Bull. **54 [2]**, 201–205 (1975)

[3.218] H.T. Larker, L. Hermansson, and J. Adlerborn: *Hot-Isostatic Pressing and Its Applicability to Silicon Carbide and Boron Carbide.* Ind. Ceram. **8 [1]**, 17–19 (1988)

[3.219] W.A. Kaysser: *Present State of Modeling of Hot Isostatic Pressing.* In: *Hot Isostatic Pressing: Theory and Applications*. R.J. Schaefer and M. Linzer (eds.). (ASM International Materials Park OH, 1991), pp. 1–13

[3.220] L.J. Bjork and A.G. Hermansson: *Hot Isostatically Pressed Alumina–Silicon Carbide–Whisker Composites.* J. Am. Ceram. Soc. **72 [8]**, 1436–1438 (1989)

[3.221] P.A. Janeway: *HIP Plays a Unique Role in Materials Processing.* Ceram. Ind. **138 [4]**, 35–37 (1992)

[3.222] J. Adlerborn, M. Burstrom, L. Hermansson, and H.T. Larker: *Development of High Temperature High Strength Silicon Nitride by Glass Encapsulated Hot Isostatic Pressing.* Mater. Des. **8 [4]**, 229–232 (1987)

[3.223] K. Uematsu, K. Itakura, N. Uchida, K. Saito, A. Miyamoto, and T. Miyashita: *Hot Isostatic Pressing of Alumina and Examination of the Hot Isostatic Pressing Map.* J. Am. Ceram. Soc. **73 [1]**, 74–78 (1990)

[3.224] A.S. Helle, K.E. Easterling, and M.F. Ashby: *Hot-Isostatic Pressing Diagrams: New Developments.* Acta Metall. **33 [2]**, 2163–2174 (1985)

[3.225] D.G. Morris: *Bonding Processes During the Dynamic Compaction of Metallic Powders.* Mater. Sci. Eng. **57 [2]**, 187–195 (1983)

[3.226] C.L. Hoenig and C.S. Yust: *Explosive Compaction of AlN, Amorphous Si_3N_4, Boron, and Al_2O_3 Ceramics.* Am. Ceram. Soc. Bull. **60 [11]**, 1175–1224 (1981)

[3.227] T. Akashi and A.B. Sawaoka: *Dynamic Compaction of Cubic Boron Nitride Powders.* J. Am. Ceram. Soc. **69 [4]**, C-78–80 (1986)

[3.228] S. Soga, K. Kondo, A. Sawaoka, and Y. Tanaka: *Shock Compaction of TiN–TiC Powders.* J. Mater. Sci. Lett. **2 [11]**, 673–674 (1983)

[3.229] M. Bengisu and O.T. Inal: *Densification and Mechanical Properties of Shock Treated Alumina and its Composites.* J. Mater. Sci. **29 [18]**, 4824–4833 (1994)

[3.230] J.H. Adair, R.R. Wills, and V.D. Linse. *Dynamic Compaction of Ceramic Powders.* In: *Emergent Process Methods for High-Technology Ceramics,* Vol.17. R.F. Davis, H. Palmour III, and R.L. Porter (eds.). (Plenum Publication New York, 1984), pp. 639–642

[3.231] M. Bengisu, O.T. Inal, and J.R. Hellmann: *Effect of Dynamic Compaction on the Retention of Tetragonal Zirconia and Mechanical Properties of Al_2O_3–ZrO_2 Composites.* J. Am. Ceram. Soc. **73 [2]**, 346–351 (1990)

[3.232] B. Tunaboylu and J. McKittrick: *Dynamic Compaction of Al_2O_3–ZrO_2 Compositions.* J. Am. Ceram. Soc. **77 [6]**, 1605–1612 (1994)

[3.233] L.J. Kecskes, T. Kottke, and A. Niiler: *Microstructural Properties of Combustion Synthesized and Dynamically Consolidated Titanium Diboride and Titanium Carbide.* J. Am. Ceram. Soc. **73 [5],** 1274–1282 (1990)

[3.234] J.B. Holt: *The Use of Exothermic Reactions in the Synthesis and Densification of Ceramic Materials.* MRS Bull. **12 [7],** 59–64 (1987)

[3.235] M.E. Washburn and W.S. Coblenz: *Reaction-Formed Ceramics.* Am. Ceram. Soc. Bull. **67 [3],** 56–63 (1988)

[3.236] J.A. Mangels and G.J. Tennenhouse: *Densification of Reaction Bonded Silicon Nitride.* Am. Ceram. Soc. Bull. **59 [12],** 1216–1222 (1980)

[3.237] W.R. Cannon, S.C. Danforth, J.H. Flint, J.S. Haggerty, and R.A. Marra: *Sinterable Ceramic Powders from Laser Driven Reactions, Part I: Process Description and Modeling.* J. Am. Ceram. Soc. **65 [7],** 324–330 (1982)

[3.238] W.R. Cannon, S.C. Danforth, J.H. Flint, J.S. Haggerty, and R.A. Marra: *Sinterable Ceramic Powders from Laser Driven Reactions, Part II: Powder Characteristics and Process Variables.* J. Am. Ceram. Soc. **65 [7],** 330–335 (1982)

[3.239] S.T. Buljan, J.G. Baldoni, and M.L. Huckabee: *Si_3N_4–SiC Composites.* Am. Ceram. Soc. Bull. 66 [2], 347–352 (1987)

[3.240] R. Lundberg, L. Kahlman, R. Pompe, R. Carlsson, and R. Warren: *SiC–Whisker-Reinforced Si_3N_4 Composites*, Am. Ceram. Soc. Bull. **66 [2],** 330–333 (1987)

[3.241] J.S. Haggerty and Y.M. Chiang: *Reaction-Based Processing Methods for Ceramics and Composites.* Ceram. Eng. Sci. Proc. **11 [7–8],** 757–781 (1990)

[3.242] J.S. Haggerty: *Reaction Sintering.* In: *Engineered Materials Handbook,* Vol.4: *Ceramics and Glasses.* (ASM International Materials Park OH, 1991), pp. 291–296

[3.243] A. Pivkina, P.J. van der Put, Y. Frolov, and J. Schoonman: *Reaction-Bonded Titanium Nitride Ceramics.* J. Eur. Ceram. Soc. **16 [1],** 35–32 (1996)

[3.244] B. North and K.E. Gilchrist: *Effect of Impurity Doping on a Reaction-Bonded Silicon Carbide.* Am. Ceram. Soc. Bull. **60 [5],** 549–554 (1981)

[3.245] G.R. Sawyer and T.F. Page: *Microstructural Characterization of "REFEL" (Reaction Bonded) Silicon Carbides.* J. Mater. Sci. **13 [4],** 885–904 (1978)

[3.246] P.A. Willermet, R.A. Pett, and T.J. Whalen: *Development and Processing of Injection Moldable Reaction Sintered SiC Compositions.* Am. Ceram. Soc. Bull. **57 [8],** 744–747 (1978)

[3.247] R.P. Meissner and Y.M. Chiang: *Processing of Reaction-Bonded Silicon Carbide Without Residual Silicon Phase.* Ceram. Eng. Sci. Proc. **9 [7–8],** 1053–1059 (1988)

[3.248] Y.M. Chiang, J.S. Haggerty, R.P. Messner, and C. Demetry: *Reaction-Based Processing Methods for Ceramic-Matrix Composites.* Am. Ceram. Soc. Bull. **68 [2],** 420–428 (1989)

[3.249] P.D.D. Rodrigo and P. Boch: *Preparation of High Purity Mullite Ceramics.* In: *13th International Conference on Science of Ceramics.* P. Odier, F. Cabannes, and B. Cales (eds.). (Les Editions de Physique Les Unix Cedex France, 1986), pp. C1-411–416

[3.250] I.A. Aksay, D.M. Dabbs, and M. Sarikaya: *Mullite for Structural, Electronic, and Optical Applications.* J. Am. Ceram. Soc. **74 [10],** 2343–2358 (1991)

[3.251] P. Miranzo, M.I. Osendi, and J.S. Moya: *Influence of Processing Method on the Microstructure and Mechanical Properties of Mullite/ZrO_2 Composites.* In: *13th International Conference on Science of Ceramics.* P. Odier, F. Cabannes, and B. Cales (eds.). (Les Editions de Physique Les Unix Cedex France, 1986), pp. C1-417–421

[3.252] J. Rincon, G. Thomas, P. Pena, S. de Aza, and J.S. Moya: *Reaction Sintered Zirconia–Mullite Composites with CaO.* In: *13th International Conference on Science of Ceramics.* P. Odier, F. Cabannes, and B. Cales (eds.). (Les Editions de Physique Les Unix Cedex France, 1986), pp. C1-423–427

[3.253] S. Lathabai, D.G. Hay, F. Wagner, and N. Claussen: *Reaction-Bonded Mullite/Zirconia Composites.* J. Am. Ceram. Soc. **79 [1],** 248–256 (1996)

[3.254] W.S. Coblenz and D. Lewis III: *In Situ Reaction of B_2O_3 with AlN and/or Si_3N_4 to form BN-Toughened Composites.* J. Am. Ceram. Soc. **71 [12],** 1080–85 (1988)

[3.255] G.J. Zhang and Z.Z. Jin: *Reactive Synthesis of AlN/TiB_2 Composite.* Ceram. Int. **22 [2],** 143–147 (1996)

[3.256] W. Stadtbauer, W. Kladnig, and G. Gritzner: *Al_2O_3–TiB_2 Composite Ceramics.* J. Mater. Sci. Lett. **8 [10],** 1217–1220 (1989)

[3.257] M.D. Sacks, N. Bozkurt, and G. Scheiffele: *Fabrication of Mullite and Mullite-Matrix Composites by Transient Viscous Sintering of Composite Powders.* J. Am. Ceram. Soc. **74 [10],** 2428–2437 (1991)

[3.258] A. Leriche, P. Descamps, et F. Cambier. *Frittage Reaction et Applications aux Composites a Dispersoides.* In: *Ceramiques Composites a Particules, Cas du Frittage-Reaction.* F. Thevenot (ed.). (Forceram Editions Septima Paris France, 1992), pp. 9–38

[3.259] G.J. Zhang, X.M. Yue, Z.Z. Jin, and J.Y. Dai: *In-Situ Synthesized TiB_2 Toughened SiC.* J. Eur. Ceram. Soc. **16 [4],** 409–412 (1996)

[3.260] L.G. Cordone and W.E. Martinsen: *Glow-Discharge Apparatus for Rapid Sintering of Al_2O_3.* J. Am. Ceram. Soc. **55 [7],** 380 (1972)

[3.261] R.A. Dugdale: *The Application of the Glow Discharge to Material Processing.* J. Mater. Sci. **1 [2],** 160–169 (1966)

[3.262] D.L. Johnson, W.B. Sanderson, J.M. Knowlton, and E.L. Kemer: *Sintering of α-Al_2O_3 in Gas Plasmas.* In: *Advances in Ceramics,* Vol.10:

Structure and Properties of MgO and Al_2O_3 Ceramics. W.D. Kingery (ed.). (The American Ceramic Society Columbus OH, 1984), pp. 656–65

[3.263] E.L. Kemer and D.L. Johnson: *Microwave Plasma Sintering of Alumina.* Am. Ceram. Soc. Bull. **64 [8],** 1132–1136 (1985)

[3.264] J.S. Kim and D.L. Johnson: *Plasma Sintering of Alumina.* Am. Ceram. Soc. Bull. **62 [5],** 620–622 (1983)

[3.265] G. Thomas, J. Freim, and W. Martinsen: *Rapid Sintering of UO_2 in a Glow Discharge.* Trans. Am. Nucl. Soc. **17,** 177 (1973)

[3.266] G. Thomas and J. Freim: *Parametric Investigation of the Glow Discharge Technique for Sintering UO_2*. Trans. Am. Nucl. Soc. **21,** 182–183 (1975)

[3.267] P.C. Kong, Y.C. Lau, E. Pfender, K. McHenry, W. Wallenhorst, and B. Koepke: *Sintering of Fully Stabilized Zirconia Powders in RF Plasmas.* In: *Ceramic Transactions, Ceramic Powder Science II.* G.L.Messing and H. Hausner (eds.). (The American Ceramic Society Westerville OH, 1988), pp. 939–946

[3.268] R.M. Young and R. McPherson: *Accelerated Densification in Plasma Sintering Due to Thermal Diffusion.* In: *9th International Symposium on Plasma Chemistry, Symposium Proceedings* Vol.I. R. D'Agostino (ed.). (International Union of Pure and Applied Chemistry Pugnochiuso Italy, 1989), pp. 19–24

[3.269] W.H. Sutton: *Microwave Processing of Ceramic Materials.* Am. Ceram. Soc. Bull. **68 [2],** 376–386 (1989)

[3.270] D. Palaith and R. Silberglitt: *Microwave Joining of Ceramics.* Am. Ceram. Soc. Bull. **68 [9],** 1601–1606 (1989)

[3.271] H. Fukushima, T. Yamamoto, and M. Matsui: *Microwave Heating of Ceramics and its Application to Joining.* In: *Microwave Processing of Materials.* W.H. Sutton, M.H. Brooks, and I.J. Chabinsky (eds.). (Materials Research Society Pittsburgh PA, 1988), pp. 267–272

[3.272] S. Vodegel: *Microwave Densification of Alumina.* Ceram. Forum Int. **73 [1],** 44–47 (1996)

[3.273] J. McKittrick, B. Tunaboylu, and J. Katz: *Microwave and Conventional Sintering of Rapidly Solidified Al_2O_3–ZrO_2 Powders.* J. Mater. Sci. **29 [8],** 2119–2125 (1994)

[3.274] G. Desgardin, L. Mazo, L. Quemeneur, and B. Raveau: *Microwave Sintering of $BaTiO_3$ Based Ceramics.*In: *Thirteenth International Conference on Science of Ceramics.* P. Odier, F. Cabannes, and B. Cales (eds.). (Les Editions de Physique Cedex France, 1986), pp. C1-397–402

[3.275] Y.L. Tian, M.E. Bordin, and D.L. Johnson: *Microwave Sintering of Ceramics Under High Gas Pressure.* In: *Microwave Processing of Materials*. W.H. Sutton, M.H. Brooks, and I.J. Chabinsky. (Mater. Res. Soc. Pittsburgh PA, 1988), pp. 213–218

[3.276] T.T. Meek, C.E. Holcomb, and N. Dykes: *Microwave Sintering of Some Oxide Materials Using Sintering Aids.* J. Mater. Sci. Lett. **6 [9],** 1060–1062 (1987)

[3.277] R.D. Blake and T.T. Meek: *Microwave Processed Composites.* J. Mater. Sci. Lett. **5 [11],** 1097–1098 (1986)

[3.278] J.D. Katz, R.D. Blake, J.J. Petrovic, and H. Sheinberg: *Microwave Sintering of Boron Carbide.* In: *Microwave Processing of Materials.* W.H. Sutton, M.H. Brooks, and I.J. Chabinsky (eds.). (Materials Research Society Pittsburgh PA, 1988), pp. 219–226

[3.279] B. Swain: *Microwave Sintering of Ceramics.* Adv. Mater. Proc. **134 [9],** 76–82 (1988)

[3.280] Y.L. Tian, D.L. Johnson, and M.E. Brodwin: *Microwave Sintering of Al_2O_3–TiC Composites.* In: *Ceramic Transaction,* Vol.1: *Ceramic Powder Science II.* G.L. Messing, E.R. Fuller Jr., and H. Hausner (eds.). (The American Ceramic Society Westerville OH, 1988), pp. 933–938

[3.281] S.S. Park and T.T. Meek: *Characterization of ZrO_2–Al_2O_3 Composites Sintered Using 2.45 GHz Radiation.* Ceram. Eng. Sci. Proc. **11 [9–10],** 1395–1404 (1990)

[3.282] J.D. Katz, R.D. Blake, and J.J. Petrovic: *Microwave Sintering of Alumina–Silicon Carbide Composites at 2.45 and 60 GHz.* Ceram. Eng. Sci. Proc. **9 [7–8],** 725–734 (1988)

[3.283] Z.A. Munir: *Synthesis of High Temperature Materials by Self-Propagating Combustion Methods.* Am. Ceram. Soc. Bull. **67 [2],** 342–349 (1988)

[3.284] R.W. Rice and W.J. McDonough: *Intrinsic Volume Changes of Self-Propagating Synthesis.* J. Am. Ceram. Soc. **68 [5],** C-122–123 (1985)

[3.285] R.W. Rice, W.J. McDonough, G.Y. Richardson, J.M. Kunetz, and T. Schroeder: *Hot Rolling of Ceramics Using Self-Propagating High Temperature Synthesis.* Ceram. Eng. Sci. Proc. **7 [7–8],** 751–760 (1986)

[3.286] R. Rice: *Processing of Ceramic Composites.* In: *Advanced Ceramic Processing and Technology,* Vol.1. J.G.P. Binner (ed.). (Noyes Park Ridge NJ, 1990)

[3.287] D.P. Stinton, T.M. Besmann, and R.A. Lowden: *Advanced Ceramics by Chemical Vapor Deposition.* Am. Ceram. Soc. Bull. **67 [2],** 350–355 (1988)

[3.288] D.P. Stinton, T.M. Besmann, R.A. Lowden, and B.W. Sheldon: *Vapor Deposition.* In: *Engineering Materials Handbook,* Vol.4: *Ceramics and Glasses.* (ASM International Materials Park OH, 1992), pp. 215–222

[3.289] M. Hirayama and K. Shohno: *CVD-BN for Boron Diffusion in Si and Its Applications to Si Devices.* J. Electrochem. Soc. **122 [12],** 1671–1675 (1975)

[3.290] W.A. Bryant: *Fundamentals of Chemical Vapor Deposition.* J. Mater. Sci. **12 [7],** 1285–1306 (1977)

[3.291] N.J. Archer: *The Preparation and Properties of Pyrolytic Boron Nitride.* In: *High Temperature Chemistry in Inorganic and Ceramic Materials.* F.P. Glasser and P.E. Potter (eds.). (The Chemical Society London UK, 1977), pp. 167–180

[3.292] D.P. Stinton, W.J. Lackey, R.J. Lauf, and T.M. Besmann: *Fabrication of Ceramic–Ceramic Composites by Chemical Vapor Deposition.* Ceram. Eng. Sci. Proc. **5 [7–8],** 668–676 (1984)

[3.293] T. Hirai and S. Hayashi: *Preparation and Some Properties of Chemically Vapor Deposited* Si_3N_4*–TiN Composite.* J. Mater. Sci. **17,** 1320–1328 (1982)

[3.294] A.J. Caputo and W.J. Lackey: *Fabrication of Fiber-Reinforced Ceramic Composites by Chemical Vapor Infiltration.* Ceram. Eng. Sci. Proc. **5 [7–8],** 654–667 (1984)

[3.295] A.J. Caputo, D.P. Stinton, R.A. Lowden, and T.M. Besmann: *Fiber-Reinforced SiC Composites with Improved Mechanical Properties*. Am. Ceram. Soc. Bull. **66 [2],** 368–372 (1987)

[3.296] J. Hurt and D. Viechnicki: *Ultrafine Grain Ceramics from Melt Phase.* In: *Ultrafine Grain Ceramics*. J.J. Burke, N.L. Reed, and V. Weiss (eds.). (Syracuse University Press New York, 1970), pp. 273–295

[3.297] D. Viechnicki and F. Schmid: *Eutectic Solidification in the System* $Al_2O_3/Y_3Al_5O_{12}$. J. Mater. Sci. **4 [1],** 84–88 (1969)

[3.298] C. Hulse and J. Batt: *Effect of Eutectic Microstructures on the Mechanical Properties of Ceramic Oxides.* (United Aircraft Final Tech.Rep. UARL-N910803–10, May 1974)

[3.299] R.L. Ashbrook: *Directionally Solidified Ceramic Eutectics.* J. Am. Ceram. Soc. **60 [9–10],** 428–435 (1977)

[3.300] F. Schmid and D. Viechnicki: *Oriented Eutectic Microstructures in the System* Al_2O_3/ZrO_2. J. Mater. Sci. **5 [6],** 470–473 (1970)

[3.301] D.J. Rowcliffe, W.J. Warren, A.G. Elliot, and W.S. Rothwell: *The Growth of Oriented Ceramic Eutectics*. J. Mater. Sci. **4 [10],** 902–907 (1969)

[3.302] W.J. Minford, R.C. Bradt, and V.S. Stubican. *Crystallography and Microstructure of Directionally Solidified Oxide Eutectics.* J. Am. Ceram. Soc. **62 [3–4],** 154–157 (1979)

[3.303] F.L. Kennard, R.C. Bradt, and V.S. Stubican. *Eutectic Solidification of* $MgO–MgAl_2O_4$. J. Am. Ceram. Soc. **56 [11],** 566–569 (1973)

[3.304] F.L. Kennard, R.C. Bradt, and V.S. Stubican. *Directional Solidification of the* ZrO_2*–MgO Eutectic.* J. Am. Ceram. Soc. **57 [10],** 428–431 (1974)

[3.305] D.J.S. Cooksey, D. Munson, M.P. Wilkinson, and A. Hellawell. *The Freezing of Some Continuous Binary Eutectic Mixtures.* Philos. Mag. **10 [107],** 745–769 (1964)

[3.306] J.D. Hunt and K.A. Jackson: *Binary Eutectic Solidification.* Trans AIME **236 [6],** 843–852 (1966)

[3.307] C.C. Sorrell, H.R. Beratan, R.C. Bradt, and V.S. Stubican: *Directional Solidification of (Ti,Zr)Carbide–(Ti,Zr) Diboride Eutectics*. J. Am. Ceram. Soc. **67 [3],** 190–194 (1984)

[3.308] J. DerHong, K.E. Spear, and V.S. Stubican: *Directional Solidification of $SiC–B_4C$ Eutectic: Growth and Some Properties.* Mater. Res. Bull. **14 [6],** 775–783 (1979)

[3.309] V.S. Stubican and R.C. Bradt: *Eutectic Solidification in Ceramic Systems.* Ann. Rev. Mater. Sci. **11,** 267–297 (1981)

[3.310] W.B. Hillig: *Melt Infiltration Process for Making Ceramic Matrix Composites.* In: *Advanced Ceramic Processing and Technology,* Vol.1. J.G.P. Binner (ed.). (Noyes Park Ridge NJ, 1990)

[3.311] W.G. Hillig, R.L. Mehan, C.R. Morelock, V.J. DeCarlo, and W. Laskow: *Silicon/Silicon Carbide Composites.* Am. Ceram. Soc. Bull. **54 [12],** 1054–1056 (1975)

[3.312] R. Riedel and W. Dressler: *Chemical Formation of Ceramics.* Ceram. Int. **22 [3],** 233–239 (1996)

[3.313] R. Riedel, M. Seher, J. Mayer, and D.V. Szabo: *Polymer-Derived Si-Based Bulk Ceramics, Part I: Preparation, Processing and Properties.* J. Eur. Ceram. Soc. **15 [8],** 703–715 (1995)

[3.314] M.F. Godon, G. Fantozzi, M. Murat, and J.P. Disson: *Manufacture of Monolithic Ceramic Bodies from Polysilazane Precursor.* J. Eur. Ceram. Soc. **15 [6],** 591–597 (1995)

[3.315] D.P. Kim, C.G. Cofer, and J. Economy: *Fabrication and Properties of Ceramic Composites with a Boron Nitride Matrix.* J. Am. Ceram. Soc. **78 [6],** 1546–1552 (1995)

[3.316] L. Kempfer: *Forming the Pieces of the Ceramic Puzzle.* Mater. Eng. **107 [6],** 23–26 (1990)

[3.317] J.D. Currey and A.J. Kohn: *Fracture in the Crossed-Lamellar Structure of Conus Shells.* J. Mater. Sci. **11 [9],** 1615–1623 (1976)

[3.318] V.J. Laraia and A.H. Heuer: *Microstructures and Mechanical Behavior of Mollusk Shells.* In: *Proceedings of the 11th Riso International Symposium on Metallurgy and Materials Science.* J.J. Bentzen, J.B.B. Sorensen, N. Christiansen, A. Horsewell, and B. Ralph (eds.). (Riso National Laboratory Roskilde Denmark, 1990), pp. 79–96

[3.319] S.A. Wainwright, W.D. Biggs, J.D. Currey, and J.M. Gosline: *Mechanical Design in Organisms.* (Edward Arnold New York, 1976)

[3.320] R.E. Newnham and G.R. Ruschau: *Smart Electroceramics.* J. Am. Ceram. Soc. **74 [3],** 463–480 (1991)

[3.321] T.Ota, M.Takahashi, T.Hibi, M.Ozawa, S.Suzuki, Y.Hikichi, and H.Suzuki: *Biomimetic Process for Producing SiC Wood.* J. Am. Ceram. Soc. **78 [12]**, 3409–3411 (1995)

Chapter 4

[4.1] B.W. Mott: *Micro-Indentation Hardness Testing*. (Butterworths Scientific London, 1956)

[4.2] M.A. Meyers and K.K. Chawla: *Mechanical Metallurgy*. (Prentice-Hall Englewood Cliffs NJ, 1984)

[4.3] M.L. Cohen: *Calculation of Bulk Moduli of Diamond and Zincblende Solids.* Phys. Rev. B **32 [12],** 7988–7991 (1985)

[4.4] M.R. Wixom: *Chemical Preparation and Shock Wave Compression of Carbon Nitride Precursors.* J. Am. Ceram. Soc. **73 [7],** 1973–1978 (1990)

[4.5] D.W. Richerson: *Modern Ceramic Engineering*. (Marcel Dekker New York, 1982)

[4.6] A.Y. Liu and M.L. Cohen: *Prediction of New Low Compressibility Solids*. Science **245,** 841–842 (1989)

[4.7] W.E. Lee and W.M. Rainforth: *Ceramic Microstructures.* (Chapman & Hall London, 1994)

[4.8] R. Warren and V.K. Sarin: *Particulate Ceramic-Matrix Composites.* In: *Ceramic Matrix Composites*. R. Warren (ed.). (Blackie and Son London, 1992), pp. 146–166

[4.9] E. Scafe, L. Fabbri, G. Grillo, and L. di Rese: *Improved Technique for Young's Modulus Determination by Flexural Resonance.* Ceram. Eng. Sci. Proc. **13 [9–10],** 1094–1102 (1992)

[4.10] J. Krautkrämer and H. Krautkrämer: *Ultrasonic Testing of Materials.* (Springer-Verlag Berlin, 1990)

[4.11] J.N. Adami, D. Bolsch, J. Bressers, E. Fenske, and M. Steen: *Uniaxial Tension and Cyclic Tension-Compression Testing of Ceramics.* J. Eur. Ceram. Soc. **7 [4],** 227–236 (1991)

[4.12] G. Higgins, B. O'Brien, and A. Sun Wai: *Mechanical Testing of Ceramics and Ceramic Matrix Composites: An Overview.* Key Eng. Mater. **72–74,** 551–568 (1992)

[4.13] D.C. Larsen and S.L. Stuchly: *The Mechanical Evaluation of Ceramic Fiber Composites.* In: *Fiber Reinforced Ceramic Composites.* K.S. Mazdiyasni (ed.). (Noyes Park Ridge NJ, 1990)

[4.14] R. Morrell: *Mechanical Test Methods for Ceramic Matrix Composites.* Brit. Ceram. Trans. **94 [1],** 1–9 (1995)

[4.15] H.P. Kirchner: *Strengthening of Ceramics.* (Marcel Dekker New York, 1979)

[4.16] J. Mencik: *Glass Science and Technology 12: Strength and Fracture of Glass and Ceramics*. (Elsevier Amsterdam Holland, 1992)

[4.17] G.D. Quinn: *Strength and Proof Testing.* In: *Engineered Materials Handbook,* Vol.4: *Ceramics and Glasses*. (ASM International Materials Park OH, 1991), pp. 585–598

[4.18] S.M. Wiederhorn: *Fracture of Ceramics.* In: *Mechanical and Thermal Properties of Ceramics.* J.B. Wachtman, Jr. (ed.). (National Bureau of Standards Special Publication 303, May 1969), pp. 217–240

[4.19] A.A. Griffith: *The Phenomena of Rupture and Flow in Solids.* Philos. Trans. R. Soc. (London) **A 221,** 163–198 (1920)

[4.20] G.C. Eckold: *Structural Design with Advanced Ceramics.* In: *Ceramic Matrix Composites.* R. Warren (ed.). (Blackie and Son Glasgow Scotland, 1992), pp. 115–144

[4.21] W. Weibull: *Statistical Distribution Function of Wide Applicability.* J. Appl. Mech. **18 [3],** 293–297 (1951)

[4.22] S. Van der Zwaag: *The Concept of Filament Strength and the Weibull Modulus.* J. Test. Eval. **17 [5],** 292–298 (1989)

[4.23] D.G. Raheja: *Assurance Technologies.* (McGraw-Hill New York, 1991)

[4.24] S.L. Fok and J. Smart: *The Accuracy of Failure Predictions Based on Weibull Statistics.* J. Eur. Ceram. Soc. **15 [9],** 905–908 (1995)

[4.25] M. Sutcu: *Weibull Statistics Applied to Fiber Failure in Ceramic Composites and Work of Fracture.* Acta Metall. **37 [2],** 651–661 (1989)

[4.26] R.W. Davidge: *Engineering Performance Prediction for Ceramics.* Mater. Sci. Tech. **2 [9],** 902–909 (1986)

[4.27] J. Lamon: *Statistical Approaches to Failure for Ceramic Reliability Assesment.* J. Am. Ceram. Soc. **71 [2],** 106–112 (1988)

[4.28] S. Ito and H. Okuda: *Evaluation of Some Ceramics: Silicon Nitride, Silicon Carbide, and Zirconia.* In: *Fine Ceramics.* S.Saito (ed.). (Elsevier Essex England, 1985), pp. 218–226

[4.29] R.W. Davidge: *Mechanical Behavior of Ceramics.* (Cambridge University Press Cambridge England, 1979)

[4.30] A.G. Evans and R.W. Davidge: *Strength and Fracture of Stoichiometric Polycrystalline* UO_2. J. Nucl. Mater. **33 [3],** 249–260 (1969)

[4.31] H. Salmang and H. Scholze: *Keramik.* (Springer-Verlag Berlin, 1982)

[4.32] T.A. Michalske and B.C. Bunker: *Slow Fracture Model Based on Strained Silicate Structures* J. Appl. Phys. **56 [10],** 2686–2693 (1984)

[4.33] R.N. Katz, G.D. Quinn, and E.M. Lenoe: *High Temperature Static Fatigue in Ceramics.* In: *Fatigue, Environment, and High Temperature Effects.* (Plenum Press New York, 1983), pp. 221–230

[4.34] F.F. Lange, B.I. Davis, and M.G. Metcalf: *Strengthening of Polyphase Materials Through Oxidation.* J. Mater. Sci. **18 [5],** 1497–1505 (1983)

[4.35] J.L. Smialek and N.S. Jacobson: *Mechanism of Strength Degradation for Hot Corrosion of* α*-SiC.* J. Am. Ceram. Soc. **69 [101],** 741–752 (1986)

[4.36] N. Claussen: *Fracture Toughness of* Al_2O_3 *with an Unstabilized* ZrO_2 *Dispersed Phase.* J. Am. Ceram. Soc. **59 [1–2],** 49–51 (1976)

[4.37] N. Claussen: *Stress-Induced Transformations of Tetragonal* ZrO_2 *Particles in Ceramic Matrices.* J. Am. Ceram. Soc. **61 [1–2],** 85–86 (1978)

[4.38] D.J. Green, R.H.J. Hannink, and M.D. Swain: *Transformation Toughening of Ceramics*. (CRC Press Boca Raton FL, 1989)

[4.39] D.B. Marshall and J.E. Ritter: *Reliability of Advanced Structural Ceramics and Ceramic Matrix Composites–A Review*. Am. Ceram. Soc. Bull. **66 [2]**, 309–17 (1987)

[4.40] R.A. Cutler, R.J. Mayhew, K.M. Prettyman, and A.V. Virkar: *High Toughness Ce-TZP/Al2O3 Ceramics with Improved Hardness and Strength*" J. Am. Ceram. Soc. **74 [1]**, 179–186 (1991)

[4.41] Y.G. Gogotsi: *Review: Particulate Silicon Nitride Based Composites*. J. Mater. Sci. **29 [10]**, 2541–2556 (1994)

[4.42] R.K. Bordia and R. Raj: *Analysis of Sintering of a Composite with a Glass or Ceramic Matrix*. J. Am. Ceram. Soc. **69 [3]**, C-55–57 (1986)

[4.43] K. Niihara, A. Nakahira, T. Uchiyama, and T. Hirai: *High-Temperature Mechanical Properties of Al_2O_3–SiC Composites*. In: *Fracture Mechanics of Ceramics*, Vol.7. R.C. Bradt, A.G. Evans, D.P.H. Hasselman, and F.F. Lange (eds.). (Plenum Press New York, 1986), pp. 103–106

[4.44] T. Zambetakis, J.L. Guille, B. Willer, and M. Daire: *Mechanical Properties of Pressure-Sintered Al_2O_3–ZrC Composites*. J. Mater. Sci. **22**, 1135–1140 (1987)

[4.45] S.J. Burden: *Ceramic Cutting Tools*. Ceram. Eng. Sci. Proc. **3 [7–8]**, 351–359 (1982)

[4.46] D. Baril and M.K. Jain: *Evaluation of SiC Platelets as a Reinforcement for Oxide Matrix Composites*. Ceram. Eng. Sci. Proc. **12 [7–8]**, 1175–1192 (1991)

[4.47] G.H. Beall: *Design of Glass–Ceramics*. Rev. Sol. State Sci. **3 [3–4]**, 333–354 (1989)

[4.48] R.R. Tummala and A.L. Friedberg: *Strength of Glass–Crystal Composites*. J. Am. Ceram. Soc. **52 [4]**, 228–229 (1969)

[4.49] W.J. Frey and J.D. Mackenzie: *Mechanical Properties of Selected Glass–Crystal Composites*. J. Mater. Sci. **2 [1]**, 124–130 (1967)

[4.50] J. Homeny and W.L. Vaughn: *Whisker-Reinforced Ceramic Matrix Composites*. MRS Bull. **12 [7]**, 66–71 (1987)

[4.51] K. Xia and T.G. Langdon: *Review: The Toughening and Strengthening of Ceramic Materials Through Discontinuous Reinforcement*. J. Mater. Sci. **29 [20]**, 5219–5231 (1994)

[4.52] R. Warren: *Fundamental Aspects of the Properties of Ceramic-Matrix Composites*. In: *Ceramic Matrix Composites*. R. Warren (ed.). (Blackie and Son London, 1992), pp. 64–111

[4.53] K.K. Chawla: *Composite Materials*. (Springer-Verlag New York, 1987)

[4.54] B. Budiansky, J.W. Hutchinson, and A. Evans: *Matrix Fracture in Fiber-Reinforced Ceramics*. J. Mech. Phys. Solids. **34 [2]**, 167–189 (1986)

[4.55] W.B. Hillig: *Strength and Toughness of Ceramic Matrix Composites*. Ann. Rev. Mater. Sci. **17**, 341–383 (1987)

[4.56] J. Aveston, G.A. Cooper, and A. Kelly: *Single and Multiple Fracture.* In: *Proceedings of the Conference on the Properties of Fiber Composites.* (IPC Science & Technology Press Surrey England, 1971), pp. 15–26

[4.57] D.B. Marshall, B.N. Cox, and A.G. Evans: *The Mechanics of Matrix Cracking in Brittle-Matrix Fiber Composites.* Acta Metall. **33 [11],** 2013–2021 (1985)

[4.58] R.W. Davidge and J.J.R. Davies: *Ceramic Matrix Fiber Composites: Mechanical Testing and Performance.* Int. J. High Tech. Ceramics. **4 [2–4],** 341–358 (1988)

[4.59] M. Sutcu: *Statistical Fiber Failure and Single Crack Behavior in Uniaxially Reinforced Ceramic Composites.* J. Mater. Sci. **23 [3],** 928–933 (1988)

[4.60] M.D. Thouless and A.G. Evans: *Effects of Pull-Out on the Mechanical Properties of Ceramic Matrix Composites.* Acta Metall. **36 [3],** 517–522 (1988)

[4.61] M.D. Thouless, O. Sbaizero, L.S. Sigl, and A.G. Evans: *Effect of Interface Mechanical Properties on Pullout in a SiC-Fiber-Reinforced Lithium Aluminum Silicate Glass Ceramic.* J. Am. Ceram. Soc. **72 [4],** 525–532 (1989)

[4.62] A.G. Evans and D.B. Marshall: *The Mechanical Behavior of Ceramic Matrix Composites.* Acta Metall. **37 [10],** 2567–2583 (1989)

[4.63] K.M. Knowles and X.F. Yang: *Mathematical Modeling of the Strength and Toughness of Unidirectional Fiber-Reinforced Ceramics.* Ceram. Eng. Sci. Proc. **12 [7–8],** 1375–1388 (1991)

[4.64] S. Kimura and E. Yasuda: *Fabrication and Properties of Carbon/Carbon Composites.* In: *Fine Ceramics*. S. Saito (ed.). (Elsevier Essex England, 1985), pp. 205–211

[4.65] L.L. Hench: *Bioceramics: From Concept to Clinic.* J. Am. Ceram. Soc. **74 [7],** 1487–1510 (1991)

[4.66] E. Fitzer and W. Huttner: *Structure and Strength of Carbon–Carbon Composites.* J. App. Phys. D **14 [3],** 347 (1981)

[4.67] S. Takano, T. Kinjo, T. Uruno, T. Tlomak, and C.P. Ju: *Investigation of Process-Structure Performance Relationship of Unidirectionally Reinforced Carbon–Carbon Composites.* Ceram. Eng. Sci. Proc. **12 [9–10],** 1914–1930 (1991)

[4.68] G. Ziegler and W. Huttner: *Engineering Properties of Carbon–Carbon and Ceramic-Matrix Composites.* In: *Engineered Materials Handbook,* Vol.4: *Ceramics and Glasses*. (ASM International Materials Park OH, 1991) p. 836

[4.69] R. Naslain: *Fibrous Ceramic–Ceramic Composites, Processing and Properties.* In: *Thirteenth International Conference on Science of Ceramics.* P. Odier, F. Cabannes, and B. Cales. (Les Editions de Physique Cedex France, 1986), pp. C1-703–715

[4.70] P. Barron-Antolin, G.H. Schiroky, and C.A. Andersen: *Properties of Fiber-Reinforced Alumina Matrix Composites.* Ceram. Eng. Sci. Proc. **9 [7–8],** 759–766 (1988)

[4.71] R.N. Singh and A.R. Gaddipati: *Mechanical Properties of a Uniaxially Reinforced Mullite–Silicon Carbide Composite.* J. Am. Ceram. Soc. **71 [2],** C-100–103 (1988)

[4.72] D.P. Stinton, A.J. Caputo, and R.A. Lowden: *Synthesis of Fiber Reinforced SiC Composites by Chemical Vapor Infiltration.* Am. Ceram. Soc. Bull. **65 [2],** 347–350 (1986)

[4.73] E. Fitzer, D. Hegen, and H. Strohmeier: *Possibility of Gas Phase Impregnation with Silicon Carbide.* Rev. Int. Hautes Temp. Refract. **17 [1],** 23–32 (1980)

[4.74] B.F. Sorensen and R. Talreja: *Toughness of Damage Tolerant Continuous Fiber Reinforced Ceramic Matrix Composites.* J. Eur. Ceram. Soc. **15 [1],** 1047–1059 (1995)

[4.75] S. Freiman: *Brittle Fracture Behavior of Ceramics.* Am. Ceram. Soc. Bull. **67 [2],** 392–402 (1988)

[4.76] H.L. Ewalds and R.J.H. Wanhill: *Fracture Mechanics.* (Edward Arnold London and Delftse Uitgevers Maatschappij b.v. Delft Netherlands, 1985)

[4.77] G. Vekinis, M.F. Ashby, and P.W.R. Beaumont: *R-Curve Behavior of* Al_2O_3 *Ceramics.* Acta Metall. **38 [6],** 1151–1162 (1990)

[4.78] S.D. Conzone, W.R. Blumenthal, and J.R. Varner: *Fracture Toughness of* TiB_2 *and* B_4C *Using the Single Edge Preckracked Beam, Indentation Strength, Chevron Notched Beam, and Indentation Strength Methods.* J. Am. Ceram. Soc. **78 [8],** 2187–2192 (1995)

[4.79] K.E. Amin: *Toughness, Hardness, and Wear.* In: *Engineered Materials Handbook,* Vol.4: *Ceramics and Glasses*. (ASM International Materials Park OH, 1991), pp. 599–609

[4.80] L.M. Barker: *Short Rod* K_{Ic} *Measurements of* Al_2O_3. In: *Fracture Mechanics of Ceramics,* Vol.3. (R.C. Bradt, D.P.H. Hasselman, and F.F. Lange (eds.). (Plenum Press New York, 1978), pp. 483–494

[4.81] K.P.R. Reddy, E.H. Fontana, and J.D. Helfinstine: *Fracture Toughness Measurement of Glass and Ceramic Materials Using Chevron-Notched Specimens.* J. Am. Ceram. Soc. **71 [6],** C-310–313 (1988)

[4.82] D. Munz, R.T. Bubsey, and J.L. Shannon, Jr.: *Fracture Toughness Determination of* Al_2O_3 *Using Four-Point-Bend Specimens with Straight-Through and Chevron Notches.* J. Am. Ceram. Soc. **63 [5–6],** 300–305 (1980)

[4.83] C. Rief and K. Kromp: *Fracture Toughness Testing.* Int. J. High. Tech. Ceram. **4 [2–4],** 301–317 (1988)

[4.84] G. Orange, H. Tanaka, and G. Fantozzi: *Fracture Toughness of Pressureless Sintered Silicon Carbide: A Comparison of* K_{Ic} *Measurement Methods.* Ceram. Int. **13 [3],** 159–165 (1987)

[4.85] G.R. Anstis, P. Chantikul, B.R. Lawn, and D.B. Marshall: *A Critical Evaluation of Indentation Techniques for Measuring Fracture Toughness:I, Direct Crack Measurement.* J. Am. Ceram. Soc. **64 [9],** 533–538 (1981)

[4.86] C.B. Ponton and R.D. Rawlings: *Vickers Indentation Fracture Toughness Part 1, Review of Literature and Formulation of Standardized Indentation Toughness Equations.* Mater. Sci. Tech. **5 [9],** 865–872 (1989)

[4.87] R.F. Krause, Jr: *Rising Fracture Toughness from the Bending Strengths of Indented Alumina Beams* J. Am. Ceram. Soc. **71 [5],** 338–343 (1988)

[4.88] S. Srinivasan and R.D. Scattergood: *R-Curve Measurements in PSZ Ceramics.* J. Mater. Res. **5 [7],** 1490–1495 (1990)

[4.89] R.M. Anderson and L.M. Braun: *Technique for the R-Curve Determination of Y-TZP Using Indentation-Produced Flaws.* J. Am. Ceram. Soc. **73 [10],** 3059–3062 (1990)

[4.90] P. Chantikul, G.R. Anstis, B.R. Lawn, and D.B. Marshall: *A Critical Evaluation of Indentation Techniques for Measuring Fracture Toughness:II, Strength Method.* J. Am. Ceram. Soc. **64 [9],** 539–543 (1981)

[4.91] M.V. Swain and N. Claussen: *Comparison of K_{Ic} Values for Al_2O_3–ZrO_2 Composites Obtained from Notched-Beam and Indentation Strength Techniques.* J. Am. Ceram. Soc. **66 [2],** C-27–29 (1983)

[4.92] R.von Mises: *Mechanik der Festen Korper im Plastich Deformablen Zustand.* Nachr. Ges. Wiss. Göttingen **582,** (1913)

[4.93] E.A. Barringer and H.K. Bowen: *Ceramic Powder Processing.* Ceram. Eng. Sci. Proc. **5 [5–6],** 285–297 (1984)

[4.94] G.C. Garvie, R.H.J. Hannink, and R.T. Pascoe: *Ceramic Steel?* Nature, London, **258 [5537],** 703–704 (1975)

[4.95] W.R. deBoskey and H. Hahn: *Opaque Lightweight Armor–Final Report.* (Defense Technical Information Center Report AD822526, November 1967)

[4.96] N. Tamari, T. Ogura, M. Kinoshita, and Y. Toibana: *Fabrication of SiC Whisker–Si_3N_4 Composite Materials and their Physical Properties.* Bulletin of the Government Industrial Research Institute Osaka Japan, **33 [2],** 129–134 (1982)

[4.97] R.A.J. Sambell, D.H. Bowen, and D.C. Phillips: *Carbon Fiber Composites with Ceramic and Glass Matrices, Part 1: Discontinuous Fibers.* J. Mater. Sci., **7 [6],** 663–673 (1972)

[4.98] E. Fitzer: *Fiber Reinforced Ceramic Composites. Whisker and Fiber Toughened Ceramics.* R.A. Bradley, D.E. Clark, D.C. Larsen, and J.O. Stiegler (eds.). (The American Society for Metals Metals Park OH, 1989), pp. 165–192

[4.99] R.T. Bhatt: *The Properties of Silicon Carbide Fiber Reinforced Silicon Nitride Composites.* In: *Whisker and Fiber Toughened Ceramics*. R.A.

Bradley, D.E. Clark, D.C. Larsen, and J.O. Stiegler (eds.). (The American Society for Metals Metals Park OH, 1989), pp. 199–207

[4.100] R.R. Wahi and B. Ilschner: *Fracture Behavior of Composites Based on Al_2O_3–TiC.* J. Mater. Sci. **15 [4],** 875–885 (1980)

[4.101] T.I. Mah and M.G. Mendiratta: *Fracture Toughness and Strength of Si_3N_4–TiC Composites.* Am. Ceram. Soc. Bull. **60 [11],** 1229–1240 (1981)

[4.102] S.T. Buljan, J.G. Baldoni, and M.L. Huckabee: *Si_3N_4–SiC Composites.* Am. Ceram. Soc. Bull. **66 [2],** 347–352 (1987)

[4.103] L. Sigl, P. Mataga, B.J. Dalgleish, R.M. McMeeking, and A.G. Evans: *On the Toughness of Brittle Materials Reinforced with a Ductile Phase.* Acta Metall. **36 [4],** 945–953 (1988)

[4.104] F. Erdogan and P.F. Joseph: *Toughening of Ceramics Through Crack Bridging by Ductile Particles.* J. Am. Ceram. Soc. **72 [2],** 262–270 (1989)

[4.105] R.H.J. Hannink and M.V. Swain: *Progress in Transformation Toughening of Ceramics.* Ann. Rev. Mater. Sci. **24,** 359–408 (1994)

[4.106] Y. Ikuma, W. Komatsu, and S. Yaegashi: *ZrO_2-Toughened MgO and Critical Factors in Toughening Ceramic Materials by Incorporating Zirconia.* J. Mater. Sci. Lett. **4 [1],** 63–66 (1985)

[4.107] M. Ruhle, N. Claussen, and A.H. Heuer: *Transformation and Microcrack Toughening as Complementary Processes in ZrO_2-Toughened Al_2O_3.* J. Am. Ceram. Soc. **69 [3],** 195–197 (1986)

[4.108] J. Wang and R. Stevens: *Review-Zirconia Toughened Alumina (ZTA) Ceramics.* J. Mater. Sci. **24 [10],** 3421–3440 (1990)

[4.109] J.S. Moya and M.S. Osendi: *Microstructure and Mechanical Properties of Mullite/ZrO_2 Composites.* J. Mater. Sci. **19 [9],** 2909–2914 (1984)

[4.110] T. Watanabe and K. Shobu: *Mechanical Properties of Hot Pressed TiB_2–ZrO_2 Composites.* J. Am. Ceram. Soc. **68 [2],** C-34–36 (1985)

[4.111] P.F. Becher and T.N. Tiegs: *Toughening Behavior Involving Multiple Mechanisms: Whisker Reinforcement and Zirconia Toughening.* J. Am. Ceram. Soc. **70 [9],** 651–654 (1987)

[4.112] A.C. Solomah, W. Reichert, V. Rondinella, L. Esposito, and E. Toscano: *Mechanical Properties, Thermal Shock Resistance, and Thermal Stability of Zirconia-Toughened Alumina-10 vol% Silicon Carbide Whisker–Ceramic Matrix Composite.* J. Am. Ceram. Soc. **73 [3],** 740–743 (1990)

[4.113] F.F. Lange: *Transformation Toughening Part 5: Effect of Temperature and Alloy on Fracture Toughness.* J. Mater. Sci. **17 [1],** 255–262 (1982)

[4.114] B. Budianski, J.W. Hutchinson, and J.C. Lambropoulos: *Continuum Theory of Dilatant Transformation Toughening in Ceramics.* Int. J. Sol. Struct. **19 [4],** 337–355 (1983)

[4.115] W. Pompe, H.A. Bahr, G. Gille, and W. Kreher: *Increased Fracture Toughness of Brittle Materials by Microcracking in an Energy*

Dissipative Zone at the Crack Tip. J. Mater. Sci. Lett. **13 [12],** 2720–2723 (1978)

[4.116] K.T. Faber: *Microcracking Contributions to the Toughness of ZrO_2-Based Ceramics.* In: *Advances in Ceramics,* Vol.12: *Science and Technology of Zirconia II.* N. Claussen, M. Ruhle, and A.H. Heuer (eds.). (The American Ceramic Society Columbus OH, 1984), pp. 293–305

[4.117] A.G. Evans and K.T. Faber: *Crack-Growth Resistance of Microcracking Brittle Materials.* J. Am. Ceram. Soc. **67 [4],** 255–260 (1984)

[4.118] M. Ruhle, A.G. Evans, R.M. McMeeking, P.G. Charalambides, and J.W. Hutchinson: *Microcrack Toughening in Alumina/Zirconia.* Acta Metall. **35 [11],** 2701–2710 (1987)

[4.119] A.G. Evans: *Microfracture from Thermal Expansion Anisotropy: I. Single Phase Systems.* Acta Metall. **26 [12],** 1845–1853 (1978)

[4.120] B.L. Karihaloo: *Contribution of $t \rightarrow m$ Phase Transformation to the Toughening of ZTA.* In: *Proceedings of the 11th Riso International Symposium on Metallurgy and Materials Science.* J.J. Bentzen, J.B.B. Sorensen, N. Christiansen, A. Horsewell, and B. Ralph (eds.). (Riso National Laboratory Roskilde Denmark, 1990), pp. 359–364

[4.121] A.V. Virkar and R.L.K. Matsumoto: *Ferroelastic Domain Switching as a Toughening Mechanism in Tetragonal Zirconia.* J. Am. Ceram. Soc. **69 [10],** C-224–226 (1986)

[4.122] B.S. Li, Y.S. Cheng, K.J. Bowman, and I.W. Chen: *Domain Switching as a Toughening Mechanism in Tetragonal Zirconia.* J. Am. Ceram. Soc. **71 [7],** C-362–364 (1988)

[4.123] G.V. Srinivasan, J.F. Jue, S.Y. Kuo, and A.V. Virkar: *Ferroelastic Domain Switching in Polydomain Tetragonal Zirconia Single Crystals.* J. Am. Ceram. Soc. **72 [11],** 2098–2103 (1989)

[4.124] A.V. Virkar and R.L.K. Matsumoto: *Toughening Mechanism in Tetragonal Zirconia Polycrystal (TZP) Ceramics.* In: *Advances in Ceramics* Vol.24 B: *Science and Technology of Zirconia III.* S. Somiya, N. Yamamoto, and H. Yanagida (eds.). (The American Ceramic Society Westerville OH, 1988), pp. 653–662

[4.125] F.F. Lange: *The Interaction of a Crack Front With a Second Phase Dispersion.* Phil. Mag. **22 [179],** 983–992 (1970)

[4.126] D.J. Green: *Fracture Toughness Predictions for Crack Bowing in Brittle Particulate Composites.* J. Am. Ceram. Soc. **66 [1],** C-4–5 (1983)

[4.127] K.T. Faber and A.G. Evans: *Crack Deflection Processes-I.Theory.* Acta Metall. **31 [4],** 565–576 (1983)

[4.128] H. Liu, K.L. Weisskopf, and G. Petzow: *Crack Deflection Process for Hot Pressed Whisker Reinforced Ceramic Composites.* J. Am. Ceram. Soc. **72 [4],** 559–563 (1989)

[4.129] D.H. Carter and G.F. Hurley: *Crack Deflection as a Toughening Mechanism in SiC–Whisker-Reinforced $MoSi_2$.* J. Am. Ceram. Soc. **70 [4],** C-79–81 (1987)

[4.130] K.T. Faber and A.G. Evans: *Intergranular Crack Deflection Toughening in Silicon Carbide.* J. Am. Ceram. Soc. **66 [6],** C-94–96 (1983)

[4.131] P.F. Becher, C.H. Hsueh, P. Angelini, and T.N. Tiegs: *Toughening Behavior in Whisker Reinforced Ceramic Matrix Composites.* J. Am. Ceram. Soc. **71 [12],** 1050–1061 (1988)

[4.132] M. Bengisu: *Densification and Mechanical Properties of Whisker and/or Zirconia Toughened Alumina, Effect of Shock Treatment and Consolidation Method.* Ph.D. Thesis. (New Mexico Institute of Mining and Technology Socorro NM, 1992)

[4.133] J.K. Wells and P.W.R. Beaumont: *The Toughness of a Composite Containing Short Brittle Fibers.* J. Mater. Sci. **23 [4],** 1274–1278 (1988)

[4.134] M.C. Shaw and K.T. Faber: *Temperature Dependent Toughening in Whisker Reinforced Ceramics.* In: *Materials Science Research,* Vol.21: *Ceramic Microstructures '86, Role of Interfaces.* J.A. Pask and A.G. Evans (eds.). (Plenum Press New York, 1987), pp. 775–794

[4.135] S. Iio, M. Watanabe, M. Matsubara, and Y. Matsuo: *Mechanical Properties of Alumina/Silicon Carbide Whisker Composites.* J. Am. Ceram. Soc. **72 [10],** 1880–1884 (1989)

[4.136] R. Hayami, K. Ueno, I. Kondou, N. Tamari, and O. Kamigato: *Si_3N_4–SiC Whisker Composite Material.* In: *Materials Science Research,* Vol.20: *Tailoring Multiphase and Composite Ceramics.* R.E. Tressler, G.L. Messing, C.G. Pantano, and R.E. Newnham (eds.). (Plenum Press New York, 1988), pp. 663–674

[4.137] T. Kandori, Y. Ukyo, and S. Wada: *Directly HIP'ed SiC Whisker Reinforced Silicon Nitrides.* In: *Whisker and Fiber Toughened Ceramics.* R.A. Bradley, D.E. Clark, D.C. Larsen, and J.O. Stiegler (eds.). (The American Society for Metals International Metals Park OH, 1989), pp. 125–129

[4.138] T.N. Tiegs: *Properties of SiC Whisker Reinforced Oxide Matrix Composites.* In: *Whisker and Fiber Toughened Ceramics.* R.A. Bradley, D.E. Clark, D.C. Larsen, and J.O. Stiegler (eds.). (The American Society for Metals Metals Park OH, 1989), pp. 105–108

[4.139] S.K. Douglas, P.W.R. Beaumont, and M.F. Ashby: *A Model for the Toughness of Epoxy–Rubber Particulate Composites.* J. Mater. Sci. **15 [5],** 1109–1123 (1980)

[4.140] R.C. Leuth: *Determination of Fracture Toughness Parameters for Tungsten Carbide–Cobalt Alloys.* In: *Fracture Mechanics of Ceramics,* Vol.2. R.C. Bradt, D.P.H. Hasselman, and F.F. Lange (eds.). (Plenum Press New York, 1974), pp. 791–806

[4.141] V.D. Krstic, P.S. Nicholson, and R.G. Hoagland: *Toughening of Glasses by Metallic Particles.* J. Am. Ceram. Soc. **64 [9],** 499–504 (1981)

[4.142] P.L. Swanson, C.J. Fairbanks, B.R. Lawn, Y.W. Mai, and B.J. Hockey: *Crack Interface Grain Bridging as a Fracture Resistant Mechanism in*

Ceramics: I, Experimental Study on Alumina. J. Am. Ceram. Soc. **70 [4],** 279–289 (1987)

[4.143] Y.W. Mai and B.R. Lawn: *Crack Interface Grain Bridging as a Fracture Resistance Mechanism in Ceramics: II, Theoretical Fracture Mechanics Model.* J. Am. Ceram. Soc. **70 [4],** 289–294 (1987)

[4.144] G. Vekinis, M.F. Ashby, and P.W.R. Beaumont: *R-Curve Behavior in* Al_2O_3 *Ceramics.* Acta Metall. Mater. **38 [6],** 1151–1162 (1990)

[4.145] H. Wieninger, K. Kromp, and R.F. Pabst: *Crack Resistance Curves of Alumina and Zirconia at Room Temperature.* J. Mater. Sci. **21 [2],** 411–418 (1986)

[4.146] J.C. Hay and K.W. White: *Grain Bridging Mechanisms in Monolithic Alumina and Spinel.* J. Am. Ceram. Soc. **76 [7],** 1849–1854 (1995)

[4.147] B.R. Lawn, L.M. Braun, S.J. Bennison, and R.E. Cook: *Reply to "Comment on Role of Grain Size in the Strength and R-Curve Properties of Alumina".* J. Am. Ceram. Soc. **76 [7],** 1900–1901 (1993)

[4.148] A.G. Evans and R.M. McMeeking: *On the Toughening of Ceramics by Strong Reinforcements.* Acta Metall. **34 [12],** 2435–2441 (1986)

[4.149] R. Morrell: *Handbook of Properties of Technical and Engineering Ceramics, Part 1.* (Her Majesty's Stationery Office London, 1985)

[4.150] K.T. Faber and A.G. Evans: *Crack Deflection Processes-II.Experiment.* Acta Metall. **31 [4],** 577–584 (1983)

[4.151] G.J. Zhang, X.M. Yue, Z.Z. Jin, and J.Y. Dai: *In-Situ Synthesized* TiB_2 *Toughened SiC.* J. Eur. Ceram. Soc. **16 [4],** 409–412 (1996)

[4.152] Y.G. Gogotsi: *Review: Particulate Silicon Nitride Based Composites.* J. Mater. Sci. **29 [10],** 2541–2556 (1994)

[4.153] M. Bengisu, O.T. Inal, and O. Tosyali: *On Whisker Toughening in Ceramic Materials.* Acta Metall. Mater. **39 [11],** 2509–2517 (1991)

[4.154] D.P.H. Hasselman: *Thermal Stress Resistance of Engineering Ceramics.* Mater. Sci. Eng. **71,** 251–264 (1985)

[4.155] R.C. Hendricks, G. McDonald, and R.L. Mullen: *Residual Stress in Plasma-Sprayed Ceramic Turbine Tip and Gas-Path Seal Specimen.* Ceram. Eng. Sci. Proc. **4 [9–10],** 802–809 (1983)

[4.156] M.M. Schwartz: *Ceramic Joining* (ASM International Materials Park OH, 1990)

[4.157] R. Danzer: *Properties of Ceramics During High Temperature Applications.* Ceram. Forum Int. **70 [6],** 280–286 (1993)

[4.158] F. Sudreau, C. Olagnon, and G. Fantozzi: *Lifetime Prediction of Ceramics: Importance of the Test Method.* Ceram. Int. **20 [2],** 125–135 (1994)

[4.159] W.D. Kingery, H.K. Bowen, and D.R. Uhlmann: *Introduction to Ceramics*. (John Wiley & Sons New York, 1976)

[4.160] F. Mignard, C. Olagnon, and G. Fantozzi: *Acoustic Emission Monitoring of Damage Evaluation in Ceramics Submitted to Thermal Shock.* J. Eur. Ceram. Soc. **15 [7],** 651–653 (1995)

[4.161] O. Sbaizero: Prove di Resistenza Agli Shock Termici. Ceramurgia **24 [4],** 143–150 (1994)

[4.162] W.P. Rogers and A.F. Emery: *Contact Thermal Shock Test of Ceramics.* J. Mater. Sci. **27 [1],** 146–152 (1992)

[4.163] A.G. Evans: *Fatigue in Ceramics.* Int. J. Fract. **16 [6],** 485–498 (1980)

[4.164] S. Horibe: *A New Method for Tension-Compression Fatigue Testing of Ceramic Materials.* J. Mater. Sci. Lett. **9 [7],** 745–747 (1990)

[4.165] F. Guiu, M.J. Reece, and D.A.J. Vaughan: *Cyclic Fatigue of Ceramics.* J. Mater. Sci. **26 [12],** 3275–3286 (1991)

[4.166] C.J. Gilbert, R.H. Dauskart, and R.O. Ritchie: *Behavior of Cyclic Fatigue Cracks in Monolithic Silicon Nitride.* J. Am. Ceram. Soc. **78 [9],** 2291–2300 (1995)

[4.167] D. Lewis and R.W. Rice: *Comparison of Static, Cyclic, and Thermal-Shock Fatigue in Ceramic Composites.* Ceram. Eng. Sci. Proc. **3 [9–10],** 714–721 (1982)

[4.168] L. Ewart and S. Suresh: *Crack Propagation in Ceramics Under Cyclic Loads.* J. Mater. Sci. **22 [4],** 1173–1192 (1987)

[4.169] S. Suresh and J.R. Brockenbrough: *Theory and Experiments of Fracture in Cyclic Compression: Single Phase, Transforming Ceramics, and Ceramic Composites.* Acta Metall. **36 [6],** 1455–1470 (1988)

[4.170] D.L. Lewis III: *Cyclic Mechanical Fatigue in Ceramic–Ceramic Composites, An Update.* Ceram. Eng. Sci. Proc. **4 [9–10],** 874–881 (1983)

[4.171] J.W. Holmes, T. Kotil, and W.T. Foulds: *High Temperature Fatigue of SiC Fiber Reinforced Si_3N_4 Ceramic Composites.* In: *Symposium on High Temperature Composites, Proc. Am. Soc. for Composites*. (Technomic Lancaster PA, 1989)

[4.172] D.C. Phillips: *Long-Fiber Reinforced Ceramics.* In: *Ceramic Matrix Composites.* R. Warren (ed.). (Blackie and Son London, 1992), pp. 167–196

[4.173] G. Choi and S. Horibe: *The Environmental Effect on Cyclic Fatigue Behavior in Ceramic Materials.* J. Mater. Sci. **30 [6],** 1565–1569 (1995)

[4.174] R. Raj: *Fundamental Research in Structural Ceramics for Service near 2000°C.* J. Am. Ceram. Soc. **76 [9],** 2147–2173 (1993)

[4.175] S. Suresh, L.X. Han, and J.J. Petrovic: *Fracture of Si_3N_4–SiC Whisker Composites Under Cyclic Loads.* J. Am. Ceram. Soc. **71 [3],** C-158–161 (1988)

[4.176] D.S. Jacobs and I.W. Chen: *Cyclic Fatigue in Ceramics: A Balance Between Crack Shielding Accumulation and Degredation.* J. Am. Ceram. Soc. **78 [5],** 513–520 (1995)

[4.177] T. Soma, M. Matsuda, M. Matsui, and I. Oda: *Cyclic Fatigue Testing of Ceramic Materials.* Int. J. High Tech. Ceram. **4 [2–4],** 289–299 (1988)

[4.178] H.N. Ko: *Fatigue Strength of Sintered Si_3N_4 Under Rotary Bending.* J. Mater. Sci. Lett. **6 [2],** 175–177 (1987)

[4.179] M. Reece and F. Guiu: *Repeated Indentation Method for Studying Cyclic Fatigue in Ceramics.* J. Am. Ceram. Soc. **73 [4],** 1004–1013 (1990)

[4.180] W.R. Cannon and T.G. Langdon: *Review, Creep of Ceramics, Part 1 Mechanical Characteristics.* J. Mater. Sci. **18 [1],** 1–50 (1983)

[4.181] Y. Maehara and T.G. Langdon: *Review–Superplasticity in Ceramics.* J. Mater. Sci. **25 [5],** 2275–2286 (1990)

[4.182] F. Wakai: *Superplasticity of Ceramics.* Ceram. Int. **17 [3],** 153–63 (1991)

[4.183] W.F. Smith: *Principles of Materials Engineering.* (McGraw-Hill New York, 1986)

[4.184] S. Deng and R. Warren: *Creep Properties of Single Crystal Oxides Evaluated by a Larson–Miller Procedure.* J. Eur. Ceram. Soc. **15 [6],** 513–520 (1995)

[4.185] M.F. Ashby: *A First Report on Deformation-Mechanism Maps.* Acta Metall. **20 [7],** 887–897 (1972)

[4.186] H.J. Frost and M.F. Ashby: *Deformation-Mechanism Maps.* (Pergamon Press Oxford England, 1982)

[4.187] R.A. Verrall, R.J. Fields, and M.F. Ashby: *Deformation Mechanism Maps for LiF and NaCl.* J. Am. Ceram. Soc. **60 [5–6],** 211–216 (1977)

[4.188] J.D. French, J. Zhao, M.P. Harmer, H.M. Chan, and G.A. Miller: *Creep of Duplex Microstructures.* J. Am. Ceram. Soc. **77 [11],** 2857–2865 (1994)

[4.189] S. Suresh and J.R. Brockenbrough: *A Theory for Creep by Interfacial Flaw Growth in Ceramics and Ceramic Composites.* Acta Metall. **38 [1],** 55–68 (1990)

[4.190] M.R. Notis, R.H. Smoak, and V. Krishnamachari: *Interpretation of Hot Pressing Kinetics by Densification Mapping Techniques.* In: *Materials Science Research,* Vol.10: *Sintering and Catalysis*. G.C. Kuczynski (ed.). (Plenum Press New York, 1975), pp. 493–507

[4.191] A.H. Chokshi and J.R. Porter: *Creep Deformation of an Alumina Matrix Composite Reinforced with Silicon Carbide Whiskers.* J. Am. Ceram. Soc. **68 [6],** C-144–145 (1985)

[4.192] A.G. Evans and B.J. Dalgleish: *Some Aspects of the High Temperature Performance of Ceramics and Ceramic Composites.* In: *Superalloys, Supercomposites, and Superceramics*. J.K. Tien and T. Caulfield (eds.). (Academic Press San Diego CA, 1989), pp. 697–720

[4.193] J.M. Birch, B. Wilshire, and D.J. Godfey: *Deformation and Fracture Processes During Creep of Reaction Bonded and Hot Pressed Silicon Nitride*. Proc. Brit. Ceram. Soc. **25,** 141–154 (1978)

[4.194] S.M. Wiederhorn, D.E. Roberts, T.J. Chuang, and L. Chuck: *Damage Enhanced Creep in a Siliconized Silicon Carbide: Phenomenology.* J. Am. Ceram. Soc. **71 [7],** 602–608 (1988)

[4.195] W.J. Kim, J. Wolfenstine, G. Frommeyer, O.A. Ruano, and O.D. Sherby: *Superplastic Behavior of Iron Carbide.* Scr. Metall. **23 [9],** 1515–1520 (1989)

[4.196] I.-W. Chen and L.A. Xue: *Development of Superplastic Structural Ceramics.* J. Am. Ceram. Soc. **73 [9],** 2585–2609 (1990)

[4.197] C.A. Johnson, R.C. Bradt, and J.H. Hoke: *Transformational Plasticity in* Bi_2O_3. J. Am. Ceram. Soc. **58 [1–2],** 37–40 (1975)

[4.198] L.A. Xue and R. Raj: *Superplastic Deformation of Zinc Sulfide Near its Transformation Temperature (1020°C).* J. Am. Ceram. Soc. **72 [10],** 1792–1796 (1989)

[4.199] T.G. Langdon: *Superplastic Ceramics, They're Not A Stretch of the Imagination Anymore.* JOM **42 [7],** 8–13 (1990)

[4.200] S.L. Hwang and I.W. Chen: *Superplastic Forming of SiAlON Ceramics.* J. Am. Ceram. Soc. **77 [10],** 2575–2585 (1994)

[4.201] R.W. Rice: *Micromechanics of Microstructural Aspects of Ceramic Wear.* Ceram. Eng. Sci. Proc. **6 [7–8],** 940–958 (1985)

[4.202] C.A. Brookes and A.K. Parry: *Some Fundamental Aspects of the Mechanical Wear of Hard Ceramic Crystals due to Sliding.* Mater. Sci. Eng. **A105/106,** 143–150 (1988)

[4.203] E.A. Almond: *Overview: Indentation Phenomena and Wear of Surfaces and Edges.* Mater. Sci. Tech. **2 [7],** 641–646 (1986)

[4.204] J.R. Alcock and O.T. Sorensen: *Slurry Abrasion Resistance of Engineering Ceramics.* Br. Ceram. Trans. **95 [1],** 30–34 (1996)

[4.205] H. Kong and M.F. Ashby: *Wear Mechanisms in Brittle Solids.* Acta Metall. **40 [11],** 2907–2920 (1992)

[4.206] D. Holz, R. Janssen, K. Friedrich, and N. Claussen: *Abrasive Wear of Ceramic-Matrix Composites.* J. Eur. Ceram. Soc. **5 [4],** 229–232 (1989)

[4.207] M. Bohmer and E.A. Almond: *Mechanical Properties and Wear Resistance of a Whisker-Reinforced Zirconia-Toughened Alumina.* Mater. Sci. Eng. **A105/106,** 105–116 (1988)

[4.208] F. Thevenot et P. Homerin: *Composites Alumine–Zircone.* In: *Ceramiques Composites a Particules, Cas du Frittage-Reaction.* F. Thevenot (ed.). (Forceram Editions Septima Paris France, 1992), pp. 39–60

[4.209] C.Cm. Wu, R.W. Rice, C.P. Cameron, L.E. Dolhert, J.H. Enloe, and J. Block: *Diamond Pin-On-Disk Wear of* Al_2O_3 *Matrix Composites and Nonoxides.* Ceram. Eng. Sci. Proc.: **12 [7–8],** 1485–1499 (1991)

[4.210] S.W. Lee, S.M. Hsu, and M.C. Shen: *Ceramic Wear Maps: Zirconia.* J. Am. Ceram. Soc. **76 [8],** 1937–1947 (1993)

[4.211] C.M. Wu and R.W. Rice: *Porosity Dependence of Wear and Other Mechanical Properties on Fine-Grain* Al_2O_3 *and* B_4C. Ceram. Eng. Sci. Proc. **6 [7–8],** 995–1011 (1985)

[4.212] C.M. Wu, R.W. Rice, D. Johnson, and B.A. Platt: *Grain Size Dependence of Wear in Ceramics.* Ceram. Eng. Sci. Proc. **6 [7–8],** 1012–1021 (1985)

[4.213] M.G. Gee and E.A. Almond: *Effects of Test Variables in Wear Testing of Ceramics.* Mater. Sci. Tech. **4 [10],** 877–884 (1988)

[4.214] S.M. Wiederhorn: *Erosion of Ceramics.* In: *Proceedings of Corrosion/Erosion of Coal Conversion Systems Materials Conference.* A.V. Levy (ed.). (National Association of Corrosion Engineering Houston TX, 1979), pp. 444–479

[4.215] J.F. Vedder: *Microcraters in Glass and Minerals.* Earth Planet. Sci. Lett. **11 [3],** 291–296 (1971)

[4.216] P.S. Follansbee, G.B. Sinclair, and J.C. Williams: *Modeling of Low Velocity Particulate Erosion in Ductile Materials by Spherical Particles.* Wear **74 [1],** 107–122 (1981–82)

[4.217] K.G. Budinski: *Engineering Materials.* Reston Publishing Reston Virginia, 1983)

[4.218] G.L. Sheldon and I. Finnie: *On Ductile Behavior of Nominally Brittle Materials During Erosive Cutting.* Trans ASME J. Eng. Ind. **88 [4],** 387–392 (1966)

[4.219] L. Murugesh, S. Srinivasan, and R.O. Scattergood: *Models and Material Properties for Erosion of Ceramics.* J. Mater. Eng. **13 [1],** 55–61 (1991)

[4.220] A.G. Evans, M.E. Gulden, and M.E. Rosenblatt: *Impact Damage in Brittle Materials in the Elastic-Plastic Response Regime.* Proc. Roy. Soc. (London) **A 361 [1706],** 343–365 (1978)

[4.221] J.E. Ritter: *Erosion Damage in Structural Ceramics.* Mater. Sci. Eng. **71,** 195–201 (1985)

[4.222] S. Srinivasan and R.O. Scattergood: *Erosion of Mg-PSZ by Solid Particle Impact.* Adv. Ceram. Mater. **3 [4],** 345–352 (1988)

[4.223] J.L. Routbort: *Erosion of Composite Ceramics.* Ceram. Acta **6 [1],** 5–13 (1994)

[4.224] S. Wada: *Effects of Hardness and Fracture Toughness of Target Materials and Impact Particles on Erosion of Ceramic Materials.* Key Eng. Mater. **71,** 51–74 (1992)

[4.225] J.L. Routbort and R.O. Scattergood: *Solid Particle Erosion of Ceramics and Ceramic Composites.* Key Eng. Mater. **71,** 23–50 (1992)

[4.226] S. Bugliosi, D. Cuppini, and G.E. D'Errico: *Come Si Erodono I Ceramici* Riv. Mecc. Oggi **11 [4],** 43–47 (1996)

[4.227] W.J. Tomlinson and S.J. Matthews: *Cavitation Erosion of Structural Ceramics.* Ceram. Intern. **20 [3],** 201–209 (1994)

[4.228] D.D. Pollock: *Physical Properties of Materials for Engineers,* Vol.I (CRC Press Boca Raton FL, 1984)

[4.229] D.R. Gaskell: *Introduction to Metallurgical Thermodynamics.* (McGraw-Hill New York, 1981)

[4.230] W.F. Hammetter: *Thermophysical Properties.* In: *Engineered Materials Handbook,* Vol.4: *Ceramics and Glasses.* (ASM International Materials Park OH, 1991), pp. 610–616

[4.231] W.J. Parker, R.J. Jenkins, C.P. Butler, and G.L. Abbott: *Flash Method of Determining Thermal Diffusivity, Heat Capacity, and Thermal Conductivity.* J. App. Phys. **32 [9],** 1679–1684 (1961)

[4.232] R. Taylor: *Construction of Apparatus for Heat Pulse Thermal Diffusivity Measurements from 300–3000 K.* J. Phys. E: Sci. Instr. **13 [11]**, 1193–1199 (1980)

[4.233] J.C. Anderson, K.D. Leaver, R.D. Rawlings, and J.M. Alexander: *Materials Science*. (Chapman Hall London, 1990)

[4.234] G. Aliprandi and F. Savioli: *Introduzione ai Ceramici Avanzati,* Vol.1. (ENEA Rome, 1989)

[4.235] L.H. Van Vlack: *Physical Ceramics for Engineers.* (Addison-Wesley Reading MA, 1964)

[4.236] J. Francl, W.D. Kingery: *Thermal Conductivity: IX, Experimental Investigation of Effect of Porosity on Thermal Conductivity.* J. Am. Ceram. Soc. **37 [2]**, 99–107 (1954)

[4.237] G.S. Sheffield and J.R. Schorr: *Comparison of Thermal Diffusivity and Thermal Conductivity Methods.* Am. Ceram. Soc. Bull. **70 [1]**, 102–106 (1991)

[4.238] D. Fournier, L. Pottier, A.C. Boccara, G. Savignat, P. Boch, J. Poirier, G. Provost: *Caratterizzazione Non-Distruttiva di Refrattari Tramite Tecniche Fototermiche.* Ceramurgia **26 [2]**, 94–97 (1996)

[4.239] A.W. Sleight: *Isotropic Negative Thermal Expansion.* Ann. Rev. Mater. Sci. **28,** 29–43 (1998)

[4.240] R.J. Beals and R.L. Cook: *Directional Dilatation of Crystal Lattices at Elevated Temperatures.* J. Am. Ceram. Soc. **40 [8]**, 279–284 (1957)

[4.241] J. Hlavac: *The Technology of Glass and Ceramics*. (Elsevier New York, 1983)

[4.242] W.L. Wolfe: *Properties of Optical Materials.* In: *Handbook of Optics.* W.G. Driscoll and W. Vaughan (eds.). (Mc-Graw–Hill New York, 1978)

[4.243] T.D. Taylor: *Structure and Properties of Glasses.* In: *Engineered Materials Handbook,* Vol.4: *Ceramics and Glasses*. (ASM International Materials Park OH, 1991), pp. 564–569

[4.244] M.W. Barsoum: *Fundamentals of Ceramics.* (Mc-Graw–Hill New York, 1997)

[4.245] *Handbook of Optical Constants of Solids.* E.D. Palik (ed.). (Academic Press Orlando FL, 1985)

[4.246] *CRC Handbook of Chemistry and Physics*. R.C. Weast (ed.). (CRC Press Boca Raton FL, 1986)

[4.247] M. Francon: *Isotropic and Anisotropic Media, Application of Anisotropic Materials to Interferometry.* In: *Advanced Optical Techniques*. A.C.S. VanHeel (ed.). (North Holland Amsterdam, 1967), pp. 23–70

[4.248] G.R. Streatfield: *Ceramic Color*. Br. Ceram. Trans. J. **89 [5]**, 177–180 (1990)

[4.249] *Mastering Color, The Kodak Library of Creative Photography*. (Kodak Mitchell Beazley Publishers–Salvat Editores Barcelona, 1985)

[4.250] C. Kittel: *Introduction to Solid State Physics* (John Wiley & Sons New York, 1986)

[4.251] *Solid State Physics Source Book*. S.P. Parker (ed.). (McGraw-Hill New York, 1988)

[4.252] C.C. Wang, S.A. Akbar, W. Chen, and V.D. Patton: *Review: Electrical Properties of High-Temperature Oxides, Borides, Carbides, and Nitrides*. J. Mater. Sci. **30 [7]**, 1627–1641 (1995)

[4.253] R.W. Schwartz: *Electronic and Magnetic Ceramics*. In: *Characterization of Ceramics*. R.E. Loehman (ed.). (Butterworth-Heinemann Stoneham MA, 1993)

[4.254] D.D. Pollock: *Electrical Conduction in Solids*. (American Society for Metals Metals Park OH, 1985)

[4.255] *Ceramic Materials for Electronics*. R.C. Buchanan (ed.). (Marcel Dekker New York, 1986)

[4.256] A.J. Moulson and J.M. Herbert: *Electroceramics*. (Chapman & Hall London, 1990)

[4.257] M. Ura: *Ceramic Substrates*. In: *Fine Ceramics*. S. Saito (ed.). (Elsevier Essex England, 1985), pp. 243–251

[4.258] H.L. Tuller and P.K. Moon: *Fast Ion Conductors: Future Trends*. Mater. Sci. Eng. **B1 [2]**, 171–191 (1988)

[4.259] P. McGeehin and A. Hooper: *Review: Fast Ion Conduction Materials*. J. Mater. Sci. **12 [1]**, 1–27 (1977)

[4.260] M. Nagai: *Ionic Conductors*. In: *Fine Ceramics*. S. Saito (ed.). (Elsevier Essex England, 1985), pp. 297–306

[4.261] K. Onnes: *Report on the Researches Made in the Leiden Cryogenic Laboratory Between the Second and Third International Congress of Refrigeration*. Supp. **34 b** (1913)

[4.262] J.G. Bednorz and K.A. Müller: *Possible High T_c Superconductivity in the Ba–La–Cu–O System*. Z. Phys. B. **64,** 189–193 (1986)

[4.263] J.H. Sharp: *A Review of the Crystal Chemistry of Mixed Oxide Superconductors*. Br. Ceram. Trans. J. **89 [1]**, 1–7 (1990)

[4.264] M.K. Wu, J.R. Ashburn, C.J. Torng, P.H. Hor, R.L. Meng, L. Gao, Z.J. Huang, Y.Q. Wang, and C.W. Chu: *Superconductivity at 93K in a New Mixed Phase Y–Ba–Cu–O Compound System at Ambient Pressure*. Phys. Rev. Lett. **58 [9]**, 908–910 (1987)

[4.265] T.P. Sheahen: *Introduction to High-Temperature Superconductivity*. Plenum Press New York, 1994)

[4.266] J.D. Doss: *Engineer's Guide to High-Temperature Superconductivity*. (John Wiley & Sons New York, 1989)

[4.267] D.M. Kroeger and A. Goyal: *Overview: Critical Current and Microstructure in Oxide Superconductors*. JOM **44 [10]**, 42–47 (1992)

[4.268] J. Bardeen, L.N. Cooper, and J.R. Schrieffer. Phys. Rev. **106,** 162 (1957)
J. Bardeen, L.N. Cooper, and J.R. Schrieffer: *Theory of Superconductivity*. Phys. Rev. **108,** 1175–1204 (1957)

[4.269] P.B. Allen: *Status of Theory for Superconductivity in the High-Tc Cuprates*. In: *High Temperature Superconductors II*. D.W. Capone II,

W.H. Butler, B. Batlogg, and C.W. Chu (eds.). (Materials Research Society Pittsburgh PA, 1988), pp. 3–5

[4.270] D.O. Welch: *Theoretical Questions Raised by High T_c.* JOM **40 [1],** 8–9 (1988)

[4.271] A.M. Wolsky, R.F. Giese, and E.J. Daniels: *The New Superconductors: Prospects for Applications.* Sci. American **Feb.,** 61–69 (1989)

[4.272] D. Larbalestier, G. Fisk, B. Montgomery, and D. Hawkswoth. *High-Field Superconductivity.* Physics Today **March,** 24–33 (1986)

[4.273] J. Narayan: *Microstructure and Properties of High T_c Superconductors.* JOM **41 [1],** 18–23 (1989)

[4.274] R. Ramesh, A. Inam, T. Sands, and C.T. Rogers: *Thin Film Y–Ba–Cu–O High T_c Superconductors: Structure-Property Relationships.* Mater. Sci. Eng. **B14 [2],** 188–213 (1992)

[4.275] V.Z. Kresin and S.A. Wolf: *Fundamentals of Superconductivity.* (Plenum Press New York, 1990)

[4.276] H.K. Liu, S.X. Dou, A.J. Bourdillon, and C.C. Sorrell: *A Comparison of the Stability of $Bi_2Sr_2CaCu_2O_{8+y}$ with $YBa_2Cu_3O_{6.5+y}$ in Various Solutions.* Supercond. Sci. Technol. **1,** 194–197 (1988)

[4.277] Z.Z. Sheng and A.M. Hermann: *Superconductivity in the Rare-Earth Free Tl–Ba–Cu–O System Above Liquid Nitrogen Temperature.* Nature (London) **332 [6159],** 138–141 (1988)

[4.278] U. Chowdry and A.W. Sleight: *Ceramic Substrates for Microelectronic Packaging.* Ann. Rev. Mater. Sci. **17,** 323–340 (1987)

[4.279] L. Solymar and D. Walsh: *Lectures on the Electrical Properties of Materials.* (Oxford University Press Oxford England, 1984)

[4.280] W. Heywang and H. Thormann: *Tailoring of Piezoelectric Ceramics.* Ann. Rev. Mater. Sci. **14,** 27–47 (1987)

[4.281] S. Nanamatsu, M. Kimura, K. Doe, S. Matsushita, and N. Yamada: *A New Ferroelectric: $La_2Ti_2O_7$.* Ferroelectrics **8,** 511–513 (1974)

[4.282] M. Serridge and T.R. Licht: *Piezoelectric Accelerometer and Vibration Preamplifier Handbook.* (Brüel & Kjær Nærum Denmark, 1987)

[4.283] T. Mitsui, I. Tatsuzaki, and E. Nakamura: *An Introduction to the Physics of Ferroelectrics.* (Gordon and Breach New York, 1976)

[4.284] M.E. Lines and A.M. Glass: *Principles and Applications of Ferroelectrics and Related Materials.* (Oxford University Press Oxford England, 1979)

[4.285] B.M. Park and S.J. Chung: *Optical, Electron Microscopic, and X-Ray Topographic Studies of Ferroic Domains in Barium Titanate Crystals Grown from High Temperature Solution.* J. Am. Ceram. Soc. **77 [12],** 3193–3201 (1994)

[4.286] Y. Furuhata and G. Toda: *Ferroelectric and Electrooptic Materials.* In: *Fine Ceramics.* S.Saito (ed.). (Elsevier Essex England, 1985), pp. 261–275

[4.287] J.C. Burfoot and G.W. Taylor: *Polar Dielectrics.* (University of California Berkeley CA, 1979)

[4.288] G.H. Haertling: *Electro-Optic Ceramics and Devices.* In: *Electronic Ceramics.* L.M. Levinson (ed.). (Marcel Dekker New York, 1988), pp. 371–492

[4.289] F.E. Luborsky, J.D. Livingston, and G.Y. Chin: *Magnetic Properties of Metals and Alloys.* In: *Physical Metallurgy Part II.* R.W. Cahn and P. Haasen (eds.). (North Holland Amsterdam Holland, 1983)

[4.290] *Physical Metallurgy.* R.W. Cahn and P. Haasen (eds.). (North Holland Amsterdam Holland, 1983)

[4.291] R.W. Hanks: *Materials Engineering Science, An Introduction.* Harcourt Brace & World New York, 1970)

[4.292] L.M. Sheppard: *Corrosion-Resistant Ceramics for Severe Environments.* Am. Ceram. Soc. Bull. **70 [7],** 1146–1166 (1991)

[4.293] Y. Hara: *Application of Fine Ceramics in Industrial Fields.* Corros. Eng. **38 [7],** 527–537 (1989)

[4.294] W.B. White: *Theory of Corrosion of Glass and Ceramics.* In: *Corrosion of Glass, Ceramics and Ceramic Superconductors.* D.E. Clark and B.K. Zoitos (eds.). (Noyes Park Ridge NJ, 1992), pp. 2–50

[4.295] R. Divakar, S.G. Seshadri, and M. Srinivasan: *Electromechanical Techniques for Corrosion Rate Determination in Ceramics.* J. Am. Ceram. Soc. **72 [5],** 780–784 (1989)

[4.296] J.P. Day: *A Study of the Chemical Reactivity in Ceramic Heat Exchangers.* Trans. ASME **A 101 [2],** 270–274 (1979)

[4.297] R.H. Jones, C.H. Henager, Jr., P.P. Trzaskoma, N.S. Stoloff, T.P. Moffat, and B.D. Lichter: *Environmental Effects on Advanced Materials.* JOM **40 [12],** 18–30 (1988)

[4.298] T.A. Michalske, B.C. Bunker, and S.W. Freiman: *Stress Corrosion of Ionic and Mixed Ionic/Covalent Solids.* J. Am. Ceram. Soc. **69 [10],** 721–724 (1986)

[4.299] A.G. Metcalfe and G.K. Shmitz: *Mechanisms of Stress Corrosion in E Glass Filaments.* Glass Tech. **13 [1],** 5–16 (1972)

[4.300] H.C. Cao, B.J. Dalgleish, C.H. Hsueh, and A.G. Evans: *High-Temperature Stress Corrosion Cracking in Ceramics.* J. Am. Ceram. Soc. **70 [4],** 257–264 (1987).

[4.301] G. Choi and S. Horibe: *The Environmental Effect on Cyclic Fatigue Behavior in Ceramic Materials.* J. Mater. Sci. **30 [6],** 1565–1569 (1995)

[4.302] A.H. Heuer and V.L.K. Lou: *Volatility Diagrams for Silica, Silicon Nitride, and Silicon Carbide and their Application to High-Temperature Decomposition and Oxidation.* J. Am. Ceram. Soc. **73 [10],** 2789–2803 (1990)

[4.303] F. Bregani: *Corrosione ad Alta Temperatura di Materiali e Rivestimenti Ceramici.* Ceramurgia **24 [4],** 151–162 (1994)

[4.304] W. Genthe and H. Hausner: *Corrosion of Aluminum Oxide in Acids and Caustic Solutions.* Ceram. Forum Int. **67 [1/2]**, 6–10 (1990)

[4.305] T. Sato, K. Haryu, T. Endo, and M. Shimada: *High Temperature Oxidation of Hot-Pressed Aluminum Nitride by Water Vapor.* J. Mater. Sci. **22 [6]**, 2277–2280 (1987)

[4.306] P.T.B. Shaffer and T.J. Mroz: *Aluminum Nitride.* In: *Handbook of Advanced Ceramic Materials.* (Advanced Refractory Technologies Buffalo NY, 1993)

[4.307] E.L. Courtright: *Engineering Property Limitations of Structural Ceramics and Ceramic Composites above 1600°C.* Ceram. Eng. Sci. Proc. **12 [9–10]**, 1725–1744 (1991)

[4.308] A. Lipp, K.A. Schwetz, and K. Hunold: *Hexagonal Boron Nitride: Fabrication, Properties, and Applications.* J. Eur. Ceram. Soc. **5 [1]**, 3–9 (1989)

[4.309] P.T.B. Shaffer and A. Goel: *Silicon Nitride.* In: *Handbook of Advanced Ceramic Materials.* (Advanced Refractory Technologies Buffalo NY, 1993)

[4.310] T. Sato, Y. Kanno, T. Endo: *Corrosion of SiC, Si_3N_4, and AlN in Molten K_2SO_4 and K_2CO_3 Salts.* Yogyo Kyokaishi **94 [1]**, 123–128 (1986)

[4.311] M.E. Gulden and A.G. Metcalf: *Stress Corrosion of Silicon Nitride.* J. Am. Ceram. Soc. **59 [9–10]**, 391–396 (1976)

[4.312] N.J. Tinghe, J. Son, and R.M. Hu: *Corrosion Reactions in SiC Ceramics.* Ceram. Eng. Sci. Proc. **8 [7–8]**, 505–511 (1987)

[4.313] D.W. McKee and D. Chatterji: *Corrosion of Silicon Carbide in Gases and Alkaline Melts.* J. Am. Ceram. Soc. **59 [9–10]**, 441–444 (1976)

[4.314] G.W. Hallum and T.P. Herbell: *Effect of High-Temperature Hydrogen Exposure on Sintered α-SiC.* Adv. Ceram. Mater. **3 [2]**, 171–175 (1988)

[4.315] G.I. Rudd., S.H. Garofalini, and D.A. Hensley: *Atomic Force Microscopy and X-Ray Photoelectron Spectroscopy Investigation of the Onset of Reactions on Alkali Silicate Glass Surfaces.* J. Am. Ceram. Soc. **76 [10]**, 2555–2560 (1995)

[4.316] L. Holland: *The Properties of Glass Surfaces*. (Chapman & Hall London, 1966)

[4.317] ASTM standard C225–85 (1990 Vol.15.02)

[4.318] ASTM standard C724–81 (1988 Vol.15.02)

[4.319] ASTM standard C650–83 (1988 Vol.15.02)

[4.320] D.E. Clark and B.K. Zoitos: *Corrosion Testing and Characterization.* In: *Corrosion of Glass, Ceramics and Ceramic Superconductors*. D.E. Clark and B.K. Zoitos (eds.). (Noyes Park Ridge NJ, 1992), pp. 51–102

Chapter 5

[5.1] N.E. Promisel: *The Concept of Characterization.* In: *Characterization of Materials in Research.* J.J. Burke and W. Weiss (eds.). (Syracuse University Press Syracuse NY, 1975), pp. 3–10

[5.2] *ASTM Specification C373, ASTM Standards, Part 13.* (American Society for Testing and Materials Philadelphia, 1969)

[5.3] D.W. Richerson: *Modern Ceramic Engineering.* (Marcel Dekker New York, 1982)

[5.4] R. Morrell: *Handbook of Properties of Technical and Engineering Ceramics, Part 1.* (Her Majesty's Stationery Office London, 1985)

[5.5] R.E. Smallman and K.H.G. Ashbee: *Modern Metallography.* (Pergamon Press Oxford England, 1969)

[5.6] E. Weidmann: *Preparation of Ceramic Specimens for Reflected Light Microscopy.* Met. Prog. **122 [6],** 51–54 (1982)

[5.7] J.H. Richardson: *Optical Microscopy for Material Sciences.* (Marcel Dekker New York, 1971)

[5.8] A.H. Heuer, V. Lou, L. Ogbugji, and T.E. Mitchell: *Lattice Resolution Studies of Engineering Ceramics.* J. Microsc. et Spectr. Electr. **2 [6],** 475–480 (1977)

[5.9] J.P. Jernot and J.L. Chermant: *Mesures de Base en Analyse D'Image.* In: *Caractérisation des Poudres et des Céramiques.* J.L. Chermant (ed.). (Forceram Hermes France, 1991), pp. 40–56

[5.10] J.I. Goldstein, D.E. Newbury, P. Echlin, D.C. Joy, C. Fiori, and E. Lifshin: *Scanning Electron Microscopy and X-Ray Microanalysis.* (Plenum Press NY, 1981)

[5.11] J.B. Bindell: *Elements of Scanning Electron Microscopy.* Adv. Mater. Proc. **143 [3],** 20–27 (1993)

[5.12] N. Baumgarten: *SEM for Imaging Specimens in their Natural States.* American Laboratory **June,** 66–70 (1990)

[5.13] L.E. Murr: *Electron and Ion Microscopy and Microanalysis.* Marcel Dekker New York, 1982)

[5.14] G. Thomas and M.J. Goringe: *Transmission Electron Microscopy of Materials.* (John Wiley & Sons New York, 1979)

[5.15] W. Braue, R.W. Carpenter, and D.J. Smith: *High-Resolution Interface Analysis of SiC-Whisker-Reinforced* Si_3N_4 *and* Al_2O_3 *Ceramic Matrix Composites.* J. Mater. Sci. **25 [6],** 2949–2957 (1990)

[5.16] O. Ünal, J.J. Petrovic, and T.E. Mitchell: *Mechanical Properties of Hot Isostatically Pressed* Si_3N_4 *and* Si_3N_4*/SiC Composites.* J. Mater. Res. **8 [3],** 626–634 (1993)

[5.17] S. Horiuchi: *Application of Atomic-Resolution Electron Microscopy to Ceramic Materials.* Am. Ceram. Soc. Bull. **64 [12],** 1590–1593 (1985)

[5.18] G. Van Tendeloo and T. Krekels: *HREM of Ceramic High* T_c *Superconductors.* J. Eur. Ceram. Soc. **16 [3],** 367–378 (1996)

[5.19] D.J. Gould: *Ion-Machining Ceramics for Transmission Electron Microscopy.* Metallography **6,** 539–542 (1973)

[5.20] H.J. Kleebe, W. Braue, H. Schmidt, G. Pezzotti, and G. Ziegler: *Transmission Electron Microscopy of Microstructures in Ceramic Materials.* J. Eur. Ceram. Soc. **16 [3],** 339–351 (1996)

[5.21] P. Gise: *Scanning Probe Microscopy: Material Analysis with an Added Dimension.* JOM **44 [12],** 8–10 (1992)

[5.22] G. Binnig and H. Rohrer: *Scanning Tunneling Microscope.* Helv. Phys. Acta. **55,** 726 (1982)

[5.23] D.A. Bonnell and D.R. Clarke: *Scanning Tunneling Microscopy and Spectroscopy of Ceramics: Silicon Carbide and Zinc Oxide.* J. Am. Ceram. Soc. **71 [8],** 629–637 (1988)

[5.24] C.S. Chang, N.J. Zheng, I.S.T. Tsong, Y.C. Wang, and R.F. Davis: *Scanning Tunneling Microscopy of Cubic Silicon Carbide Surfaces.* J. Am. Ceram. Soc. **73 [11],** 3264–3268 (1990)

[5.25] D.A. Bonnell: *Some Theoretical and Experimental Considerations in Scanning Tunneling Microscopy of Ceramic Materials.* In: *Ceramic Transactions,* Vol.5. (American Ceramic Society Westerville OH, 1989) pp. 315–326

[5.26] G. Binnig, C.F. Quate, and C. Gerber: *Atomic Force Microscope.* Phys. Rev. Lett. **56,** 930 (1986)

[5.27] *Scanning Probe Microscopy.* (Digital Instruments Technical Documentation B1, v3-6/97)

[5.28] G.I. Rudd., S.H. Garofalini, and D.A. Hensley: *Atomic Force Microscopy and X-Ray Photoelectron Spectroscopy Investigation of the Onset of Reactions on Alkali Silicate Glass Surfaces.* J. Am. Ceram. Soc. **76 [10],** 2555–2560 (1995)

[5.29] R. Levi-Setti, J.M. Chabala, and Y.L. Wang: *Aspects of High Resolution Imaging with a Scanning Ion Microscope.* Ultramicroscopy **24,** 97–114 (1988)

[5.30] J.R.A. Cleaver, E.C.G. Kirk, R.J. Young, and H. Ahmed: *Scanning Ion Beam Techniques for the Examination of Microelectronic Devices.* J. Vac. Sci. Tech. **B6 [3],** 1026–1029 (1988)

[5.31] C.G. Pantano and J.F. Kelso: *Chemical Analysis of Fracture Surfaces.* In: *Fractography of Ceramic and Metal Surfaces.* J.J. Mecholsky, Jr. and S.R. Powell Jr. (eds.). (American Society for Testing and Materials Philadelphia PA, 1984), pp. 139–156

[5.32] T.B. Rymer: *Electron Diffraction.* (Chapman & Hall London, 1970)

[5.33] *Vacuum Products Catalogue.* (Kurt J. Lesker Co. Clairton PA, 1993)

[5.34] B.D. Cullity: *Elements of X-Ray Diffraction.* (Addison-Wesley Reading MA, 1978)

[5.35] M.F.C. Ladd and R.A. Palmer: *Structure Determination by X-Ray Crystallography.* (Plenum Press New York, 1977)

[5.36] M.M. Woolfson: *An Introduction to X-Ray Crystallography*. (Cambridge at the University Press Cambridge England, 1970)

[5.37] R.H. Doremus: *Glass Science*. (John Wiley & Sons New York, 1994)

[5.38] J.W. Jeffery: *Methods in X-Ray Crystallography*. (Academic Press London, 1971)

[5.39] *Characterization of Ceramics*. R.E. Loehman (ed.). (Butterworth-Heinemann Stoneham MA, 1993)

[5.40] L. Neergard: *Nondestructive Evaluation*. In: *Characterization of Ceramics*. R.E. Loehman (ed.). (Butterworth-Heinemann Stoneham MA, 1993)

[5.41] L.A. Lott and D.C. Kunerth: *NDE Testing and Inspection*. In: *Engineered Materials Handbook,* Vol.4: *Ceramics and Glasses*. (ASM International Materials Park OH, 1991), pp. 617–627

[5.42] R.F. Firestone: *NDE: Improving Reliability of Advanced Ceramics*. Am. Ceram. Soc. Bull. **68 [6],** 1177–1186 (1989)

[5.43] D.R. Black, H.E. Burdette, R.C. Dobbyn, R.D. Spal, and M. Kuriyama: *Advanced X-Ray Imaging Methods for Ceramics*. Am. Ceram. Soc. Bull. **70 [6],** 1004–1009 (1991)

[5.44] R.N. Yancey and S.J. Klima: *High Resolution Computed Tomography of Advanced Composite and Ceramic Materials*. Ceram. Eng. Sci. Proc. **12 [9–10],** 2029–2041 (1991)

[5.45] R.A. Armistead: *CT:Quantitative 3-D Inspection.* Adv. Mater. Proc. **133 [3],** 42–48 (1988)

[5.46] W.A. Ellingson, P.S. Wong, S.L. Dieckman, J.L. Ackerman, and L. Garrido: *Magnetic Resonance Imaging: A New Characterization Technique for Advanced Ceramics.* Am. Ceram. Soc. Bull. **68 [6],** 1180–1186 (1989)

[5.47] S. Karunanithy and M. Falk: *Nondestructive Characterization of Slip Cast Fiber- and Whisker-Reinforced Alumina.* Ceram. Eng. Sci. Proc. **10 [7–8],** 998–1004 (1989)

[5.48] P.S. Wang, S.G. Malghan, S.J. Dapkunas, K.F. Hens, and R. Raman: *"NMR Characterization of Injection Moulded Alumina Green Compacts, Part II:*T_2*-Weighted Proton Imaging.* J. Mater. Sci. **30 [4],** 1069–1074 (1995)

[5.49] L.A. Lott and D.C. Kunerth: *NDE Testing and Inspection*. In: *Engineered Materials Handbook,* Vol.4: *Ceramics and Glasses*. (ASM International Materials Park OH, 1991), pp. 617–627

[5.50] C.L. Vorres, D.E. Yuhas, and L.W. Kessler: *Scanning Laser Acoustic Microscopy: A New Tool For Nondestructive Testing*. In: *Novel NDE Methods for Materials*. B.R. Rath (ed.). (The Metallurgical Society of AIME Warrendale Pennsylvania, 1983), pp. 77–82

[5.51] J. Winterfeld, H.D. Tietz, and T. Palm: *Non-Destructive Testing of Volume Defects in Structural Ceramics*. Ceram. Forum Int. **71 [3],** 96–100 (1994)

[5.52] P.K. Khandelwal, R.R. Kinnick, and P.W. Heitmman: *Photoacoustic Microscopy of Ceramic Turbine Blades*. Am. Ceram. Soc. Bull. **64 [8],** 1112–1115 (1985)

[5.53] S. Sathish: *Scanning Acoustic Microscopy of SCS-6 Silicon Carbide Fiber.* J. Am. Ceram. Soc. **79 [1],** 209–212 (1996)

[5.54] M. Suganuma: *Scanning Acoustic Microscopy of Indentation Damage in Y_2O_3-Stabilized Tetragonal Zirconia*. J. Am. Ceram. Soc. **78 [11],** 2889–2896 (1995)

[5.55] D.B. Williams: *Practical Analytical Electron Microscopy in Materials Science*. (Verlag Chemie International Deerfield Beach FL, 1984)

[5.56] K. Ewsuk: *Consolidation of Bulk Ceramics*. In: *Characterization of Ceramics*. R.E. Loehman (ed.). (Butterworth-Heinemann Stoneham MA, 1993)

[5.57] Y. Bando, K. Kurashima, and S. Nakano: *Modern Applications of a New 300 KV Field Emission Transmission Electron Microscope to the Study of Advanced Materials*. J. Eur. Ceram. Soc. **16 [3],** 379–384 (1996)

[5.58] A.L. Albee and L. Ray: *Correction Factors for Electron Probe Microanalysis of Silicates, Oxides, Carbonates, Phosphates, and Sulfates*. Anal. Chem. **42 [12],** 1408–1414 (1970)

[5.59] J. Homeny, W.L. Vaughn, and M.K. Ferber: *Silicon Carbide Whisker/Alumina Matrix Composites: Effect of Whisker Surface Treatment on Fracture Toughness*. J. Am. Ceram. Soc. **73 [2],** 394–402 (1990)

[5.60] G.H. Campbell, M. Rühle, B.J. Dalgleish, and A.G. Evans: *Whisker Toughening: A Comparison Between Aluminum Oxide and Silicon Nitride Toughened with Silicon Nitride*. J. Am. Ceram. Soc. **73 [3],** 521–530 (1990)

[5.61] P.H. Holloway: *Surface and Interface Characterization of Advanced Materials*. Ceram. Eng. Sci. Proc. **3 [9–10],** 431–449 (1982)

[5.62] M.N. Rahaman, Y. Boiteux, L.C. deJonghe: *Surface Characterization of Silicon Nitride and Silicon Carbide Powders*. Am. Ceram. Soc. Bull. **65 [8],** 1171–1176 (1986)

[5.63] L.R. Pederson and L.S. Dake: *X-Ray Photoelectron Spectroscopy of Electronic Ceramic Materials: Complex Chromites and Manganites.* In: *Ceramic Transaction,* Vol.5: *Advanced Characterization Techniques for Ceramics*. W.S. Young, G.L. McVay, and G.E. Pike (eds.). (The American Ceramic Society Westerville OH, 1989), pp. 43–54

[5.64] G. Selvaduray, M.B. Lange, and D. Sullivan: *Chemical Characterization of Ceramic Substrates for Microelectronic Packaging*. In: *Ceramic Transaction,* Vol.5: *Advanced Characterization Techniques for Ceramics.* W.S. Young, G.L. McVay, and G.E. Pike (eds.). (The American Ceramic Society Westerville OH, 1989), pp. 55–61

[5.65] J.J. Brennan and S.R. Nutt: *SiC-Whisker-Reinforced Glass–Ceramic Composites: Interfaces and Properties*. J. Am. Ceram. Soc. **75 [5],** 1205–1216 (1992)

[5.66] W.C. Johnson, D.F. Stein, and R.W. Rice: *Analysis of Grain-Boundary Impurities and Fluorite Additives in Hot-Pressed Oxides by Auger Electron Spectroscopy*. J. Am. Ceram. Soc. **57 [8],** 342–344 (1974)

[5.67] M.F. Dilmore, D.E. Clark, and L.L. Hench: *Aqueous Corrosion of Lithia–Alumina–Silicate Glasses*. Am. Ceram. Soc. Bull. **57 [11],** 1040–1048 (1978)

[5.68] D.E. Clark, M.F. Dilmore, E.C. Ethridge, and L.L. Hench: *Aqueous Corrosion of Soda–Silica and Soda–Lime–Silica Glass*. J. Am. Ceram. Soc. **59 [1–2],** 62–65 (1976)

[5.69] R.J. Blattner: *Characterizing the Top 10000 Å of Ceramics*. In: *Ceramic Transaction Vol.5, Advanced Characterization Techniques for Ceramics*. W.S. Young, G.L. McVay, and G.E. Pike (eds.). (The American Ceramic Society Westerville OH, 1989), pp. 31–42

[5.70] S.W. Novak, H. Kanber, and R. Lowden: *Characterization of Materials Using SIMS Image Depth Profiling*. In: *Ceramic Transaction,* Vol.5: *Advanced Characterization Techniques for Ceramics*. W.S. Young, G.L. McVay, and G.E. Pike (eds.). (The American Ceramic Society Westerville OH, 1989), pp. 102–111

[5.71] R. Venkatesh: *Interface Engineering of Alumina Fiber/Glass Composites*. Ph.D. Thesis. (New Mexico Insitute of Mining and Technology Socorro NM, 1993)

[5.72] R.H. Scott, P.F.S. Jackson, and A. Strasheim: *Application of Laser Source Mass Spectroscopy to Analysis of Geological Materials*. Nature (London) **232 [5313],** 623–624 (1971)

[5.73] L.L. Murr: *Electron and Ion Microscopy and Microanalysis*. (Marcel Dekker New York, 1982)

[5.74] W.K. Chu, J.W. Mayer, and M.A. Nicolet: *Backscattering Spectrometry*. (Academic Press New York, 1978)

[5.75] L.A. Boatner and C.W. White: *The Application of Rutherford Backscattering and Channeling Techniques to the Characterization of Ceramics*. In: *Ceramic Transaction,* Vol.5: *Advanced Characterization Techniques for Ceramics*. W.S. Young, G.L. McVay, and G.E. Pike (eds.). (The American Ceramic Society Westerville OH, 1989), pp. 112–127

[5.76] D.B. Williams: *Practical Analytical Electron Microscopy in Materials Science*. (Verlag Chemie International Deerfield Beach FL, 1984)

[5.77] A. Greiner: *Darstellung von Si_3N_4/SiC Verbundkeramiken mit Nanokristallinen SiC-Einschlussen in Mikrokristallinen Si_3N_4-Kornern*. M.S. Thesis. (Faculty of Chemistry University of Stuttgart Germany, 1993)

[5.78] L.L. Hench: *Characterization of Glass*. In: *Characterization of Materials in Research*. J.J. Burke and V. Weiss (eds.). (Syracuse University Press Syracuse NY, 1975), pp. 211–251

[5.79] M.I. Baraton, L. Boulanger, M. Cauchetier, V. Lorenzelli, M. Luce, T. Merle, P. Quintard, and Y.H. Zhou: *Nanometric Boron Nitride Powders: Laser Synthesis, Characterization, and FT-IR Surface Study*. J. Eur. Ceram. Soc. **13 [4]**, 371–378 (1994)

[5.80] D.A. Sagala and S. Koyasu: *Infrared Reflection of $Ba(Mg_{1/3}Ta_{2/3})O_3$ Ceramics*. J. Am. Ceram. Soc. **76 [10]**, 2433–2436 (1993)

[5.81] L. Bergman and R.J. Nemanich: *Raman Spectroscopy for Characterization of Hard, Wide-Bandgap Semiconductors: Diamond, GaN, GaAlN, AlN, BN*. Ann. Rev. Mater. Sci. **26,** 551–579 (1996)

[5.82] L.E. McNeil, M. Grimsditch, and R.H. French: *Vibrational Spectroscopy of Aluminum Nitride*. J. Am. Ceram. Soc. **76 [5]**, 1132–1146 (1993)

[5.83] D.J. Kim, H.J. Jung, and I.S. Yang: *Raman Spectroscopy of Tetragonal Zirconia Solid Solutions.* J. Am. Ceram. Soc. **76 [8]**, 2106–2108 (1993)

[5.84] P.S. Wang, S.G. Malghan, S.J. Dapkunas, K.F. Hens, and R. Raman: *NMR Characterization of Injection Moulded Alumina Green Compacts Part I-Nuclear Spin-Spin Relaxation*. J. Mater. Sci. **30 [4]**, 1059–1064 (1995)

Chapter 6

[6.1] M.R. Pascucci and R.N. Katz: *Modern Day Applications of Advanced Ceramics*. Interceram **42 [2]**, 71–78 (1993)

[6.2] R. Morrell: *Ceramics in Modern Engineering.* In: *Engineering Applications of Ceramic Materials.* M.M. Schwartz (ed.). (American Society for Metals Metals Park OH, 1985), pp. 3–12

[6.3] L.B. Sibley and M. Zlotnick: *Considerations for Tribological Applications of Engineering Ceramics.* Mater. Sci. Eng. **71,** 283–293 (1985)

[6.4] A. Bennett: *Requirements for Engineering Ceramics in Gas Turbine Engines*. Mater. Sci. Tech. **2 [9]**, 895–899 (1986)

[6.5] A. Krauth and K. Berroth: *Engineering Ceramics for Industrial Applications; Wear-, Heat-, and Automotive Technology*. In: *Engineering Applications of Ceramic Materials.* M.M. Schwartz (ed.). (American Society for Metals Metals Park OH, 1985), pp. 308–313

[6.6] CoorsTek Internet site: *http://www.coorstek.com/coorstek/ = coorstek.asp?menu prodserv&reference=/ccprodserv/prodserv1.htm*

[6.7] D. Zeus: *How the Use of Advanced Ceramics as Tribomaterial Has Affected the Evolution of Mechanical Seals*. Ceram. Forum Int. **68 [1–2]**, 36–45 (1991)

[6.8] Y. Hara: *Application of Fine Ceramics in Industrial Fields*. Corros. Eng. **38 [7]**, 527–537 (1989)

[6.9] C. Toy, T. Baykara: *21.Yuzyilin Malzemesi, Seramikler*. Bilim ve Teknik **27 [317]**, 6–15 (1994)

[6.10] S. Parrott: *Engineering Ceramics for Aggressive Environments*. Metals & Mater. **6 [4]**, 207–210 (1990)

[6.11] F. Muilwijk and J.P.P. Tholen: *Ceramic Membranes, Applications and Properties*. In: *Euro Ceramics Vol.3*. G. Ziegler and H. Hausner (eds.). (Trans Tech Zürich Switzerland, 1993) pp. 3.596–3.599

[6.12] *Ceramics in Military Applications*. Ceramic Fact Sheets. The American Ceramic Society Web Site: *http://www.acers.org/info/facts/facts.html*

[6.13] J.E. Sheenan, K.W. Buesking, and B.J. Sullivan: *Carbon–Carbon Composites*, Ann. Rev. Mater. Sci. **24,** 19–44 (1994)

[6.14] R.K. Bart and J.C. Lindberg: *Ceramic Bodyguards*. Adv. Mater. & Proc. **132 [3]**, 69–72 (1987)

[6.15] G. Savage: *Ceramic Armor*. Met. Mater. **6 [8]**, 487–492 (1990)

[6.16] D.J. Viechnicki, M.J. Slavin, and M.I. Kliman: *Development and Current Status of Armor Ceramics*. Am. Ceram. Soc. Bull. **70 [6]**, 1035–1039 (1991)

[6.17] *Modern Metal Cutting* (Sandvik , 1995)

[6.18] V.E. Annamalai, T. Sornakumar, C.V. Gokularathnam, and R. Krishnamurthy: *Cutting Tool Application of Ceria–Zirconia*. J. Eur. Ceram. Soc. **12 [3]**, 227–233 (1993)

[6.19] S.J. Burden: *Ceramic Cutting Tools*. Ceram. Eng. Sci. Proc. **3 [7–8]**, 351–359 (1982)

[6.20] G. Geiger: *Ceramic Coatings Enhance Material Performance*. Am. Ceram. Soc. Bull. **71 [10]**, 1470–1481 (1992)

[6.21] D. W. Richerson: *Modern Ceramic Engineering*. (Marcel Dekker New York, 1982)

[6.22] M.P. Groover: *Fundamentals of Modern Manufacturing*. (Prentice-Hall Upper Saddle River NJ, 1996)

[6.23] M.C. Shaw: *Principles of Abrasive Processing*. (Oxford University Press Oxford England, 1996)

[6.24] E.P. DeGarmo, J.T. Black, and R.A. Kohser: *Materials and Processes in Manufacturing*. (Prentice-Hall Upper Saddle River NJ, 1997)

[6.25] A. Parker: *Ceramic Applications in Reciprocating Engines*. Met. Mater. **6 [1]**, 14–20 (1990)

[6.26] *Technical Ceramics*. (NGK Spark Plug Co. Technical publication 00-2075(5)-9211)

[6.27] P.E.D. Morgan: *Structuring Chemical Technology to Produce Cost-Effective Ceramic Products on the Large Scale*. Am. Ceram. Soc. Bull. **72 [7]**, 65–70 (1993)

[6.28] J.M. Tulliani, L. Mantanaro, C. Borello, and P.P. Demaestri: *Filtri Ceramici per il Controllo Delle Emissioni di Motori Diesel: Presente e Futuro*. Ceramurgia **25 [2]**, 81–90 (1995)

[6.29] *Diesel Oxidation Catalysts.* (Degussa AG Pamphlet PT 105-7-5-396 H Germany)

[6.30] M.R. Montierth and J.P. Day: *Supporti Catalitici per il Controllo delle Emissioni da Autoveicoli.* Ceramurgia **24 [1],** 9–15 (1994)

[6.31] *Automotive Emission Control Catalysts.* (Degussa AG Pamphlet PT 92-10-5-396 H Germany)

[6.32] *Motronic-Technische Unterrichtung* (Bosch 2.Ausgabe Germany, 1985)

[6.33] S. Nunomura, J. Nakayama, H. Abe, O. Kamigaito, and K. Matsusue: *Mechanical Properties*. In: *Advanced Technical Ceramics*. S.Somiya (ed.). (Academic Press San Diego CA, 1989), pp. 223–258

[6.34] M. Demirsoy: *Otomotiv Endustrisinde Seramik Kullanimindaki Teknolojik Gelismeler.* In: *3.cu Otomotiv ve Yan Sanayii Semp. ve Sergisi Cilt 1.* (TMMOB Ankara Turkiye, 1991), pp. 115–123

[6.35] *The Morgan Crucible Company: A Leader in Ceramics Technology*. In: *Automotove Technology International*. (Sterling London, 1998)

[6.36] Mercedes-Benz Internet site: *mercedesbenz.com*

[6.37] D.H. Wright: *Testing Automotive Materials and Components.* (Society of Automotive Engineers Warrendale PA, 1993)

[6.38] *Automobil Revue 96*. (Hallwag AG Bern Switzerland, 1996)

[6.39] P.A. Badkar: *Alumina Ceramics for High Temperature Applications*. Key Eng. Mater. **56–57,** 45–58 (1991)

[6.40] M. Ferraris and C. Badini: *Materiali Compositi a Matrice Vetrosa e Vetroceramica*. Ceramurgia **24 [2],** 51–57 (1994)

[6.41] P. Dietrichs: *Ceramic Linings for Waste Incinerators.* Ceram. Forum Int. **68 [3],** 92–96 (1991)

[6.42] C.G. Marvin: *Refractories-An Overview.* Am. Ceram. Soc. Bull. **66 [7],** 1101–1102 (1987)

[6.43] R.A. Penty and J.W. Bjirklie: *Silicon Carbide for High Temperature Heat Exchangers*. Ceram. Eng. Sci. Proc. **3 [1–2],** 120–127 (1982)

[6.44] G. Fisher: *Refractory Uses-Practicality of High Technology Ceramics.* Am. Ceram. Soc. Bull. **66 [7],** 1103–1108 (1987)

[6.45] L.M. Sheppard: *Corrosion-Resistant Ceramics for Severe Environments.* Am. Ceram. Soc. Bull. **70 [7],** 1146–1466 (1991)

[6.46] D.E. Mahagin and R.E. Dahl: *Nuclear Applications of Boron and the Borides.* In: *Boron and Refractory Borides*. V.I. Matkovich (ed.). (Springer-Verlag Berlin, 1977), pp. 613–632

[6.47] L.H. Van Vlack: *Physical Ceramics for Engineers*. (Addison-Wesley Reading MA, 1964)

[6.48] F. Thevenot: *Boron Carbide-A Comprehensive Review*. J. Eur. Ceram. Soc. **6 [4],** 205–225 (1990)

[6.49] Hj. Matzke: *Ceramics for Energy Production and Environmental Applications with Emphasis on Nuclear Energy*. Ceram. Acta **6 [1],** 31–50 (1994)

[6.50] C.E. Johnson, R.G. Clemmer, and G.W. Hollenberg: *Solid Breeder Materials*. J. Nucl. Mater. **103–104,** 547–554 (1981)

[6.51] *Ceramic Applications in Medicine. Ceramic Fact Sheets.* (The American Ceramic Society Web Site: http://www.acers.org/info/facts/facts.html)

[6.52] G. Heimke: *Alumina.* In: *Concise Encyclopedia of Medical and Dental Materials*. D. Williams (ed.). (Pergamon Press Oxford England, 1990), pp. 28–34

[6.53] C. Kasperk, R. Ewers, B. Simons, and R. Kasperk: *Algae-Derived (Phycogene) Hydroxylapatite; A Comparative Histological Study*. Int. J. Oral Maxillofacial Surg. **17,** 319–324 (1988)

[6.54] *Algipore-hydroxy apatite*. (Handout No.1504.2.IX.94 La Friatec AG Mannheim Germany, 1994)

[6.55] M. Ferraris and C. Badini: *Materiali Compositi a Matrice Vetrosa e Vetroceramica.* Ceramurgia **24 [2],** 51–57 (1994)

[6.56] G. Heimke: *Recent Developments in Bioceramics*. In: *Euro-Ceramics*, Vol.3: *Engineering Ceramics*. G. DeWith, R.A. Terpstra, and R. Metselaar (eds.). (Elsevier Applied Science London, 1989), pp. 3.1–3.10

[6.57] J.R. Kelly: *Ceramics in Restorative and Prosthetic Dentistry.* Ann. Rev. Mater. Sci. **27,** 443–468 (1997)

[6.58] G. Gomez-Roman, W. Schulte, B. D'Hoedt, and D. Axman-Krcmar: *The Frialit-2 Implant System: Five-Year Clinical Experience in Single-Tooth and Immediately Postextraction Applications*. Int. Journ. Oral Maxillofacial Implants **12 [3],** 299–309 (1997)

[6.59] L.L. Hench: *Bioceramics: From Concept to Clinic*. J. Am. Ceram. Soc. **74 [7],** 1487–510 (1991)

[6.60] P.S. Christel: *Zirconia-Toughened Ceramics*. In: *Concise Encyclopedia of Medical and Dental Materials*. D. Williams (ed.). (Pergamon Press Oxford England, 1990), pp. 375–379

[6.61] K. Inamori: *Biological Applications*. In: *Advanced Technical Ceramics*. S. Somiya (ed.). (Academic Press San Diego CA, 1989), pp. 209–222

[6.62] M.S. Judd and D.L. Eaton: *Controlled-Pore Diatomite Ceramics for Biotechnology*. Am. Ceram. Soc. Bull. **69 [4],** 674–676 (1990)

[6.63] *Ceramic Applications in Electronics.* (Ceramic Fact Sheets The American Ceramic Society Web Site: *http://www.acers.org/info/facts/facts.html*)

[6.64] *Aluminum Nitride*. (Form A-14065A 1/90 Carborundum)

[6.65] M. Ura: *Ceramic Substrates*. In: *Fine Ceramics.* S. Saito (ed.). (Elsevier Essex England, 1985), pp. 243–251

[6.66] P.T.B. Shaffer and T.J. Mroz: *Aluminum Nitride.* In: *Handbook of Advanced Ceramic Materials*. (Advanced Refractory Technologies Buffalo NY, 1993)

[6.67] A.J. Moulson and J.M. Herbert: *Electroceramics*. (Chapman & Hall London, 1990)

[6.68] Y. Furuhata and G. Toda: *Ferroelectric and Electrooptic Materials*. In: *Fine Ceramics*. S. Saito (ed.). (Elsevier Essex England, 1985), pp. 261–275

[6.69] M.E. Lines and A.M. Glass: *Principles and Applications of Ferroelectrics and Related Materials*. (Oxford University Press Oxford England, 1979)

[6.70] R.W. Schwartz: *Electronic and Magnetic Ceramics*. In: *Characterization of Ceramics*. R.E. Loehman (ed.). (Butterworth-Heinemann Stoneham MA, 1993)

[6.71] K. Wakino: *Piezoelectric and Pyroelectric Ceramics*. In: *Fine Ceramics*. S. Saito (ed.). (Elsevier Essex England, 1985), pp. 251–260

[6.72] *Solid State Physics Source Book*. S.P. Parker (ed.). (McGraw-Hill New York, 1988)

[6.73] M. Miyayama and H. Yanagida: *Ceramic Semiconductors: Gas Sensors*. In: *Fine Ceramics*. S. Saito (ed.). (Elsevier Essex England, 1985), pp. 292–297

[6.74] C. Gresgovich and S. Duclos: *Ceramic Scintillators*. Ann. Rev. Mater. Sci. **27**, 69–88 (1997)

Chapter 7

[7.1] P.M. Rice: *Pottery Analysis*. (The University of Chicago Press Chicago, 1987)

[7.2] W. Ryan and C. Radford: *Whitewares, Production, Testing, and Quality Control*. (Pergamon Press Oxford England, 1987)

[7.3] W.D. Kingery, H.K. Bowen, and D.R. Uhlmann: *Introduction to Ceramics*. (John Wiley & Sons New York, 1976)

[7.4] *Materials Handbook*. Ceram. Ind. **Jan.,** (1991)

[7.5] *Encyclopedia Britannica*. (Encyclopedia Britannica Chicago, 1994)

[7.6] P. Dickin: *CADCAM at Wedgewood*. Am.Ceram.Soc.Bull. **72 [12],** 71–74 (1993)

[7.7] *Plant Report: Florida Tile Industries*. Am.Ceram.Soc.Bull. **73 [1],** 30–33 (1994)

[7.8] *Commercial Glasses, Advances in Ceramics, Vol.18*. D.C. Boyd and J.F. McDowell (eds.). (The American Ceramic Soc. Columbus OH, 1986), pp. 67–72

[7.9] R.A. Eppler: *Selecting Ceramic Pigments*. Am. Ceram. Soc. Bull. **66 [11],** 1600–1604 (1987)

[7.10] F.M. Lea: *The Chemistry of Cement and Concrete*. (Edward Arnold London 1970)

[7.11] M.F. Ashby and D.H.R. Jones: *Engineering Materials 2*. (Pergamon Press Oxford England, 1986)

[7.12] H.F.W. Taylor: *Cement Chemistry*. (Academic Press London, 1990)

[7.13] S.P. Shah and C. Ouyang: *Fracture Mechanics for Failure of Concrete*. Ann. Rev. Mater. Sci. **24,** 293–320 (1994)

[7.14] E.M. Gartner and D.F. Myers: *New Cements and Novel Cement Matrices*. In: *Ceramic Transactions*, Vol.16: *Advances in Cementitious*

Materials. S. Mindess (ed.). (The American Ceramic Society Westerville OH, 1991) pp. 621–637.

[7.15] J.D. Gilchrist: *Fuels, Furnaces and Refractories*. (Pergamon Press Oxford England, 1977)

[7.16] E.P. DeGarmo, J.T. Black, and R.A. Kohser: *Materials and Processes in Manufacturing*. (Prentice-Hall Upper Saddle River NJ, 1997)

[7.17] R.H. Doremus: *Glass Science*. (John Wiley & Sons New York, 1994)

[7.18] R.M. Brick, A.W. Pense, and R.B. Gordon: *Structure and Properties of Engineering Materials*. (McGraw-Hill New York, 1977)

[7.19] T.D. Taylor: *Structure and Properties of Glasses*. In: *Engineered Materials Handbook,* Vol.4: *Ceramics and Glasses*. (ASM International Materials Park OH, 1991), pp. 564–569.

[7.20] I. Cooke and P. Mayes: *Introduction to Innovation and Technology Transfer.* (Artech House Norwood MA, 1996)

[7.21] M.P. Groover: *Fundamentals of Modern Manufacturing*. (Prentice-Hall Upper Saddle River NJ, 1996)

Subject Index

Printing: Mercedes-Druck, Berlin
Binding: Buchbinderei Lüderitz & Bauer, Berlin